Steam Plant
Operation

Steam Plant Operation

Everett B. Woodruff
Herbert B. Lammers
Thomas F. Lammers

Seventh Edition

McGraw-Hill

New York   San Francisco   Washington, D.C.   Auckland   Bogotá
Caracas   Lisbon   London   Madrid   Mexico City   Milan
Montreal   New Delhi   San Juan   Singapore
Sydney   Tokyo   Toronto

Library of Congress Cataloging-in-Publication Data

Lammers, Thomas F.
    Steam plant operation / Thomas F. Lammers.—7th ed.
        p.   cm.
    First-fourth eds. by E. B. Woodruff and H. B. Lammers. Fifth ed. by E. B. Woodruff
    and T. F Lammers. Sixth ed. by T. F. Lammers
    Includes index.
    ISBN 0-07-036150-9 (alk. paper)
    1. Steam power plants.—I. Woodruff, Everett B. (Everett Bowman), 1900–1982
    —II. Lammers, Herbert B. 1902–1981
    1900–1982. Steam-plant operation. II. Title.
  TJ405.L36   1998
  621.1—dc21   97-46056
                                                             CIP

McGraw-Hill

*A Division of The **McGraw·Hill** Companies*

    3 4 5 6 7 8 9 0 DOC/DOC 9 0 3 2 1 0 9

ISBN 0-07-036150-9

The sponsoring editors for this book were Hal Crawford and Robert Esposito, the editing supervisor was Ruth W. Mannino, and the production supervisor was Pamela Pelton. It was set in Century Schoolbook by Dina John of McGraw-Hill's New York desktop publishing department.

Printed and bound by R. R. Donnelley and Sons Company.

To my wife, Marianne, and to my children,
Stephen and Michelle, with all my love and
gratitude.

Contents

Preface

I am very pleased to have the opportunity to complete the seventh edition of this book, a book whose six previous editions have assisted thousands of people for over 60 years in the understanding of steam power plants. The presentation of material in the form of illustrations and clear descriptions has helped many to gain the fundamental knowledge of the many complex systems found in these plants. This knowledge has allowed them to further their careers in this field and, where required for an operating license, has assisted many in the passing of an operator's examination.

Presented in a practical and easily understood format, as in previous editions, the complex systems found in power plants are described, and the means by which they are operated and maintained are defined so that the operation is safe, reliable, economic, and, very important, performed in an environmentally acceptable manner. Thus, the combustion processes of solid, liquid, and gaseous fuels are described, as well as the boilers that are designed to handle this combustion and to produce the steam necessary for the particular power plant, whether it be for a process, for heating, or for the production of electricity.

As society accepts the technological advances of this computer age, it is often forgotten that it is the reliable operation of power plants, particularly steam power plants, that provide the energy source for these advancements. Throughout the world, 80 to 90 percent of the electricity produced results from steam. The fuel energy for this is predominantly from the combustion of fossil fuels, with coal still providing a very significant margin over oil, natural gas, and other fuels requiring combustion. *Steam Plant Operation* continues to present the sytems necessary to produce this power—boilers, combustion equipment, steam turbines, pumps, condensers, etc.—that are required for efficient operation, as well as the environmental control equipment to control plant emissions within regulated bounds.

Since power plant equipment remains in operation for many decades, some older equipment descriptions are retained in this edition; they continue to remain in operation, and they illustrate operating fundamentals. In this edition, this information is complemented with the principles of operation on modern fluidized bed boilers, which can handle hard-to-burn solid fuels and control emissions of nitrogen oxides (NO_x) and sulfur dioxides (SO_2). In addition, cogeneration facilities are introduced, and these facilities join gas turbine technology with a steam power plant to form a combined facility. Where necessary, updated material is introduced to reflect recent trends in power plant system design.

There still remain critical decisions on the handling of municipal solid waste, as its disposal remains a problem in the United States and worldwide. Waste-to-energy plants remain a viable solution to this problem; however politics have delayed many facilities. Recycling has become an important part of this solution, and the advantages and disadvantages of recycling are presented as well as their integration with waste-to-energy plants. The use of material recycling facilities and the various systems that are used for their operation are described.

This edition no longer includes the descriptions of the venerable steam engine of which few remain in operation today. The information included in previous editions assisted those who took examinations where questions on the subject were still asked. The elimination of this material allows the addition of current technology on other subjects.

It is a privilege to be able to continue the tradition of this book by presenting material in an easily understood format. This idea was originally established in earlier editions by my father, Herbert Lammers, and his friend and co-author, Everett Woodruff. These two practical men devoted much of their lives to the safe and efficient operation of steam power plants. As a result, *Steam Plant Operation* has assisted many in the basic understanding of steam-plant technology.

Finally, without the contribution of illustrations and information by many suppliers and designers of power plant equipment, this book would not have been possible. I am sincerely grateful to all who assisted me in this project—the seventh edition of *Steam Plant Operation*.

Thomas F. Lammers

Steam and Its Importance

In today's modern world, all societies are involved to various degrees with technological breakthroughs that are attempting to make our lives more productive and more comfortable. These technologies include sophisticated electronic devices, the most prominent of which are computer systems. Many of the systems in our modern world depend on a reliable and relatively inexpensive energy source—*electricity*.

With the availability of electricity providing most of the industrialized world a very high degree of comfort, the source of this electricity and the means for its production are often forgotten. It is the power plant that provides this critical energy source, and in the United States approximately 90 percent of the electricity is produced from power plants that use steam as an energy source, with the remaining 10 percent of the electricity produced primarily by hydroelectric power plants. In other parts of the world, similar proportions are common for their electric production.

The power plant is a facility that transforms various types of energy into electricity or heat for some useful purpose. The energy input to the power plant can vary significantly, and the plant design to accommodate this energy is drastically different for each energy source. The forms of this input energy can be as follows:

1. The *potential energy* of an elevated body of water, which, when used, becomes a hydroelectric power plant.
2. The *chemical energy* that is released from the hydrocarbons contained in fossil fuels such as coal, oil, or natural gas, which becomes a fossil fuel fired power plant.
3. The *solar energy* from the sun, which becomes a solar power plant.

4. The *fission or fusion energy* that separates or attracts atomic particles, which becomes a nuclear power plant.

With any of these input sources, the power plant's output can take
various forms:

1. Heat for a process or for heating
2. Electricity that is subsequently converted into other forms of energy
3. Energy for transportation such as for ships

In these power plants, the conversion of water to steam is the pre-
dominate technology, and this book will describe this process and the
various systems and equipment that are used commonly in today's
operating steam power plants.

Each power plant has many interacting systems, and in a steam
power plant these include fuel and ash handling, handling of combustion air and the products of combustion, feedwater and condensate,
steam, environmental control systems, and the control systems that
are necessary for a safe, reliable, and efficiently run power plant. The
seventh edition of *Steam-Plant Operation* continues to blend descriptions and illustrations of both new and older equipment, since both
are in operation in today's power plants. One noticeable change in
this edition is the elimination of the discussion on steam engines.
These wonderful mechanical devices, which were so critical to industry throughout the world for many decades as they powered machinery, have been nearly totally replaced, most often by electric motors.

1.1   The Use of Steam

Steam is a critical resource in today's industrial world. It is essential
for the production of paper and other wood products, for the preparation and serving of foods, for the cooling and heating of large buildings, for driving equipment such as pumps and compressors, and for
powering ships. However, its most important priority remains as the
primary source of power for the production of electricity.

Steam is extremely valuable because it can be produced anywhere
in the world by using the heat that comes from the fuels that are
available in the area. Steam also has unique properties that are
extremely important in producing energy. Steam is basically recycled,
from steam to water and then back to steam again, all in a manner
that is nontoxic in nature.

The steam plants of today are a combination of complex engineered
systems that work to produce steam in the most efficient manner that
is economically feasible. Whether the end product of this steam is

electricity, heat, or a steam process required to develop a needed product such as paper, the goal is to have that product produced at the lowest cost possible, and this often is related to the heat required to produce the steam or the actual end product.

In every situation, however, the steam power plant must first obtain heat. This heat must come from an energy source, and this varies significantly, often based on the plant's location in the world. These sources of heat could be

1. A fossil fuel—coal, oil, or natural gas

2. A nuclear fuel such as uranium

3. Other forms of energy, which can include waste heat from exhaust gases of gas turbines; bark, wood, bagasse, vine clippings, and other similar waste fuels; by-product fuels such as carbon monoxide (CO), blast furnace gas (BFG), or methane (CH_4); municipal solid waste (MSW); sewage sludge; geothermal energy; and solar energy

Each of these fuels contains potential energy in the form of a heating value, and this is measured in the amount of British thermal units (Btus) per each pound or cubic feet of the fuel (i.e., Btu/lb or Btu/ft^3) depending on whether the fuel is a solid or a gas. (*Note:* A British thermal unit is about equal to the quantity of heat required to raise one pound of water one degree Fahrenheit.)

This energy must be released, and with fossil fuels, this is done through a carefully controlled combustion process. In a nuclear power plant that uses uranium, the heat energy is released by a process called *fission*. In both cases the heat is released and then transferred to water. This can be done in various ways, such as through tubes that have the water flowing on the inside. As the water is heated, it eventually changes its form by turning into steam. As heat is continually added, the steam reaches the desired temperature and pressure for the particular application.

The system in which the steam is generated is called a *boiler*. Boilers can vary significantly in size. A relatively small one supplies heat to a building, and other industrial-sized boilers provide steam for a process. Very large systems produce enough steam at the proper pressure and temperature to result in the generation of 1300 megawatts (MW) of electricity in an electric utility power plant. Such a large power plant would provide the electric needs for over 1 million people.

Small boilers that produce steam for heating or for a process are critical in their importance in producing a reliable steam flow, even though it may be saturated steam at a pressure of 200 psig and a steam flow of 5000 lb/h. This then can be compared with the large utility boiler that produces 10 million pounds of superheated steam per hour at pressures and temperatures exceeding 3800 psig and 1000°F. To the operator of either size plant, reliable, safe, and efficient operation is of the utmost importance. The capacity, pressure, and temperature ranges of boilers and their uniqueness of design reflect their applications and the fuel that provides their source of energy.

Not only must the modern boiler produce steam in an efficient manner to produce power (heat, process, or electricity) with the lowest operational cost that is practical, but also it must perform in an environmentally acceptable way. Environmental protection is a major consideration in all modern steam generating systems, where low-cost steam and electricity must be produced with a minimum impact on the environment. Air pollution control that limits the emissions of sulfur dioxide (SO_2) and other acid gases, particulates, and nitrogen oxides (NO_x) is a very important issue for all combustion processes.

For example, low NO_x burners, combustion technology, and supplemental systems have been developed for systems fired by coal, oil, or natural gas. These systems have met all the requirements that have been imposed by the U.S. Clean Air Act, and as a result, NO_x levels have been reduced by 50 to 70 percent from uncontrolled levels.

1.2   The Steam-Plant Cycle

The simplest steam cycle of practical value is called the *Rankine cycle,* which originated around the performance of the steam engine. The steam cycle is important because it connects processes that allow heat to be converted to work on a continuous basis. This simple cycle was based on dry saturated steam being supplied by a boiler to a power unit such as a turbine that drives an electric generator. The steam from the turbine exhausts to a condenser, from which the condensed steam is pumped back into the boiler. It is also called a *condensing cycle,* and a simple schematic of the system is shown in Fig. 1.1.

This schematic also shows heat (Q_{in}) being supplied to the boiler and a generator connected to the turbine for the production of electricity. Heat (Q_{out}) is removed by the condenser, and the pump supplies energy (W_p) to the feedwater in the form of a pressure increase to allow it to flow through the boiler.

A higher plant efficiency is obtained if the steam is initially superheated, and this means that less steam and less fuel are required for a specific output. If the steam is reheated and passed through a second turbine, cycle efficiency also improves, and moisture in the steam is reduced as it passes through the turbine. This moisture reduction minimizes erosion on the turbine blades.

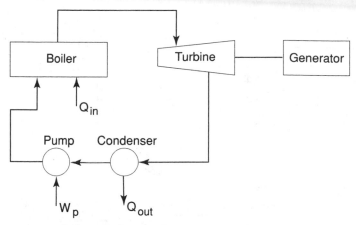

Figure 1.1 Schematic diagram for a Rankine cycle.

By the addition of regenerative feedwater heating, the original Rankine cycle was improved significantly. This is done by extracting steam from various stages of the turbine to heat the feedwater as it is pumped from the condenser back to the boiler to complete the cycle. It is this cycle concept that is used in modern power plants, and the equipment and systems for it will be described in this book.

1.3 The Power Plant

The steam generator or boiler is a major part of the many systems that comprise a steam power plant. A typical pulverized-coal-fired utility power plant is shown schematically in Fig. 1.2. The major systems of this power plant can be identified as

1. Coal receipt and preparation
2. Coal combustion and steam generation
3. Environmental protection
4. Turbine generator and electric production
5. Condenser and feedwater system
6. Heat rejection, including the cooling tower

In this example, the fuel handling system stores the coal supply, prepares the fuel for combustion by means of pulverization, and then transports the pulverized coal to the boiler. A forced-draft (FD) fan supplies the combustion air to the burners, and this air is preheated in an air heater, which improves the cycle efficiency. The heated air is also used to dry the pulverized coal. A primary air fan is used to supply heated air to the pulverizer for coal drying purposes and is the

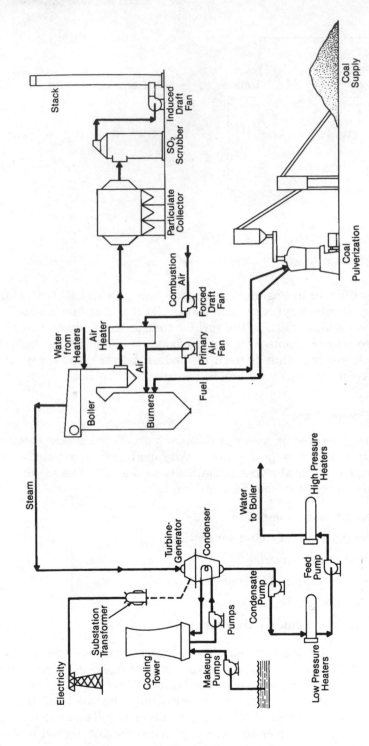

Figure 1.2 Schematic of a typical pulverized-coal-fired utility power plant. Reheater, ash and reagent handling, and sludge disposal are not shown. (*Babcock & Wilcox, a McDermott company.*)

source of the primary air to the burners as the fuel-air mixture flows from the pulverizers to the burners. The fuel-air mixture is then burned in the furnace portion of the boiler.

The boiler recovers the heat from combustion and generates steam at the required pressure and temperature. The combustion gases are generally called *flue gas,* and these leave the boiler, economizer, and finally the air heater and then pass through environmental control equipment. In the example shown, the flue gas passes through a particulate collector, either an electrostatic precipitator or a bag filterhouse, to a sulfur dioxide (SO_2) scrubbing system, where these acid gases are removed, and then the cleaned flue gas flows to the stack through an induced-draft (ID) fan. Ash from the coal is removed from the boiler and particulate collector, and residue is removed from the scrubber.

Steam is generated in the boiler under carefully controlled conditions. The steam flows to the turbine, which drives a generator for the production of electricity and for distribution to the electric system at the proper voltage. Since the power plant has its own electrical needs, such as motors, controls, and lights, part of the electricity generated is used for these plant requirements.

After passing through the turbine, the steam flows to the condenser, where it is converted back to water for reuse as boiler feedwater. Cooling water passes through the condenser, where it absorbs the rejected heat from condensing and then releases this heat to the atmosphere by means of a cooling tower. The condensed water then returns to the boiler through a series of pumps and heat exchangers, called *feedwater heaters,* and this process increases the pressure and temperature of the water prior to its reentry into the boiler, thus completing its cycle from water to steam and then back to water.

The type of fuel that is burned determines to a great extent the overall plant design. Whether it be the fossil fuels of coal, oil, or natural gas, biomass, or by-product fuels, considerably different provisions must be incorporated into the plant design for systems such as fuel handling and preparation, combustion of the fuel, recovery of heat, fouling of heat-transfer surfaces, corrosion of materials, and air pollution control. Refer to Fig. 1.3, where a comparison is shown of a natural gas–fired boiler and a pulverized-coal-fired boiler, each designed for the same steam capacity, pressure, and temperature. This comparison only shows relative boiler size and does not indicate the air pollution control equipment that is required with the coal-fired boiler, such as an electrostatic precipitator and an SO_2 scrubber system. Such systems are unnecessary for a boiler designed to burn natural gas.

In a natural gas–fired boiler, there is minimum need for fuel storage and handling because the gas usually comes directly from the

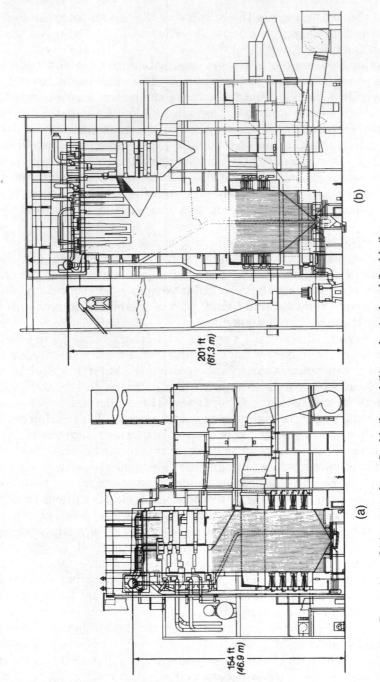

201 ft
(61.3 m)

(b)

154 ft
(46.9 m)

(a)

Figure 1.3 Comparison of (*a*) a natural gas–fired boiler and (*b*) a pulverized-coal-fired boiler, each producing the same steam capacity, pressure, and temperature. (*Babcock & Wilcox, a McDermott company.*)

pipeline to the boiler. In addition, only a relatively small furnace is required for combustion. Since natural gas has no ash, there is no fouling in the boiler because of ash deposits, and therefore the boiler design allows heat-transfer surfaces to be more closely spaced. The combination of a smaller furnace and the closer spacing results in a more compact boiler design. The corrosion allowance is also relatively small, and the emissions control required relates primarily to the NO_x that is formed during the combustion process. The boiler designed for natural gas firing is therefore a relatively small and economical design.

The power plant becomes much more complex when a solid fuel such as coal is burned. Coal and other solid fuels have a high percentage of ash that is not combustible, and this ash must be a factor in designing the plant. A coal-fired power plant must include extensive fuel handling, storage, and preparation facilities; a much larger furnace for combustion; and wider spaced heat-transfer surfaces. Additional components are also required:

1. Sootblowers, which are special cleaning equipment to reduce the impact of fouling and erosion
2. Air heaters, which provide air preheating to dry fuel and enhance combustion
3. Environmental control equipment such as electrostatic precipitators, bag filterhouses, and SO_2 scrubbers
4. Ash handling systems to collect and remove ash
5. Ash disposal systems including a landfill

The units shown in Fig. 1.3 are designed for the same steam capacity, but one is designed for natural gas firing and the other is designed for pulverized coal firing. Although the comparison of the two units shows only a relative difference in the height of the units, both the depth and the width of the coal-fired unit are proportionately larger as well.

1.4   Utility Boilers for Electric Power

Both in the United States and worldwide, the majority of electric power is produced in steam power plants using fossil fuels and steam turbines. Most of the electric production comes from large electric utility plants, although the newer plants are much smaller and owned and operated by independent power producers (IPPs).

Until the 1980s, the United States and other Western nations developed large electrical networks, primarily with electric utilities.

Over the past several decades in the United States, the incremental demand of about 2 percent has been met through independent power producers (IPPs). However, the United States is not dependent on this IPP capacity. The average electricity reserve margin is 20 percent. This allows the opportunity to investigate the possible changes of established institutions and regulations, to expand wheeling of power to balance regional supply, and to demand and satisfy these low incremental capacity needs in less expensive ways. (*Note: Wheeling* is the sale of power across regions and not restricted to the traditional local-only supply.)

Many developing countries do not have this luxury. In fact, their electric supply growth is just meeting demand, and in many cases, the electric supply growth is not close to meeting demand. Power outages are frequent, and this has a serious impact on the local economy.

As an average for large utility plants, a kilowatt-hour (kWh) of electricity is produced for each 8500 to 9500 Btus that are supplied from the fuel, and this results in a net thermal efficiency for the plant of 36 to 40 percent. These facilities use steam-driven turbine generators that produce electricity up to 1300 MW, and individual boilers are designed to produce steam flows ranging from 1 million to 10 million lb/h. Modern plants use cycles that have, at the turbine, steam pressures ranging from 1800 to 3500 psi and steam temperatures from 950 to 1000°F and at times over 1000°F.

In the United States, approximately 3500 billion kWh of electricity is generated from the following energy sources:

Coal	51%
Oil	2
Natural gas	16
Nuclear	20
Hydroelectric	8
Geothermal and others	3
Total	100%

Therefore, nearly 70 percent of the electric production results from steam generators that use the fossil fuels of coal, oil, or natural gas. A portion of the energy from natural gas powers gas turbine cogeneration plants that incorporate a steam cycle. Since nuclear plants also use steam to drive turbines, when added to the fossil fuel plant total, almost 90 percent of electricity production comes from steam power plants, which certainly reflects the importance of steam.

The overwhelmingly dominant fossil fuel used in modern U.S. power plants is coal, since it is the energy source for over 50 percent of the electric power produced. There are many types of coal, as discussed in Chap. 4, but the types most often used are bituminous, sub-bituminous, and lignite. Although it is expected that natural gas or fuel oil will be the fuel choice for some future power plants, such as gas turbine combined-cycle facilities, coal will remain the dominant fuel for the production of electricity.

It is the belief of some in the power industry that the approximate 20 percent of the electricity that is now produced in the United States from nuclear power will be reduced to about 10 percent over the next several decades. If this occurs, the majority, if not all, of this power will be replaced with coal-fired units. This additional coal-fired capacity may come from reactivated coal-fired plants that are currently in a reserve status, as well as new coal-fired units.

On a worldwide basis, a similar pattern is present as in the United States, with coal being the predominant fuel for the production of electricity:

Coal	44%
Oil	10
Natural gas	8.5
Nuclear	17
Hydroelectric	20
Other	0.5
Total	100.0%

Although many newer and so-called sophisticated technologies often get the headlines for supplying the future power needs of the world, electric power produced from generated steam, with the use of fossil fuels or with the use of nuclear energy, results in the production of 80 to 90 percent of the world's energy requirements. Therefore, steam continues to have a prominent role in the world's economic future.

1.4.1 Coal-fired boilers

Coal is the most abundant fuel in the United States and in many other parts of the world. The benefit of its high availability, however, is offset by the fact that it is the most complicated fuel to burn. Many problems occur with the systems required to combust the fuel efficiently and effectively as well as the systems that are required to handle the ash that remains after combustion. Even with similar coals, designs vary from even one boiler designer because of operating experience and testing. For different boiler designers, significant differences in design are apparent because of the designers' design philosophy and the experience gained with operating units.

The environmental control aspects of coal firing also present complexities. These include both social and political difficulties when trying to locate and to obtain a permit for a coal-fired plant that has atmospheric, liquid, and solid emissions that have to be taken into consideration in the plant design. Also, as noted previously, there are a wide variety of coals, each with its own characteristics of heating value, ash, sulfur, etc., that have to be taken into account in the boiler design and all its supporting systems. For example, coal ash can vary from 5 to 25 percent by weight among various coals. Of the total operating costs of a coal-fired plant, approximately 60 to 80 percent of the costs are for the coal itself.

The large coal-fired power plant utilizes pulverized coal firing, as described in detail in Chaps. 2 and 5. An example of a medium-sized modern pulverized-coal-fired boiler is shown in Fig. 1.4 and incorporates low NO_x burners to meet current emission requirements (see Chap. 5). This unit is designed to produce 1,250,000 lb/h of steam at 2460 psig and 1005°F/1005°F (superheat/reheat). This unit has the coal burners in the front wall and, as part of the NO_x control system, has secondary air ports above the burners. This unit has a two gas pass, three air pass tubular air heater (see Chap. 2). The forced-draft (FD) fan also takes warm air from the top of the building (above the air heater) by means of a vertical duct. This design of the combustion air intake improves the air circulation within the building as well as using all available heat sources for improving plant efficiency. The environmental control equipment is not shown in this illustration.

A larger pulverized-coal-fired boiler is shown in Fig. 1.5. This illustration shows a boiler system and its environmental control equipment that produces approximately 6,500,000 lb/h of steam for an electrical output of 860 MW. This is a radiant-type boiler that is designed to produce both superheated and reheated steam for use in the turbine. For air heating, it incorporates a regenerative air heater instead of a tubular air heater. For environmental control, it uses a dry scrubber for the capture of SO_2 and a baghouse for the collection of particulates. The boiler shown is designed for indoor use (see building enclosing equipment), but depending on location, many boilers and their auxiliary systems are designed as outside installations.

1.4.2   Oil- and gas-fired boilers

The use of oil and gas as fuels for new utility boilers has declined except for certain areas of the world where these otherwise critical fuels are readily available and low in cost. Large oil-producing countries are good examples of places where oil- and gas-fired boilers are installed. In other areas of the world, their use as fuels for utility boilers has declined for various reasons: high cost, low availability, and

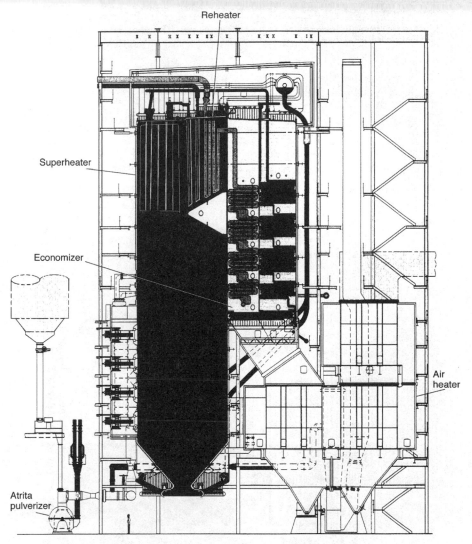

Figure 1.4 Medium-sized pulverized-coal-fired boiler producing 1,250,000 lb/h of steam at 2460 psig and 1005°F/1005°F (superheat/reheat). (*DB Riley, Inc.*)

government regulations. However, there have been significant improvements in combined cycle systems that have made the use of oil and more often natural gas in these systems more cost-effective. In addition, plants that have these gas turbine cycles are more easily sited than other types of power plants because of their reduced environmental concerns. However, in the majority of cases, they depend on a critical fuel, natural gas, whose availability for the long term may be limited.

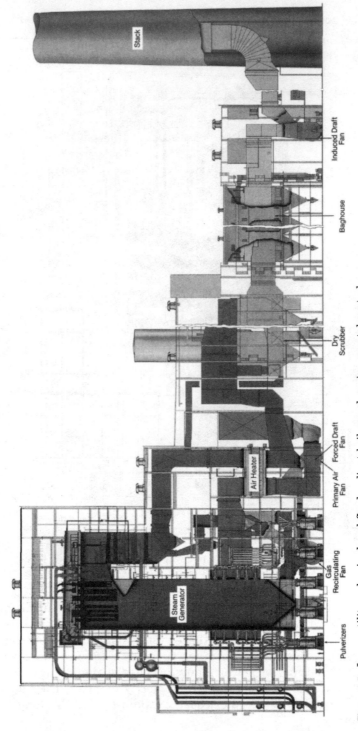

Figure 1.5 Large utility pulverized-coal-fired radiant boiler and environmental control systems that produces steam for a plant output of 860 MW. (*Babcock & Wilcox, a McDermott company.*)

1.4.3 Steam considerations

The reheat steam cycle is used on most fossil fuel–fired utility plants. In this cycle, high-pressure superheated steam from the boiler passes through the high-pressure portion of the turbine, where the steam reduces in pressure, and then this lower-pressure steam returns to the boiler for reheating. After the steam is reheated, it returns to the turbine, where it flows through the intermediate- and low-pressure portions of the turbine. The use of this cycle increases the thermal efficiency of the plant, and the fuel costs are therefore reduced. In a large utility system, the reheat cycle can be justified because the lower fuel costs offset the higher initial cost of the reheater, piping, turbine, controls, and other equipment that is necessary to handle the reheated steam.

1.4.4 Boiler feedwater

When water is obtained from sources that are either on or below the surface of the earth, it contains, in solution, some scale-forming materials, free oxygen, and in some cases, acids. These impurities must be removed because high-quality water is vital to the efficient and reliable operation of any steam cycle. Good water quality can improve efficiency by reducing scale deposits on tubes, it minimizes overall maintenance, and it improves the availability of the system. All of this means lower costs and higher revenues.

Dissolved oxygen attacks steel, and the rate of this attack increases significantly as temperatures increase. By having high chemical concentrations or high solids in the boiler water and feedwater, boiler tube deposition can occur, and solids can be carried over into the superheater and finally the turbine. This results in superheater tube failures because of overheating. Deposits and erosion also occur on the turbine blades. These situations are serious maintenance problems and can result in plant outages for repairs. The actual maintenance can be very costly; however, this cost can be greatly exceeded by the loss of revenues caused by the outage that is necessary to make the repairs.

As steam-plant operating pressures have increased, the water treatment systems have become more important to obtaining high availability. This has led to more complete and refined water treatment facilities.

1.5 Industrial and Small Power Plants

Various industries require steam to meet many of their needs: heating and air conditioning; turbine drives for pumps, blowers, or compressors; drying and other processes; water heating; cooking; and cleaning. This so-called industrial steam, because of its lower pressure and temperature as compared with utility requirements, also can be used to generate electricity. This can be done directly with a turbine for electric production only or as part of a cogeneration system, where a turbine is used for electric production and low-pressure steam is extracted from the turbine and used for heating or for some process. The electricity that is produced is used for in-plant requirements, with the excess often sold to a local electric utility.

Another method is a combined cycle system, where a gas turbine is used to generate electric power and a heat recovery system is added using the exhaust gas from the gas turbine as a heat source. The generated steam flows to a steam turbine for additional electric generation, and this cogeneration results in an improvement in the overall efficiency. The steam that is generated also can be used as process steam either directly or when extracted from the system, such as an extraction point within the turbine.

One of the most distinguishable features of most industrial-type boilers is a large saturated water boiler bank between the steam drum and the lower drum. Figure 1.6 shows a typical two-drum design. This particular unit is designed to burn pulverized coal or fuel oil, and it generates 885,000 lb/h of steam. Although not shown, this boiler also requires environmental control equipment to collect particulates and acid gases contained in the flue gas.

The boiler bank serves the purpose of preheating the inlet feedwater to the saturated temperature and then evaporating the water while cooling the flue gas. In lower-pressure boilers, the heating surface that is available in the furnace enclosure is insufficient to absorb all the heat energy that is needed to accomplish this function. Therefore, a boiler bank is added after the furnace and superheater, if one is required, to provide the necessary heat-transfer surface.

As shown in Fig. 1.7, as the pressure increases, the amount of heat absorption that is required to evaporate water declines rapidly, and the heat absorption for water preheating and superheating steam increases. See also Table 1.1 for examples of heat absorption at system pressures of 500 and 1500 psig.

The examples shown in the table assume that the superheat is constant at 1000°F, which is 100° higher than the saturated temperature for the particular pressure (see Chap. 3).

It is also common for boilers to be designed with an economizer and/or an air heater located downstream of the boiler bank in order to reduce the flue gas temperature and to provide an efficient boiler cycle.

It is generally not economical to distribute steam through long steam lines at pressures below 150 psig because, in order to minimize the pressure drop that is caused by friction in the line, pipe sizes

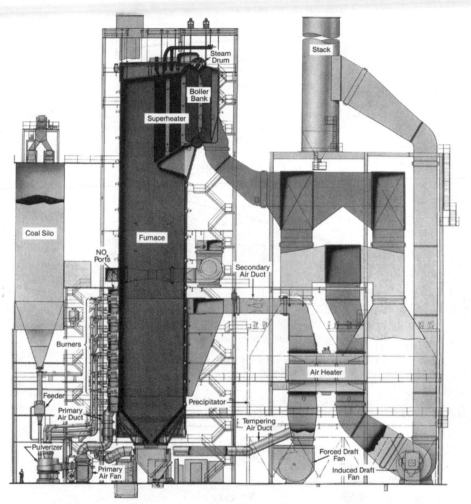

Figure 1.6 Large industrial-type pulverized-coal- and oil-fired two-drum boiler. (*Babcock & Wilcox, a McDermott company.*)

increase with the associated cost increase. In addition, for the effective operation of auxiliary equipment such as sootblowers and turbine drives on pumps, boilers should operate at a minimum pressure of 125 psig. Therefore, few plants of any size operate below this steam pressure. If the pressure is required to be lower, it is common to use pressure-reducing stations at these locations.

For an industrial facility where both electric power and steam for heating or a process are required, a study must be made to evaluate the most economical choice. For example, electric power could be purchased from the local utility and a boiler could be installed to meet the heating or process needs only. By comparison, a plant could be

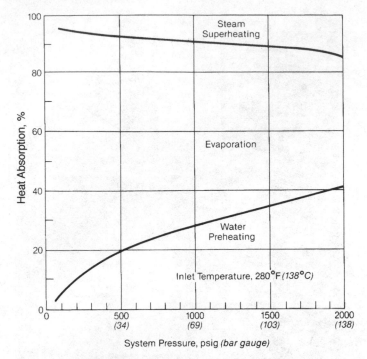

Figure 1.7 Effect of steam pressure on evaporation in industrial boilers. (*Babcock & Wilcox, a McDermott company.*)

TABLE 1.1 Heat Absorption Percentages for Water Preheating, Evaporation, and of Steam Superheating

	500 psig	1500 psig
Water preheating	20%	34%
Evaporation	72	56
Steam superheating	8	10
TOTAL	100%	100%

installed where both electricity and steam are produced from the same system.

1.5.1 Fluidized bed boilers

There are various ways of burning solid fuels, the most common of which are in pulverized-coal-fired units and stoker-fired units. These designs for boilers in the industrial size range have been in operation for many years and remain an important part of the industrial boiler base for the burning of solid fuels. These types of boilers and their features continue to be described in this book.

Although having been operational for over three decades, but not with any overall general acceptance, the fluidized bed boiler is becoming more popular in modern power plants because of its ability to handle hard-to-burn fuels with low emissions. As a result, this unique design can be found in many industrial boiler applications and in small utility power plants, especially those operated by independent power producers (IPPs). Because of this popularity, this book includes an expansion of the features of some of the many designs available and the operating characteristics of each.

In fluidized bed combustion, fuel is burned in a bed of hot particles that are suspended by an upward flow of fluidizing gas. The fuel is generally a solid fuel such as coal, wood chips, etc. The fluidizing gas is a combination of the combustion air and the flue gas products of combustion. When sulfur capture is not required, the fuel ash may be supplemented by an inert material such as sand to maintain the bed. In applications where sulfur capture is required, limestone is used as the sorbent, and it forms a portion of the bed. Bed temperature is maintained between 1550 and 1650°F by the use of a heat-absorbing surface within or enclosing the bed.

As stated previously, fluidized bed boilers feature a unique concept of burning solid fuel in a bed of particles to control the combustion process, and the process controls the emissions of sulfur dioxide (SO_2) and nitrogen oxides (NO_x). These designs offer versatility for the burning of a wide variety of fuels, including many that are too poor in quality for use in conventional firing systems.

The state of fluidization in a fluidized bed boiler depends mainly on the bed particle diameter and the fluidizing velocity. There are two basic fluid bed combustion systems, the bubbling fluid bed (BFB) and the circulating fluid bed (CFB), and each operates in a different state of fluidization.

At relatively low velocities and with coarse bed particle size, the fluid bed is dense with a uniform solids concentration, and it has a well-defined surface. This system is called the *bubbling fluid bed* (BFB) because the air in excess of that required to fluidize the bed passes through the bed in the form of bubbles. This system has relatively low solids entrainment in the flue gas.

With the *circulating fluid bed* (CFB) design, higher velocities and finer bed particle size are prevalent, and the fluid bed surface becomes diffuse as solids entrainment increases and there is no defined bed surface. The recycle of entrained material to the bed at high rates is required to maintain bed inventory.

It is interesting that the BFB and CFB technologies are somewhat similar to stoker firing and pulverized coal firing with regard to fluidizing velocity, but the particle size of the bed is quite different. Stoker firing incorporates a fixed bed, has a comparable velocity, but has a much coarser particle size than that found in a BFB. For pulverized-coal firing, the velocity is comparable with a CFB, but the particle size is much finer than that for a CFB.

Bubbling fluid bed (BFB) boiler. Of all the fluid bed technologies, the bubbling bed is the oldest. The primary difference between a BFB boiler and a CFB boiler design is that with a BFB the air velocity in the bed is maintained low enough that the material that comprises the bed (e.g., fuel, ash, limestone, and sand), except for fines, is held in the bottom of the unit, and the solids do not circulate through the rest of the furnace enclosure.

For new boilers, the BFB boilers are well suited to handle high-moisture waste fuels, such as sewage sludge, and also the various sludges that are produced in pulp and paper mills and in recycle paper plants. The features of design and the uniqueness of this technology, as well as the CFB, are described in Chap. 2. Although the boiler designs are different, the objectives of each are the same, and the designs are successful in achieving them.

Circulating fluid bed (CFB) boiler. The CFB boiler provides an alternative to stoker or pulverized coal firing. In general, it can produce steam up to 1,000,000 lb/h at 1850 psig and 1000°F. It is generally selected for applications with high-sulfur fuels, such as coal, petroleum coke, sludge, and oil pitch, as well as for wood waste and for other biomass fuels such as vine clippings from large vineyards. It is also used for hard-to-burn fuels such as waste coal culm, which is a fine residue generally from the mining and production of anthracite coal. Because the CFB operates at a much lower combustion temperature than stoker or pulverized-coal firing, it generates approximately 50 percent less NO_x as compared with stoker or pulverized coal firing.

The use of CFB boilers is rapidly increasing in the world as a result of their ability to burn low-grade fuels while at the same time being able to meet the required emission criteria for nitrogen oxides (NO_x), sulfur dioxide (SO_2), carbon monoxide (CO), volatile organic compounds (VOC), and particulates. The CFB boiler produces steam economically for process purposes and for electric production.

The advantages of a CFB boiler are reduced capital and operating costs that result primarily from the following:

1. It burns low-quality and less costly fuels.

2. It offers greater fuel flexibility as compared with coal-fired boilers and stoker-fired boilers.

3. It reduces the costs for fuel crushing because coarser fuel is used as compared with pulverized fuel. Fuel sizing is slightly less than that required for stoker firing.

4. It has lower capital costs and lower operating costs because pollution control equipment is not required.

1.5.2 Combined cycle and cogeneration systems

In the 1970s and 1980s, the role of natural gas in the generation of electric power in the United States was far less than that of coal and oil. The reasons for this included

1. Low supply estimates of natural gas that projected it to last for less than 10 years
2. Natural gas distribution problems that threatened any reliable fuel delivery
3. Two OPEC (Organization of Petroleum Exporting Countries) oil embargoes that put pressure on the domestic natural gas supply
4. Concerns that natural gas prices would escalate rapidly and have an impact on any new exploration, recovery, and transmission

For these reasons, a Fuel Use Act was enacted in the late 1970s that prohibited the use of natural gas in new plants.

This situation has changed dramatically because now the electric power industry is anticipating a continuing explosive growth in the use of natural gas. The reasons for this growth are

1. Continued deregulation of both natural gas and electric power
2. Environmental restrictions that limit the use of coal in many areas of the country
3. Continued perception of problems with the use of oil as a fuel for power plants because of greater dependence on foreign oil
4. Rapidly advancing gas turbine and combined cycle technology with higher efficiencies and lower emissions
5. Easier financing of power projects because of shorter schedules and more rapid return on investments

Perhaps the greatest reason for the growth is the current projection of natural gas supplies. Where before the natural gas supply was expected to last approximately 10 years, the current estimate is between 70 and 100 years based on the current production and use levels. Although these optimistic estimates are very favorable, they could promote a far greater usage, which could seriously deplete this critical resource in the future, far sooner than expected. Therefore, careful long-term plans must be incorporated for this energy source.

Advancements in combustion technology have encouraged the application of natural gas to the generation of electric power. The gas turbine is the leader in combustion improvements. By using the most advanced metallurgy, thermal barrier coatings, and internal air cooling technology, the present-day gas turbines have higher outputs, higher reliability, lower heat rates, lower emissions, and lower costs. At present in the United States, nearly all new power plants that are fired by natural gas use gas turbines with combined cycles.

Combined cycles (or cogeneration cycles) are a dual-cycle system. The initial cycle burns natural gas, and its combustion gases pass through a gas turbine that is connected to an electric generator. The secondary cycle is a steam cycle that uses the exhaust gases from the gas turbine for the generation of steam in a boiler. The steam generated flows through a steam turbine that is connected to its electric generator. Figure 1.8 shows a block diagram of a cogeneration system.

The interest in the combined cycle for power plants has resulted from the improved technology of gas turbines and the availability of natural gas. The steam cycle plays a secondary role in the system because its components are selected to match any advancement in technology such as the exhaust temperatures from gas turbines.

The recovery of the heat energy from the gas turbine exhaust is the responsibility of the boiler, which for this combined cycle is called the *heat-recovery steam generator* (HRSG). As the exhaust temperatures

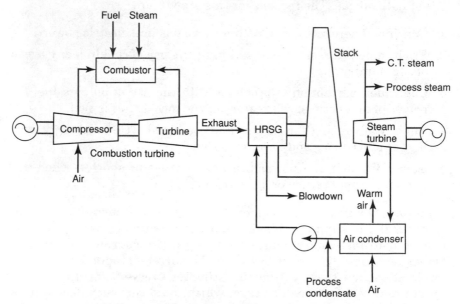

Figure 1.8 Diagram of a cogeneration system using a gas turbine and a steam cycle. (*Westinghouse Electric Corp.*)

from the more advanced gas turbines have increased, the design of the HRSG has become more complex.

The standard configuration of the HRSG, as shown in Fig. 1.9, is a vertically hung heat-transfer tube bundle with the exhaust gas flowing horizontally through the steam generator and with natural circulation for the water and steam. If required to meet emission regulations, selective catalytic reduction (SCR) elements for NO_x control (see Chap. 12) are placed between the appropriate tube bundles.

The advantages of gas turbine combined cycle power plants are the following:

1. Modular construction results in the installation of large, high-efficiency, base-loaded power plants in about 2 years.

2. Rapid, simple cycle startup of 5 to 10 minutes from no load to full load, which makes it ideal for peaking or emergency backup service.

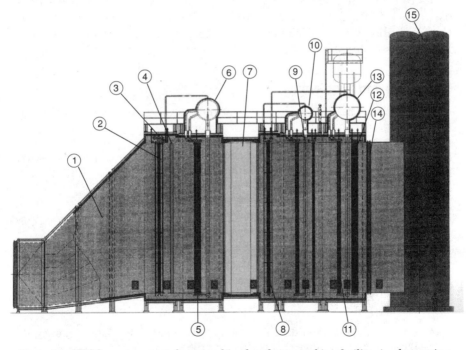

Figure 1.9 HRSG arrangement for a combined cycle gas turbine facility. 1 = hot casing. 2 = top-supported high-pressure superheater. 3 = desuperheating spray piping for control of steam temperature. 4 = ammonia injection grid for NO_x control system. 5 = high-pressure evaporator designed for natural circulation. 6 = high-pressure drum equipped with drum internals for control of steam purity. 7 = selective catalytic reduction (SCR) elements, which may be required to meet U.S. government regulations for NO_x emissions. 8 = high-pressure economizer tubes. 9 = low-pressure evaporator designed for natural circulation. 10 = low-pressure drum with drum internals. 11 = low-pressure economizer tubes. 12 = deaerator evaporator. 13 = integral deaerator drum. 14 = condensate preheater. 15 = stack with emission monitoring taps. (*DB Riley, Inc.*)

3. High exhaust temperatures and gas flows enable the efficient use of heat-recovery steam generators for the cogeneration of steam and power.

4. Low NO_x and CO emissions.

1.6 Summary

Steam is generated for many useful purposes from relatively simple heating systems to the complexities of a fossil fuel–fired or nuclear-fired electric utility power plant. All types of fuels are burned, and many different combustion systems are used to burn them efficiently and reliably.

This book will describe the various systems and equipment of a steam power plant that are so important to everyday life, whether it be for the generation of electricity, for heating, or for a process that leads to a product. The environmental control systems that are a necessary part of a modern plant are also thoroughly described because their reliability and efficiency are necessary to the successful operation of these plants.

Questions and Problems

1.1 Why are the study and understanding of steam power plants so important?

1.2 Describe the various forms of energy input to a power plant. Provide examples of the plant output that uses this energy.

1.3 Provide a list of the major uses of steam in industry.

1.4 What are the sources of heat that are used to generate steam?

1.5 What is a British thermal unit (Btu)?

1.6 Why is air pollution control equipment so important in the production of steam?

1.7 Provide a sketch and describe the operation of the Rankine steam cycle. What is the advantage of superheating the steam?

1.8 What are the major systems of a coal-fired power plant? Provide a brief description of each.

1.9 What is the purpose of an air heater?

1.10 Why is condensing the steam from the turbine and returning it to the boiler so important?

1.11 Why is a natural gas–fired boiler far less complex than a boiler that burns coal?

1.12 For a coal-fired or other solid fuel–fired boiler, what additional systems are necessary to account for the ash contained in the fuel?

1.13 What percentage of the total electric production results from steam power plants?

1.14 Why would a coal-fired plant be more difficult to obtain an operating permit for as compared with a plant fired with natural gas? Provide ideas on how this can be overcome.

1.15 For large utility boilers burning natural gas and oil, why have their use declined except for certain parts of the world?

1.16 Why are water treatment systems important to a well-operated power plant?

1.17 For most industrial-type boilers, what is the most distinguishing feature of this design? What is its purpose?

1.18 From an environmental point of view, what are the advantages of a fluidized bed boiler?

1.19 Name the two types of fluidized bed boilers and briefly describe their characteristics.

1.20 Describe a combined cycle system that uses a gas turbine. What are the advantages of this system? What is the single most important disadvantage?

2

Boilers

2.1 The Boiler

A boiler (or steam generator, as it is commonly called) is a closed vessel in which water, under pressure, is transformed into steam by the application of heat. Open vessels and those generating steam at atmospheric pressure are not considered to be boilers. In the furnace, the chemical energy in the fuel is converted into heat, and it is the function of the boiler to transfer this heat to the water in the most efficient manner.

A steam electric power plant is a means for converting the potential chemical energy of fuel into electrical energy. In its simplest form it consists of a boiler supplying steam to a turbine, and the turbine driving an electric generator.

The *ideal* boiler includes

1. Simplicity in construction, excellent workmanship, materials conducive to low maintenance cost, and high availability

2. Design and construction to accommodate expansion and contraction properties of materials

3. Adequate steam and water space, delivery of clean steam, and good water circulation

4. A furnace setting conducive to efficient combustion and maximum rate of heat transfer

5. Responsiveness to sudden demands and upset conditions

6. Accessibility for cleaning and repair

7. A factor of safety that meets code requirement

In general, the boiler must be conservatively designed to ensure reliable performance over the life of the plant. This conservative design is required because of all the variables that occur over the life of the plant, such as the use of different fuels, degradation of performance over time, and the occurrence of upset conditions.

The term *boiler setting* was applied originally to the brick walls enclosing the furnace and heating surface of the boiler. As the demand grew for larger-capacity steam generating units, the brick walls gave way to air-cooled refractory walls and then to water-cooled tube walls. The term *boiler setting* is used here to indicate all the walls that form the boiler and furnace enclosure and includes the insulation and lagging of these walls.[1]

A boiler should be designed to absorb the maximum amount of heat released in the process of combustion. This heat is transmitted to the boiler by *radiation, conduction,* and *convection,* the percentage of each depending on the boiler design.

Radiant heat is heat radiated from a hot to a cold body and depends on the temperature difference and the color of the body that receives the heat. Absorption of radiant heat increases with the furnace temperature and depends on many factors but primarily on the area of the tubes exposed to the heat.

Conduction heat is heat that passes from the gas to the tube by physical contact. The heat passes from molecule of metal to molecule of metal with no displacement of the molecules. The amount of absorption depends on the conductivity or heat-absorption qualities of the material through which the heat must pass.

Convection heat is heat transmitted from a hot to a cold body by movement of the conveying substance. In this case, the hot body is the boiler flue gas; the cold body is the boiler tube containing water.

In designing a boiler, each form of heat transmission is given special consideration. In the operation of a boiler unit, all three forms of heat transmission occur simultaneously and cannot readily be distinguished from each other.

Considerable progress has been made in boiler design from the standpoint of safety, efficiency of the fuel-burning equipment, and efficiency of the heat transferred. More and more emphasis is being placed on efficiency, flexibility, and boiler availability. Boiler designs are being developed not only for the traditional utility and industrial applications but also for plants designed for cogeneration of electricity and process steam. Boilers are also designed to burn low-grade coal, such as lignite, or to burn *municipal solid waste* (MSW) in the form of

---

[1]The definition is taken from *Steam: Its Generation and Use,* Babcock & Wilcox, a McDermott company.

mass burning or *refuse-derived fuel* (RDF) (see Chap. 13). The newer boilers are designed to be fully automated; their design also must take into account the environmental control equipment that is mandatory under regulations (see Chap. 12).

Boilers are built in a variety of sizes, shapes, and forms to fit conditions peculiar to the individual plant and to meet varying requirements. With increasing fuel cost, greater attention is being given to improvement of the combustion efficiency. Many boilers are designed to burn multiple fuels in order to take advantage of the fuel most economically available.

Increased boiler "availability" has made units of increased capacity practical, and this has resulted in lower installation and operating costs. For the small plant, all boilers preferably should be of the same type, size, and capacity, since standardization of equipment makes possible uniform operating procedures, reduces spare parts stock to a minimum, and contributes to lower overall costs.

The types of applications are many. Boilers are used to produce steam for heating, process, and power generation and to operate turbines, pumps, etc. This text is concerned with boilers used in stationary practice, although marine boilers and their systems have many of the same characteristics.

2.2   Fundamentals of Steam Generation

2.2.1   Boiling

The process of boiling water to make steam is a phenomenon that is familiar to all of us. After the boiling temperature is reached (e.g., 212°F at an atmospheric pressure of 14.7 psia), instead of the water temperature increasing, the heat energy from the fuel results in a change of phase from a liquid to a gaseous state, i.e., from water to steam. A steam-generating system, called a *boiler,* provides a continuous process for this conversion.

A kettle boiler, as shown in Fig. 2.1, is a simple example of such a device where a fixed quantity of water is heated. The heat raises the water temperature, and for a specific pressure, the boiling temperature (also called *saturation temperature*) is reached, and bubbles begin to form. As heat continues to be applied, the temperature remains constant, and steam flows from the surface of the water. If the steam were to be removed continuously, the water temperature would remain the same, and all the water would be evaporated unless additional water were added. For a continuous process, water would be regulated into the vessel at the same flow rate as the steam being generated and leaving the vessel.

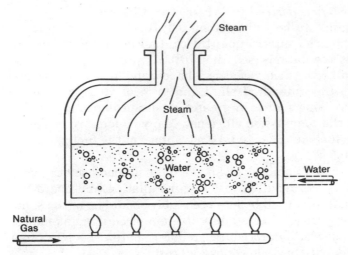

Figure 2.1   A kettle boiler. (*Babcock & Wilcox, a McDermott company.*)

2.2.2 Circulation

For most boiler or steam generator designs, water and steam flow through tubes where they absorb heat, which results from the combustion of a fuel. In order for a boiler to generate steam continuously, water must circulate through the tubes. Two methods are commonly used: (1) natural or thermal circulation and (2) forced or pumped circulation. These methods are shown in Fig. 2.2.

Natural circulation.   For natural circulation, no steam is present in the unheated tube segment identified as *AB*. With the input of heat, a steam-water mixture is generated in the segment *BC*. Because the steam-water mixture in segment *BC* is less dense than the water segment *AB*, gravity causes the water to flow down in segment *AB* and the steam-water mixture in *BC* to flow up into the steam drum. The rate of circulation depends on the difference in average density between the unheated water and the steam-water mixture.

The total circulation rate depends on four major factors:

1. *Height of boiler.* Taller boilers result in a larger total pressure difference between the heated and unheated legs and therefore can produce larger total flow rates.

2. *Operating pressure.* Higher operating pressures provide higher-density steam and higher-density steam-water mixtures. This

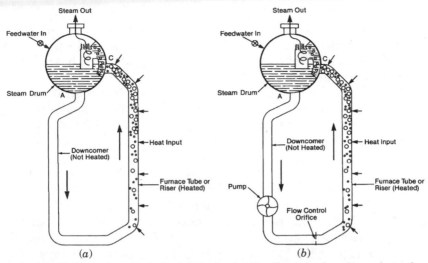

Figure 2.2 Boiler water circulation methods. (*a*) Simple natural or thermal circulation loop. (*b*) Simple forced or pumped circulation loop. (*Babcock & Wilcox, a McDermott company.*)

reduces the total weight difference between the heated and unheated segments and tends to reduce flow rate.

3. *Heat input.* A higher heat input increases the amount of steam in the heated segments and reduces the average density of the steam-water mixture, thus increasing total flow rate.

4. *Free-flow area.* An increase in the cross-sectional or free-flow area (i.e., larger tubes and downcomers) for the water or steam-water mixture may increase the circulation rate.

Boiler designs can vary significantly in their circulation rates. For each pound of steam produced per hour, the amount of water entering the tube can vary from 3 to 25 lb/h.

Forced circulation. For a forced circulation system, a pump is added to the flow loop, and the pressure difference created by the pump controls the water flow rate. These circulation systems generally are used where the boilers are designed to operate near or above the critical pressure of 3206 psia, where there is little density difference between water and steam. There are also designs in the subcritical pressure range where forced circulation is advantageous, and some boiler designs are based on this technology.

2.2.3 Steam-water separation

The steam-water mixture is separated in the steam drum. In small, low-pressure boilers, this separation can be accomplished easily with a large drum that is approximately half full of water and having natural gravity steam-water separation.

In today's high-capacity, high-pressure units, mechanical steam-water separators are needed to economically provide moisture-free steam from the steam drum (see Sec. 2.5). With these devices in the steam drum, the drum diameter and its cost are significantly reduced.

At very high pressures, a point is reached where water no longer exhibits the customary boiling characteristics. Above this critical pressure (3206 psia), the water temperature increases continuously with the addition of heat. Steam generators are designed to operate at these critical pressures, but because of their expense, generally they are designed for large-capacity utility power plant systems. These boilers operate on the "once-through" principle, and steam drums and steam-water separation are not required.

2.3 Fire-Tube Boilers

Fire-tube boilers are so named because the products of combustion pass through tubes or flues, which are surrounded by water. They may be either *internally* fired (Fig. 2.3) or *externally* fired (see Fig. 2.5). Internally fired boilers are those in which the grate and combustion chamber are enclosed within the boiler shell. Externally fired boilers are those in which the setting, including furnace and grates, is separate and distinct from the boiler shell. Fire-tube boilers are classified as vertical tubular or horizontal tubular.

The vertical fire-tube boiler consists of a cylindrical shell with an enclosed firebox (Figs. 2.3 and 2.4). Here tubes extend from the *crown sheet* (firebox) to the upper tube sheet. Holes are drilled in each sheet to receive the tubes, which are then rolled to produce a tight fit, and the ends are beaded over.

In the vertical *exposed-tube* boiler (Fig. 2.3), the upper tube sheet and tube ends are above the normal water level, extending into the steam space. This type of construction reduces the moisture carryover and superheats the steam leaving the boiler. However, the upper tube ends, not being protected by water, may become overheated and leak at the point where they are expanded into the tube sheet by tube expanders during fabrication. The furnace is water-cooled and is formed by an extension of the outer and inner shells that is riveted to the lower tube sheet. The upper tube sheet is riveted directly to the shell. When the boiler is operated, water is carried some distance

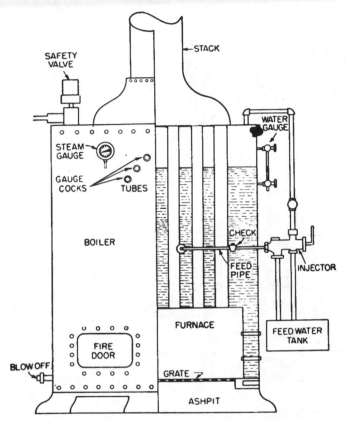

Figure 2.3 Sectional view of vertical fire-tube boiler, exposed-tube type.

below the top of the tube sheet, and the area above the water level is steam space. This design is seldom used today.

In *submerged-tube* boilers (Fig. 2.4), the tubes are rolled into the upper tube sheet, which is below the water level. The outer shell extends above the top of the tube sheet. A cone-shaped section of the plate is riveted to the sheet so that the space above the tube sheet provides a smoke outlet. Space between the inner and outer sheets comprises the steam space. This design permits carrying the water level above the upper tube sheet, thus preventing overheating of the tube ends. This design is also seldom used today.

Since vertical boilers are portable, they have been used to power hoisting devices and operate fire engines and tractors as well as for stationary practice and still do in some parts of the world. They range in size from 6 to 75 bhp; tube sizes range from 2 to 3 in in diameter;

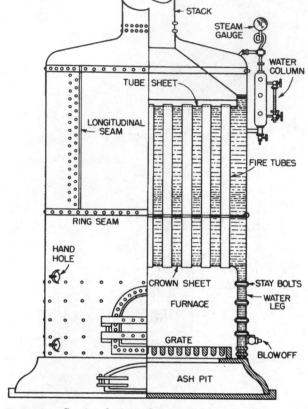

Figure 2.4 Sectional view of vertical fire-tube boiler, sub-merged-tube type.

pressures to 100 psi; diameters from 3 to 5 ft; and height from 5 to 10 ft. With the exposed-tube arrangement, 10 to 15°F of superheat may be obtained.

Vertical fire-tube boilers are rapid steamers, their initial cost is low, and they occupy little floor space. Boilers of this type usually employ a standard base. Combustion efficiency is improved when the boiler is elevated and set on a refractory base to obtain added furnace volume. This is especially important if bituminous coal is to be burned and smoke is to be reduced to a minimum. If the boiler is stoker-fired, either raise the boiler or pit the stoker for the required setting height.

Horizontal fire-tube boilers are of many varieties, the most common being the *horizontal-return tubular* (HRT) boiler (Fig. 2.5). This boiler has a long cylindrical shell supported by the furnace sidewalls and is set on saddles equipped with rollers to permit movement of the boiler as it expands and contracts. It also may be suspended from hangers

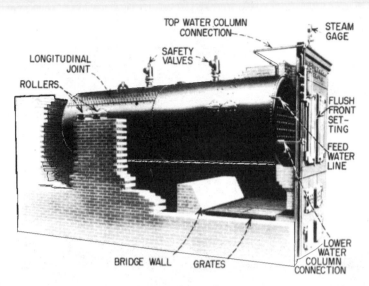

Figure 2.5 Horizontal-return tubular boiler and setting. (*Zurn Industries, Inc.*)

(Fig. 2.6) and supported by overhead beams. Here the boiler is free to move independently of the setting. Expansion and contraction do not greatly affect the brick setting, and thus maintenance is reduced.

In the original designs of this boiler, the required boiler shell length was secured by riveting (see Fig. 2.5) several plates together. The seam running the length of the shell is called a *longitudinal joint* and is of butt-strap construction. Note that this joint is above the fire line to avoid overheating. The *circumferential joint* is a lap joint.

Today's design of a return tubular boiler (see Fig. 2.7) has its plates joined by fusion welding. This type of construction is superior to that of a riveted boiler because there are no joints to overheat. As a result, the life of the boiler is lengthened, maintenance is reduced, and at the same time higher rates of firing are permitted. Welded construction is used in modern boiler design.

The boiler setting of Fig. 2.6 includes grates (or stoker), bridge wall, and combustion space. The products of combustion are made to pass from the grate, over the bridge wall (and under the shell), to the rear end of the boiler. Gases return through the tubes to the front end of the boiler, where they exit to the breeching or stack. The shell is bricked in slightly below the top row of tubes to prevent overheating of the longitudinal joint and to keep the hot gases from coming into contact with the portion of the boilerplate that is above the waterline.

The conventional HRT boiler is set to slope from front to rear. A blowoff line is connected to the underside of the shell at the rear end

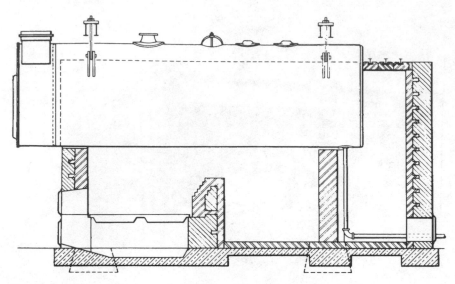

Figure 2.6 Horizontal-return tubular boiler and setting, overhanging front.

of the boiler to permit drainage and removal of water impurities. It is extended through the setting, where blowoff valves are attached. The line is protected from the heat by a brick lining or protective sleeve. Safety valves and the water column are located as shown in Fig. 2.5. A *dry pipe* is frequently installed in the top of the drum to separate the moisture from the steam before the steam passes to the steam outlet.

Still another type of HRT boiler is the horizontal four-pass forced-draft packaged unit (Fig. 2.7), which can be fired with natural gas or fuel oil. In heavy oil-fired models, the burner has a retractable nozzle for ease in cleaning and replacing. It is this type of design that is the most common fire-tube boiler found in today's plants.

Gases from the combustion chamber reverse at the rear to pass downward to the tubes directly beneath the chamber. Again they reverse to pass through the tube bank above the combustion chamber and reverse and pass through the top tube section to the stack, thus making a four-pass unit.

Such units are available in sizes of 15 to 800 bhp (approximately 1000 to 28,000 lb/h) with pressures of 15 to 250 psi. Some units are designed for nearly 50,000 lb/h. These units are compact, requiring a minimum of space and headroom, are automatic in operation, have a low initial cost, and do not need a tall stack. For these reasons, they find application and acceptance in many locations. Because of their compactness, however, they are not readily accessible for inspection and repairs. Larger fire-tube boilers tend to be less expensive and use

Figure 2.7 Horizontal four-pass fire-tube package boiler designed for natural gas firing. (*Cleaver-Brooks, a Division of Aqua-Chem, Inc.*)

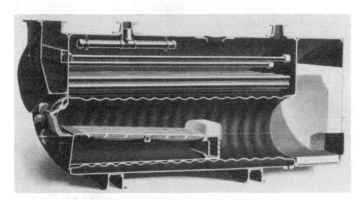

Figure 2.8 Scotch marine boiler.

simpler controls than water-tube units; however, the large shells of these fire-tube boilers limit them to pressures less than 250 psi.

The Scotch marine boiler (Fig. 2.8) is a horizontal fire-tube return tubular unit in which the combustion chamber is surrounded by water. It consists of a cylindrical shell containing the firebox and tubes. The tubes surround the upper portion of the firebox and are rolled into tube sheets on each end of the boiler. Combustion gases pass to the rear of the furnace, returning through the tubes to the front, where they are discharged to the stack.

The water-cooled furnace and limited furnace volume of the Scotch marine boiler make smokeless combustion difficult when firing bituminous coal unless overfire air or steam jets are used.

Scotch marine boilers are self-contained, do not require a setting, and are internally fired. They are portable packaged units requiring a minimum of space and headroom. Units of this type are found in marine and stationary service burning coal, oil, and natural gas; however, this design has limited use in today's modern boiler designs. For coal burning, the long grates make cleaning of fires and handling of ashes very difficult.

Fire-tube boilers serve in most industrial plants where saturated steam demand is less than 50,000 lb/h and pressure requirements are less than 250 psig. With few exceptions, nearly all fire-tube boilers made today are packaged designs that can be installed and in operation in a short period of time.

Fire-tube boilers were designed originally for hand firing of coal, wood, and other solid fuels; however, today nearly all are designed for oil or natural gas firing and are generally similar in design to the unit shown in Fig. 2.7. Solid fuel firing can be accommodated if there is enough space underneath the boiler to add firing equipment and the required furnace volume to handle the combustion of the fuel. The burning of solid fuels also requires environmental control equipment for particulate removal and possibly for SO_2 removal depending on local site requirements. These added complexities and costs basically have eliminated fire-tube boilers for consideration when firing solid fuels.

2.4 Water-Tube Boilers

A water-tube boiler is one in which the products of combustion (called *flue gas*) pass around tubes containing water. The tubes are interconnected to common water channels and to the steam outlet. For some boilers, baffles to direct the flue gas flow are not required. For others, baffles are installed in the tube bank to direct the flue gas across the heating surfaces and to obtain maximum heat absorption. The baffles may be of refractory or membrane wall construction, as discussed later. There are a variety of boilers designed to meet specific needs, so care must be exercised in the selection, which should be based on plant requirements, fuel considerations, and space limitations. Water-tube boilers generally may be classified as straight tube and bent tube.

The electric steam boiler (Fig. 2.9) provides steam at high pressure. It is a packaged unit generating steam for heating and process. The small units (1000 to 10,000 lb/h of steam) operate at low voltage, while the larger units (7000 to 100,000 lb/h of steam) operate at

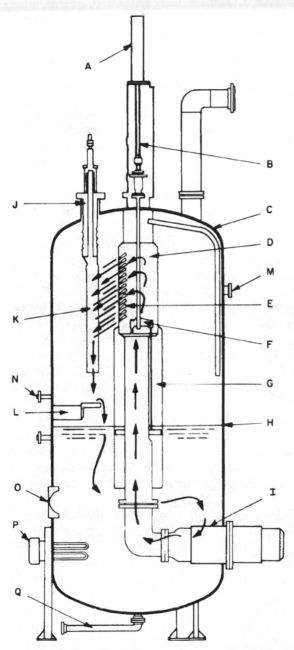

Figure 2.9 Electrode high-voltage steam boiler, interior view.
A = Control cylinder. B = Control cylinder rod. C = Boiler
shell. D = Jet column. E = Jets. F = Control linkage. G =
Control sleeve. H = Water level. I = Circulating pump. J =
Insulator. K = Electrode. L = Counterelectrode. M = Safety
valve. N = Water level control. O = Manhole. P = Standby
heater. Q = Tank drain.

13,800 V. Such units find application in educational institutions, commercial and office buildings, hospitals, processing plants, etc.

Operation is as follows: Water from the lower part of the boiler shell (C) is pumped to the jet column (D) and flows through the jets (E) to strike the electrodes (K), thus creating a path for the electric current. As the unevaporated portion of the water flows from the electrode to the counterelectrode (L), a second path for current is created. Regulation of the boiler output is accomplished by hydraulically lifting the control sleeve (G) to intercept and divert the streams of water from some or all of the jets (E); this prevents the water from striking the electrode. The control sleeve (G) is moved by the lift cylinder (B), which is positioned by the boiler pressure and load control system to hold the steam pressure constant or to limit the kilowatt output to a desired level.

To shut the boiler off, it is only necessary to stop the pump. A proportioning-type feedwater regulator (not shown) is used to maintain a constant water level. Water failure simply causes the boiler to cease operation with no overheating or danger involved.

The advantages of the electric steam generator are compactness, safety of operation, absence of storage tanks, the ability to use electric power during off-peak periods, and responsiveness to demand. Its disadvantages lie in the use of high voltage, high power costs, and availability, if it must be used during other than off-peak periods. When such an installation is contemplated, all factors, including the initial cost, need to be considered.

The boiler shown in Fig. 2.10 is an early design of a water-tube boiler that was designed to burn coal on a stoker. It has vertical inclined tubes and has four drums. The upper drums are set on saddles fastened to horizontal beams, the center drum is suspended from an overhead beam by slings, and the lower drum (mud drum) hangs free, suspended from the tubes.

Water enters the right-hand drum (top) and flows down the vertical bank of tubes to the lower (mud) drum. It then moves up the inclined bank of tubes because of natural circulation as a result of heating from the flue gases, passing through the center drum and returning to its point of origin. Steam is made to pass around a baffle plate. In the process, most of the entrained moisture is removed before the steam enters the circulators. The steam in passing to the steam drum through the circulators receives a small degree of superheat, 10 to 15°F, by the time it enters the steam drum.

Another older design is the cross-drum straight-tube sectional-header boiler shown in Fig. 2.11, arranged for oil firing and equipped with an interdeck superheater. It is designed with a steep inclination of the main tube bank to provide rapid circulation. Boilers of this general type fit into locations where headroom was a criterion. In Figs.

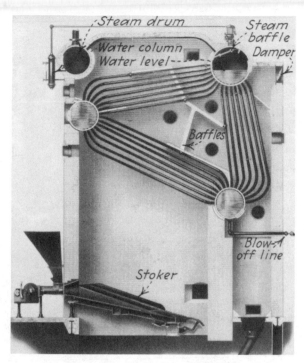

Figure 2.10   Water-tube boiler, four-drum type. (*DB Riley, Inc.*)

2.10 and 2.11, note the refractory furnace with these designs, and compare these designs with the water-cooled furnace designs of boilers described in other figures and as designed today.

The Stirling boiler shown in Fig. 2.12 was basically the next step in boiler design and had several design features to meet various space and headroom limitations. In this design, three drums are set transversely, interconnected to the lower drum by tubes slightly inclined. The tubes are shaped and bent at the ends so that they enter the drum radially.

The upper drums are interconnected by steam circulators (top) and by water circulators (bottom). Note that the center drum has tubes leaving, to enter the rear tube bank. This is done to improve the water circulation. The heating surface is then a combination of water-wall surface, boiler tubes, and a small amount of drum surface. The interdeck superheater and economizer likewise contain heating surface.

The furnace is water-cooled as compared with refractory lined. Downcomers from the upper drum supply water to the sidewall headers, with a steam-water mixture returning to the drum from the wall tubes. Feedwater enters the economizer, where it is initially heated

Figure 2.11   Cross-drum straight-tube boiler.
(*Babcock & Wilcox, a McDermott company.*)

and then enters the left (top) drum and flows down the rear bank of tubes to the lower (mud) drum. Steam generated in the first two banks of boiler tubes returns to the right and center drums; note the interconnection of drums, top and bottom. Finally, all the steam generated in the boiler and waterwalls reaches the left-hand drum, where the steam is made to pass through baffles or a steam scrubber or a combination of the two, to reduce the moisture content of the steam before it passes to the superheater. The superheater consists of a series of tube loops. The steam then passes to the main steam line or steam header.

The upper drums are supported at the ends by lugs resting on steel columns. The lower drum is suspended from the tubes and is free to move by expansion, imposing no hardship on the setting. The superheater headers are supported (at each end) from supports attached to steel columns overhead.

The flue gas resulting from combustion passes over the first bank of tubes, through the superheater, and down across the second pass; the flue gas then reverses in direction and flows up through the third

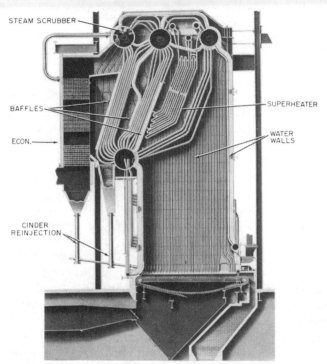

STEAM SCRUBBER

BAFFLES

ECON.

SUPERHEATER

WATER
WALLS

CINDER
REINJECTION

Figure 2.12 Stirling boiler, four-drum type fired with a spreader stoker. (*Babcock & Wilcox, a McDermott company.*)

pass. Note the baffles that direct the flow of the flue gas. On leaving the boiler, the flue gas enters the economizer, traveling down (in a counterflow direction to the water flow) through the tubes to the exit.

Most of the steam is generated in the waterwalls and the first bank of boiler tubes, since this heating surface is exposed to radiant heat. This unit is fired with a spreader stoker. It can also be fired with gas, oil, or a combination of the two, providing flexibility in operation. Fly ash (containing unburned carbon particles) is collected at the bottom of the third pass and from the economizer and is reinjected back onto the grate through a series of nozzles located in the rear wall, thus improving the boiler efficiency and lowering fuel costs. Over-fire air is introduced through nozzles in the rear walls and sidewalls to improve combustion.

Although the three- and four-drum designs, as well as the cross-drum design, shown in Figs. 2.10 to 2.12, have been replaced with the modern two-drum and one-drum designs, many of these older units are still in operation today.

There are many varieties of packaged boilers. The smaller units are completely factory assembled and ready for shipment. The larger units are of modular construction with final assembly and erection done in the field.

The FM package boiler (Fig. 2.13) is available in capacities from 10,000 to 200,000 lb/h of steam, with steam pressures of 525 to 1050 psi and steam temperatures to 825°F. Superheaters, economizers, and air preheaters can be added, with operating and economic design consideration.

The packaged unit includes burners, soot blowers, forced draft fan, controls, etc. The steam drum is provided with steam separating devices to meet steam purity requirements.

The features of the FM boiler are (1) furnace waterwall cooling including sidewalls, rear wall, roof, and floor, which eliminates the need for refractory and its associated maintenance; (2) a gas-tight setting preventing gas leaks; (3) a steel base frame that supports the entire boiler; (4) an outer steel lagging permitting an outdoor installation; (5) drum internals providing high steam purity; (6) tube-bank access ports providing ease in inspection; and (7) soot blowers providing boiler bank cleaning.

The shop-assembled unit can be placed into service quickly after it is set in place by connecting the water, steam, and electric lines; by making the necessary flue connection to the stack; and depending on the size of the unit, by connecting the forced-draft fan and the associated duct.

Figure 2.13 Type FM integral furnace package boiler for fuel oil and natural gas firing. (*Babcock & Wilcox, a McDermott company.*)

Flue gases flow from the burner to the rear of the furnace (i.e., through the radiant section), reversing to pass through the super-heater and convection passes. These units are oil, gas, or combination fired. A combustion control system accompanies the boiler installation.

The advantages of package boilers over field-erected boilers are (1) lower cost, (2) proven designs, (3) shorter installation time, and (4) generally a single source of responsibility for the boiler and necessary auxiliaries. However, package boilers are limited by size because of shipping restrictions and generally can fire only gas and oil. Nevertheless, where shipping permits, package boilers are designed for capacities up to 600,000 lb/h and steam pressures to 1800 psig and steam temperatures to 900°F.

Figure 2.14 shows an MH series package boiler in the process of shop assembly. These units are oil and gas fired and can be designed for capacities ranging from 60,000 to 200,000 lb/h with steam pressures to 1000 psi and steam temperatures to 800°F. Designs vary between manufacturers, as do the design pressures and temperatures.

The SD steam generator (Fig. 2.15) has been developed to meet demands of power and process steam in a wide range of sizes. This

Figure 2.14 MH series package boiler during shop assembly. (*DB Riley, Inc.*)

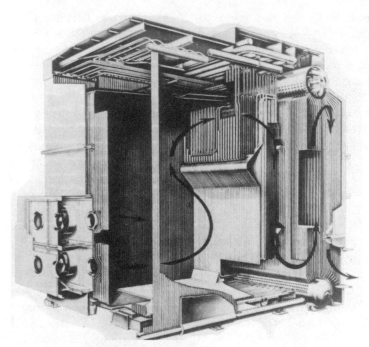

Figure 2.15 SD type steam generator. (*Foster Wheeler Corp.*)

unit is available in capacities to 800,000 lb/h of steam, steam pressures to 1600 psi, and steam temperatures to 960°F. It is a complete
waterwall furnace construction, with a radiant and convection
superheater. If economics dictate, an air preheater or economizer
(or combination) can be added. The unit shown is for a pressurized
furnace designed for oil, gas, or waste fuels such as coke-oven or
blast furnace gases.

The combination radiant-convection superheaters provide a relatively constant superheat temperature over the normal operating
range. When superheat control is necessary, a spray-type desuperheater (attemperator) is installed. This is located in an intermediate
steam temperature zone that ensures mixing and rapid, complete
evaporation of the injected water.

The rear drum and lower waterwall headers are bottom supported,
permitting upward expansion. Wall construction is such that the unit
is pressure-tight for operating either with a balanced draft or as a
pressurized unit. The steam drum is equipped with chevron dryers
and horizontal separators to provide dry steam to the superheater.

The prominent nose at the top of the furnace ensures gas turbulence and good distribution of gases as they flow through the boiler.

The unit is front fired and has a deep furnace, from which the gases pass through the radiant superheater, down through the convection superheater, up and through the first bank of boiler tubes, and then down through the rear bank of tubes to the boiler exit (to the right of the lower drum).

The outer walls are of the welded fin-tube type; baffles are constructed of welded fin tubes; the single-pass gas arrangement reduces the erosion in multiple-pass boilers that results from sharp turns in the gas stream.

High-temperature water (HTW) boilers.   High-temperature water (HTW) boilers provide hot water under pressure for space heating of large areas such as buildings. Water is circulated at pressures up to 450 psig through the system. The water leaves the HTW boiler at subsaturated temperatures (i.e., below the boiling point at that pressure) up to 400°F. (Note that the boiling temperature at 450 psig is approximately 458°F.) Sizes generally range up to 60 million Btu/h for package units and larger for field-erected units, and the units are designed for oil or natural gas firing. Most units are shop assembled and shipped as packaged units. The large units are shipped in component assemblies, and it is necessary to install refractory and insulation after the pressure parts are erected.

The high-pressure water can be converted into low-pressure steam for process. For example, with a system operating at a maximum temperature of 365°F, the unit is capable of providing steam at 100 psi.

A *high-temperature water system* is defined as a fluid system operating at temperatures above 212°F and requiring the application of pressure to keep the water from boiling. Whereas a steam boiler operates at a fixed temperature that is its saturation temperature, a water system, depending on its use, can be varied from an extremely low to a relatively high temperature.

The average water temperature within a complete system will vary with load demand, and as a result, an expansion tank is used to provide for expansion and contraction of the water volume as its average temperature varies. To maintain pressure in the system, steam pressurization or gas pressurization is used in the expansion tank. For the latter, air or nitrogen is used.

The pressure is maintained independent of the heating load by means of automatic or manual control. Firing of each boiler is controlled by the water temperature leaving the boiler.

The hot-water system is advantageous because of its flexibility. For the normal hot-water system there are no blowdown losses and little or no makeup, installation costs are lower than for a steam-heating system, and the system requires less attention and maintenance. The system can be smaller than an equivalent steam system because of the huge water-storage capacity required by a steam system; peak loads and pickup are likewise minimized, with resulting uniform firing cycles and higher combustion efficiency.

The high-temperature water system is a closed system. When applied to heating systems, the largest advantage is for the heating of multiple buildings. For such applications, the simplicity of the system helps reduce the initial cost.

Only a small amount of makeup is required to replace the amount of water that leaks out of the system at valve stems, pump shafts, and similarly packed points. Since there is little or no free oxygen in the system, return-line corrosion is reduced or eliminated, which is in contrast to wet returns from the steam system, wherein excessive maintenance is frequently required. Feedwater treatment can be reduced to a minimum.

Comparison of fire- and water-tube boilers. Fire-tube boilers ranging to 800 bhp (approximately 28,000 lb/h) and oil- and gas-fired water-tube boilers with capacities to 200,000 lb/h are generally shop assembled and shipped as one package. The elimination of field-assembly work, the compact design, and standardization result in a lower cost than that of comparable field-erected boilers.

Fire-tube boilers are preferred to water-tube boilers because of their lower initial cost and compactness and the fact that little or no setting is required. They occupy a minimum of floor space. Tube replacement is also easier on fire-tube boilers because of their accessibility. However, they have the following inherent disadvantages: the water volume is large and the circulation poor, resulting in slow response to changes in steam demand and the capacity, pressure, and steam temperature are limited.

Packaged boilers may be of the fire-tube or water-tube variety. They are usually oil or gas fired. Less time is required to manufacture packaged units; therefore, they can meet shorter project schedules. Completely shop assembled, they can be placed into service quickly. The packaged units are automatic, requiring a minimum of attention, and hence reduce operating costs. In compacting, however, the furnace and heating surfaces are reduced to a minimum, resulting in high heat-transfer rates, with possible overheating and potential increased maintenance and operating difficulties. Thus caution must be exercised in the selection of packaged units because the tendency toward compactness can be carried too far. Such compactness also makes the units somewhat inaccessible for repairs. Because these units operate at high ratings and high heat transfer, it is important to provide optimal water conditioning at all times; otherwise, overheating and damage to the boiler may result.

Water quality is critical for successful boiler operation. Impurities in the water can quickly destroy the boiler and its components. Poor water quality can damage or plug water-level controls and cause unsafe operating conditions. Water treatment (refer to Chap. 11) is provided by a variety of equipment: deaerators for the removal of oxygen, water softeners, chemical additives, and boiler blowdown packages. In all cases, boiler water must be analyzed to determine its composition and the type of water treatment that is required.

Water-tube boilers are available in various capacities for high-pressure and high-temperature steam. The use of tubes of small diameter results in rapid heat transmission, rapid response to steam demands, and high efficiency. Water-tube boilers require elaborate settings, and initial costs are generally higher than those of fire-tube boilers in the range for which such units are most frequently designed. However, when this capacity range is exceeded for high-pressure and high-temperature steam, only the water-tube boiler is available. Air filtration, which plagued the earlier water-tube boiler, has now been minimized in the design by means of membrane tube water walls, improved expansion joints, and casings completely enclosing the unit. Feedwater regulation is no longer a problem when the automatic feedwater regulator is used. In addition, water-tube boilers are capable of burning any economically available fuel with excellent efficiency, whereas packaged units must use liquid or gaseous fuels to avoid fouling the heating surfaces.

Thus, in selecting a boiler, many factors other than first cost are to be considered. Important are availability, operating and maintenance costs, fuel costs, space, and a host of other factors. Most important perhaps are fuel costs. During the life of the equipment, we can expect fuel costs to be many times the cost of the boiler and associated equipment.

2.5   Steam-Water Separation

In the past, difficulty with carry-over and impurities in the steam was frequently encountered (*carry-over* is the passing of water and impurities to the steam outlet). Efforts were directed to reduce carry-over to a minimum by separating the water from the steam through the installation of baffles and the dry pipe. Both measures met with some success. The dry pipe ran the length of the drum, the ends being closed and the upper side of the pipe being drilled with many small holes. The top center of this pipe was connected to the steam outlet. Steam entering through the series of holes was made to change direction before entering the steam outlet, and in the process the water was separated from the steam. The bottom of the dry pipe contained a drain, which ran below the normal water level in the drum. The dry pipe was installed near the top of the drum so as not to require removal for routine inspection and repairs inside the drum.

The dry pipe proved to be fairly effective for small boilers but unsuited for units operating at high steam capacity. Placing a baffle ahead of the dry pipe offered some slight improvement in steam quality but was still not considered entirely satisfactory.

Modern practice requires high-purity steam for process, for the superheater, and for the turbine. An important contribution to increased boiler capacity and high rating is the fact that the modern boiler is protected by clean, high-quality feedwater. The application of both external and internal feedwater treatment is supplemented by the use of steam scrubbers and separators that are located in the steam drum.

Steam drums are used on recirculating boilers that operate at sub-critical pressures. The primary purpose of the steam drum is to separate the saturated steam from the steam-water mixture that leaves the heat-transfer surfaces and enters the drum. The steam-free water is recirculated within the boiler with the incoming feedwater for further steam generation. The saturated steam is removed from the drum through a series of outlet nozzles, where the steam is used as is or flows to a superheater for further heating.

The steam drum is also used for the following:

1. To mix the saturated water that remains after steam separation with the incoming feedwater

2. To mix the chemicals that are put into the drum for the purpose of corrosion control and water treatment

3. To purify the steam by removing contaminants and residual moisture

4. To provide the source for a blowdown system where a portion of the water is rejected as a means of controlling the boiler water chemistry and reducing the solids content

5. To provide a storage of water to accommodate any rapid changes in the boiler load.

The most important function of the steam drum, however, remains as the separation of steam and water. Separation by natural gravity can be accomplished with a large steam-water surface inside the drum. This is not the economical choice in today's design because it results in larger steam drums, and therefore the use of mechanical separation devices is the primary choice for separation of steam and water.

Efficient steam-water separation is of major importance as it obtains high-quality steam that is free of moisture. This leads to the following key factors in efficient boiler operation:

1. It prevents the carry-over of waterdroplets into the superheater, where thermal damage could result.

2. It minimizes the carry-under of steam with the water that leaves the drum, where this residual steam would reduce the circulation effectiveness of the boiler.

3. It prevents the carry-over of solids. Solids are dissolved in the waterdroplets that may be entrained in the steam if not separated properly. By proper separation, this prevents the formation of deposits in the superheater and ultimately on the turbine blades.

Boiler water often contains contaminants that are primarily in solution. These contaminants come from impurities in makeup water, treatment chemicals, and leaks within the condensate system such as the cooling water. Impurities also occur from the reaction of boiler water and contaminants with the materials of the boiler and of the equipment prior to entering the boiler. The steam quality of a power plant depends on proper steam-water separation as well as the feedwater quality, and this is a major consideration to having a plant with high availability and low maintenance costs. Even low levels of solids in the steam can damage the superheater and turbine, causing significant outages, high maintenance costs, and loss of production revenues.

Prior to the development of quality steam-water separators, gravity alone was used for separation. Because the steam drum diameter requirements increased significantly, the use of a single drum became uneconomical, and therefore it became necessary to use multiple smaller drums. Figure 2.12 is an example of this showing three steam drums. Although there are units of this design still in operation, they are no longer common.

The cyclone separators illustrated in cross-sectional elevation in Figs. 2.16 and 2.17 overcome many of the shortcomings previously mentioned for the baffle and dry pipe. Depending on the size of the boiler, there is a single or double row of cyclone steam separators with scrubbers running the entire length of the drum. Baffle plates are located above each cyclone, and there is a series of corrugated scrubber elements at the entrance to the steam outlet. Water from the scrubber elements drains to a point below the normal water level and is recirculated in the boiler.

In the installation shown in Fig. 2.17, operation is as follows: (1) The steam-water mixture from the risers enter the drum from behind

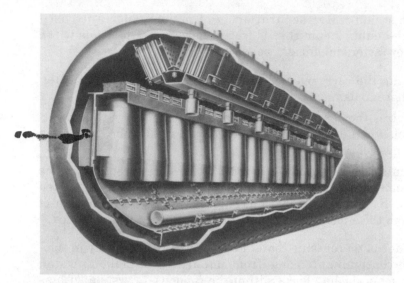

Figure 2.16 Single-row arrangement of cyclone steam separators with secondary scrubbers. (*Babcock & Wilcox, a McDermott company.*)

Figure 2.17 Double-row arrangement of cyclone steam separators for primary steam separation with secondary scrubber elements at top of drum. (*Babcock & Wilcox, a McDermott company.*)

the baffle plate before entering the cyclone; the cyclone is open at top and bottom. (2) Water is thrown to the side of the cyclone by centrifugal force. (3) Additional separation of water and steam occurs in the passage of steam through the baffle plates. (4) On entering the scrubber elements, water is also removed with steam passing to the steam outlet. Separators of this type can reduce the solids' carry-over to a very low value depending on the type of feedwater treatment used, the rate of evaporation, and the concentration of solids in the water. The cyclone and scrubber elements are removable for cleaning and inspection and are accessible from manways that are located in the ends of the steam drum.

The combination of cyclone separators and scrubbers provides the means for obtaining steam purity corresponding to less than 1.0 part per million (ppm) solids content under a wide variation of operating conditions. This purity is generally adequate in commercial practice; however, the trend to higher pressures and temperatures in steam power plants imposes a severe demand on steam-water separation equipment.

2.6 Principles of Heat Transfer

The design of a furnace and boiler is largely related to heat transfer. Heat flows from one body to another by virtue of a temperature difference, and it always flows in the direction of higher temperature to lower temperature. This heat flow takes place through the use of one or more of three methods: conduction, convection, and radiation. At times, all three methods of heat transfer are involved, supplementing or complementing each other.

1. *Conduction.* In a solid body, the flow of heat is by means of conduction and is accomplished through the transfer of kinetic energy from one molecule to another, even though the substance as a whole is at rest. Fluidized bed boilers, with their high solids content, use a high percentage of this form of heat transfer.

2. *Convection.* In liquids or gases, heat may be transferred from one point to another through the movement of substance. For example, the hot flue gas transfers its heat to a tube, where the absorbed heat is transferred to the fluid that flows through the tubes, in most boiler designs this being water and steam.

3. *Thermal radiation.* Every substance, solid, liquid, or gas, is capable of emitting electromagnetic waves, by which thermal energy may be transmitted. In boiler and furnace design, radiation of gas to a solid is very important. Some, but not all, gases absorb and

radiate heat. The radiation of the gases originates from the oscillations of charged atoms within the molecules of the gases. Some basic gases such as hydrogen, nitrogen, and oxygen are comprised of molecules in which the atoms are completely neutral, and these gases cannot transmit or absorb radiation. However, gases such as carbon dioxide (CO_2) and vapor such as water vapor (H_2O) do have charged atoms and can radiate and absorb considerable thermal energy. Since flue gas in a boiler has considerable quantities of these two gases, it therefore possesses considerable radiating power.

2.7 Superheaters

Steam that has been heated above the saturation temperature corresponding to its pressure is said to be *superheated*. This steam contains more heat than does saturated steam at the same pressure (see Chap. 3), and the added heat provides more energy for the turbine for conversion to electric power.

A superheater surface is that surface which has steam on one side and hot gases on the other. The tubes are therefore dry with the steam that circulates through them. Overheating of the tubes is prevented by designing the unit to accommodate the heat transfer required for a given steam velocity through the tubes, based on the desired steam temperature. To accomplish this, it is necessary to have the steam distributed uniformly to all the superheater tubes and at a velocity sufficient to provide a scrubbing action to avoid overheating of the metal. Carry-over from the steam drum must be at a minimum.

Superheaters are referred to as *convection, radiant,* or *combination* types. The convection superheater is placed somewhere in the gas stream, where it receives most of its heat by convection. As the name implies, a radiant superheater is placed in or near the furnace, where it receives the majority of its heat by radiation.

The conventional convection-type superheater uses two headers into which seamless tubes are rolled or welded. The headers are baffled so that the steam is made to pass back and forth through the connecting tubes, which carry their proportioned amount of steam, the steam leaving at the desired temperature. The headers are small, and access to the tubes is achieved by removing handhole caps similar to those for boiler-tube access.

With either the radiant or the convection-type superheater, it is difficult to maintain a uniform steam-outlet temperature, so a combination superheater is often installed (Fig. 2.18). The *radiant* section is shown above the screen tubes in the furnace; the *convection* section lies between the first and second gas passages. Steam leaving the

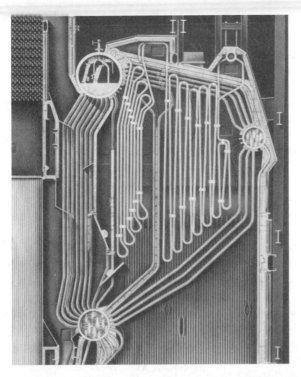

Figure 2.18 Combination radiant- and convection-type superheater. (*DB Riley, Inc.*)

boiler drum first passes through the convection section, then to the radiant section, and finally to the outlet header. Even this arrangement may not produce the desired results in maintaining a constant steam temperature within the limits prescribed, and so a bypass damper, shown at the bottom of the second pass of the boiler, is sometimes used. A damper of this type can be operated to bypass the gas or a portion of the gas around the convection section, thus controlling the final steam-outlet temperature for various boiler capacities. Figures 2.22 and 2.25 also are examples of combination superheaters for modern utility-type boilers.

For many applications a constant superheat temperature is required over a load range. Consequently, sufficient surface must be installed to ensure that this temperature can be obtained at the minimum load condition. For higher loads, unless the temperature is controlled, the superheat temperature will exceed limits because of the amount of installed surface. For this reason, a desuperheater (or attemperator) is used to control the superheat temperature. This is accomplished by mixing water (in a spray-type attemperator) with the steam in adequate amounts at an intermediate stage of the super-heater. This ensures that the maximum steam temperature is not exceeded. (A submerged-surface type of attemperator is often used and is installed in the lower drum where two-drum boilers are used.)

The final exit-steam temperature is influenced by many factors, such as gas flow, gas velocity, gas temperature, steam flow and velocity, ash accumulation on furnace walls and heat-transfer surfaces, method of firing, burner arrangement, type of fuel fired, etc.

In the *convection* superheater, steam temperature increases with the capacity, whereas in the *radiant* superheater, steam temperature decreases with the capacity. For maximum economy, a constant superheat temperature is desirable. Moreover, uniform steam temperatures are desired to meet the design requirements of a steam turbine or a particular process.

Overheating of the superheater tubes must be prevented in the design, and this requires uniform steam flow and steam velocity through the tubes. This is accomplished in a variety of ways: by spacing the takeoffs from the steam drum to the superheater, by installing baffles in the superheater header, by placing ferrules in the tubes at the steam entrance to the tubes, or by other means. Care must be exercised to obtain uniform flow without an excessive pressure drop through the superheater because this has an impact on the design pressure of the boiler and thus its costs.

A reheater (or reheat superheater) is used in utility applications for the reheating of steam after it leaves the high-pressure portion of a turbine. Thus a superheater and reheater incorporated into the boiler design will improve the overall plant efficiency.

Some modern boilers may have twin furnaces, one containing the superheater and the other the reheater section. The superheater is usually a combination radiant and convection section, with various types of arrangements. Constant steam temperature is obtained with the use of an attemperator or other methods, as described later.

Various methods are used to support the superheater tubes, as shown in Figs. 2.18 and 2.19, as well as those shown in other illustrations in this book. Superheater tubes vary in size generally from 1 to 3 in in diameter, and steam temperatures range to 1050°F.

Superheated steam has many advantages: It can be transmitted for long distances with little heat loss; condensation is reduced or eliminated; superheated steam contains more heat (energy), and hence less steam is required; and erosion of turbine blading is reduced because of the elimination of moisture in the steam.

The design of the superheater in a boiler is very important to the overall boiler design. Incorrect design of the superheater could result

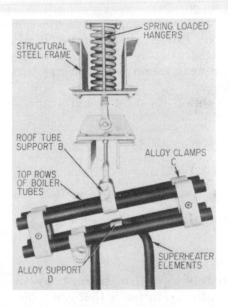

STRUCTURAL STEEL FRAME

SPRING LOADED HANGERS

ROOF TUBE SUPPORT B

ALLOY CLAMPS C

TOP ROWS OF BOILER TUBES

ALLOY SUPPORT D

SUPERHEATER ELEMENTS

Figure 2.19 Superheater support and tube clamps. (*DB Riley, Inc.*)

in excessive initial boiler costs and in future high maintenance costs. Two areas that influence the design are tube side spacing and the arrangement of heating surface.

1. *Side spacing.* The side-to-side tube spacing in a superheater is important with regard to superheater cleanliness. With oil- or coal-fired units, slagging is a potential problem, and superheaters usually are designed with wider side spacing than gas-fired units.

2. *Surface arrangements.* Superheaters can be designed as parallel flow, counterflow, or a combination of the two arrangements. A counterflow arrangement results in the least amount of surface; however, for high-temperature requirements, this could be the highest-cost superheater arrangement. This is possible because of high alloy metal requirements as a result of the hottest steam in contact with the hottest gas. Usually, with high-temperature superheaters a combination of parallel and counterflow results is the most economical arrangement. For low-temperature superheaters, a counterflow arrangement is often the economical choice.

2.8 Superheat Steam Temperature Control

High turbine efficiency over a wide load range depends a great deal on having a constant steam temperature over that load range.

Therefore, it is necessary to design a boiler that can provide this constant steam temperature over the load range, which can be 50 percent of full load or more. For plants that use a reheat cycle, it is also necessary that the boiler provides reheat steam temperatures that are also constant over the control range.

There are three basic methods to obtain constant steam temperature:

1. Attemperation
2. Flue gas bypass (or flue gas proportioning)
3. Flue gas recirculation

Other methods are also used, including tilting burners (see Fig. 2.38).

2.8.1   Attemperation

This is the method used predominately to control steam temperatures. The basic theory of control by attemperation is that if a superheater is made large enough to give the desired steam temperature at low loads, it will give a steam temperature at high loads that is higher than that desired if there is no method of control. However, if some means of reducing this temperature at high loads is used, a constant steam temperature over the range from low to full load may be obtained. This is the function of attemperation, to reduce the steam temperature at high loads in order that a constant steam temperature to the turbine or to some process results. There are two types of attemperators: (1) drum-type and (2) spray-type.

Drum-type designs are sometimes used on some industrial boiler applications of two-drum boiler designs, whereas spray-type designs generally are used on all units that require attemperation.

Drum-type attemperator. This type uses tubes through which the steam flows, and these tubes are located in a drum (usually the lower drum) and surrounded by water at a lower temperature than the steam. Heat is transferred from the steam to the water in the drum, and the temperature of the steam is thus reduced.

Spray-type attemperator. This type uses the principle that if water is sprayed into steam, the water will evaporate, forming steam, with the temperature of the final mixture lower than the initial steam temperature. However, if solids or impurities are present in the spray water, they are then entrained in the final steam mixture, which is not acceptable. Therefore, this method of attemperation is usually limited to those situations where pure water is available.

Either of the preceding methods of attemperation is used at an intermediate point in the flow of steam through the superheater or after the superheater. Unless the boiler operates at low superheat temperatures, an attemperator located after the superheater is not desirable because the highest metal temperatures possible would exist, and this would require higher alloys for the superheater tubes and thus higher costs.

For the superheater design where the attemperator is located at some intermediate point, the section of the superheater ahead of the attemperator is called the *primary superheater,* whereas the section located after the attemperator is called the *secondary superheater.* Refer to Fig. 2.41 for a typical illustration of this arrangement.

2.8.2   Flue gas bypass

The control of the superheat steam temperature by means of flue gas bypass is relatively simple. Consider a superheater over which a certain quantity of flue gas is flowing and from which steam leaves at a certain temperature. If the steam flow through the superheater, the flue gas temperature, and the steam temperature entering the superheater are constant, and if the flue gas quantity flowing over the superheater is decreased, the heat transferred from the flue gas to the superheater is decreased, with a consequent decrease in the steam temperature leaving the superheater.

The variation in flue gas quantities that flow over the superheater is controlled by dampers. The flue gas that is bypassed around the superheater usually passes over the economizer, boiler, or air heater surface so that its heat content is reduced.

This method of control has two disadvantages:

1. Slow reaction speed

2. Possible fouling, warping, and sticking of dampers

The use of flue gas bypass systems generally is combined with a spray-type attemperator to provide an optimal design for steam temperature control. These systems are also generally found on large utility-type boilers in modern boiler designs.

2.8.3   Flue gas recirculation

Superheat temperature control by attemperation involves the concept that the superheater must be large enough so that at the lowest control load the desired superheat temperature would be obtained. This therefore requires attemperation at high loads.

Superheat temperature control by flue gas bypass involves the fact that superheater absorption at a given load is controlled by varying the flue gas over the superheater. Flue gas recirculation is a system where the heat available to the superheater is regulated by controlling the heat absorption of the furnace. If the heat absorption requirement of the superheater increases, furnace absorption is caused to decrease, which increases the heat available to the superheater. If the heat absorption requirement of the superheater decreases, furnace absorption is caused to increase, which decreases the heat available to the superheater.

This control of furnace heat absorption is obtained by recirculating flue gas to the furnace, with the flue gas normally recirculated from a point after the economizer and prior to the air heater. Furnace heat absorption is primarily a function of the flue gas temperature throughout the furnace, because heat is mainly transferred by radiation. Therefore, by the introduction of flue gas recirculation into the furnace, this reduces furnace absorption by changing the flue gas temperature in the furnace. The boiler shown in Fig. 2.41 is designed with flue gas recirculation in combination with an attemperator.

2.9   Heat-Recovery Equipment

In the boiler heat balance, the greatest loss results from heat loss in the exit flue gases. In order to operate a boiler unit at maximum efficiency, this loss must be reduced to an absolute minimum. This goal is accomplished by installing economizers and air preheaters.

Theoretically, it is possible to reduce the exit flue gas temperature to that of the incoming air. Certain economic limitations prevent carrying the temperature reduction too far, since the costs of the added investment to accomplish this goal may more than offset any savings obtained. Furthermore, if reduction in temperature is carried below the dew point (the temperature at which condensation occurs), corrosion problems may be experienced. Therefore, savings resulting from the installation of heat-recovery apparatus must be balanced against added investment and maintenance costs.

An *economizer* is a heat exchanger located in the gas passage between the boiler and the stack, designed to recover some of the heat from the products of combustion. It consists of a series of tubes through which water flows on its way to the boiler. Economizers may be *parallel-flow* or *counterflow* types or a combination of the two. In parallel-flow economizers, the flue gas and water flow in the same direction; in counterflow economizers, they flow in opposite directions. For parallel flow, the hottest flue gases come into contact with the coldest feedwater; for counterflow, the reverse occurs. Counterflow units are considered to be more efficient, resulting in increased heat absorption with less heat-transfer surface. The gas side of the economizer is usually of single-pass construction. In operation, feedwater enters at one end of the economizer and is directed through a system of tubes and headers until it enters the steam drum at a higher temperature. Economizers are referred to as *return tubular* because the water is made to pass back and forth through a series of return bends. Typical locations of an economizer are shown in Figs. 2.25 and 2.43.

The original economizers were constructed of cast iron, whereas today steel is used. Access to the tubes was made by removing handhole covers, the tubes being rolled or welded into the headers as shown in Fig. 2.20. These economizers were used when the intention was to hold the pressure drop to a minimum and when feedwater conditions were such as to necessitate internal inspection and cleaning.

Instead of using headers, economizers using flanged joints similar to those shown in Fig. 2.21 were frequently constructed. Such units had the advantage of using a minimum number of return-bend fittings, of not requiring handhole fittings and gaskets, and of being free from expansion difficulties. A number of takeoffs to the steam drum provided uniform water distribution to the drum without disturbing the water level.

However, designs have evolved, and the modern economizer consists of a continuous coil of tubes welded onto inlet and outlet headers. This construction has the advantage of eliminating gaskets, handholes, etc.; it also permits acid cleaning of tubes, which was not possible with previous designs.

Tubes range in size from 1 to 2 in in diameter. The size of the economizer is influenced by many factors, such as cost, space availability, type of fuel, and whether or not an air preheater is to be installed.

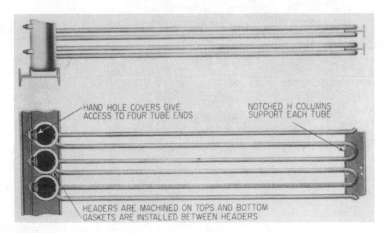

HAND HOLE COVERS GIVE
ACCESS TO FOUR TUBE ENDS

NOTCHED H COLUMNS
SUPPORT EACH TUBE

HEADERS ARE MACHINED ON TOPS AND BOTTOM
GASKETS ARE INSTALLED BETWEEN HEADERS

Figure 2.20 Construction of economizer. (*DB Riley, Inc.*)

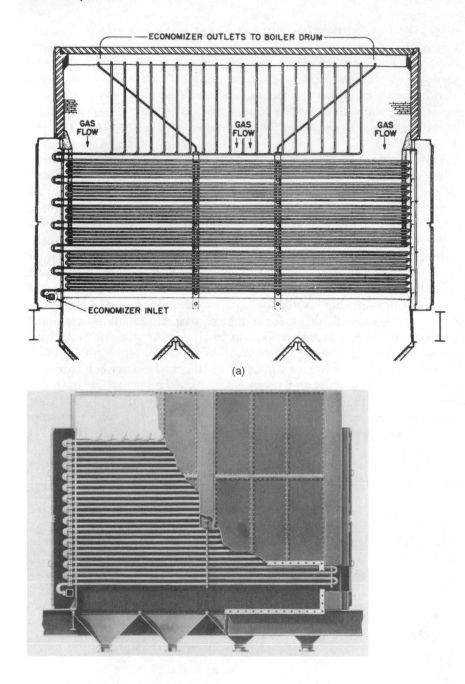

Figure 2.21 Return-bend economizers. (*a*) Continuous-tube type. (*b*) Loop-tube type. (*Babcock & Wilcox, a McDermott company.*)

When both an economizer and an air preheater are to be installed, consideration must be given to preventing the exiting flue gas temperature from dropping below the dew point. An example of the arrangement of the superheater, reheater, and economizer in a large modern utility coal-fired boiler is shown in Fig. 2.22.

In utility and many industrial power plants, economizers and air preheaters are both installed to obtain maximum efficiency. For the modern plant, typical improvements in efficiencies are as follows: boiler efficiency, 74 percent; boiler and economizer, 82 percent; and boiler, economizer, and air preheater, 88 percent. Savings in fuel costs result from these higher efficiencies.

The *air preheater* (or air heater) consists of plates or tubes having hot gases on one side and air on the other. The heat in the flue gas leaving the boiler or economizer is recovered by the incoming air, thereby reducing the flue gas temperature and increasing efficiency. There are generally two types of air preheaters, *tubular* and *regenerative*. The tubular type consists of a series of tubes (Fig. 2.23) through which the flue gases pass, with air passing around the outside of the tube. In the illustration shown, baffles are arranged to make the preheater a four-pass unit for the airflow. Tubes are expanded into tube sheets at the top and bottom, the entire assembly being enclosed in a steel casing. Note the air-bypass dampers

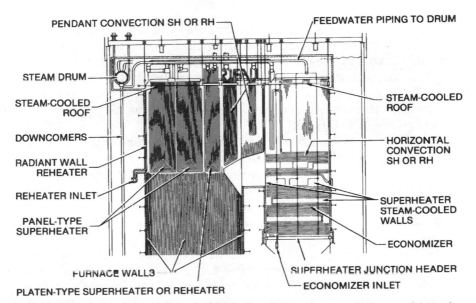

Figure 2.22 Arrangement of superheater, reheater, and economizer of a large pulverized-coal-fired steam generator. (*ABB Combustion Engineering Systems.*)

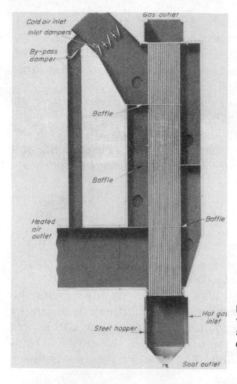

Figure 2.23 Tubular air heater with a four-air-pass baffle arrangement. (*Babcock & Wilcox, a McDermott company.*)

that are used to ensure that the exit flue gas temperature does not fall below a minimum temperature. At low loads, air is bypassed to maintain this minimum temperature.

The regenerative air preheater shown in Fig. 2.24 transfers heat in the flue gas to the combustion air through heat-transfer surface in a rotor that turns continuously through the flue gas and airstreams at slow speeds (1 to 3 rpm).

To increase the service life of the heat-transfer surface elements, design consideration is given to the following: (1) excess temperature at the hot end—by the use of scale-resistant steel; (2) corrosion at the cold end—by greater sheet thickness, low-alloy steel, enameled sheets, glazed ceramics, and honeycomb blocks made of ceramics; and (3) danger of clogging—by enlarged flue gas passage cross sections and enameled sheets.

Units are equipped with soot blowers that use superheated steam or compressed air. Washing and fire-extinguishing devices consist of a series of spray nozzles mounted in the housing. Thermocouples are mounted at the cold end and are close to the heating surfaces and in

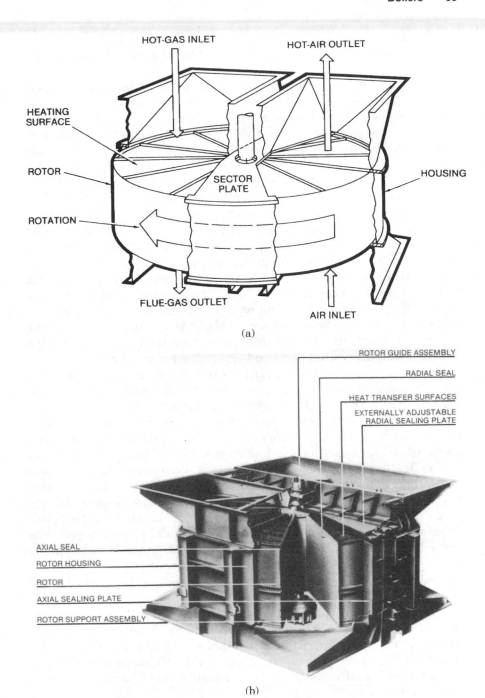

Figure 2.24 Regenerative air preheater, bisector design for vertical air and flue gas flow. (*a*) Cutaway drawing. (*b*) Photograph of unit. (*ABB Air Preheater, Inc.*)

the flue gas and air ducts. They serve to monitor any falling below the acid dew point of the flue gases and also to give early warning of danger of fire.

Air heaters have been accepted as standard equipment in power-plant design and are justified because they increase plant efficiency. The degree of preheat used depends on many factors, such as furnace and boiler design, type of fuel and fuel-burning equipment, and fuel cost. Preheated air accelerates combustion by producing more rapid ignition and facilitates the burning of low-grade fuels. In the process it permits the use of low excess air, thereby increasing efficiency. When pulverized coal is burned, preheated air assists in drying the coal, increasing pulverizer mill capacity, and accelerating combustion.

For stoker firing, depending on the type of stoker and the type of fuel burned, care must be taken not to operate with too high a preheated air temperature. Too high a temperature may damage the grates. Difficulty also may be experienced with matting of the fuel bed and clinkers. The degree of preheating is determined by the kind of fuel, the type of fuel-burning equipment, and the burning rate or grate-heat release. Preheated air at 350°F is usually considered the upper limit for stokers; for pulverized coal, high-temperature preheated air is tempered when it enters and leaves the pulverizer.

For the air preheater, a low air inlet or low exit gas temperature or a combination of the two may result in corrosion when fuels containing sulfur are burned should the metal temperature fall below the dew point. Two dew points need to be considered: the water dew point, which occurs at approximately 120°F, and the flue gas dew point, which varies with the quantity of sulfur trioxide in the flue gas and with other factors. The acid dew point occurs at a higher temperature than the water dew point. The metal temperature is considered to be approximately the average of the air-gas temperature at any given point. Corrosion may be prevented by preheating the air before it enters the preheater, by bypassing a portion of the air around the preheater, and by using alloys or corrosion-resistant metals. Steam coil air heaters are used when required to preheat the air prior to the air entering the air heater, and the steam coil air heater is located after the forced-draft (FD) fan.

The use of an air preheater increases the overall unit efficiency from 2 to 10 percent. The amount of increase depends on the unit location, the steam capacity, and whether or not an economizer is also installed. While air preheaters increase the efficiency, this increase must be evaluated against the added cost of installation, operation, and maintenance.

Figure 2.25 shows a large coal-fired boiler for utility use and highlights those portions of the boiler which have just been reviewed.

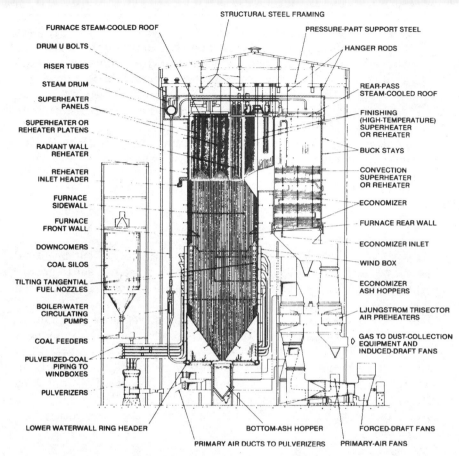

STRUCTURAL STEEL FRAMING
FURNACE STEAM-COOLED ROOF
DRUM U BOLTS
RISER TUBES
STEAM DRUM
SUPERHEATER PANELS
SUPERHEATER OR REHEATER PLATENS
RADIANT WALL REHEATER
REHEATER INLET HEADER
FURNACE SIDEWALL
FURNACE FRONT WALL
DOWNCOMERS
COAL SILOS
TILTING TANGENTIAL FUEL NOZZLES
BOILER-WATER CIRCULATING PUMPS
COAL FEEDERS
PULVERIZED-COAL PIPING TO WINDBOXES
PULVERIZERS
LOWER WATERWALL RING HEADER

PRESSURE-PART SUPPORT STEEL
HANGER RODS
REAR-PASS STEAM-COOLED ROOF
FINISHING (HIGH-TEMPERATURE) SUPERHEATER OR REHEATER
BUCK STAYS
CONVECTION SUPERHEATER OR REHEATER
ECONOMIZER
FURNACE REAR WALL
ECONOMIZER INLET
WIND BOX
ECONOMIZER ASH HOPPERS
LJUNGSTROM TRISECTOR AIR PREHEATERS
GAS TO DUST-COLLECTION EQUIPMENT AND INDUCED-DRAFT FANS

BOTTOM-ASH HOPPER
FORCED-DRAFT FANS
PRIMARY AIR DUCTS TO PULVERIZERS   PRIMARY-AIR FANS

Figure 2.25 Side elevation of large pulverized-coal-fired steam generator for high subcritical pressure operation. (*ABB Combustion Engineering Systems.*)

2.10 Furnace Design Considerations

The furnace portion of a boiler provides a place for the combustion of the fuel, contains the combustion gases, and then directs those gases to the heating surfaces of the boiler. In nearly all modern boilers, furnaces are water-cooled enclosures and therefore absorb heat from combustion and cool the combustion gases (flue gas) before they enter the convection heating surfaces of the boiler.

The heating surface that forms the walls of a water-cooled furnace is often considered to be the most expensive saturated surface in a boiler because only one side of this surface absorbs heat. Therefore, a good, economical design must be the smallest furnace allowable to burn the fuel completely, with consideration given to containing and directing the flue gas to the heating surfaces of the boiler.

Although the size of the furnace is often controlled by space limitations and by boiler convection pass requirements, it is primarily set by the fuel-burning requirements. Proper fuel burning requires that the burners be far enough from the walls and floor to prevent carbon buildup and flame impingement and that enough flame travel distance is provided to ensure complete combustion. If multiple fuels are used, each fuel-burning characteristic must be known in order to evaluate its impact on furnace size.

When designing a furnace, or when evaluating a boiler for a particular application, the following guidelines are important:

1. A few large burners are less expensive than more smaller burners.
2. A high, narrow boiler is generally less expensive than a low, wide one.

Therefore, a boiler design should incorporate the largest practical burner size, and the width and depth of the furnace are set to ensure the proper clearances. The overall space limitations, if any, are a major factor in the design.

The selection of the number of burners must consider the planned operation of the boiler in order to meet the operating requirements at various loads. This is often referred to as the *turn-down ratio*. For example, if operation is planned for 20 percent of full load, the turn-down ratio is 5:1. The burners would have to be sized so that they could function properly and efficiently at these lower loads. This often requires taking burners out of service in order to operate effectively.

2.11 Furnace Construction

Internally fired boilers, such as the firebox and packaged types, are self-contained and require no additional setting. Externally fired boilers require special consideration in terms of furnace construction, particularly since each installation is designed to meet specific plant requirements and space availability.

The horizontal-return tubular boiler (see Fig. 2.5) is supported by the furnace walls. It is mounted on lugs set on rollers, permitting the boiler to move longitudinally. An improved method of installation is shown in Fig. 2.6. Here, the refractory walls do not carry the weight of the boiler.

Expansion and contraction for water-tube boilers are taken care of in a number of ways: (1) by suspending the drums and headers from slings attached to overhead columns, (2) by supporting the drum at the end, on columns or overhead beams, and (3) by anchoring the lower drum at the floor level, permitting expansion upward.

In the past, refractory arches frequently were installed in furnaces equipped with chain-grate stokers. Their primary purpose was to assist in maintaining stable ignition with a reduction in smoke emission. Such arches were difficult to maintain, resulting in frequent replacement that required outages of boiler units. These arches have largely been replaced by water-cooled arches or by small snub-nose refractory arches, also water-cooled. Over-fire air jets are provided to improve combustion efficiency.

The low-head three-drum boiler (Fig. 2.26) is shown in the process of being erected. The top drums are supported at the ends, resting on steel columns or beams, and the lower drum hangs suspended from the tubes. Note that the boiler is set high to provide adequate furnace volume. This unit had a capacity of 20,000 lb/h of steam and was suitable for coal, oil, or gas firing, but it has been replaced with more modern designs.

The vertical boiler (Fig. 2.27) is a two-drum three-gas-pass water-tube boiler, with side waterwalls. The steam drum is supported on steel beams, while the lower (mud) drum is suspended from the

Figure 2.26 Low-head boiler during construction. (*The Bigelow-Liptak Corp.*)

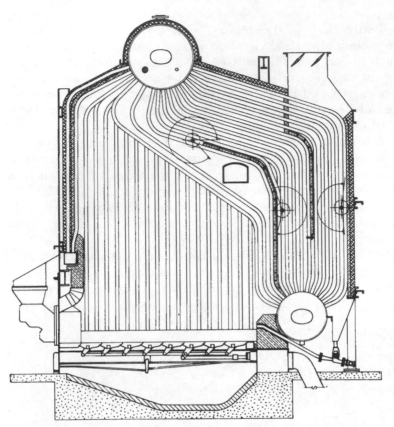

Figure 2.27 Vertical boiler with a dump grate spreader stoker. (*Union Iron Works Co.*)

inclined-vertical tubes. It is fired by a spreader stoker. (Refer to Chap. 5.) Note the cinder reinjection at the rear of the furnace. Its purpose is to improve the boiler efficiency by burning the unburned combustibles that have fallen into the hopper at the boiler outlet.

Until the 1920s, any increased steam requirements were met with increasing the number of boilers in a system. In order to reduce costs, attempts were made to increase the size of the boilers. These boilers were basically refractory-lined furnace designs, and methods of firing coal, primarily on stokers, were no longer adequate.

The use of pulverized coal became the answer to the requirements of high combustion rates and higher boiler steam capacities. The refractory furnace was no longer adequate, and the water-cooled furnaces were developed. These furnace designs eliminated the problem with the rapid deterioration of refractory walls because of the molten ash (slag) that formed on the hot walls. The water-cooled walls also lowered the temperature of the gases leaving the furnace. This not only improved the heat-absorption capability of the boiler but also reduced the accumulation of ash (slagging) in the convection heating surfaces of the boiler.

Water-cooled furnace walls not only reduced maintenance on the furnace and fouling of the convection heating surfaces (and thus minimized forced boiler outages), but the wall also absorbed heat, which helped to generate more steam. As a result, boiler tube bank surface was reduced because of this additional steam-generating surface in the furnace. In order to obtain a higher cycle efficiency, feedwater and steam temperatures were increased, with an increase in steam pressure, and this further reduced boiler tube bank surface. However, it was replaced with an additional superheater surface.

As a result, boilers designed for steam pressures above 1200 psig consist of basically furnace water walls, superheaters, and supplemental heat recovery equipment of economizers (for heating feedwater) and air heaters (for heating combustion air). Boilers designed for lower pressures have a considerable amount of steam-generating surface in boiler banks in addition to the water-cooled furnace.

Most modern boiler furnaces have walls that are water cooled. This not only reduces maintenance on the furnace walls but also serves to reduce the temperature of the gas entering the convection bank to the point where slag deposits and superheater corrosion can be controlled by soot blowers.

Furnace wall tubes are spaced on close centers to obtain maximum heat absorption. Tangent tube construction, used on earlier designs, has been replaced with membrane walls (see Fig. 2.32) in which a steel bar or membrane is welded between adjacent tubes.

The two-drum boiler shown in Fig. 2.28 is of waterwall construction with membrane walls. A radiant-type superheater is located in the furnace. The boiler is fired primarily by coal, wood, bark, or other solid fuels on a traveling-grate spreader stoker, with auxiliary burners (gas or oil) located in the rear wall providing flexibility for supplementary fuel firing.

The single-elevation top support ensures an even downward expansion without differential stresses or binding. The drum rests on overhead steel beams, and the superheater is hung from slings. Boilers of this type are largely prefabricated, with the furnace walls built in panel sections. Later the panels are welded together to form the membrane wall furnace sections. These units are carefully built under controlled shop conditions, for ease of erection, requiring a minimum amount of time to assemble.

This is a single-gas-pass boiler, and therefore, no baffling is required. There are no local areas of high-velocity products of combustion to cause tube erosion. When required, cinder return from the last pass of the boiler is by gravity to the rear of the grate. Over-fire air

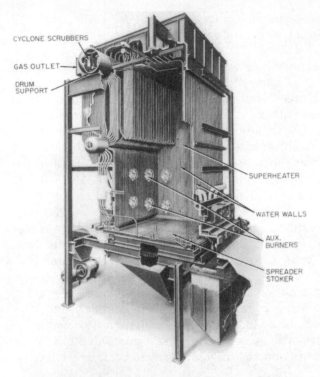

Figure 2.28 The Sterling SS boiler with a traveling-grate spreader stoker. (*Babcock & Wilcox, a McDermott company.*)

jets are provided to improve combustion. These units are generally available in capacities of 60,000 to 400,000 lb/h of steam, pressures of 160 to 1050 psi, and temperatures to 900°F.

The *Stirling Power Boiler* (SPB) shown in Fig. 2.29 is designed with a *controlled combustion zone* (CCZ) furnace to provide better mixing of the fuel and air. This design is used primarily in the firing of waste fuels, particularly bark and refuse-derived fuel (RDF), a solid fuel that is processed from municipal solid waste (MSW). Figure 2.29 shows a boiler designed to burn RDF on a spreader stoker, with auxiliary natural gas burners, in a waste-to-energy plant.

Furnace heat release is expressed in Btu per hour per cubic foot (Btu/h/ft³) of furnace volume. The permissible heat release varies with design, depending on whether the furnace is refractory-lined or water-cooled, the extent of water cooling, heat transfer, and the type of fuel burned. High furnace heat release is usually accompanied by high furnace temperatures. When coal low in ash-fusion temperature is being burned, the ash adheres to the refractory surface, causing erosion and spalling. The ash also may adhere to the heating sur-

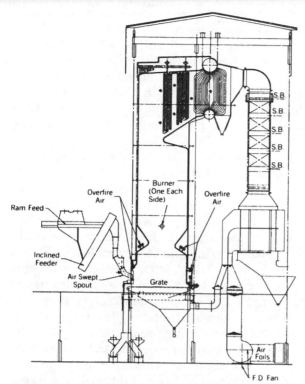

Figure 2.29 Controlled-combustion-zone (CCZ) furnace design. (*Babcock & Wilcox, a McDermott company.*)

faces, reducing the heat transfer and frequently fouling the gas passages with a loss in boiler capacity and efficiency. For the refractory-lined furnace, high furnace heat release is more severe than for waterwall installations.

As noted previously, because refractory walls were unable to meet the severe service conditions to which they were subjected, waterwalls were introduced, even for the smaller boiler units. Excessive maintenance and outage of equipment are thus avoided, and the addition of waterwalls increases the boiler capacity for a given furnace size.

The first application of furnace water cooling was the installation of the water screen when burning pulverized coal. This screen consisted of a series of tubes located above the ashpit and connected to the boiler water-circulating system. Its purpose was to reduce the temperature of the ash below its fusion point; thus slagging was prevented.

The waterwall was added next. In replacing the refractory walls, the added heating surface increased the boiler output, and with the elimination of refractory maintenance, boiler availability was improved.

The amount of water cooling that can be applied is determined in part by combustion conditions to be experienced at low steam capacity, since excessive cooling reduces stability of ignition and combustion efficiency. Therefore some furnaces are partially water cooled, or the waterwall is partially insulated; each design is based on experience.

Details of wall construction are illustrated in Figs. 2.30 and 2.31; tube construction with full and partial stud tubes is shown. The studs are used to anchor the refractory in place, while tie bars hold the tubes in line.

Various types of wall blocks are used. The choice is determined by their individual capacity for heat conductivity and by the varying conditions to which they are exposed in different parts of the furnace. The blocks may be rough faced or smooth, of bare metal or refractory faced. Depending on known heat-transfer coefficients, blocks are applied to meet design specifications and to limit the heat input to the tubes in order to prevent overheating and other problems.

Special attention must be given to wall sections subjected to flame impingement, to tube bends, and to division walls and slag screens subject to the blast action of the flame. Special refractory materials provide protection against molten slag and erosion. The arrangement of studs and the extent of refractory covering are modified to meet the specific requirements of the individual furnace and the type of fuel burned. In operation, any excess refractory is washed away until a state of thermal equilibrium is reached because of the cooling effect of

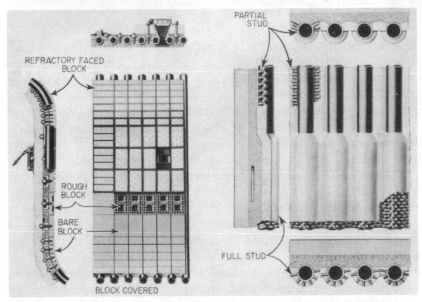

Figure 2.30 Water-cooled furnace wall construction. (*Babcock & Wilcox, a McDermott company.*)

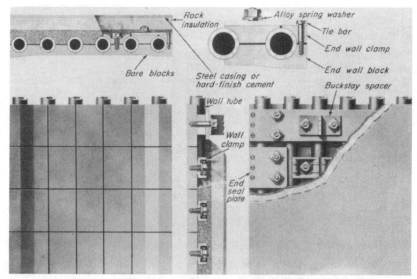

Figure 2.31 Block-covered wall showing method of clamping blocks on tubes. (*Babcock & Wilcox, a McDermott company.*)

the studs. Fully studded tubes are used to assist ignition and to promote complete combustion for sections of the furnace where maximum temperatures are desired. Partially studded tubes are usually used in cooler zones of the furnace and where more rapid heat absorption is advantageous.

Over the years, efforts have been directed to the reduction of air infiltration into the boiler setting in order to improve unit efficiency while maintaining boiler capacity. The use of waterwalls with welded outer casings has reduced this leakage considerably. The pressurized furnace was the next step. It uses an all-welded casing behind the tube enclosure, the insulation being located behind the casing.

However, on pressurized units, flue gas can still leak through the walls to cause overheating of the inner and outer casings. Such leakage causes flue gas and fly ash to enter the casings. The flue gas may be saturated with sulfur, resulting in corrosion of the casings. On balanced-draft designed units, air infiltration was a problem that reduced boiler efficiency. The membrane wall construction was developed to solve these problems (Fig. 2.32). Tightness is accomplished by welding a bar between the tubes, insulation being placed behind the tubes, with casing or lagging on the outside.

For boilers such as those shown in Figs. 2.10 and 2.11, the bulk of the heat absorbed was the result of convection and conduction; only the lower rows of tubes received heat by radiation. The number of

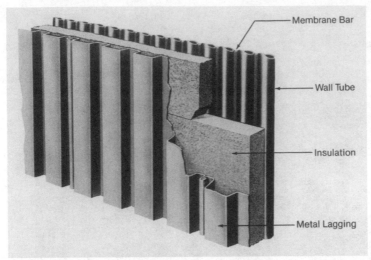

Figure 2.32 Membrane wall construction with block insulation and metal lagging. (*Babcock & Wilcox, a McDermott company.*)

square feet of heating surface was then used to determine the capacity of the unit, approximately 10 ft² of heating surface being considered capable of generating 34.5 lb/h of steam "from and at 212°F" feedwater temperature. (See Sec. 3.8 for an explanation of this measurement.) Where waterwalls comprise the greater portion of the heating surface, receiving most of the heat by radiation, the previous standard cannot be applied. Therefore, for modern units, boiler performance and steam capacity are calculated by the designer based on design performance data and experience with similar units in the field.

When pulverized coal is being fired, difficulty may be experienced with deposits of furnace slag. This is especially troublesome when the coal contains ash having a low fusion temperature. The slag becomes very hard and difficult to remove, especially when it is attached to the brickwork. Furthermore, a portion of the refractory is frequently removed along with the slag, thus increasing the maintenance cost.

Furnaces can be designed to burn coal of any fusion range. If the ash is removed in the dry state, the unit is referred to as a *dry-bottom* furnace. Or for low-fusion-ash coal, the unit may be designed to remove the ash in liquid form; the unit is then called a *wet-bottom* furnace. The liquid ash can be removed on a continuous basis. Here the molten ash collects on the furnace floor, is made to flow over a weir located in the floor of the furnace, and drops into a bath of water below. Later the ash is removed from the hopper hydraulically. Or the molten ash may be permitted to remain and collect on the furnace floor to be tapped off at intervals. On being discharged, the molten ash encounters a jet of high-velocity water; the chilling of the ash causes it to break up into a fine granular form for ease of disposal.

Wet-bottom furnaces (often called *slag-tap furnaces*) have been used for both pulverized coal and cyclone firing systems. Cyclone furnaces were developed to burn crushed coal and to form a molten sticky slag layer. They were designed to burn coals that were not well suited to pulverized coal firing.

The use of slag-tap boilers designed for pulverized coal firing declined in the early 1950s when there were significant design improvements in dry-bottom units and these minimized the ash-deposition problems. However, slag-tap boilers with cyclone furnaces continued to be used until the mid-1970s, when environmental restrictions imposed limits on NO_x emissions that mandated that NO_x be controlled. The slag-tap boilers operated at high furnace temperatures, and this resulted in high NO_x formation. Cyclone furnaces still remain in operation; however, other means for controlling NO_x formation have to be incorporated as part of the boiler design.

2.12   Industrial and Utility Boilers

Boilers generally are classified into two categories, industrial and utility.

Industrial boilers are used primarily to provide steam to processes or manufacturing activities and are designed to meet the following criteria:

1. Steam pressures that are controlled by the specific process, most often lower than utility needs
2. High reliability with minimum maintenance
3. Use of one or more locally inexpensive fuels, including process by-products and waste fuels
4. Low initial capital cost and low operating cost
5. The operating pressures range from 150 to 1800 psi with saturated or superheated steam

With the advent of independent power producers (IPPs) as compared with large electric utilities, many of today's new power plants are significantly smaller in size as compared with utilities and burn waste fuels, but they maintain the high-availability objectives of their utility counterparts.

Utility boilers are used primarily to generate electricity in large power stations and are designed to optimize overall thermodynamic efficiency with the highest possible availability. Units have very high steam flow rates, with superheated steam outlet pressures from 1800 to 3850 psig at 1050°F.

Boilers are designed to meet various requirements for capacity and for space limitations. Thus there are many packaged units, shop-assembled in various capacities and pressures. Others are of modular construction, adapted to space limitations. Many are field erected yet have many of the component parts shop-assembled in modular form to reduce field construction costs. Boilers with pressures as high as 3850 psi and steam temperatures to 1050°F are available in many different designs. The large units are single-purpose boilers, providing steam to a single turbine having single- or double-reheat arrangements. Designs include twin furnaces for steam-and-reheat temperature control, forced water circulation, once-through boilers, pressurized furnaces, and many other innovations.

Although many units are designed to operate at steam temperatures of 1050°F, operating experience has found that the boilers firing high-sulfur coals are susceptible to corrosion because of the higher tube metal temperatures, and this has resulted in high maintenance costs and lower availability. Therefore, many newly designed utility plants design for steam temperatures of no higher than 1000°F. This lower temperature permits the use of lower-cost alloys in the superheater tubes, the steam piping, and turbine and also provides a higher safety margin for minimizing corrosion in the boiler.

An example of a large utility pulverized-coal-fired once-through boiler is shown in Fig. 2.33. This unit is designed for a steam flow of 6,415,000 lb/h at 3785 psi and 1005°F with reheat conditions of 5,200,000 lb/h at 569 psi and 1005°F.

These utility boilers get to be quite large. Units have been designed to burn low-grade Texas lignite coal, the boiler being 310 ft high and 200 ft deep. The boiler is of a height equivalent to a 31-story building and consumes nearly 550 tons of coal per hour. (See Figs. 12.2 and 12.3.)

The PFI integral-furnace boiler (Fig. 2.34) is a standard bent-tube bottom-supported two-drum unit arranged with a gas-tight membrane wall furnace and a bare-tube boiler for pressurized or balanced-draft operation, with a completely water-cooled furnace. Maximum shop subassembly facilitates field erection and lowers construction costs. The superheater is an inverted-loop design and is fully drainable because the headers are located at the bottom of the unit. By comparison, note the superheater arrangement in Fig. 2.36, which is not drainable. The unit is designed for capacities from 100,000 to 500,000 lb/h of steam, pressures to 1150 psi, and steam temperatures to 950°F. The fuel is usually oil or natural gas, singly or in combination; gaseous waste fuels also can be used.

These units provide steam for heating, power, or process. They can be located outdoors if desired and require a minimum of space. A relatively constant steam temperature over a wide load range can be

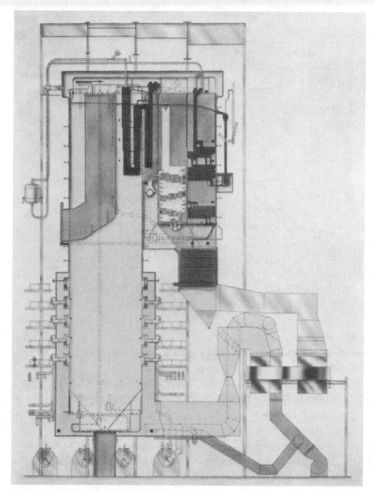

Figure 2.33 Large pulverized-coal-fired once-through boiler. (*Foster Wheeler Corp.*)

obtained, and an air preheater or economizer can be adapted readily to the unit if economics so dictate.

Figure 2.35 shows a unit designed to produce 200,000 lb/h of steam at 625 psi and 750°F. It is equipped with waterwalls, a radiant superheater, an economizer, a tubular air preheater, and a mechanical dust collector. It is fired by a continuous-ash-discharge spreader stoker and auxiliary burners. It is designed to burn pine and gum bark, oil, natural gas, or coal. The boiler is supported from overhead steel columns as shown, and its expansion is downward.

The VU-60 boiler (Fig. 2.36) is a pressurized furnace unit using bare waterwall tubes with floor tubes covered as shown. A front-firing

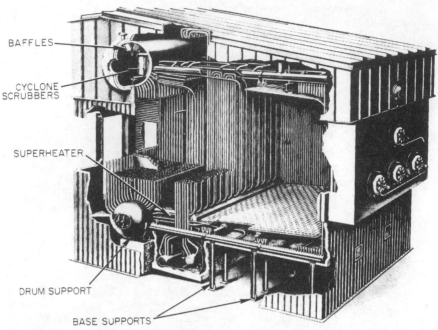

BAFFLES

CYCLONE
SCRUBBERS

SUPERHEATER

DRUM SUPPORT

BASE SUPPORTS

Figure 2.34 Type PFI integral furnace boiler. (*Babcock & Wilcox, a McDermott company.*)

arrangement is shown, although tangential firing is available, if desired. It is designed to burn oil, gas, and waste heating fuels and can deliver over 1 million lb of steam per hour.

As an alternative to horizontal burners, tangential firing systems are used in some boiler designs. In this design, both fuel and combustion air are projected from the corners of the furnace along a line tangent to a small circle (Fig. 2.37). A rotary motion is imparted to the flame body, which spreads and fills the furnace area and creates a high degree of turbulence for effective mixing that enhances complete combustion.

These burners are also capable of tilting, which creates an effective means for controlling superheat and reheat steam temperatures. As shown in Fig. 2.38, fuel and air nozzles tilt in unison to raise and lower the flame in the furnace to control furnace heat absorption and thus heat absorption in the superheater and reheater.

A steam generator available for industrial and utility use in a variety of sizes is shown in Fig. 2.39. It is a single-gas-pass unit, which eliminates turns, baffles, and pockets where concentrations of fly ash might accumulate. This type of construction also eliminates high velocities across the heating surfaces, which are conducive to tube erosion.

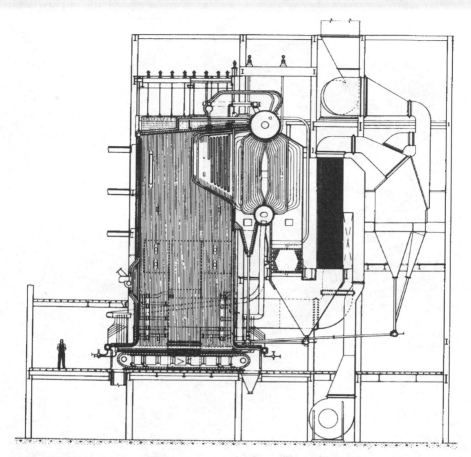

Figure 2.35 Type VU-40 boiler equipped with a traveling-grate spreader stoker. (*ABB Combustion Engineering Systems.*)

The prominent nose at the top of the furnace ensures the flue gas transition from the furnace, thus providing for good distribution of flue gases as they flow through the superheater and boiler bank. This unit is available in capacities to 500,000 lb/h of steam at 1500 psi and temperatures to 960°F. Units such as this can be oil, gas, or pulverized coal fired or combination fired. The unit shown is a pulverized-coal-fired boiler.

Primary air is introduced at the fuel inlet from the pulverizers and is part of the air-fuel mixture; secondary air, at the burner exit; and tertiary air, through the center of the burner, all three combining to produce a turbulent ignition with a minimum of excess air.

The waterwall tubes are fin tube welded. Wall construction is such that the unit is pressure-tight for operating either with a balanced draft or a pressurized furnace. The drum is equipped with horizontal

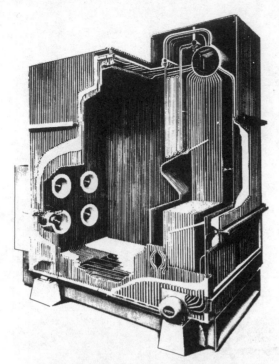

Figure 2.36 Type VU-60 gas- and oil-fired boiler with horizontal burners. (*ABB Combustion Engineering Systems.*)

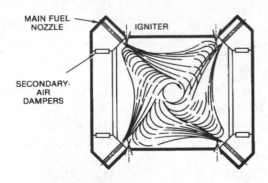

MAIN FUEL
NOZZLE

IGNITER

SECONDARY-
AIR
DAMPERS

Figure 2.37 Tangential firing pattern. (*ABB Combustion Engineering Systems.*)

steam separators to provide dry steam to the superheater. Units such as this operate at high efficiency.

A Turbo furnace designed for pulverized coal firing is shown in Fig. 2.40. This is an open-pass design, incorporating both a radiant and a convection type of superheater. Its unique feature, however, is the furnace design. The furnace envelope is constricted in a venturi section just above the furnace level. The burners are angled downward into

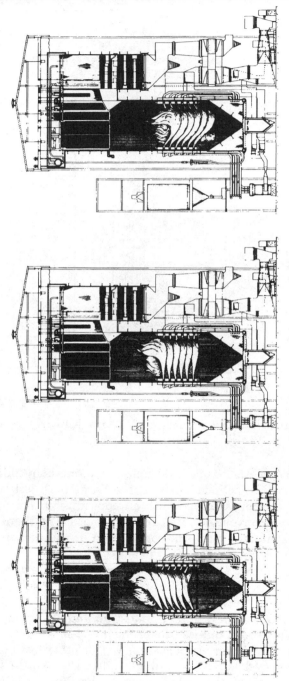

Figure 2.38 Selective furnace utilization and steam temperature control are accomplished by tilting nozzles in a tangentially fired system. (*ABB Combustion Engineering Systems.*)

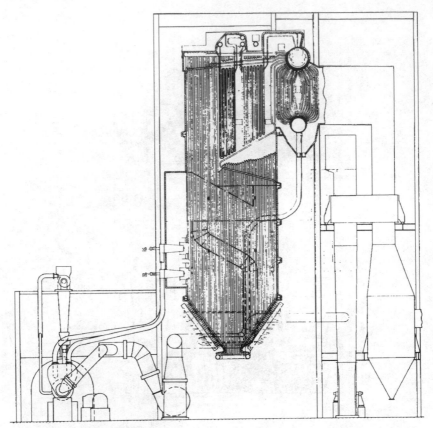

Figure 2.39 Pressurized furnace design industrial boiler for pulverized-coal or combination firing. (*Foster Wheeler Corp.*)

the lower section of the furnace. Because the flame is initially directed downward by the inclined burners, the flame path is lengthened, and since each burner is paired with an opposing burner, individual burner flames are convoluted in a turbulent mass. The thorough mixing results in complete, uniform combustion of the fuel in an area that is centered within the furnace envelope.

For the lower furnace the burning gases accelerate upward through the venturi restriction, with further turbulent mixing completing the combustion process.

The Turbo furnace is used in both industrial and utility applications, with sizes available to produce steam for a 600-MW turbine, and is designed to burn a wide variety of coals as well as other fossil fuels.

Another pulverized-coal-fired radiant boiler is shown in Fig. 2.41. Proceeding from the water-cooled furnace, the flue gases pass over the secondary superheater (which is of the radiant type) to the reheater superheater, to the primary superheater, to the economizer

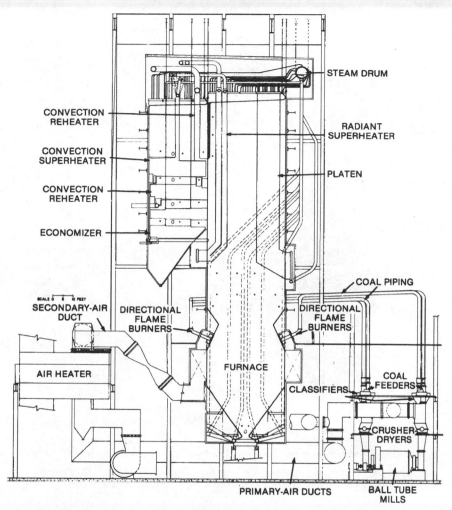

Figure 2.40 Turbo-furnace-type utility steam generator designed for pulverized-coal firing. (*DB Riley, Inc.*)

and regenerative air preheater, and finally to the stack after they pass through the environmental control systems (see Chap. 12). Steam from the boiler drum flows through the primary superheater and then is conveyed through the secondary superheater. Prior to entering the secondary superheater, the steam passes through an attemperator, where water is added to maintain a constant steam temperature at the secondary superheater outlet before flowing to the turbine. After passing through the high-pressure stages of the turbine, the steam is returned to the boiler, where it passes through the reheat superheater and then returns to the low-pressure stage of the turbine.

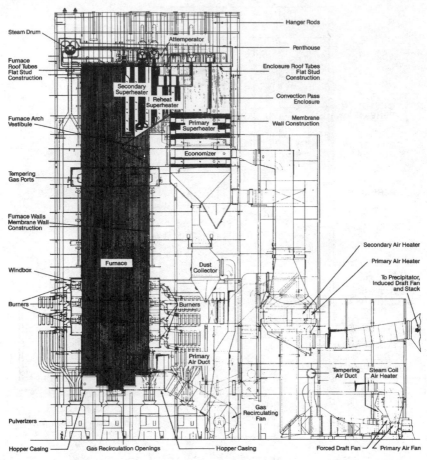

Figure 2.41 A pulverized-coal-fired utility radiant boiler using gas recirculation for steam temperature control. (*Babcock & Wilcox, a McDermott company.*)

A gas-recirculating duct is connected at the base of the furnace for the introduction of flue gas to control combustion conditions in the bottom of the furnace as well as the furnace-outlet gas temperature, if desired. High-preheated air (from the air preheater) is provided for drying the coal in the pulverizer, with a tempering arrangement to control pulverizer outlet air temperature. Units such as this can be designed for various capacities, pressures, and steam temperatures. These boilers are designed for steam flows ranging from 300,000 to 1 million lb/h, pressures from 1800 to 2400 psig, and superheat temperatures over 1000°F.

The utility boiler shown in Fig. 2.42 has a capacity of 1,620,000 lb/h at 2525 psig and 1005/1005°F (superheat and reheat temperatures). It contains a combination radiant and convection superheater, a reheater

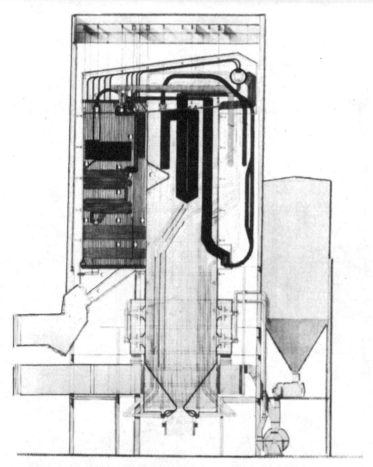

Figure 2.42   Pulverized-coal-fired utility boiler. (*DB Riley, Inc.*)

section, and an economizer. Reheat-temperature control is obtained by gas-flow-proportioning dampers. This boiler uses natural circulation of water. It also utilizes opposed firing, which creates extreme turbulence and is conducive to operating efficiently at low excess air.

A *natural-circulation* unit is one in which the pumping head is provided by the difference in density between the saturated liquid in the downcomers and the steam-water mixture in the heated risers. A steam drum is required to provide the recirculated saturated liquid to the unheated downcomers and saturated steam to the boiler outlet or to the superheater. For the *controlled-circulation* unit, the system uses a pump to ensure sufficient pumping head to obtain the proper cooling of the furnace parts. As with a natural circulation unit, a steam drum is provided to separate the water and the steam.

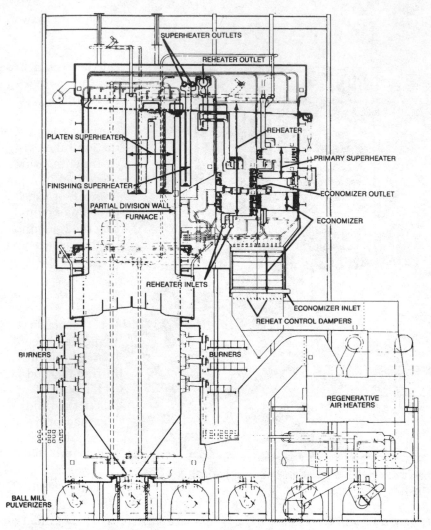

Figure 2.43 Once-through type of steam generator for large utility applications. (*Foster Wheeler Corp.*)

A *once-through* boiler unit is considered to be one that does not use recirculation at full load. Such a unit is shown in Fig. 2.43. It has a capacity of approximately 5 million lb/h of steam at 3800 psi and 1010/1010°F. Eighteen pulverizers are required to process the coal for this unit.

In operation, feedwater flows through the economizer and upper partial division walls in the furnace. The heated fluid from these walls is routed through an external downcomer to supply the first enclosure pass in the furnace section. The furnace enclosure is cooled in this manner by several series-connected passes. Full mixing of the fluid between these passes is achieved as a means of reducing unbalance.

After leaving the furnace circuitry, the fluid (a steam-water mixture) is heated in the convection pass enclosure and the roof circuit, with partial mixing between the two sections. From the roof circuit outlet, the steam then is routed to the superheaters for final heating to full steam temperature.

The furnace is arranged for opposed firing. Pulverized coal mixes with preheated air in the burner zone, where combustion takes place. The flue gas flows upward through the platen and finishing superheaters. The primary superheater and reheater are installed in a parallel-pass arrangement with proportioning damper control of the flue gas flowing over the reheater. The economizer is installed partly in one of the parallel passes and partly in the section following the rejoining of the parallel passes. Flue gases are then directed through two regenerative air heaters to the flue-gas-cleaning equipment and to the stack.

2.13   Considerations for Coal Firing

For years, coal has been burned effectively on stokers of various types and as pulverized coal. Boilers designed for stoker firing have been in the industrial size classification, however; both industrial and utility boilers are designed for pulverized-coal firing. Each firing method has its benefits and each has limitations, as described in this book.

A third method of burning coal and other solid fuels is the fluidized bed boiler. This rapidly developing technology has gained operating experience and operator acceptance over the past years, and it is now commonly accepted as an alternative to stoker and pulverized-coal firing. Because of its uniqueness and relatively recent acceptance, this technology is described in detail later in this chapter. This section will describe the basic difference between stoker and pulverized-coal firing.

2.13.1   Spreader stokers

Spreader stokers of the traveling-grate type are probably the most popular stoker application with industrial boilers because of their relative simplicity, low maintenance, and ability to respond to load variations. Spreader stokers can operate effectively with coals having a wide range of moisture, volatile, and ash constituents. These types of stokers are readily adaptable for bark firing alone or in combination with coal. They are also utilized in the burning of municipal solid waste (MSW) in the form of refuse-derived fuel (RDF). (See Chap. 5 for a description.)

Spreader stokers are associated with high particulate emissions. This necessitates effective fly ash reinjection systems to maintain optimal efficiency unless the furnace design permits a more complete burnout of the combustibles. In addition to fly ash reinjection, suitable fly-ash-collecting devices such as mechanical dust collectors, electrostatic precipitators, and baghouses (see Chap. 12) are necessary to maintain low stack emissions. Scrubbers for the removal of sulfur dioxide in the flue gas also may be required where the operating permit requires such equipment. As explained later, this is one reason why fluidized bed boilers have become an attractive alternative.

With spreader stokers, a substantial amount of combustion air is introduced into the furnace above the grate as over-fire air. These high-pressure jet streams of over-fire air induce a swirling, turbulent motion above the grate area, thus eliminating the escape of vaporized volatile matter and simultaneously ensuring rapid combustion. Lack of over-fire air results in increased smoke density at the boiler or stack exit and higher carbon loss and therefore a loss of efficiency.

2.13.2 Pulverized-coal firing

Stoker firing is extremely sensitive to fuel size, segregation, and moisture. These fuel factors coupled with grate operating factors such as bed depth, grate speed, plenum-chamber air pressure, and over-fire air and stoker variables such as rate of feed and fuel trajectory cause stoker operation to be subject to high variability. In contrast, pulverized-coal firing was developed to provide for higher reliability. In addition, pulverized-coal firing helps achieve minimum carbon loss, lower excess air requirements, and therefore higher operating efficiency. However, the practical lower limit of pulverized-coal-fired boilers is a steam flow of approximately 200,000 lb/h.

Pulverized-coal firing requires the installation of high-efficiency fly ash collecting systems to cope with a high concentration of particles under 30 μm (a micron or one-millionth of a meter). For spreader stokers, concentration of particles under 30 μm is less than half of that found with pulverized-coal firing.

Of all the emissions from a coal-fired plant, emphasis is placed on the visible constituents because of attention drawn to the plume from the stack and the complaints that result. However, the effective control of both particulates and acid gases is mandatory to meet stringent operating permit limitations. Therefore, highly reliable and efficient air pollution control equipment is required, as described in Chap. 12.

2.13.3 Stoker versus pulverized-coal firing

When selecting coal-firing equipment, the advantages and disadvantages of each firing system must be considered. Some of the major advantages and disadvantages of stoker-fired units as compared with pulverized-coal-fired units are the following:

Advantages

Lower cost

Lower particulate carry-over with a larger percentage of particles greater than 10 μm

Simpler particulate removal equipment

Generally less overall space

Lower power requirements

Lower maintenance requirements

Less sophisticated auxiliary equipment and controls

Disadvantages

Limited steam capacity range of about 75,000 to 400,000 lb/h

Efficiency 4 to 7 percent lower

Limited coal size flexibility

Limited choice of coal types

Limited load swing and pickup capability

Higher excess air

2.14   Pressurized versus Balanced-Draft Boiler Designs

The trend of furnace design has changed over the past decades. Although smaller industrial units that are fired with gas or oil are usually of the pressurized-furnace design, i.e., a design incorporating only a forced-draft fan, the trend of the utility industry and larger industrial boilers today is to have balanced-draft boilers, i.e., designs incorporating both a forced-draft and an induced-draft fan.

The addition of environmental control systems downstream from the boiler has raised the total system resistance. Under conditions of increased resistance, a pressurized-furnace design, with its forced-draft-fan system and its already high pressures, would develop excessively high discharge pressures with a resulting higher flue gas and fly ash leakage from the boiler. Consequently, a balanced-draft system with both forced-draft and induced-draft fans, which results in a lower furnace operating pressure, is the preferred choice.

The important factor in the trend to balanced-draft systems is leakage through penetrations and seals from the pressurized furnace to the ambient air. The problem is magnified in units burning pulverized coal because fly ash penetrates into both the penthouse and the ambient air. The difficulty in providing long-term tightness at penetrations, joints, and seals results in excessive leakage. Although both balanced-draft and pressurized units must provide penetrations to allow tubes to pierce the walls and roof, infiltration of ambient air into a balanced-draft unit does not result in the problems caused by outward leakage of flue gas and fly ash.

The advantage of the pressurized unit is that the induced-draft fan is eliminated. In this case, the forced-draft fan must be larger, but it handles clean air at ambient temperatures instead of hot fly-ash-laden flue gas. By eliminating the induced-draft (ID) fan, considerable floor area is saved, as well as the use of flues that direct the flue gas to and from the fan. Without the ID fan, the associated fan maintenance is also eliminated. The pressure design is advantageous for units designed to burn oil or natural gas, especially of the packaged boiler type. Any openings in the boiler setting, such as doors, must be pressurized to prevent the escape of flue gas.

The advantage of balanced-draft systems is that the unit operates with a furnace pressure slightly below atmospheric, making operating conditions and furnace observations simpler. Thus there is an infiltration of ambient air into the furnace and convection passes instead of an outward leaking of flue gas and fly ash. However, this infiltration has to be minimized because it reduces boiler efficiency.

2.15   Fluidized Bed Boilers

Fluidized bed boilers have been used as a combustion device for over 30 years. These combustors are noted for their capability to burn, in an environmentally acceptable manner, low-grade fuels, which can range from wet biomass sludges to high-ash coal wastes, as well as conventional fuels. This ability to burn a wide range of fuels is a result of the turbulent mixing that is found in the fluid beds, resulting in good heat and mass transfer. These characteristics not only improve the combustion process but also allow the burning to take place at temperatures ranging between 1500 and 1600°F instead of normal combustion temperatures of 3000 to 3500°F.

The ability to burn materials at this lower temperature is important because it makes it possible to burn low-quality fuels that do not have a high enough heating value to support combustion at high temperatures. Therefore, fuels with lower fusion temperatures (refer to Chap. 4) can be handled, and sorbents such as limestone can be added directly into the combustion area with the fuel for the purpose of controlling sulfur emissions. The lower combustion temperatures also mean lower emissions of nitrogen oxides, NO_x (see Chap. 12), which are atmospheric pollutants.

As the heating value of the fuel increases, a point is reached where a fluid bed boiler with water-cooled walls and possibly in-bed tubes are required. As a general rule, a high-ash fuel with a low moisture content (less than 30 percent) and a heating value above 2500 Btu/lb can be burned in a water-cooled fluid bed boiler, while fuels below 2500 Btu/lb require a refractory-lined fluid bed combustor in order to support a higher combustion temperature.

A fluid bed system can be designed to handle a wide range of fuel heating values that require varying degrees of heat extraction. The controllable combustor variables are bed temperature, the heat-transfer coefficient, the effective heat-transfer area, and excess air. Each of these can be varied for a fixed combustor design in different combinations to handle a wide range of fuels and boiler steam flow conditions while maintaining good control of combustion and emissions.

Fluidized bed boilers can burn fuels cleanly. Since sulfur is found in many solid fuels, sulfur dioxide emissions can be controlled by introducing a sulfur sorbent such as limestone ($CaCO_3$) directly into the combustor. The overall chemical reaction is

$$CaCO_3 + SO_2 + \tfrac{1}{2}O_2 \rightarrow CaSO_4 + CO_2$$

resulting in a dry $CaSO_4$ (calcium sulfate) product that is removed from the system with the ash. Reduction of SO_2 emissions by 90 percent or greater is achieved by this technique. Low combustion temperatures also result in low emissions of nitrogen oxides.

2.15.1 The process

The fluidized bed combustion technology has distinct advantages for burning solid fuels and recovering the heat energy to produce steam. The process features a mixture of particles that are suspended in an upward-flowing air–flue gas stream, and this mixture results in properties that are similar to a fluid. Combustion takes place in the bed with high heat transfer to the furnace and, very important, low combustion temperatures. The major benefits of this process are (1) fuel flexibility, enabling the unit to use fuels with low heating values (i.e., waste fuels), and (2) reduced air pollution emissions. The process is suitable for both industrial applications and the production of electric power.

2.15.2 Comparison with pulverized-coal and stoker firing

Pulverized-coal firing.   In a pulverized-coal-fired boiler, the combustion process consists of burning fine fuel particles (70 percent less than 200 mesh) that are suspended in air and combustion gases. The zone in the furnace around the burners is the hottest zone, with temperatures in the range of 3000 to 3500°F.

Stoker firing.   With the use of a stoker, the fuel particles are much larger than those used with pulverized-coal firing. For bituminous coal, the fuel size is approximately 1 to $1\frac{1}{4}$ in. Depending on the type of stoker, where some of the coal is burned in suspension, most of the coal is burned as a mass on some type of moving grate with the air passing through the fixed bed of coal. Temperatures in the fuel bed can exceed 3000°F.

Fluid bed combustion. For a fluid bed combustor, the size of the coal falls between the coal size requirements of pulverized-coal and stoker firing. Coal is crushed to less than $\frac{1}{4}$ in, and it is fed into the lower portion of a fluid bed furnace. The solids are maintained at a temperature of 1500 to 1600°F in an upwardly moving stream of air and combustion gases.

As coal is fed into the bed, it is quickly heated to its ignition temperature, at which it ignites and becomes a part of the burning mass. The flow of air and coal to the dense fluid bed is controlled so that a continuous amount of heat is released to meet the steam load demand. Due to the long residence time of the fuel, and because of the high heat-transfer process within the bed due to conduction (as compared with convection), the coal is burned efficiently in the fluidized bed process at temperatures considerably lower than in the pulverized-coal- and stoker-firing methods.

The fuel particles remain in the dense bed until they are carried along by the combustion gases or removed with the bed drain solids as ash residue. As the fuel particles burn, their size reduces below a value where the terminal and gas velocities are equal, and then they become entrained in the gases.

There are two types of fluidized bed combustors that generally can be described as follows:

1. *Bubbling fluid bed (BFB).* In these units, combustion occurs mostly in the bed because of a lower air velocity in the bed and a larger fuel size.

2. *Circulating fluid bed (CFB).* These units are designed to have more particles blown from the bed than for a bubbling bed. The particles are collected by a particle separator and recirculated in the furnace.

Both the BFB and CFB technologies allow the burning of fuels at approximately 1600°F as compared with approximately 2600°F with conventional firing methods such as pulverized-coal or stoker firing. This characteristic occurs because fluid bed boilers use conduction heat transfer in the furnace as compared with radiation heat transfer, which is present with conventional firing. The high solids content and their close contact allow this to occur. The heat transfer in terms of

Btu per hour per square foot ($Btu/h/ft^2$) is actually higher with conduction than it is with radiation, which results in an economical boiler design as well as obtaining the advantages of handling hard-to-burn fuels and reducing emissions.

2.15.3 Advantages of fluidized bed combustion

The reduction in SO_2 and NO_x emissions is the primary advantage of this technology. It is possible to burn high-sulfur coals and achieve low SO_2 emissions without the addition of costly scrubbers for sulfur removal. This process also can burn low-grade fuels, i.e., those having a low heating value, that may be very difficult to burn as a pulverized fuel or on a stoker.

The fluidized bed boiler is designed to operate with a bed operating temperature range of 1500 to 1600°F, and this low operating temperature results in the following operational advantages.

Reduced emissions of SO_2 and NO_x. Because of lower operating temperatures, an inexpensive material such as limestone ($CaCO_3$) can be added to the bed and acts as a sorbent to remove SO_2 from flue gas. When limestone is added to the bed, a reaction occurs in the furnace between the lime in the limestone and the SO_2 in the combustion gas. SO_2 emissions can be reduced by more than 90 percent with this process.

There is a reaction between nitrogen and oxygen at temperatures above 2700°F that results in the formation of nitric oxide (NO_x). The rate of this reaction decreases rapidly as the temperature is reduced. With an operating bed temperature of between 1500 and 1600°F in a fluidized bed boiler, the amount of NO_x found is less than in pulverized-coal- and stoker-fired systems, which operate at higher temperatures. NO_x emissions can be reduced even further, if necessary, by other techniques added to the overall system.

Fuel flexibility. The ability to use a large variety of fuels offers the following advantages. In general, a CFB unit offers greater fuel flexibility than a BFB unit.

1. *Low heating value.* The combustion process allows the burning of fuels having very low heating values, thus having the ability to burn waste fuels such as waste coal. This capability results from the rapid heating of the fuel particles by the large mass of hot bed material and the long residence time that the fuel spends in the bed. Both offset the effects of a lower combustion temperature.

2. *Fuel ash.* Not only does the combustion process result in lower emissions, but also the lower combustion temperature allows the burning of fuels that have high fouling and slagging characteristics at temperatures below their ash fusion temperature, where the problems are created. As a result, there is a significant reduction in the operating problems commonly associated with boilers when using these fuels.

3. *Fuel preparation.* For coals having ash content, the fluidized bed boiler offers an advantage in fuel preparation over pulverized-coal-fired systems. These fuels require greater installed pulverizer capacity, and pulverizers require frequent maintenance. The fluid bed boiler requires only crushed coal of less than $\frac{1}{4}$ in, and this is easier and less costly to prepare.

2.15.4 Descriptions of fluidized bed boilers

Fluidized bed boilers are designed to handle the following types of fuels:

Coal	Biomass	Sludge	Waste
Bituminous	Wood	Papermill	Refuse-derived fuel (RDF)
Lignite	Bark	Deinking	Tire chips
Gob (waste coal)	Sawdust	Sewage	
	Cardboard		

There are two types of fluidized bed boilers designed to burn these fuels, the bubbling fluid bed (BFB) and the circulating fluid bed (CFB), and they have the following general characteristics, with variations incorporated by the specific designer.

Bubbling fluid bed (BFB). A schematic of this boiler is shown in Fig. 2.44, which shows the main features of a BFB boiler. The process features a mixture of particles that are suspended in an upwardly flowing stream mixture of air and combustion gases, and this results in fluid-like properties. Within the bed there is an intimate mixing of the air and fuel that produces optimal combustion. There is a bed level that can be seen easily, and there is a distinct transition between the bed and the space above the bed; this is generally called the *freeboard area.*

The BFB boiler is a two-stage combustion system. Solid-fuel particles burn within the bed. Volatiles from the fuel and very fine fuel particles leave the bed and are burned in the freeboard area of the

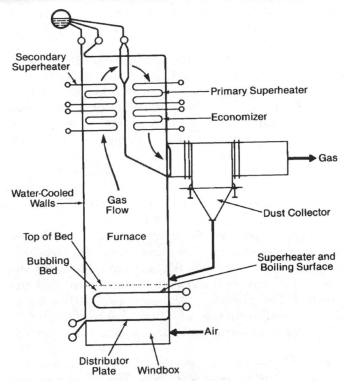

Figure 2.44 Schematic of bubbling fluidized bed (BFB) boiler. (*Babcock & Wilcox, a McDermott company.*)

bed. Secondary air is also injected into the freeboard area for the purpose of optimizing combustion in the second stage of the combustion process.

BFB boilers operate at bed temperatures between 1500 and 1600°F. This low temperature results in a relatively low uncontrolled NO_x emission, since the rate of the formation of NO_x decreases rapidly as the combustion temperature is reduced. With the addition of postreduction techniques (i.e., systems located after the boiler), such as selective noncatalytic reduction (SNCR) and selective catalytic reduction (SCR) (see Chap. 12), even lower NO_x emissions can be achieved.

The low bed operating temperature also provides the proper environment for SO_2 capture. A relatively inexpensive material, such as limestone or dolomite, is used as the bed material. At normal bed operating temperatures, this bed material acts as an excellent sorbent for the removal of SO_2 from the flue gas. If SO_2 removal is not required, then the bed material is usually sand. For SO_2 removal, experience in the operation of BFB and CFB boilers has shown that SO_2 removal with a BFB boiler is not quite as good as with a CFB boiler.

For most high-moisture and low-heating-value fuels, the bed temperature can be maintained easily in the 1500 to 1600°F range with the absorption of heat by the wall tubes of the lower furnace. However, with low-moisture and high-heating-value fuels, additional heat-transfer surface is required to maintain the 1500 to 1600°F bed temperature range. This heat-transfer surface is installed as in-bed tube bundles. However, these in-bed tube bundles can be subject to high rates of erosion. It is for this reason that BFB technology is generally used for high-moisture and low-heating-value fuels such as sludges and high-moisture wood and similar fuels. The BFB design is relatively simple and has a lower cost than the CFB unit.

For all fuels, the bed temperature is maintained uniform within 25°F because of the vigorous mixing of gas and solids. For coal-fired BFB boilers, a recycle system is normally incorporated where the solids in the flue gas, when leaving the economizer, are separated and recycled back to the bed. This maximizes the combustion efficiency as well as the capture of the sulfur in the coal. Bed level is controlled by draining and cooling an appropriate amount of material from the operating bed.

Figure 2.45 shows a bubbling bed system designed to produce steam for the generation of 160 MW of electricity. The boiler produces 1,100,000 lb/h of steam at 1800 psig and 1005°F when firing bituminous coal. Multiple-feed nozzles are used to inject fuel and limestone into the bottom of the combustor. By using a relatively large plan area, overall velocity in the lower combustor is relatively low, which creates a dense bubbling bed several feet high. Above the bed is a taller upper combustor. As the flue gas flows upward through this area, larger solids drop back into the bed, while finer particles are carried out and into the convective pass. Once through the convective pass, these fines are collected in a mechanical dust collector. Final particulate removal occurs in a baghouse or precipitator.

Evaporation, superheat, and reheat, if required, in a bubbling bed are accomplished by a combination of waterwall tubes, a convective pass surface, and often a heating surface immersed in the bubbling bed itself. The location of the heating surface is optimized to provide rapid startups, correct bed temperatures for optimal SO_2 capture, and ease of steam temperature control over wide load ranges.

Circulating fluid bed (CFB). A schematic of a CFB boiler design is shown in Fig. 2.46, where the main features of this CFB boiler are identified. The dense bed does not contain tube bundles. The required

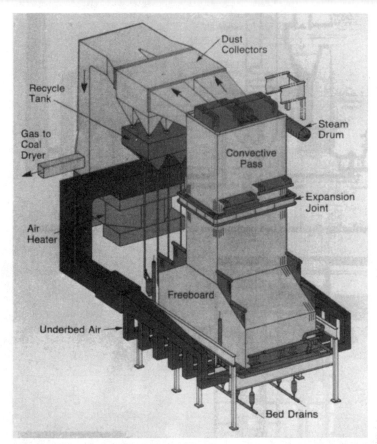

Figure 2.45 Bubbling fluidized bed (BFB) boiler design to produce steam for 160-MW unit. (*ABB Combustion Engineering Systems.*)

heat-transfer surface consists of the furnace enclosure (waterwalls) and internal division walls located across the boiler width. The elimination of the in-bed tube bundle is possible because of the large quantity of solids that are recycled internally and externally around the furnace. It is the method of collecting and recycling of solids that is the primary difference in designs of CFB boilers. Various designs of particle collectors and cyclones are used for this purpose. Furnace temperatures remain uniform because the mass flow rate of the recycled solids is many times the mass flow rate of the flue gas. The heat transferred to the furnace walls provides the heat absorption necessary to maintain the required bed temperature range of 1500 to 1600°F. CFB boilers are generally used to burn coal, tire chips, and waste fuels.

Circulating fluid bed (CFB) combustion system designs include the following major components: a refractory-lined lower combustor sec-

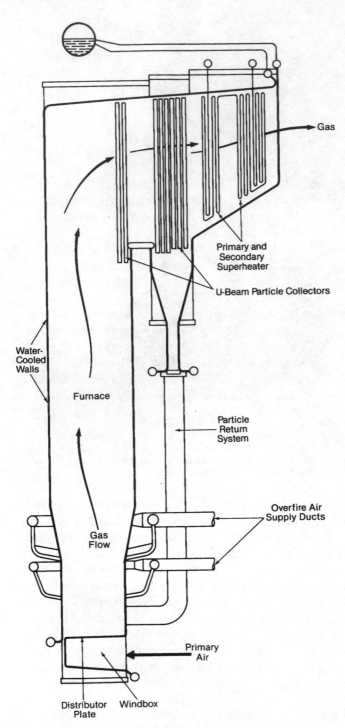

Figure 2.46 Schematic of circulating fluidized bed (CFB) boiler. (*Babcock & Wilcox, a McDermott company.*)

tion with fluidizing nozzles on the floor; an upper combustor water-wall section; a transition piece that includes a hot solids separator and solids reentry downcomer, which may be all or partially lined with refractory; and a convective boiler section. Primary air is supplied several feet above the lower combustor section to impart a staged-firing effect.

A CFB combustion system design is shown in Fig. 2.47 and Fig. 2.48. In this system, fuel and limestone (for sulfur capture) are fed into the lower portion of the combustion chamber in the presence of fluidizing air. The turbulent environment causes the fuel and sorbent to mix quickly and uniformly with the bed material. Fluidizing air causes the fuel, limestone, and bed material to circulate and rise throughout the combustion chamber and enter the hot cyclone collector. Hot flue gases and fly ash are separated from coarse solids in the cyclone. The solids, including any unburned fuel, are then reinjected into the combustion chamber through the nonmechanical loop seal. The continuous circulation of solids through the system provides longer fuel residence time, which results in very efficient fuel combustion. The limestone in the fuel mixture reacts with sulfur dioxide and other sulfur compounds in the flue gas and is removed with the ash as an inert dry solid, calcium sulfate. The relatively low combustion temperature (approximately 1600°F) and the introduction of secondary air at various levels above the grid provide for staged combustion that limits the formation of nitrogen oxides (NO_x).

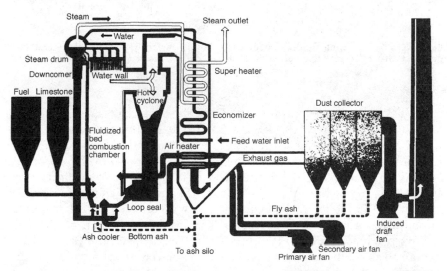

Figure 2.47 Circulating fluidized bed (CFB) combustion system with a hot cyclone collector. (*Foster Wheeler Pyropower, Inc.*)

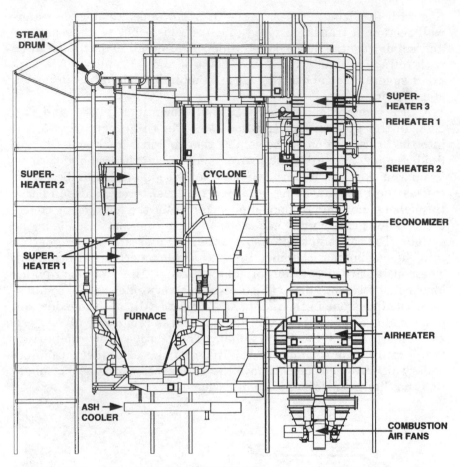

Figure 2.48 Circulating fluidized bed (CFB) boiler design with hot cyclone collector design, either refractory lined or water cooled. (*Foster Wheeler Pyropower, Inc.*)

A CFB boiler system similar to that shown in Figs. 2.47 and 2.48 is designed and operates to burn anthracite culm, a waste by-product of the coal mining industry in Pennsylvania. This previously unusable energy source is now used to provide steam for a nearby coal-drying facility and for the production of 80 MW of electricity. In the burning of the culm, the culm banks will be leveled (currently 200 to 300 ft high), and the inert ash will be used to fill in abandoned strip mines, thus restoring the landscape. Each of the two boilers produces 355,000 lb/h of steam at 1500 psig and 955°F. This design uses a hot cyclone to collect and recirculate the solids to the combustion chamber.

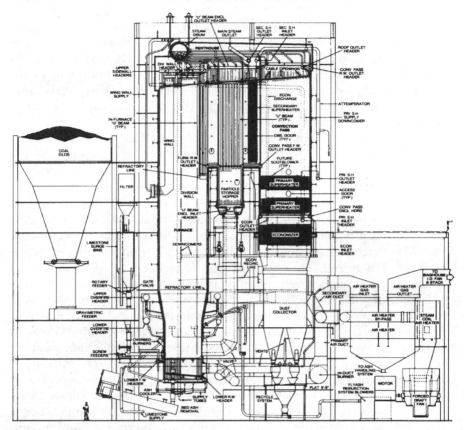

Figure 2.49 Circulating fluid bed (CFB) boiler, Ebensburg Power Company. (*Babcock & Wilcox, a McDermott company.*)

Figure 2.49 shows another design of a CFB boiler that burns bituminous waste from Pennsylvania. The use of this energy source eliminates a waste coal pile and provides advantages similar to those described above. The boiler is designed to produce 465,000 lb/h of steam at 1550 psig and 955°F. The steam is used for both heating and processing to a nearby industrial complex and for the production of approximately 50 MW of electricity.

This design incorporates a U-beam particle collector, which is a series of obstructions in the flue gas path that cause velocity to slow and particles to deflect and drop out of the flow. The U-beam separator incorporates a particle storage hopper between the separator and the return leg. In addition, two staggered rows of U-beams are located in the furnace. As part of the overall solids recirculation system, a dust collector is utilized after the economizer for final solids removal.

A solids reinjection device, called by some designers an *L-valve* or *loop seal,* is an important part of the CFB design. It is a simple non-mechanical hydraulic barometric seal against the combustor shell. Most are large-diameter refractory-lined pipes with no internals. Flow of these solids may or may not be controlled through valving to help maintain a constant bed temperature and solids inventory.

Ash is removed from the bottom of the combustor to control the solids inventory and bed quality. Because of the large amount of ash in a fluidized bed boiler, erosion can be a problem, and proper designs must be incorporated to minimize the effect of erosion. The abrasive nature of the ash alone from coal waste, for example, is cause for significant erosion. The addition of limestone for sulfur capture adds to this erosion problem. Sand is used in some bubbling bed designs and in CFBs where sulfur content of the fuel is low and limestone addition is not required. Thus the addition of sand as a bed material can create an even more severe erosion problem.

Nearly all CFB designs use staging in order to minimize NO_x emissions. Primary air is the fluidizing air, and it enters through distribution ports on the combustion floor, while secondary air enters through ports in the combustion chamber. This staging of combustion, together with low combustion temperatures, results in low NO_x emissions. Final particulate control is maintained with the use of conventional electrostatic precipitators or fabric filters, as described in Chap. 12.

There are various CFB boiler designs, each based on the designer's research and operating experience and each having unique advantages. Several of the more common designs are as follows, and each has the goal of ensuring the proper solids content in the unit.

Hot cyclone separator. As shown in Fig. 2.48, this design collects primary solids prior to the convection pass in a hot cyclone separator for recycle to the furnace. Secondary cycles are usually recycled from a baghouse. Alternative designs are also utilized where a water-cooled cyclone is used instead of a hot refractory-lined cyclone separator, and it becomes part of the overall heat-transfer surface of the boiler.

Cold cyclone separator. This design collects primary solids after the convection pass for recycle to the furnace, and therefore all recycled solids pass through the convection pass.

Impact separator. The design shown in Fig. 2.49 collects primary solids prior to the convection pass for recycle to the furnace, with secondary solids recycled from a multiclone dust collector.

The designers of CFB boilers continue to strive to improve their product based on the experience and testing of operating units and on research projects. Therefore, designs continue to evolve, and one such design is shown in Fig. 2.50. This design uses an internal recirculation system for solids control.

The boiler shown is coal fired. Approximately 60 to 70 percent of the combustion air is introduced through the bottom of the bed. The bed material consists of coal, lime, spent sorbent, sand (when limestone is not used), and ash. The bottom of the bed is supported by water-cooled membrane walls with air nozzles for air distribution. The coal and limestone are fed into the lower bed by gravimetric feeders (see Chap. 5). In the presence of fluidizing air, coal and limestone are mixed under the turbulent conditions and react as a fluid. Bed temperatures are in the range of 1500 to 1600°F.

The remainder of the combustion air is admitted as over-fire air at two levels at the top of the lower furnace in both the front and rear walls. This admission of primary and secondary air results in staged combustion, which limits the formation of NO_x.

The fluidizing air velocity is such that it entrains the solid particles and carries them through the combustor shaft. The distribution of solids in a CFB furnace has a high-density region that exists in the bottom of the furnace and a lower density region that exists in the shaft of the furnace. A gradual transition exists between these two regions.

The entrained solids and gas mixture passes through in-furnace U-beam separators, where approximately 75 percent of the solids, which include unburned carbon and unutilized calcium oxide (CaO), are separated and returned to the furnace internally. The remaining 25 percent of the solids is separated by external U-beam separators and returned to the furnace. All collected solids return to the lower furnace by falling as a curtain along the rear wall.

The fines collected by the secondary separator, in this case a multiclone dust collector, are also recirculated to the lower furnace to minimize carbon loss and to increase the use of limestone. If a solid fuel has a low ash content, or if it is low in sulfur, sand is used to maintain the furnace solids inventory. Solids are removed from the bottom of the unit through a bed drain system, where the solids are cooled with or without recovery of the heat from the solids. The purpose of draining the bed material from the furnace is to control the bed solids inventory and to remove oversized material that accumulates during operation. Since the drained material is at bed temperature (approximately 1600°F), it contains a considerable amount of heat. Therefore, the material is cooled to an acceptable temperature before disposal into the ash system.

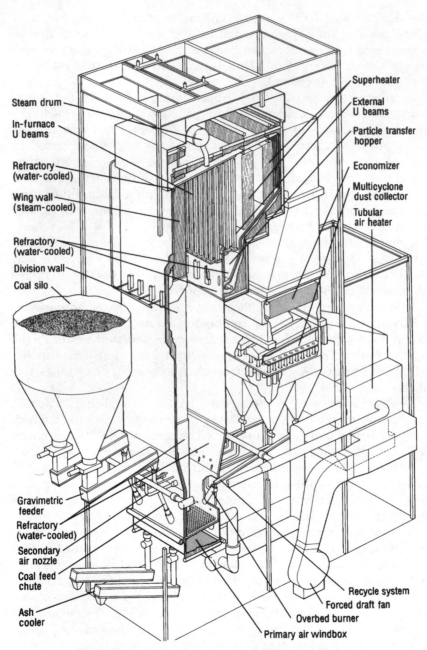

Steam drum

In-furnace
U beams

Refractory
(water-cooled)

Wing wall
(steam-cooled)

Refractory
(water-cooled)

Division wall

Coal silo

Gravimetric
feeder

Refractory
(water-cooled)

Secondary
air nozzle

Coal feed
chute

Ash
cooler

Superheater

External
U beams

Particle transfer
hopper

Economizer

Multicyclone
dust collector

Tubular
air heater

Recycle system

Forced draft fan

Overbed burner

Primary air windbox

Figure 2.50 Circulating fluid bed (CFB) boiler design with an internal primary recycle system. (*Babcock & Wilcox, a McDermott company.*)

The solids separation system is a major part of all CFB boiler designs, and it is perhaps the most distinguishing difference between the various designs. Both hot and cold refractory-lined cyclones are used, as well water-cooled cyclones that become part of the boiler circulation system. The design shown in Fig. 2.50 incorporates a U-beam separator design, and a schematic of this is shown in a plan view in Fig. 2.51.

This boiler design has two stages of primary solids separators: in-furnace and external separators. A particle-transfer hopper is located at the bottom of the U-beams, and the separated solids are discharged from this hopper directly into the furnace. The falling solids in the hopper form a pressure seal between the furnace and the solids-transfer hopper to prevent any gas-solid mixture from bypassing the U-beams.

The combination of the primary collector in the U-beam and the multiclone dust collector provides a high collection efficiency, and therefore a high solids loading results. This is necessary for a good design on a CFB boiler.

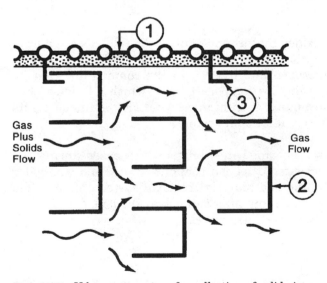

Figure 2.51 U-beam separators for collection of solids in a circulating fluid bed (CFB) boiler. 1 = Sidewall membrane panel. 2 = U-beam. 3 = Seal baffle. (*Babcock & Wilcox, a McDermott company.*)

2.15.5   Emissions control with fluidized bed boilers

SO_2.   When fuels with a sulfur content are burned, most of the sulfur is oxidized to SO_2, which then becomes part of the flue gas. When limestone ($CaCO_3$) is added to the bed, the limestone undergoes a transformation called *calcination* where lime (CaO) is formed and reacts with the SO_2 in the flue gas to form calcium sulfate ($CaSO_4$). $CaSO_4$ is chemically stable at fluid bed temperatures, and it is removed from the system as a solid with the ash for disposal.

In a CFB unit, a solids collector is located immediately after the furnace and recycles all the captured solids to the bed. Some designs also have recycling of solids from the baghouse located downstream. This method results in additional sulfur capture and improved combustion efficiency. Because of this recycling, CFB designs generally have a higher SO_2 capture rate than BFB units.

An SO_2 removal efficiency of approximately 90 percent can be obtained in a CFB unit. If removal requirements greater than this are mandated, the amount of limestone required increases rapidly, which significantly increases operating costs. This could result in the use of a pulverized-coal-fired unit with scrubbers as the more economical choice instead of a CFB unit.

NO_x.   The nitrogen oxides that are present in the flue gas come from two sources, the oxidation of nitrogen compounds in the fuel and the reaction between nitrogen and oxygen in the combustion air (see Chap. 12). Because of the low temperatures at which a fluidized bed boiler works, NO_x formation is minimized, and NO_x emission limits often can be met without additional systems.

Particulates.   The ash contained in solid fuel is released during the combustion process. Some of this ash remains in the fluid bed boiler and is discharged by the bed material removal system, often called the *bed drain system*. The remaining ash leaves the boiler in the flue gas, where it is collected by a fabric filter.

2.15.6   Bed material

There is a wide variation in ash properties in fuels, and therefore ash is not usually depended on to form a stable bed. Therefore, another inert material, such as sand, must be added to the system at the proper size requirements. When sulfur removal is a requirement, limestone is used as a sand replacement to achieve SO_2 capture.

2.15.7 Heat transfer

In conventional boilers such as those burning pulverized coal, a portion of the fuel ash is carried with the flue gas through the furnace and boiler heating surfaces. This ash is less than 10 lb of inert solids per 1000 lb of flue gas, i.e., the quantity of ash is less than 1 percent of the amount of flue gas. The majority of the heat transfer from the flue gas to the furnace water walls is by radiation.

In a CFB boiler, the amount of solids in the flue gas leaving the furnace may exceed 5000 lb of solids per 1000 lb of flue gas, i.e., the quantity of solids is greater than 500 percent (or 5 times) of the amount of flue gas. As a result of the high solids content, additional heat-transfer characteristics are considered in the design. The majority of the heat transfer to the in-bed tubes of a bubbling bed unit and to the furnace walls of a CFB unit is by solids and flue gas conduction, with a small amount by solids and flue gas radiation. In a conventional boiler, flue gas radiation is the most important and solids convection is the least important to overall heat transfer.

The high solids concentration is very significant because, for the same temperature, the heat transfer in a fluidized bed boiler furnace is considerably higher than the heat transfer in a conventional furnace. However, because the furnace temperatures in a fluid bed boiler are between 1500 and 1600°F as compared with 3000 and 3500°F for a pulverized-coal- or a stoker-fired boiler, the overall heat transfer for each system is comparable.

BFB heat transfer. A BFB boiler is divided into three zones for the purposes of heat transfer:

1. Bubbling bed or dense bed

2. Disengaging

3. Upper furnace or freeboard

CFB heat transfer. A CFB boiler does not utilize an in-bed heating surface, but its heat absorption is totally from the furnace waterwalls and the internal furnace division walls. The heat transfer comes from two distinct zones in the furnace: (1) the dense bed and (2) the remainder of the furnace.

Heat balance. In conventional firing such as pulverized coal or stokers, only a small amount of heat is removed from the combustion gases in the zone of maximum heat release. As a result, the flue gas temperature becomes very high (3000°F) before it is cooled by the furnace waterwalls.

In fluidized bed combustion, heat is removed from the zone of maximum heat release at a much higher rate. Therefore, the temperature of the flue gas is limited to the design level of 1500 to 1600°F.

For BFB boiler designs, when required for fuels with a relatively high heating value, heating surface is placed in the bed of hot solids and burning fuel to maintain the temperature. For CFB designs, the large quantity of solids that circulate in the system removes heat from the active combustion zone and transfers it to the heating surface throughout the furnace.

2.16 Modern Coal-Fired Plants—Summary

Modern coal-fired utility plants require about 1 lb of coal to produce 10 lb of steam, and the combustion of this 1 lb of coal results in approximately 10 lb of flue gas. This 10 lb of steam can generate 1 kW of electricity.

In the 1940s, electrical generating units in the 50-MW range were considered large. By today's standards, units 10 times larger are considered only average in size.

A modern 500-MW steam turbine requires nearly 4 million pounds per hour of steam. To supply the heat required, nearly 5000 tons of coal must be burned every day, and this requires about 80 carloads of coal each day just to satisfy this one single average-sized boiler.

At today's rate of electricity generation, the United States has about 250 such plants in operation, consuming approximately 500 million tons of coal each year. Industrial coal-fired boiler applications add about 40 percent to this figure. Coal-fired plants account for more than 50 percent of the total electricity production, and these coal-fired plants spend approximately 85 percent of their total generating costs on the coal itself. It is for this reason that plant operators strive to achieve the highest efficiency that is reasonably possible.

Because of the high cost of fuel, very few, if any, medium- and large-size boilers are in operation today without heat-recovery equipment. Heat recovery for a stoker-fired boiler consists of either an economizer or an air heater or both. When the economizer water outlet temperature is at least 50°F below the boiler drum saturation temperature and the economizer is designed for an exit gas temperature of about 350°F, an economizer may be the only choice.

Air heaters are not normally used with small stoker-fired industrial boilers because of the added power requirements, increased equipment costs, overall space requirements, and the limit on under-grate air temperature. However, on larger-sized stoker-fired boilers where the size of the economizer is limited by the higher feedwater temperature, it is necessary to provide both an air heater and an economizer to produce the most efficient boiler system.

Pulverized-coal-fired boilers include an air heater along with an economizer as a means of heat recovery because of the pulverizer's requirement for hot primary air for coal drying. The use of an air heater also permits designing for lower stack temperatures than would be possible with an economizer alone because special corrosion-resistant materials can be utilized in the air heater's cold-end sections. This results in improvements in overall efficiency.

2.17 Process Steam and Its Application

In the steam power plants that generate only electric power and do not use any process steam, the economical thermal efficiency for a fossil fuel plant is approximately 40 percent, and for a nuclear plant it is approximately 33 percent (see Fig. 2.63). Therefore, more than 50 percent of the heat released from the fuel must be transferred to the environment in some manner. The majority of this heat loss in a fossil fuel plant is from the flue gases that flow out of the stack and from the condenser cooling water.

This overall plant efficiency could be improved significantly by operating steam plants where some steam is extracted from the cycle at a sufficient pressure for use in an industrial process or in space heating. An overall efficiency of 65 percent would be possible with this arrangement, and combined systems have been common for years at certain locations. However, in most electric power plant locations, the demand for process steam has not been sufficient to use a combined cycle like this.

There are situations where this has become attractive, such as cogeneration, biomass, and waste-to-energy installations, where district heating and other process steam applications are viable options to the plant design and economics of the project.

2.18 Combined Cycle and Cogeneration Systems

The waste heat from a power system such as a gas turbine can serve as the heat source for a heat-recovery system that is part of a steam turbine cycle. This combination of two types of systems is called a *combined cycle system*, and it can increase the overall plant efficiency to approximately 50 percent, as compared with 40 percent for a typical power plant cycle using a fossil fuel. If such a system also were to use the produced steam as process steam, the overall energy use could be even higher.

Waste heat has been utilized for years by many industries to obtain the most from their heating investment. Unique combined cycle systems have been developed by the steel making, oil refining, pulp and paper, and food processing industries in order to reduce their operating costs.

The goal of more efficiently converting fuel energy into electrical and mechanical power led to the combining of plant cycles that have two thermodynamic power cycles. The gas turbine has improved greatly in both availability and reliability, and because of this, a combined cycle system generally refers to a plant that consists of a gas turbine, a heat-recovery system that converts heat energy to steam energy, and a steam turbine. The heat-recovery system incorporates a waste-heat boiler that is now commonly called a *heat-recovery steam generator* (HRSG). This cycle application also can be referred to as a *cogeneration system.*

Figure 2.52 shows a schematic of a simple combined cycle system that consists of a gas turbine generator, a heat-recovery steam generator (HRSG), a steam turbine generator, a condenser, and auxiliary systems. If necessary to meet low NO_x emissions, NO_x-reduction systems can be added.

The gas temperature leaving the gas turbine is in the range of 950 to 1050°F. The gas turbine can be operated independently of the steam cycle with the installation of a gas bypass to the stack located after the gas turbine.

The major advantages of a combined cycle system is the improved thermal efficiency and low environmental emissions with the use of natural gas as a fuel. It also offers the following:

1. A shorter schedule, with the operation of the gas turbine in about 1 year and the addition of the steam cycle generally within an additional year.

2. The gas turbine portion of the system can be used by itself for rapid startup and in meeting peak power requirements of the electric system. The steam system requires approximately an hour to be in operation from a cold start.

3. Low capital cost as compared with other firing methods such as coal firing.

However, the disadvantages of the system are significant and must be seriously evaluated. These include the use of higher-cost natural gas, the potential for the unavailability of this critical resource,

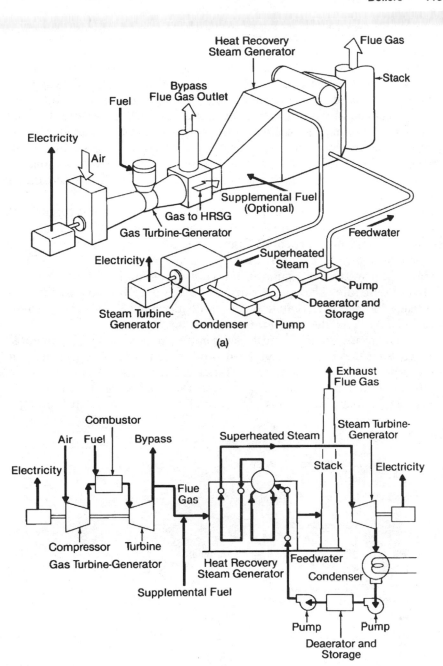

Figure 2.52 Schematics of combined cycle system with a gas turbine and a steam cycle incorporating a heat-recovery steam generator (HRSG). (*a*) Equipment configuration schematic. (*b*) Flow schematic. (*Babcock & Wilcox, a McDermott company.*)

potentially higher maintenance costs of the system, and lower availability.

The gas turbine combined cycle system is used primarily for electricity production, but it also can be part of a cogeneration system where the heat-recovery portion of the system also can be used to furnish process steam as well as electricity. A gas turbine system without a combined steam cycle results in a low efficiency of approximately 30 percent. Steam could be used for heating or in a process, and such systems can obtain over 60 percent of total energy use as compared with the near 50 percent of a combined cycle system that is designed to produce electricity only.

A simple gas turbine system is shown in Fig. 2.53, and it consists of an air compressor, a combustor where the fuel (generally natural gas) is burned, and a gas turbine that drives an electric generator. It is basically a simple system, and when compared with a coal-fired steam cycle system, it results in low capital cost and a relatively short delivery schedule. This gas turbine system is used by some utilities to increase their electric capacity in small increments, especially when the capacity is needed to fill intermittent operation.

Because of its use of natural gas, complex and costly environmental control systems are not required, and as a result, the permitting process for the location of these facilities is shorter, since the plants are more acceptable to the public. This acceptability also leads to a shorter schedule for installation and operation of such a system. On the nega-

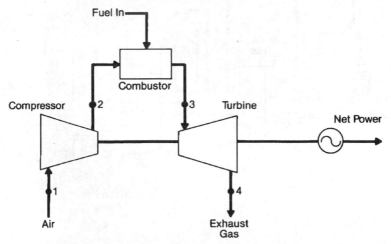

Figure 2.53 Simple gas turbine power cycle. (*Babcock & Wilcox, a McDermott company.*)

tive side, these plants use natural gas, a critical resource for home heating and other specific applications. The viability of the gas turbine cycle will be acceptable only as long as there is a surplus of natural gas and this can be directed toward the production of electricity.

The simple gas turbine cycle is similar to the jet engine cycle. Air is compressed and mixed with the fuel (natural gas) in the combustor, where it is burned. The high-temperature combustion products (approximately 2300°F) flow through the turbine and produce work by the expansion of the gases as they cool. A portion of this work produced by the turbine is used to drive the compressor, which is on the same shaft, and the remainder is available to produce electric power by means of an electric generator that is coupled to the turbine. The exhaust gases from the turbine are then vented to the atmosphere as waste heat. The electricity output of the gas turbines is currently in the range of 1 to 150 MW; however, larger units are being developed.

As noted previously, the cycle efficiency of a gas turbine system is relatively low at 25 to 30 percent because of the high exhaust gas temperatures and also because a significant portion of the turbine output is used to drive the compressor, which is on the same shaft. A major advantage of the gas turbine cycle, as compared with a steam cycle, is its ability to operate at much higher temperatures in the turbine. Gas turbines generally operate with gas temperatures entering the turbine of between 1800 and 2200°F. The newer designs operate with temperatures near 2350°F, which increases the thermal efficiency of the cycle. However, because the temperature of the exhaust gases leaving the gas turbine is high (approximately 1000°F), such systems are incorporated with a steam cycle in a combined cycle system, and thus this otherwise wasted heat is utilized.

Thermal efficiency is important in all power plants because it relates to the effectiveness of converting heat energy into work, which is electrical output in a power plant. Thus the higher the efficiency, the lower are the operating costs because of lower fuel costs. Efficiency is defined as follows:

$$\text{Thermal efficiency, \%} = \frac{\text{net work produced}}{\text{total heat input}} \times 100$$

For a combined cycle, the thermal efficiency is the work produced by the two cycles divided by the total heat supplied.

As noted previously, the combined cycle is a gas turbine plant that is supplemented with a steam plant that uses the exhaust gases from the outlet of the gas turbine and passes them through a steam generator, which is often called a *waste-heat boiler*. A schematic of this com-

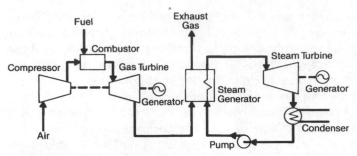

Figure 2.54 Simple combined cycle system. (*Babcock & Wilcox, a McDermott company.*)

bined system is shown in Fig. 2.54. The steam generator uses the hot turbine exhaust gas as a heat source for a steam turbine cycle. Electricity is produced from generators attached to both the gas turbine and the steam turbine. The steam generator recovers the exhaust heat from the gas turbine, and even though it is a waste-heat boiler, it is today commonly called a *heat-recovery steam generator* (HRSG).

Figures 2.55 and 2.56 show an advanced combined cycle concept that is designed to perform any operational duty from peaking to base load operation. It is designed for a gross plant output of over 250 MW, and it has a unique single-shaft gas turbine–steam turbine arrange-

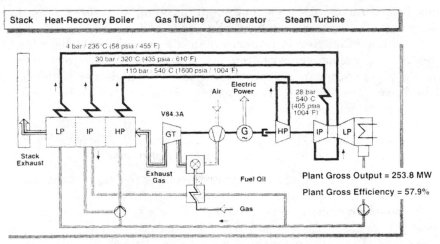

Figure 2.55 Single-shaft design concept for 253-MW combined cycle system. (*Siemens Power Corp.*)

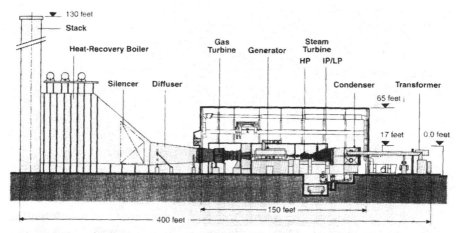

Figure 2.56 Plant layout for 253-MW single-shaft combined cycle system design concept. (*Siemens Power Corp.*)

ment. The common generator is solidly coupled to the gas turbine. A synchronous clutch is used for the steam turbine connection to the generator to provide maximum operating flexibility.

2.19 Nuclear Steam Generation

The energy produced from nuclear power plants results in the production of over 20 percent of the total electric power produced in the United States. Worldwide, electric production from nuclear power is also approximately 20 percent. Nuclear power, therefore, is an extremely important source of energy because it is second only to coal as a fuel energy source for the production of electricity (see Chap. 1).

The nuclear steam generator produces steam and therefore is a boiler. The furnace for burning conventional fossil fuels is replaced by a reactor, which contains a core of nuclear fuel.

The heart of the reactor is the *core*. The core contains uranium fuel, which, as a result of the fissioning of the uranium, generates heat to produce steam. The core is encased in a pressure vessel, which in turn is enclosed by shielding and a reactor building, all of which contain the radiation emitted from the nuclear reaction.

In commercial reactors for utility applications, the core consists of a number of fuel elements, each fuel element containing fuel rods that encapsulate the uranium dioxide (UO_2) fuel pellets. These fuel elements are arranged so that a self-sustaining nuclear chain reaction takes place. The fuel utilizes the fissionable isotope of uranium ^{235}U, and in most reactor designs it is enriched to approximately 3 percent, with the remaining 97 percent consisting of the uranium isotope ^{238}U. ^{238}U is a fertile isotope of uranium that absorbs neutrons, and it eventually transforms into another element, plutonium, which can be used as a nuclear fuel, usually mixed with uranium. There are also some reactor designs that use only natural uranium, i.e., uranium that has not been enriched in the isotope ^{235}U.

When a neutron strikes the nucleus of the fissionable isotope ^{235}U, the nucleus splits and releases a large amount of heat together with the release of additional neutrons that maintain the fissioning process; thus there is a chain reaction. The liberated neutrons travel at a high rate of speed. Since slow-moving neutrons are more effective in splitting nuclei of ^{235}U than are fast-moving neutrons, the neutron velocity must be slowed. This is accomplished by the use of a moderator.

The moderator does the slowing down. Various materials are used as moderators, such as graphite, ordinary water, or heavy water (water that contains heavy hydrogen instead or ordinary hydrogen). The moderator can slow speeding neutrons without absorbing them. In commercial light-water reactors, ordinary water serves as both the moderator and the coolant.

The control rods contain substances that absorb neutrons readily. They are arranged so that they may be inserted or withdrawn from within the fuel core as required to control the chain reaction. When the control rods are inserted into the reactor core, they absorb neutrons so that the chain reaction is slowed or stopped. As the rods are withdrawn, the neutrons become active again, and the nuclear chain reaction starts up again. Thus the control rods are used to raise or lower the power output of the reactor.

Another component of a reactor is the coolant. The function of the coolant is to remove the heat developed in the core, which can be used to produce steam to generate electricity. The coolant may be ordinary water, heavy water, a gas, or a liquid such as liquid sodium.

In one power reactor system, called a *pressurized-water reactor* (PWR) system, water is used as both the moderator and the coolant. The water is kept under pressure in the reactor vessel and the primary system. From the reactor vessel, the water is pumped to a heat exchanger (steam generator), which converts the water to steam in a secondary piping system. The steam is then used to power a turbine generator. The schematic shown in Fig. 2.57 shows a PWR system that is fueled by slightly enriched uranium in the form of uranium oxide pellets held in zirconium-alloy tubes in the

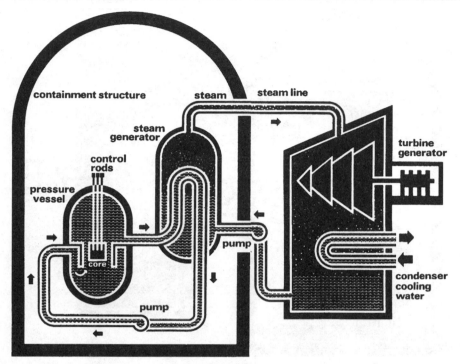

Figure 2.57 Pressurized-water reactor (PWR) nuclear power plant. (*American Nuclear Society.*)

core. Water is pumped through the core to transfer heat to the steam generator. The coolant water is kept under pressure in the primary system through the core to prevent boiling, and it transfers its heat to the water in the steam generator (the secondary system) to make the steam.

In another power reactor system, called a *boiling-water reactor* (BWR) system, water is again used as both moderator and coolant, but here the water is allowed to boil within the reactor vessel. The steam thus generated then passes directly to the turbine generator. The schematic shown in Fig. 2.58 shows a BWR system that also uses a reactor fueled by slightly enriched uranium in the form of uranium oxide pellets held in zirconium-alloy tubes in the core. Since there is no secondary system with this design, the turbine portion of the plant is designed to handle any radioactivity carried by the steam, and this requires special shielding and containment structures.

Figure 2.59 shows the arrangement of a *liquid-metal fast breeder reactor* (LMFBR) nuclear plant design that pumps molten sodium in

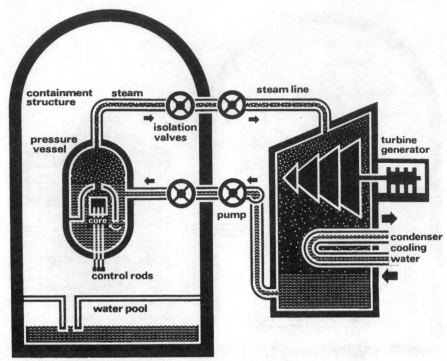

Figure 2.58 Boiling-water reactor (BWR) nuclear power plant. (*American Nuclear Society.*)

the primary loop through the reactor core containing the fuel. This sodium in the primary loop collects the heat and transfers it to a secondary liquid sodium loop in the heat exchanger, from which it is pumped to the steam generator, where steam is generated and used to power the turbine generator.

In addition to producing electricity, this type of reactor also produces more fissionable material than it consumes, which results in the name *breeder reactor*. When irradiated, certain nonfissionable materials may be transformed into material that is fissionable. An LMFBR begins operation with a core of fissionable ^{235}U surrounded by nonfissionable ^{238}U. During operation, the ^{238}U is bombarded by high-velocity neutrons and transmuted to fissionable plutonium-239 (^{239}Pu). The plutonium is extracted periodically and fabricated into a new fuel. This design uses fast neutrons as compared with the slow neutrons that resulted from the moderator of the PWR and BWR designs.

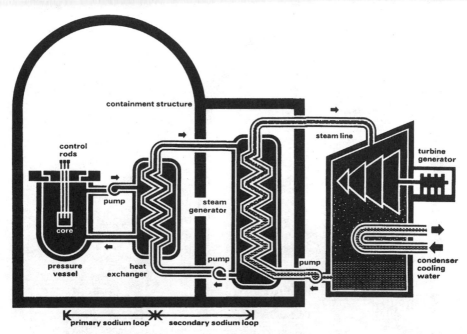

Figure 2.59 Liquid metal fast breeder reactor (LMFBR) nuclear power plant. (*American Nuclear Society.*)

Another type of nuclear system is shown in Fig. 2.60 and uses helium gas as a coolant. This system is called a *high-temperature gas-cooled reactor* (HTGR). The HTGR shown is a type of reactor that is fueled by uranium carbide particles distributed in graphite in the core. Helium gas is used as a coolant to transfer the heat from the core to the steam generator. Steam is generated in this secondary cycle and is used to drive a turbine generator in a conventional turbine cycle.

Figure 2.61 shows still another design that uses heavy water as both the coolant and moderator, and this design is called the *CANDU pressurized heavy water reactor* (PHWR). In this design, the calandria, or reactor vessel, is a cylindrical tank filled with heavy water (deuterium oxide) moderator at low temperature and pressure. Hundreds of pressure tubes (fuel channels) penetrate the calandria, and fuel bundles containing natural uranium fuel are inserted in the pressure tubes. Pressurized heavy water coolant is pumped past the uranium fuel, and the heat of fission is transferred to the coolant. The

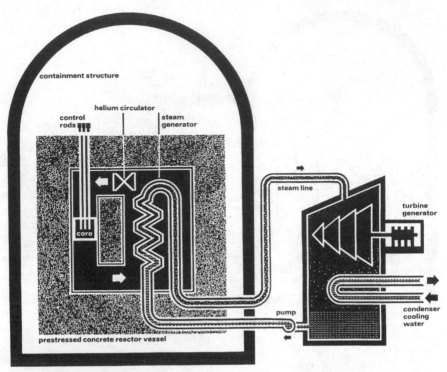

Figure 2.60 High-temperature gas-cooled reactor (HTGR) nuclear power plant. (*American Nuclear Society.*)

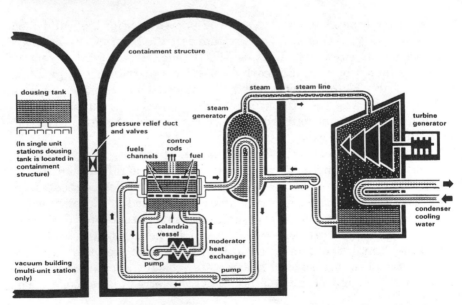

Figure 2.61 CANDU pressurized heavy water reactor nuclear power plant. (*American Nuclear Society.*)

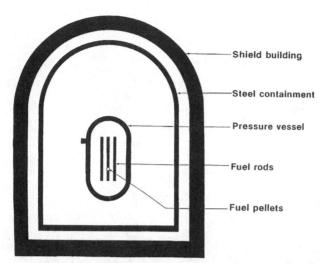

Shield building

Steel containment

Pressure vessel

Fuel rods

Fuel pellets

Figure 2.62 Barriers against radioactive release in nu clear power plant. Shield building: reinforced concrete structure ≥3 ft. Steel containment: essentially leak-tight shell of steel plate. Pressure vessel: height, ≤75 ft; diameter, 20 ft; walls, 9 in thick. Fuel rods: zirconium alloy about 12 ft long; diameter, $\frac{1}{2}$ in. Fuel pellets: dense ceramic pellets in which most of the fission products remain bound. (*American Nuclear Society.*)

coolant flows to the steam generators, where it gives up its heat to ordinary light water to produce steam that drives a conventional turbine generator.

A by-product of nuclear energy is the release of radioactivity during the fission process. The nuclear plant is designed to prevent the release of radioactivity by having a series of barriers that prevent its release. Figure 2.62 shows the various barriers. The fuel rods contain the fission products. If these rods were to leak, the series of additional barriers, the primary system including the pressure vessel, and the design of the reactor building prevent the release of radioactive products. Monitors within the reactor system will alert the operators when levels become too high. If this occurs, the nuclear system is shut down, and the leaking fuel assembly is identified and replaced.

Figure 2.63 shows a schematic comparison of a fossil fuel–fired power plant with various nuclear power plant system designs and the plant efficiencies for each type of design.

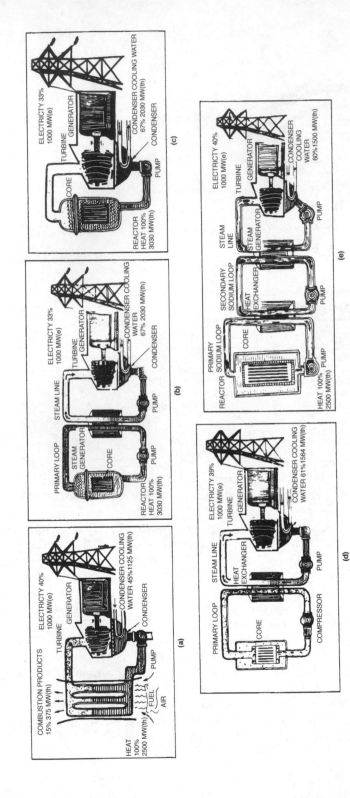

Figure 2.63 Comparison of fossil fuel power plant with nuclear power plant systems. (*a*) Fossil fuel power plant. (*b*) Nuclear power plant: boiling-water reactor (BWR). (*c*) Nuclear power plant: pressurized-water reactor (PWR). (*d*) Nuclear power plant: high-temperature gas-cooled reactor (HTGR) and gas-cooled fast reactor (GCFR). (*e*) Nuclear power plant: liquid metal fast breeder reactor (LMFBR). (*American Nuclear Society.*)

Questions and Problems

2.1   What is a boiler?

2.2   What are the requirements of a good boiler?

2.3   Define a boiler setting.

2.4   What is meant by heat being transmitted by radiation? By conduction? By convection?

2.5   Describe the process of boiling.

2.6   What methods are used for the circulation of water and steam through a boiler?

2.7   What is a fire-tube boiler? Describe its operational features.

2.8   In today's power plants, what type of fire-tube boiler is the most common, and how is it fired?

2.9   If a fire-tube boiler is a four-pass design, describe the features of such a design.

2.10   What is a water-tube boiler?

2.11   What are the advantages and disadvantages of package boilers?

2.12   What is a high-temperature-water (HTW) boiler? What prevents the water from boiling? For what type of application does this type of design provide advantages?

2.13   Provide some major comparisons between fire-tube and water-tube boilers, and provide the predominant advantages and disadvantages of each.

2.14   Identify the major purposes of the steam drum.

2.15   Why is steam-water separation equipment so important to good boiler operation?

2.16   Describe how a cyclone steam separator works.

2.17   What is a superheater? What advantage does it serve in the overall power plant operation? Where is the superheater located in the boiler?

2.18   What is a radiant type of superheater? A convection type superheater?

2.19 What methods are used to provide a constant steam temperature over a specified load range?

2.20 What is a reheater, and when is it used?

2.21 What two primary areas have the greatest influence on the superheater design other than flue gas temperature and quantity?

2.22 What are the three basic methods for maintaining a constant steam temperature? Which of these is the predominant method? Describe its operation.

2.23 What is an economizer? What is the advantage of using an economizer? If an air preheater is part of the overall boiler design, is the economizer located before or after the air heater with regard to the flue gas flow? Why?

2.24 Describe the various types of air preheaters and their purpose in boiler design.

2.25 In reference to the designs of superheaters, economizers, and air heaters, what is meant by the terms *parallel flow* and *counterflow*? What are the advantages of each design?

2.26 What effect, if any, does the dew point have on the corrosion of economizers and air heaters? Does sulfur in the fuel change the dew point? Why is this important?

2.27 When is a steam coil air heater required? Where is it located?

2.28 Describe the purpose of the furnace portion of the boiler, and define the more important design features of it.

2.29 How are boilers supported to allow for expansion and contraction during operation?

2.30 Why are waterwall furnaces so important to the boiler design as compared with older refractory boiler designs?

2.31 What is meant by membrane furnace wall construction? What is its advantage?

2.32 How would you classify industrial and utility boiler designs?

2.33 For coal firing, what are the three primary methods for the combustion of coal? Describe the major characteristics of each, including their advantages and disadvantages.

2.34 What is the difference between a pressurized and a balanced-draft boiler design?

2.35 Describe the fluidized bed combustion process and how it compares with pulverized-coal and stoker firing.

2.36 Name the major advantages of fluidized bed combustion.

2.37 What are the two types of fluidized bed boilers? Describe their general characteristics.

2.38 For a CFB boiler design, what component is the most distinguishing design feature, and what is its purpose?

2.39 How are the emissions of SO_2, NO_x, and particulates controlled in a fluidized bed boiler?

2.40 What is used for the bed material in a fluidized bed boiler? Why?

2.41 Why is a combined cycle system important? Describe such a system that uses a gas turbine generator. Define its advantages and disadvantages.

2.42 In a nuclear power plant, briefly describe the fission process that releases heat for the generation of steam.

2.43 Describe and develop a simple schematic sketch of a pressurized water reactor (PWR) system and a boiling water reactor (BWR) system.

2.44 What systems are part of a nuclear power plant and are designed to prevent the release of radioactivity?

3

Design and
Construction
of Boilers

A boiler, which includes drums, tubes, flues, ducts, auxiliary equipment, and their associated supports, is subject to continual stress resulting from expansion and contraction and from elevated temperatures when it is in service. Boilers must be made adequately strong and with suitable materials to withstand these forces and temperatures. Boiler design requires a detailed study of the forces that are exerted on the various parts and of the temperatures to which they will be exposed. In addition, because of the corrosive properties of substances contained in fossil fuels and in water, attention must be given to the selection of the proper materials to ensure a high availability of the boiler system.

Pressure-vessel construction codes adopted by federal, state, and local jurisdictional authorities play an important part in determining safety requirements and thus the construction features of pressure vessels. The most widely used of these is the *ASME Boiler and Pressure Vessel Code,* published by the American Society of Mechanical Engineers. This code gives excellent guidance and basic requirements to the boiler designer.

3.1 Materials Used in Boiler Construction

When two forces of equal intensity are acting on the ends of a bar or plate and are acting in opposite directions, the bar or plate is said to be "in tension." If the forces are each acting toward the center, the material is said to be "in compression." When two forces are acting in

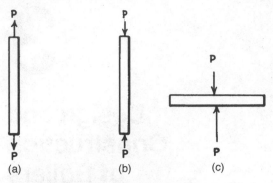

Figure 3.1 Bar subjected to the three types of stress.
(*a*) Tension. (*b*) Compression. (*c*) Shear.

the same plane, in very much the same way as the forces produced by a pair of scissors or shears, the material between these two forces is said to be "in shear" (Fig. 3.1). Depending on the location, boiler parts are subjected to the stresses resulting from these forces. *Stress* is the internal resistance that the material offers to being deformed by the external force. Stress is defined as the internal force divided by the area over which it is applied. The main importance of stress is its magnitude, but the nature of the load and the resulting stress distribution are also important. The boiler design must consider whether the loading is mechanical or thermal, whether it is steady or transient, and whether the stress pattern is uniform. As long as the material does not fail, the stress is at least equal to the external force. The external force is usually expressed in pounds per square inch (psi). For example, the force on a support rod is 4000 lb. If the rod is 1 in in diameter (*d*), the area (*A*) is equal to

$$A = \frac{\pi d^2}{4}$$

where $\pi = 3.1416$[1]

$$A = 0.7854 \times 1 \times 1 = 0.7854 \text{ in}^2$$

The force per square inch on the rod is

$$4000 \div 0.7854 = 5093 \text{ psi}$$

When a material is loaded, it is deformed or changed in shape. Up to a certain load, for any given material, the amount of this deformation is proportional to the load. In the case of a tensile stress, the

---

[1]See Appendix B.

deformation is called *elongation*. If a given bar will deflect 0.013 in with a load of 15,000 lb, it will deflect 0.026 in with a load of 30,000 lb. After a certain load has been reached, this proportion will no longer hold true; neither will the material return to its original shape when the load is removed. The more ductile the steel, the greater is the elongation before rupture. The load in pounds per square inch required to cause the material to be permanently deflected is called the *elastic limit*. The load in pounds per square inch required to cause complete failure in a piece of material is called the *ultimate strength*.

There is a close relation between the compressive and the tensile strengths of steel. However, in boiler construction, the tensile strength of steel is of foremost importance. The tensile strength of steel is determined by placing a test specimen in a testing machine and subjecting it to a measured "pull" until it fails. The elastic limit is determined by making two marks on the test specimen and noting the elongation as the force is applied. When the elongation is no longer proportional to the applied force, the elastic limit has been reached. Before the specimen actually fails, the cross-sectional area is usually much reduced as a result of the elongation. The ultimate strength is calculated from the original area. The tensile strength of steel used in boiler construction varies from 50,000 to 100,000 psi and the elastic limit from 25,000 to 55,000 psi. There is also a relation between the tensile and the shearing strengths of steel; i.e., a piece of steel that has high tensile strength also has high shearing strength. Many new alloys have been introduced over the years and a significant amount of new data on materials has become available. The designer must consult the latest code edition to ensure the use of proper material specifications and design values.

The *ductility* of metal is determined by measuring the increase in length, or total elongation, and the final area at the plane of rupture after a specimen has broken. Ductility is expressed as percent elongation or percent reduction of area.

The steel used in boiler construction must be of the highest quality, with the constituents carefully controlled. Unlike the steel used in general construction, the steel in boilers must withstand the load at elevated temperatures. Temperature has a more serious effect on the boiler than pressure. Excessive temperatures must be guarded against just as carefully as excessive pressures. Carbon steel is a suitable and economical material to about 850 to 950°F metal temperature depending on the pressure, but above this temperature the strength of carbon steel decreases very rapidly, and it becomes necessary to resort to the use of alloy steels (an *alloy* is a mixture of two or more metals). Several substances are used to produce high-temperature alloy steel, and molybdenum is one of the most important. Small percentages of molybdenum in the steel used to manufacture super-heater tubes, piping, and valves increase the ability of these parts to withstand high temperatures.

The combination of stress and high temperature results in changes to the internal structure of steel. One change is a small but permanent deformation of the steel known as *creep*. When a boiler is adequately designed and correctly operated, this change is too small to be measured over its lifetime. Another change that decreases the strength of steel is *graphitization*. This occurs as a result of an internal structure change in the steel due to a separation of the graphite.

Loads applied repeatedly to a piece of material in the form of vibration eventually will cause failure even when the load is below the determined elastic limit. When the correct materials are used and the boiler is operated properly, these factors do not cause difficulty or limit the useful life of the boiler. They are mentioned here to emphasize the importance of correct design and operation. A sophisticated vibration analysis study is often conducted for components that may be susceptible to significant vibration during operation. Material selection is then made based on this analysis.

The quality of steel is determined by chemical analysis for alloys and impurities. Carbon is reduced to approximately 0.25 percent in the boilerplate to ensure optimal ductility for welding and bending operations. Too high a percentage of carbon will cause the steel to harden and crack under the influence of pressure and temperature. The common impurities in steel are sulfur, phosphorus, manganese, and silicon. Sulfur reduces the ductility of steel and makes it difficult to work, especially when hot. Phosphorus makes steel hard and to some degree strong; it is undesirable in boilerplate because it lessens its ability to withstand vibratory forces. Manganese makes steel hard and difficult to cut and work; it is considered desirable because it combines with sulfur and lessens the effect of the sulfur. Silicon has a tendency to make steel hard, but since it does not decrease its tensile strength, it is considered desirable.

Chemical analysis is also used to control the introduction of chromium, nickel, molybdenum, and other substances used in making alloy steel for high-temperature boiler and superheater tubes.

Steel used in boiler construction was for years manufactured by the Bessemer process, which was invented in 1855 and for many years was the principal method of making steel. The development and perfection of the open-hearth process and subsequently the development of the basic-oxygen process and the basic electric-furnace process are now the primary steel manufacturing methods.

In the processing of steel, samples are analyzed and materials are added to obtain the desired steel alloy. When the analysis shows that the metal has the desired chemical content, it is poured into ingot molds. When cool, the steel ingots are taken to the rolling mill, heated, and worked into plates of the desired thickness and the shapes required for boiler construction.

Cast iron has high compressive but low tensile strength and is more resistant to corrosion than steel. In early steam boiler designs, cast iron was used extensively; however, high tensile strength is a requirement of metals used in modern boiler construction. Therefore, the use of cast iron is restricted to items such as stoker and pulverizer parts.

Copper is an unimportant material in boiler construction. This is chiefly because of its high cost and low tensile strength. However, it is used extensively in all electrical equipment found in power plants. Brass and other alloys are used in making many small boiler fittings, valves, etc. Their resistance to corrosion is their chief advantage.

3.2 Stresses in Tubes, Boiler Shells, and Drums

Tube, shell, and drum stresses will be explained by considering a seamless cylinder enclosed with heads in each end, as shown in Fig. 3.2. Pressure applied to the interior of this cylinder is distributed equally to all areas. This pressure causes the cylinder to be subjected to stress in three places.

First, the total force on each head is equal to the applied pressure (in pounds per square inch) multiplied by the area of the head in square inches. These forces A and B (Fig. 3.2a) produce tensile stress in the circumferential section (Fig. 3.2b).

Second, these same forces on the heads have a tendency to make them bulge (Fig. 3.3).

Third, stress is created along a longitudinal section MN of the cylinder (Fig. 3.4a). Assume that the cylinder is cut in half at section $ABCD$ (Fig. 3.5). Further assume that the lower half of the cylinder is replaced by a heavy plate. The remaining vessel consists of half a cylinder and a flat plate. The force on the plate equals the applied pressure multiplied by the area $ABCD$. (The rectangular surface

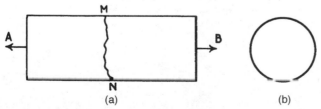

(a)   (b)

Figure 3.2 Tension in circumferential section of a shell or drum. (a) Seamless cylinder. (b) Cross section of cylinder showing metal in tension.

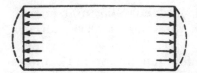

Figure 3.3 Bulging tendency of flat heads.

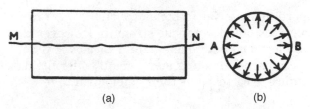

(a)   (b)

Figure 3.4 Tension in a longitudinal section of a drum or shell. (*a*) Internal forces tending to cause longitudinal failure. (*b*) Radial forces on inside of drum.

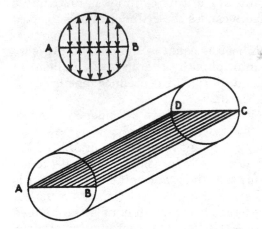

Figure 3.5 Diagrammatic illustration of transverse forces on the longitudinal section of a cylinder. The forces act on projected areas *A, B, C,* and *D* and produce tension stresses in sections *AD* and *BC*.

ABCD is known as the *projected area* of the curved surface.) This force is resisted by two strips of metal *AB* and *CD*. Each of these strips has a cross-sectional area equal to the length of the cylinder multiplied by the thickness of metal in the cylinder.

An example will show the application of this reasoning to the calculations of the stresses in an actual cylindrical shell.

Example A cylindrical shell, as in Fig. 3.2*a* and *b,* has an inside diameter of 6 ft and is 20 ft long; it is made of ½-in plate and is subjected to a pressure of 100 psi. (1) Find the stress in pounds per square inch on the metal in the circumferential section of the shell. (2) What is the stress in pounds per square inch in the cylindrical sides due to the transverse pressure?

Solution (1) Total pressure on the head is what tends to cause failure; for this reason the head must be held on the sides by additional metal.

Total pressure on the head:

$$A = \frac{\pi D^2}{4} \quad \text{(see App. B)}$$

$$A = 6 \times 6 \times 0.7854 \times 144 \text{ in}^2/\text{ft}^2 = 4071.51 \text{ in}^2$$

$$P = 4071.51 \times 100 = 407{,}151 \text{ lb total pressure}$$

The area in the metal ring (Fig. 3.2*b*) that must support this load:

$$C = \pi D \quad \text{(see App. B)}$$

$$A = 6 \times 12 \text{ in/ft} \times 3.1416 \times \tfrac{1}{2} = 113.10 \text{ in}^2 \text{ of metal}$$

The force in each square inch of metal is

$$\frac{407{,}151}{113.10} = 3600 \text{ psi}$$

(2) Projected area (refer to Fig. 3.5):

$$A = 6 \times 20 \times 144 \text{ in}^2/\text{ft}^2 = 17{,}280 \text{ in}^2$$

Total pressure exerted on this area:

$$P = 17{,}280 \times 100 = 1{,}728{,}000 \text{ lb}$$

But since this pressure is held by both sides of the shell, one side must hold half, or $1{,}728{,}000 \div 2 = 864{,}000$ lb. The area of the strip of metal that supports this pressure is 12 in/ft $\times$ 20 $\times$ ½ = 120 in². The force on this area equals $864{,}000 \div 120 = 7200$ psi.

From these calculations the force on each square inch of metal tending to pull the drum apart in a transverse section is 3600 psi. The force on the longitudinal section is 7200 psi. The force on the longitudinal section is two times as much as the force on the transverse section. This fact must be considered in the determination of plate thickness requirements.

In determining the internal design pressure of a boiler drum or shell, consideration must be given to the tensile strength of the steel used in the construction, the thickness of the plate, the dimensions of the drum or shell, the permissable factor of safety, and the efficiency of the longitudinal joint. The tensile strength of steel and its ability to withstand temperature have been increased by the use of alloys and improved methods of manufacture. Steel varies in tensile strength from 50,000 to 100,000 psi. When the tensile strength of a boilerplate is not known, it may be assumed to be 45,000 psi for wrought iron and 55,000 psi for steel when calculating the internal design pressure. The metal in a boiler must never be subjected to a pressure greater than or even approaching the elastic limit. This is the reason there is a definite limit to the pressure used in applying a hydrostatic test. For the same reason, safety valves are used to limit the pressure when a boiler is in operation.

The required thickness of boilerplates is reduced proportionally by the use of high-tensile-strength steel. Sometimes drums are designed with plates of a greater thickness in that section which is drilled to receive the tubes. This procedure, together with welded construction, results in a decrease in drum weight. The thickness of the plate must always be sufficient for the specified design pressure; in addition, there is also a minimum-thickness requirement based on drum diameter.

For a given pressure, the drum metal thickness must be increased when the drum diameter is increased. Solve the previous problem using a shell 4 ft in diameter in place of the 6-ft-diameter shell.

The maximum stress to which a boilerplate may be subjected varies from one-fourth to one-seventh the tensile strength of the material. This comparison of tensile strength to actual working stress is known as the *factor of safety* (FS). If the tensile strength of a boilerplate is 55,000 psi, the stress to which it may be subjected varies from 55,000/7 = 7857 to 50,000/4 = 12,500 psi, depending on its age, type of construction, and condition. A minimum factor of safety of 4 may be used in determining the plate thickness of new boilers. In no case may the internal design pressure be increased above that allowable for new boilers. Secondhand boilers shall have a factor of safety of at least $5\frac{1}{2}$ unless constructed according to American Society of Mechanical Engineers (ASME) rules, when the factor shall be at least 5.

Note that the actual design values are per the latest ASME code requirements that must be adhered to by the boiler designer.

The equation for setting the minimum required wall thickness is per the ASME code (Sec. VIII) and is as follows:

$$t = \frac{PR}{SE - 0.6P}$$

where t = minimum required wall thickness (in)

   P = internal design pressure (psi)

   S = allowable stress of material at design temperature (psi)
      (These values are given for a wide range of steel specifica-
      tions and operating temperatures in the *ASME Boiler and
      Pressure Vessel Code.*)

   E = efficiency of weld joints or of ligaments between openings
      (For fully x-rayed with manual penetration, E = 1.0.)

   R = inside radius of drum (in)

Example   A drum will be constructed of SA-285 Grade B carbon steel by
welding. The internal design pressure is 350 psi, and there are no unrein-
forced openings in the drum. The drum inside diameter is 48 in. What is
the minimum required wall thickness of the drum?

Solution   From the ASME code, the allowable stress for SA-285 Grade B
steel is 12,500 psi. The joint efficiency is 100 percent. Thus,

$$t = \frac{350 \times 24}{12,500 \times 1 - 0.6(12,500)}$$

$$t = 1.68 \text{ in}$$

For commercial sizes, this plate would be ordered at a thickness of 1.75 in.
If the material selected were SA-516 Grade 70 carbon steel, which has an
allowable stress of 17,500 psi, what would be the required thickness?

$$t = \frac{350 \times 24}{17,500 \times 1 - 0.6(17,500)}$$

$$t = 1.2 \text{ in}$$

For commercial size, this plate would be ordered at 1.25 in.

The better grade of carbon steel plate results in a plate thickness
0.5 in less than that required with SA-285 steel, which also results in
a lighter drum. However, the better grade of material is more costly.
This then becomes an economic evaluation by the designer in selection of the proper material for the application.

In general, today's modern boiler drums are designed using carbon
steel plate of SA-299, a 75,000-psi tensile strength material, for
drums requiring a thickness of 4 in or greater. Below this thickness,
carbon steel plate of SA-516, a 70,000-psi tensile strength material, is
used most often.

The *efficiency of a joint or tube ligament* is found by dividing the
strength of the section in question by the strength of the solid plate.
When calculations or destructive tests show that a joint or tube ligament fails when subjected to one-half as much force as the solid plate,
the efficiency is 50 percent. The efficiency of a seamless shell is 100

percent. Welded joints with the reinforcement removed flush with the surface have an efficiency of 100 percent, but when the reinforcement is not removed, the efficiency is 90 percent. When a drum is drilled for tubes in a line parallel to the axis, the efficiency of the ligament may be calculated as follows:

$$\text{Efficiency of ligament} = \frac{P - d}{P}$$

where P = pitch of the tube hole (in)
$\quad d$ = diameter of tube hole (in)

The internal design pressure of existing boiler installations may be estimated by the following formula if, for some reason, the design pressure is unknown. Usually the vital statistics of a boiler design are identified on the nameplate, which is easily located on the boiler.

$$P = \frac{\text{TS} \times t \times E}{R \times \text{FS}}$$

where P = internal design pressure on inside of drum or shell (psi)
$\quad$ TS = ultimate strength of plate (psi)
$\quad t$ = thickness of plate (in)
$\quad R$ = inside radius of drum (in) [inside diameter divided by 2]
$\quad$ FS = factor of safety (ultimate strength divided by allowable working stress)
$\quad E$ = efficiency of joints or tube ligaments (ultimate strength of joint or ligament divided by ultimate strength of plate). When more than one joint or tube ligament is involved, the lowest efficiency must be used.

Example A boiler drum 4 ft in diameter is made of 1-in plate; the tensile strength of the steel is 55,000 psi; the efficiency of the joint is 85 percent. What is the estimated design pressure of the drum?

Solution Estimated design pressure in pounds per square inch (assume FS = 6):

$$P = \frac{\text{TS} \times t \times E}{R \times \text{FS}}$$

$$P = \frac{55,000 \times 1 \times 0.85}{2 \times 12 \times 6} = 325 \text{ psi}$$

3.3 Drum and Shell Construction

The pressure parts for water-tube boilers originally were made of iron and later of steel. The steam drums of today's boilers are fabricated from thick steel plates and forgings and are joined together by welding. The development of the steam boiler has been concurrent with advances in metallurgy and continuing improvements in fabrication methods, including the welding of steel and steel alloys.

The first boilers used steam-generating tubes made of cast iron, and these were replaced with steel tubes. In the early 1900s, a manufacturing process was developed for seamless steel boiler tubes that provided both strength and reliability together with a reasonable cost.

Prior to 1930, the standard method for the joining of boiler drum plates was riveting. The thickness of drum plate was limited to about $2\frac{3}{4}$ in because there was not a satisfactory method known to obtain a tight joint in thicker plate. As a result, boiler pressures also were limited and thus plant cycle efficiencies were low. Attempts were made to forge and machine a solid ingot of steel to form a steam drum; however, this was a very expensive process.

Fusion welding was developed at this time, as well as the use of x-ray procedures to ensure the integrity of the weld. By x-ray examination along with physical tests of samples of the weld material, the weld integrity could be determined without any effect on the drum. The testing of drum integrity by x-ray examination and the development of procedures for the qualification and testing of welders were critical in the establishment of quality control to the fabrication of boilers.

Drum fabrication consists of pressing flat plate into half cylinders or, depending on the thickness, by rolling the plate into cylindrical shells. One or two longitudinal welds are required depending on the method used. The drum length is achieved by welding circumferentially the required number of shell sections.

The heads (i.e., drum end enclosures) of modern boiler drum designs are generally hemispherically shaped and are hot formed by pressing or spinning flat plate in dies. The heads are then attached to the drum cylinders by circumferential welding. Automatic welding techniques are used to join shell courses and heads. The thickness in most cases is determined by the pressure that is to be used, as explained in Sec. 3.2. The drum must be strong enough to withstand the mechanical load of itself and the water as well as the expansion stresses.

The heads in return tubular boilers and some other fire-tube boilers are called *tube sheets*; they are welded to the cylindrical section of the shell. The heads of water-tube boiler drums are made from boilerplate by use of a die. The head is dished, and the manhole and flange around the outside are all formed in a single operation. The head is prepared for welding by making the flange equal to the diameter of the shell.

Steam, on being liberated from the surface of the hot water in a boiler drum or shell, has a tendency to carry particles of water along with it. This water may contain impurities in the form of soluble salts and insoluble particles that are very objectionable.

Some of the design features that decrease the tendency of a boiler to carry over water with the steam are (1) adequate circulation so that the steam generated near the furnace will be quickly carried to the steam drum, (2) sufficient surface area of the water in the steam drum to liberate the steam from the water without excessive agitation, and (3) provisions in the steam drum for retaining the water while allowing the steam to flow into the steam outlet.

Older fire-tube boilers were sometimes fitted with steam domes that consisted of a small drum riveted to the highest point of the shell. Steam was taken from the top of this dome, thus providing an opportunity for the water to separate from the steam. A center vertical row of tubes was sometimes omitted from an HRT boiler to improve circulation and thereby reduce carry-over.

Several methods have been used to reduce carry-over from water-tube boilers. One of the simplest but least effective of these was the dry pipe, which was located at the highest point in the drum and connected to the steam-outlet nozzle. The upper side of this pipe contains many holes, through which the steam must flow as it leaves the drum. Another method was to install a water-steam separator in the upper portion of the drum to remove the particles of water, and this is the method used today in modern boiler design, as explained in Sec. 2.5. In some cases older designed boilers were provided with separate drums in which these separators were installed. Refer to Fig. 2.12 for an example of this design.

3.4   Welded Construction

Welding is a joining of one or more materials by the application of heat that is localized in the region of the joint with or without the addition of filler metal. For most welding processes, the induced heat raises the temperature of the materials above the melting point, which creates a weld pool. The heat source is then removed, which allows the pool to solidify and create a metallurgical bond between the joined surfaces. There are many welding processes, but the most widely used for joining pressure parts is fusion welding with the addition of filler metal.

All welding performed on power boilers and pressure vessels must be done by code-qualified personnel, who must follow qualified procedures. The personnel must be trained in using the correct welding position and the correct filler material and when to apply the necessary pre-heat treatment prior to welding, as well as the stress-relief heat treatment after the welding has been completed.

Fusion welding has been universally adopted in the construction of boiler drums. The construction of high-pressure boilers by use of riveted joints required an extreme thickness of plate, and increased thickness was limited because of the difficulty in obtaining a tight joint. Welded joints relieve the operators of the inconvenience caused by joint leakage and of the necessity for caulking. Welding also lessens the possibility of caustic embrittlement, which is partially attributed to boiler water entering the joint, evaporating, and thus producing a solution containing a large amount of impurities. For these reasons, riveted joints are not used in current-day designs.

Rigid specifications cover the application of fusion welding to the pressure parts of boilers. This work must be done in the manufacturing plant or at the construction site with special equipment operated by a qualified welder who must pass a test that meets specific welding requirements. The completed work must be subjected to rigid inspection and tests. It is not permissible to field weld the pressure parts of a boiler unless provisions are made to stress relieve the welded portion. Pressure part welding in the field also must conform to the requirements of the ASME code, where qualification tests and inspection requirements must be met.

Terms used to describe the welding process are defined as follows. The *throat* of a weld is the joint where the filler material is of minimum thickness. The *fillet weld* is approximately triangular in cross section with the throat in a plane that is 45° to the surface joined. A *double-welded butt joint* is formed by adding welding metal from both sides of the joint and reinforcement on both sides. A *single-welded butt joint* is formed by adding filler metal and reinforcement on only one side.

In the fabrication of a fusion-welded drum, the carbon content of materials used must not exceed approximately 0.3 percent. Carbon contents above this are considered high carbon, and these materials cannot be used as welded pressure parts because they are not structurally sound. The plates to be welded may be cut to size or shape by machining, by shearing, or by flame cutting. When flame cutting is used, the edges must be uniform and free from slag.

In the design of boiler drums it is frequently necessary to join plates of unequal thickness to account for the lower ligament efficiency caused by tube holes in the portion of the drum. A two-drum boiler design is an example of this. When welded joints are used, this is permissible within the following limits: The offset of the plates from each other must not exceed one-quarter of the plate thickness, with a maximum permissible offset of $\frac{1}{8}$ in for the longitudinal joints and $\frac{1}{4}$ in for the girth joints. In all cases when plates of unequal thickness are joined together, the thicker plate must be reduced by tapering until the plates are equal in thickness at the joint. The design of welded joints must be such that bending stresses are not imposed directly on the welded joint. Backing strips are used when one side of the joint is inaccessible for welding. The filler metal is then added from only one side of the joint. In this case, the reinforcing metal must be at least $\frac{1}{16}$ in, although it may be machined off if so desired.

After fusion welding a drum or shell (including any connections or attachments), it is always necessary that the assembly be stress relieved. Stress relieving is accomplished by heating the drum uniformly to a temperature of 1100 to 1200°F. The temperature to which the drum is heated for stress relieving is varied to meet the characteristics of the metal used in the construction. The drum to be stress relieved must be brought up slowly to the specified temperature and held there for a period of time equal to at least 1 h per inch of plate thickness. After this period of heating, the drum is allowed to cool slowly in a "still" atmosphere. The stress-relieving procedure must be repeated if it is necessary to weld on additional outlets or if repairs are made. Welded attachments may be locally stress relieved by heating a circumferential band around the entire vessel. The entire band must be brought up to the specified stress-relieving temperature. A similar procedure is followed in stress relieving welded pipe joints.

The welding of pressure vessels is made possible not only by the improved technique of the process but also by the advancement in the method of inspecting and testing the completed joint. The tests applied to welded pressure vessels may be divided into two classifications: the *destructive test,* in which the metal is stressed until it fails, and *nondestructive testing* (NDT), which does not injure the metal and may therefore be applied to the completed vessel.

Special test specimens are prepared for the application of destructive tests. Plates of the same thickness and material as those used in the actual construction are prepared for welding and attached to one end of the longitudinal seam. The same welding material and technique are used on both the joint and the test plates. Two specimens for tension and one for a bend test are cut from the welded test plates. One of the tension specimens is cut transversely to the welded joint and must show a tensile strength not less than the minimum requirement for the plate material used (Fig. 3.6). The other tension specimen is taken entirely from the deposited weld metal, and the tensile strength must be at least that of the minimum range of the plate welded. When the plate thickness is less than $\frac{5}{8}$ in, an all-weld-metal tension test may be omitted. The bend-test specimen is cut across the welded joint through the full thickness of the plate with a width $1\frac{1}{2}$ times the thickness of the plate. The specimen must bend without cracking until there is an elongation of at least 30 percent of the outside fibers of the weld (Fig. 3.7). Material tests, types of test speci-

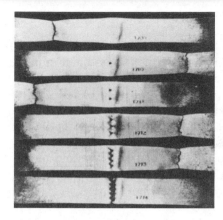

Figure 3.6 Tensile-test specimens showing that to cause failure, the area of the weld had to be reduced by 30 percent by drilling five holes. (*Babcock & Wilcox, a McDermott company.*)

Figure 3.7 Bend-test specimen with the outside fibers elongated 55 percent. (*Babcock & Wilcox, a McDermott company.*)

mens, and test methods are changed over time based on experience. All of these must be conducted in accordance with the procedures established by the American Society of Testing and Materials (ASTM) and other authorities.

X-rays provide a satisfactory nondestructive procedure for examining welds in boilerplate. This method of inspecting boilerplate has been developed to a high degree of perfection. The process consists of passing powerful x-rays through the weld and recording the intensity by means of a photographic film. The variation and intensity of the x-rays as recorded on the film show cracks, slag, or porosity. The exposed and processed films showing the condition of the welded joint are known as *radiographs.*

Both longitudinal and circumferential fusion-welded joints of boiler drums must be examined radiographically throughout their length. Welded joints are prepared for x-ray examination by grinding, chipping, or machining the welded metal to remove irregularities from the surface. A small specified amount of reinforcement, or crown, on the weld is permissible. Single-welded joints can be radiographed without removing the backing strip, provided it does not interfere with the interpretation of the radiograph.

Radiography, or x-ray, has been a primary weld quality-control technique for many years. These x-ray machines have been supplemented with the use of gamma-ray equipment, which uses a radioactive isotope. The choice of either method is usually based on the mobility of the part that has to be examined and also on the availability of equipment. X-ray units are best used in a manufacturing facility where fixed-position equipment is available and the parts to be examined can be readily moved into proper position. For field use or when a part cannot be moved easily, gamma-ray equipment is used. A significant advantage of radiography is that a permanent inspection record is made of each inspection.

Thickness gauges (penetrometers) are placed on the side of the plate to test the ability of x-rays to show defects. These gauges are rectangular strips of about the same metal as that being tested and not more than 2 percent of the thickness of the plate. They contain identification markings and holes that must show clearly at each end of the radiograph. The entire length of the welded joint is photographed. The purpose of these penetrometers is to ensure the correct exposure time of the x-ray. An image of the holes in the penetrometer appears on the x-ray film, and therefore any flaws of equal or larger size would be expected to appear. The ASME code specifies the acceptance standard as to the size of porosity and defects from slag inclusion.

Figure 3.8a shows a radiograph of a satisfactory welded joint, while Fig. 3.8b shows a joint that is unsatisfactory due to excessive porosity. Welded joints are judged unsatisfactory when the slag inclusion or porosity as shown on the radiograph exceeds the amount shown by a standard radiograph reproduction that may be obtained from the ASME.

The percentage of defective welds in modern boiler shops is very small. When the radiograph shows an excessive amount of slag inclusion or other unsatisfactory condition within the joint, the defective portion must be chipped out. The section is then rewelded, after which it must again be x-rayed.

In addition to radiography, there are other nondestructive tests that may be applied to determine the condition of boilers. These are ultrasonics, magnetic particle, and liquid-penetrant testing methods. Radiography and *ultrasonic testing* (UT) are used for volumetric examination, while magnetic particle and liquid-penetrant testing are used for surface examination.

Ultrasonic testing (UT) is the technology that developed rapidly in the nondestructive testing (NDT) of pressure parts. The ultrasonic

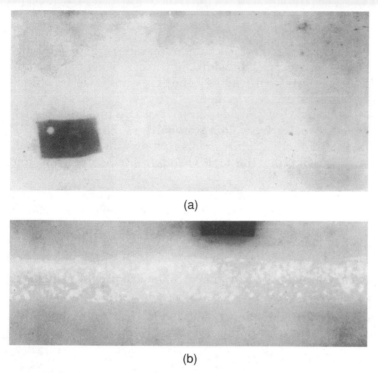

(a)

(b)

Figure 3.8 Radiographs. (*a*) Satisfactory weld. (*b*) Unsatisfactory weld as a result of slag inclusions. (*Babcock & Wilcox, a McDermott company.*)

wave that is produced is reflected by any flaw in the material being examined. The wave is displayed on an oscilloscope.

The testing of material thickness is performed by ultrasonic testing. A common cause of pressure part failure is the loss of material that results from oxidation, corrosion, or erosion. Ultrasonic testing is relatively fast and is used extensively for measuring the wall thickness of tubes or piping.

During outages for planned maintenance, a UT examination of furnace walls (and other tubes where erosion or corrosion is expected) is often made, and the thicknesses of tubes are recorded and compared with the original thicknesses or any previous thicknesses determined by UT testing. Where thinning has occurred to a point where replacement is required, cladding of the tube area or actual tube replacement is performed to reestablish tube integrity.

3.5 Structural Support Components

Pressure vessels, such as steam drums and headers, are normally supported by saddles, cylindrical support skirts, hanger lugs and brackets, ring girders, or integral support legs. Where these supports are located on the vessel, concentrated loads are imposed on its shell. Therefore, it is important that the support arrangements minimize the local stresses on the vessel. In addition, the components must provide support for the specified loading conditions and also be able to withstand the temperature requirements.

Structural elements that provide support of the pressure vessels may be attached directly by welding or bolting. They also can be attached indirectly by clips, pins, or clamps or may be completely unattached by transferring the load through surface bearing and friction.

In general, loads applied to structural components are categorized as dead loads, live loads, or transient loads and are defined as follows:

1. *Dead loads* are loads due to the force of gravity on the equipment and supports.

2. *Live loads* vary in magnitude and are applied to produce the maximum design conditions.

3. *Transient loads* are time-dependent loads that are expected to occur randomly for the life of the structural component.

Examples of these types of loads include

1. Weight of the component and its contents during operating conditions, including loads due to static and dynamic head and fluid flow

2. Weight of support components

3. Loads caused by environmental effects such as wind and snow

4. Dynamic loads including those caused by earthquake, vibration, or rapid pressure change

5. Loads from the thermal expansion of piping

6. Loads from expansion and contraction due to pressure

Boiler designs vary among manufacturers and generally reflect their experience over many years of operation. Thus boiler designs and their support are constantly evolving, with each manufacturer having a variation in the detailed design. One such arrangement is shown in Fig. 3.9, which shows a modern design of a membrane waterwall furnace using tie bars and buckstays for support.

Because superheaters and reheaters are located in areas that have high flue gas temperatures, the major support loads are generally carried by the tubes. For pendent superheaters, the major support

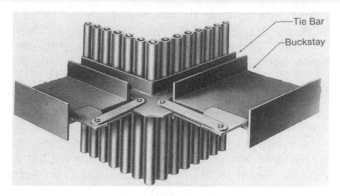

Figure 3.9 Tie-bar and buckstay arrangement at corner of furnace. (*Babcock & Wilcox, a McDermott company.*)

points are located outside the flue gas stream. Refer to Fig. 2.22 for a typical arrangement of this support method.

For horizontally arranged superheaters, the support load is usually transferred to boiler or steam-cooled enclosure tubes. Depending on the design, the support and tube arrangement must slide on each other to provide relative movement between boiler tubes and superheater tubes. Again, each boiler design can vary based on design preference and operating experience. Stringer support tubes are often used in horizontal superheaters, as well as in economizers. These are vertical tubes containing water or steam that act as supports.

3.6 Manholes, Handholes, and Fittings

The design and construction of boiler drums must allow service and maintenance personnel access to properly inspect, clean, and repair the inside of the drum and/or its internals. Therefore, manholes are made at suitable places in the drum, and these are normally in the heads. For headers, it is only necessary to gain access to tubes and to get a hand inside. In these places, handholes are provided, and they are usually placed opposite each tube in the headers of some water-tube boilers and give access to the tubes for replacement, rerolling, and cleaning.

The minimum size of a manhole opening is, for a circular manhole, 15 in in diameter, and for an elliptical manhole, 11 in by 15 in or 10 in by 16 in. The minimum size of an elliptical handhole opening is $2\frac{3}{4}$ in by $3\frac{1}{2}$ in. In actual practice, the holes must be large enough to provide the necessary access for cleaning and for tube replacement.

Manhole openings must be reinforced to compensate for the metal removed in the drum. This reinforcement may be obtained either by forming a flange around the hole from the drum metal or by using a frame welded to the drum. Manhole-reinforcing frames are made of rolled, forged, or cast steel. Woven gaskets are used on manhole covers for pressures up to 500 psi, and metallic gaskets are used for higher pressures.

In most cases, both manholes and handholes are elliptical in shape. This makes it possible for the cover plate to be placed on the inside and removed through the hole. Placing the cover plate on the inside puts the pressure against the seat. The yoke holds the cover in place but does not have to counteract the pressure. Round handhole cover plates must be placed inside the drum or header through a larger opening than the hole that they are to cover. When outside cover plates are used, the bolt must hold all the pressure. In low- and medium-pressure boilers, gaskets are used to prevent leakage around the handhole cover plates. In high-pressure boilers, the gaskets are eliminated, and the cover plates are seal welded to prevent leakage.

In addition to manholes, steam drums must have a number of connections, including those for attaching feedwater inlet, steam outlet, safety valves, water column, chemical feed, and continuous blowdown. These connections are made by welding during fabrication and before the drum is stress relieved. All welding must be stress relieved, and radiographic examinations are required.

The protruding ends of drum connections are generally flanged or prepared for either butt or socket welding. The socket welding is preferred for smaller pipe, butt welding for intermediate sizes, and flanges for the larger sizes. The steam outlet and safety valve connection on low- and moderate-pressure boiler drums are flanged, but on high-pressure boilers all connections are welded. Again, code requirements must be followed.

All boilers must have a blowoff-line connection to the lowest part of the water-circulating system, which is usually the lower (mud) drum. Waterwall and economizer headers that do not drain into the boiler must have separate drain connections.

Superheaters must have connections to both inlet and outlet headers for draining and venting. However, some superheaters have the headers located above the tubes and therefore cannot be drained.

3.7   Boiler Assembly

It is always advantageous to do as much of the assembly work as possible in the manufacturer's shop. The shop personnel not only are trained on specific jobs but have tools and equipment available for performing the work in a cost-effective manner. This advantage of shop assembly can be applied in varying degrees to all types of boilers. Relatively low capacity oil- and gas-fired boilers are usually delivered fully assembled. However, when restricted conditions make it impossible to install the complete unit, sections are shipped separately for assembly at the plant site. Fire-tube boilers are essentially contained within the shell and are readily factory assembled. Burners and other accessories are also made a part of the factory assembly. After the boilers are set on the foundation and the various services are connected, the boilers are ready for operation. Water-tube boilers with integral furnaces for burning gas and fuel oil are factory assembled and transported to the customer's plant. However, capacities are limited by the physical sizes that can be transported. These transportation limitations are reduced by designing boilers for increased output for a given physical size, making them longer but within the restrictions of width and height, or by adapting transportation facilities to this particular problem, e.g., by the use of special railroad cars and barges. As a result of these measures, package water-tube boilers are built with capacities of 350,000 lb of steam per hour. Shop-assembled units have even been made for capacities of 600,000 lb/h at pressures up to 1800 psi where barge shipments to the site are permissible.

Large power boilers must be field erected. However, both costs and erection time are reduced by modular, or "building block," design and construction. This procedure consists of designing and fabricating sections that are identical and which can be added to meet a range of boiler capacities. This results in standardization and reduced costs. The use of membrane wall panels is common, and these are sent in sections and then are field assembled and welded.

In top-supported boilers, the steam drum is securely anchored in place, and the lower portions of the boiler, including the lower drum, are allowed to expand downward when the boiler is operating. In the case of high boilers, there is considerable movement of the lower portion of the boiler as a result of temperature change. This movement of lower drums, waterwall tubes, and other pressure parts must be taken into consideration when furnace walls, stokers, ash hoppers, and other boiler equipment are installed or repaired.

Bottom-supported boilers expand upward, minimizing the problem of sealing the bottom of the furnace to prevent air leaks. This arrangement is used effectively in connection with slag-tapped furnaces and some types of stokers. It is generally more economical to support smaller units from the bottom and larger units from the top.

The upper portion of the boiler is supported by the tubes when the unit is bottom supported. The expansion resulting from increased temperature causes the drums to move upward. This movement must be considered in the design and installation of piping and of flues and ducts that connect to the boiler. Figure 2.34 is an example of a bottom-supported unit.

After the drums are in place, the tubes are installed. When headers are used, the tubes and header assembly or module are tested in the factory and shipped as a complete unit. The bent tubes are formed in the factory to the required shape and installed on the job. In the modern boiler equipped with waterwalls, many different shapes of tubes are required. Regardless of whether the tubes are assembled in the factory or on-site, the three-roll expander is almost universally used for joining the tubes to the drums and headers unless the pressure is high and thus welded attachments are required.

Tube expanding or *tube rolling* is a process of cold working the end of a tube into contact with the metal of the drum or header containing tube holes. When a tube is expanded, the outside diameter, inside diameter, and length increase, and the wall thickness decreases. A properly expanded tube will provide a pressure-tight joint of great strength and stability.

The conventional expander consists of three rolls mounted in separate compartments in a cage. The cage contains a hole that permits a tapered mandrel to be inserted between the rollers. The mandrel forces the rollers out against the tube. When the mandrel is turned, rollers are rotated as they are forced against the wall of the tube. Continued turning of the mandrel causes the end of the tube to be enlarged and the outer surface to be pressed against the tube hole. The mandrel is driven by either an electric or an air-operated motor. When thick-walled high-pressure boiler tubes are expanded, the service on the rollers is most severe.

Several methods have been developed for determining when a tube has been expanded sufficiently, but satisfactory results are obtained by relying on the judgment of expert workers. Mill scale cracking from the surface of the sheet around the tube hole has been found to be a reliable method of determining when the tube has been expanded sufficiently to make a tight joint. However, the hydrostatic test that must be conducted on the boiler is the ultimate check for tightness.

At low and moderate pressures it is satisfactory to expand the tube into the smooth seats. For high-pressure boilers and special application it is advisable to cut or expand grooves in the tube seats. This grooving results in considerable improvement in both the resistance to leakage and the holding power of the tube. The tube metal is expanded into these grooves.

3.8 Heating Surface and Capacity

The capacity of boilers has been stated in terms of square feet of heating surface, rated boiler horsepower, percentage of rating, and maximum steam-generating capacity in pounds per hour. The boiler-horsepower and percentage-rating units are used in expressing the capacity of small, simple types of boilers, and even these have become relatively obsolete today in the defining of boiler capacity. The capacity of modern boilers is expressed in pounds of steam per hour at a pressure and temperature when supplied with a specified feedwater temperature.

The *heating surface* of a boiler consists of those areas which are in contact with heated gases on one side and water on the other. The boiler code of the ASME specifies that the gas side is to be used when calculating heating surfaces. It follows that in determining the heating surface for fire-tube boilers, the inside diameter of the tubes must be taken; for water-tube boilers, the outside diameter or the tube area exposed to the flue gas is used.

As an example, unless otherwise stated, the area included as heating surface in a horizontal-return tubular boiler comprises two-thirds of the area of the shell, plus all the tube surface, plus two-thirds of the area of both heads, minus the area of the tube holes. The boiler heating surface is usually expressed in square feet. The boiler manufacturer will furnish this information on request. If, however, it becomes necessary to calculate the heating surface from the physical dimensions of the boiler, the following procedure may be used:

Example Find the heating surface of an HRT boiler 78 in in diameter, 20 ft long, containing eighty 4-in tubes (inside diameter) bricked in one-third of the distance from the top. Find the (1) area of the shell, (2) area of the tubes, and (3) area of the heads.

Solution (1) Multiply the diameter of the shell in inches by 3.1416 × the length of the shell in inches × $\frac{2}{3}$ (as only two-thirds is in actual contact with the gases) and divide by 144 (the number of square inches in 1 ft^2).

$$\frac{78 \times 3.1416 \times 240 \times 0.66}{144} = 270 \text{ ft}^2 \text{ of heating surface in shell}$$

(2) Multiply the inside diameter of the tube in inches by 3.1416 × the length of the tube in inches × the number of tubes and divide by 144 (the number of square inches in 1 ft^2).

$$\frac{4 \times 3.1416 \times 240 \times 80}{144} = 1675 \text{ ft}^2 \text{ of heating surface in tubes}$$

(3) Multiply the square of the diameter of the boiler in inches by 0.7854 × 2 (as there are two heads) × $\frac{2}{3}$ (boiler is bricked one-third from top). From this must be subtracted the area of the tube holes; so to subtract, multiply the square of the diameter of the tube in inches by 0.7854 × the number of tubes × 2 (as tubes have openings in each head). Divide this product by 144 (the number of square inches in 1 ft²). (*Note:* $\pi D^2/4 = 0.7854D^2$.)

$$\frac{(78 \times 78 \times 0.7854 \times 2 \times 0.66) - (4 \times 4 \times 0.7854 \times 80 \times 2)}{144}$$

$$= 30 \text{ ft}^2 \text{ of heating surface in heads}$$

Heating surface	Ft²
Shell	270
Tubes	1675
Heads	30
Total heating surface	1975

When extreme accuracy is unnecessary, the area of the heads may be omitted from the heating-surface calculations, or it may be assumed to account for 1.5 percent of the combined area of the heads and shell.

The *rated horsepower* of a boiler depends on its type and the number of square feet of heating surface it contains. The square feet of heating surface divided by the factor corresponding to the type of boiler will give the rated horsepower (boiler horsepower). The factors or square feet of heating surface per boiler horsepower are as follows:

Type of boiler	Ft² of surface/ boiler hp
Vertical fire-tube	14
Horizontal fire-tube	12
Water-tube	10

Example Find the rated horsepower of the HRT boiler in the previous example.

Solution The horizontal-return tubular boiler in the previous example was found to have 1975 ft² of heating surface. Thus,

$$\frac{1975}{12} = 164.6 \text{ rated boiler hp}$$

A boiler is operating at 100 percent of rating when it is producing 34.5 lb[1] of steam per hour, from feedwater at 212°F to saturated steam at 212°F (abbreviated "from and at 212°F") for each rated boiler horsepower. The rated steam output of a 164.6-hp boiler is 34.5 × 164.6 = 5678.6 lb of steam per hour from and at 212°F. Modern boilers are capable of generating more than twice this amount of steam per rated boiler horsepower. That is to say, they may be operated in excess of 200 percent of rating.

The evaporation from and at 212°F is equivalent to adding 970.3 Btu[2] (see Table C.1 in App. C) to each pound of water already at 212°F. The resulting steam would be at atmospheric pressure and dry and saturated.

The factor of evaporation is the heat added to the water in the actual boiler in Btu per pound divided by 970.3. To determine the heat added by the boiler, the feedwater temperature, boiler outlet pressure, and percentage of moisture or degrees of superheat in the outlet steam must be known. When these factors are known, the heat input per pound can be determined as explained in Sec. 3.9.

The equivalent evaporation from and at 212°F is determined by multiplying the actual steam generated in pounds per hour by the factor of evaporation. The equivalent evaporation in pounds of steam per hour divided by 34.5 equals the developed boiler horsepower. The developed boiler horsepower divided by the rated boiler horsepower equals the percentage of rating of the boiler.

The performance of modern high-capacity boilers has rendered the boiler-horsepower and percentage-rating method of evaluating boiler size and output obsolete. The feedwater is heated almost to the boiling temperature in the economizers. In fact, some economizers are built to evaporate a portion of the water and deliver steam to the boiler drum. The furnaces are constructed with waterwalls that have

---

[1]When the boiler-horsepower unit (of 34.5 lb evaporation per hour of equivalent steam from and at 212°F) was adopted, it represented the heat input normally requi by the average steam engine to produce a horsepower-hour of energy. This quantity of heat bears no relation to the requirements of modern turbines but is still sometimes used in designating the performance of small boilers.

The modern terminology is to define all boiler sizes in terms of steam capacity (pounds per hour) at a specific steam pressure and temperature when supplied with feedwater at a specific temperature, e.g., a boiler designed to produce 500,000 lb/h at 1800 psig and 950°F when supplied with feedwater at 350°F.

[2]*Note:*
h_g at 212°F = 1150.4 Btu/lb
h_f at 212°F = 180.1 Btu/lb
$\quad \Delta h$ = 970.3 Btu/lb
$\qquad$ or
h_{fg} (from steam tables) = 970.3 Btu/lb

high rates of heat transfer. The heat is transmitted to these water-wall tubes by radiation, and the convection heat transfer takes place in the boiler proper, superheater, economizer, and air preheater. Under these conditions, the boiler horsepower and percentage of rating are inadequate in measuring and stating boiler output. The manufacturers of boilers have adopted a policy of stating the boiler output in terms of pounds of steam per hour at operating pressure and temperature. When actual conditions vary from design specification, the necessary compensating corrections are made so that the actual operating results can be compared with expected performance.

In evaluating boiler performance and in making comparisons between boilers, it is helpful to consider the relative square feet of heat-absorbing surface and cubic feet of furnace volume involved. The *convection heating surface* of a water-tube boiler includes the exterior surfaces of the tubes, drums, headers, and those connections which are exposed to the flue gas. The *waterwall heating surface* includes furnace wall areas and partition walls that are composed of tubes which form a part of the boiler circulating system. The projected area of these tubes and the extended metal surfaces exposed to the heat of the furnace are included in the furnace heating surface. The *furnace volume* includes the cubic feet of space available for combustion of the fuel before the products of combustion enter the convection sections of the boiler or superheater. *Superheater and reheater surface* consists of the exterior area of the tubes through which the steam flows. All the tube area subjected to either convection or radiant heat or a combination of both is included as heating surface. In like manner, the number of square feet of heating surface is an important factor to use when evaluating economizers and air preheaters.

The size of a boiler unit is determined by the amount of heat-transfer surface in the boiler, furnace, superheater, economizer, and air heater and by the heat-release volume in the furnace. However, boiler manufacturers assign varying heat-transfer rates to the heat-transfer surfaces and heat-liberation rates to the furnace volume. The values assigned are influenced by many factors but are essentially based on experience and are increased as the design is improved. These design improvements include better circulation, more effective removal of moisture from steam leaving the drum, cleanliness factors that relate to the fuels being burned, and combustion equipment that is more effective in utilizing the furnace volume.

Perhaps the most critical technology in the design of steam-generation equipment is heat transfer. A key advancement in steam-water flow was the invention of a ribbed inside surface of a tube as compared with a tube with a smooth internal surface. By preventing the deterioration of heat transfer under many flow conditions, the internally ribbed tube made possible the use of natural circulation boilers at virtually all pressures up to the critical pressure (3206 psia). Because of the increased cost, ribbed tubes generally are used above 2200 psia.

3.9 Boiler Capacity Calculation

Steam tables and charts are required to determine the heat content of water and steam. The physical properties of steam have been determined and agreed on by qualified international committees, and these properties are tabulated in steam tables. (See App. C.) These tables are used in calculating the heat content (enthalpy) of steam and water for various pressure and temperature conditions, as well as specific volume and other steam and water properties. It must be remembered that the total heat contents of steam (enthalpies) given in the tables are the Btus required to raise the water from 32°F to the boiling point, evaporate it into steam, and further increase the temperature when superheaters are used. The 32°F is an arbitrary point taken for convenience in making the steam tables. Actually, a substance contains heat until its temperature is lowered to absolute zero. The steam discharged from a boiler drum may be dry and saturated, or it may be wet. When the boiler unit is equipped with a superheater, the steam will be superheated, i.e., heated above the saturation temperature corresponding to the pressure.

Dry and saturated steam is at the temperature corresponding to the boiler pressure (not superheated) and does not contain moisture. It is saturated with heat, since additional heat will raise the temperature above the boiling point and the removal of heat will result in the formation of moisture. When the absolute pressure (gauge plus 14.7, approximately) is known (refer to Chap. 4), the enthalpy may be read directly from the steam tables. Note that 14.7 psia is the barometric reading of pressure that is exerted by the atmosphere at sea level.

Example Steam at 450 psia, dry and saturated, is produced by a boiler when supplied with feedwater at 180°F. What is the enthalpy of both the steam and feedwater in Btu/lb, and what is the amount of heat that must be generated by the boiler in Btu/lb? If the boiler produces 120,000 lb/h of steam, what is the heat output of the boiler in Btu/h?

Solution

$$h_g \text{ at 450 psia} = 1204.6 \text{ Btu/lb} \text{(from Table C.1)}$$
$$h_f \text{ at 180°F} = \underline{147.9} \text{ Btu/lb} \text{(from Table C.2)}$$
$$\Delta h = 1056.7 \text{ Btu/lb}$$

Heat generated by boiler = 1056.7 Btu/lb

Heat output of boiler = 120,000 × 1056.7 = 126.8 × 10⁶ Btu/h

If the boiler efficiency were 82 percent, then the boiler heat input would be

$$\text{Heat input} = \frac{126.8 \times 10^6}{0.82} = 154.6 \times 10^6 \text{ Btu/h}$$

Note that the saturation temperature of the 450 psia steam is 456.28°F per Table C.1, which is the boiling point at that pressure.

When the values cannot be found directly in the steam tables, it becomes necessary to interpolate to find the required values.

Example   Steam is at a pressure of 132 psig, dry and saturated. What is the enthalpy in Btus?

Solution

132 + 14.7 = 146.7 psia pressure

Since the enthalpy of steam at 146.7 psi cannot be read directly from the values given in the steam tables, it becomes necessary to calculate the value (interpolate) from the values given.

We note that 146.7 is between 140 and 150 psi, both of which may be found in the tables:

Enthalpy h_g at 150 psia = 1194.1 Btu/lb

Enthalpy h_g at 140 psia = 1193.0 Btu/lb

The change in enthalpy for a difference of 10 psi pressure = 1.1 Btu/lb

The change in pressure to be considered is 146.7 − 140 = 6.7.

A change of 6.7 psi pressure causes a change of

$$\frac{6.7}{10} \times 1.1 = 0.74 \text{ Btu/lb}$$

The Btu/lb at 146.7 psi equals the Btu/lb at 140 psi (1193.0) plus the amount resulting from a change of 6.7 psi (0.74) = 1193.74 Btu/lb.

Wet steam is a mixture of steam and water. The heat supplied has been insufficient to evaporate all the water. The amount of moisture present in steam can be determined readily by means of a throttling calorimeter. When the pressure and quality (i.e., the percentage of steam) are known, the enthalpy can be calculated from the values given in the steam tables. The *quality* of wet steam refers to the percentage of the mixture that has been evaporated, i.e., 100 minus the percentage of moisture.

$$h_w = h_f + Xh_{fg}$$

where h_w = enthalpy of wet steam (Btu/lb)
h_f = enthalpy of saturated liquid (Btu/lb)
X = quality of steam (%)
h_{fg} = latent heat of vaporization (Btu/lb)

Example Wet steam at 150 psia, quality 95 percent, is produced by a boiler. What is the enthalpy in Btu/lb?

Solution From steam tables, h_{fg} = 863.6, h_f = 330.51. X is given as 95 percent.

$$h_w = 330.51 + 0.95 \times 863.6$$
$$h_w = 1150.93 \text{ Btu/lb}$$

Superheated steam has an enthalpy and temperature above that of dry saturated steam at the same pressure. The heat necessary for producing the superheat is applied (by use of a superheater) after the steam has been removed from the presence of water. The enthalpy is the sum of the heat in the liquid (h_f), that used for vaporization (h_{fg}), and that added in the superheater. The amount of superheat is designated either by total temperature or by the temperature above the boiling point. This temperature above the boiling point is referred to as the *degrees of superheat*. The superheated-steam tables in App. C give the enthalpy when the pressure and total temperature or degrees of superheat are given.

Example Using the example above, where 120,000 lb/h of steam is produced at 450 psia, dry and saturated with feedwater at 180°F, assume that the steam produced is superheated at 450 psia and 800°F instead of dry and saturated. Calculate the boiler heat input assuming the same boiler efficiency of 82 percent as before. Also determine the degrees of superheat in the steam. Compare the heat input of the two boilers. Why is there a difference?

Solution

$$h \text{ at 450 psia and 800°F} = 1414.3 \text{ Btu/lb} \text{ (from Table C.3)}$$
$$h_f \text{ at 180°F} = \underline{147.9} \text{ Btu/lb} \text{ (from Table C.2)}$$
$$\Delta h = 1266.4 \text{ Btu/lb}$$

Heat generated by boiler = 1266.4 Btu/lb

$$\text{Heat input} = \frac{120{,}000 \times 1266.4}{0.82} = 185.3 \times 10^6 \text{ Btu/h}$$

$$\text{Saturation temperature at 450 psia} = 456.28°\text{F}$$

$$\text{Degrees of superheat} = 800 - 456.28 = 343.72°\text{F of superheat}$$

Superheated steam has two advantages over steam that is not superheated:

1. It increases the efficiency of the steam cycle by having a higher heat content (enthalpy), i.e., Btu/lb of steam.
2. It is dryer (actually 100 percent steam) and is less likely to condense in the lower stages of the turbine.

Moisture in the turbine will cause erosion on the turbine blades, and this results in high maintenance costs and loss of power production and thus a loss of revenues that usually is far greater than the maintenance costs.

Therefore, all modern power plants (with the exception of nuclear power plants, where most use saturated steam or only slightly superheated steam) use superheated steam, and many of the large utility plants use a second superheater (called a *reheater*) that reheats the steam after it has passed through a portion of the turbine. This reheated steam then passes back to the lower-pressure stages of the turbine.

The reheated steam has a temperature generally the same as the primary steam, but the pressure is considerably lower. For example, the primary steam has a pressure of 1400 psia and a temperature of 1000°F ($h = 1493.2$ Btu/lb). After this steam has passed through a portion of the turbine, it is returned to the boiler, where, for this example, the steam pressure and temperature are 200 psia and 400°F ($h = 1210.3$ Btu/lb). The steam is reheated to a temperature of 1000°F at a pressure of 200 psia (assume no pressure drop) ($h = 1528$ Btu/lb) and returned to the low-pressure stages of the turbine. The objective of this design is to achieve a higher thermal efficiency and improve operating costs by reducing fuel costs.

In large power plants, each fraction of a percentage of improvement in the overall efficiency is worthwhile, whether the improvement comes from higher steam pressures and temperatures, lower turbine backpressures, or reduced heat losses by the addition of equipment and/or insulation. Such plants use millions of tons of coal per year, and small gains in thermal efficiency result in large savings in fuel costs. It is important to remember that of the plant's total operating costs, coal represents 60 to 80 percent. Therefore, improved cycle efficiency is very important to the reduction of operating costs.

Example A water-tube boiler has a capacity of 75,000 lb of steam per hour and operates at 435 psig and a total steam temperature of 700°F. Water enters the economizer at 235°F and leaves at 330°F. The combined efficiency is 85 percent. The areas of the heating surfaces in square feet are as follows: boiler, 5800; furnace waterwall, 750; superheater, 610; and economizer, 3450. The furnace volume is 1400 ft³. Calculate the heat transfer in Btu/h per square foot for the (1) combined boiler and furnace waterwall area, (2) superheater, and (3) economizer; (4) also calculate the heat release in the furnace in Btu/h per cubic foot of volume. Neglect blowdown from the boiler and carry-over of moisture in steam to superheater.

Solution (1) Heat-transfer rate in combined boiler and furnace waterwall areas:

$$h_g \text{ at 450 psia, saturated} = 1204.6 \text{ Btu/lb}$$

$$h_f \text{ at 330°F} = \underline{300.68 \text{ Btu/lb}}$$

$$\Delta h = 903.92 \text{ Btu/lb}$$

$$\text{Heat-transfer rate of boiler and furnace} = \frac{75,000 \text{ lb/h} \times 903.92 \text{ Btu/lb}}{5800 + 750}$$

$$= 10,350 \text{ Btu/h/ft}^2$$

(2) Heat-transfer rate in superheater:

$$h \text{ at 450 psia and 700°F} = 1359.9 \text{ Btu/lb}$$

$$h \text{ at 450 psia saturated} = \underline{1204.6 \text{ Btu/lb}}$$

$$\Delta h = 155.3 \text{ Btu/lb}$$

$$\text{Heat-transfer rate of superheater} = \frac{75,000 \times 155.3}{610}$$

$$= 19,094 \text{ Btu/h/ft}^2$$

(3) Heat-transfer rate in economizer:

$$h \text{ at 330°F} = 300.68 \text{ Btu/lb}$$

$$h \text{ at 235°F} = \underline{203.33 \text{ Btu/lb}}$$

$$\Delta h = 97.35 \text{ Btu/lb}$$

$$\text{Heat-transfer rate of economizer} = \frac{75,000 \times 97.35}{3450}$$

$$= 2116 \text{ Btu/h/ft}^2$$

(4) Heat release per furnace volume:

$$h \text{ at 450 psia and 700°F} = 1359.9 \text{ Btu/lb}$$

$$h_f \text{ at 235°F} = \underline{203.3 \text{ Btu/lb}}$$

$$\Delta h = 1156.6 \text{ Btu/lb}$$

$$\text{Heat input} = \frac{75,000 \times 1156.6}{0.85} = 102.1 \times 10^6 \text{ Btu/h}$$

$$\text{Heat release} = \frac{102.1 \times 10^6}{1400} = 72,929 \text{ Btu/h/ft}^3$$

The study of the conversion of forms of energy is called *thermodynamics*. It is the science that describes the transformation of one form of energy into another. In this book, the following will be emphasized:

1. *Chemical to thermal*; i.e., the combustion of a fuel produces heat energy.

2. *Thermal to mechanical*; i.e., the heat energy converts water to steam and the steam is used to power a turbine.

3. *Mechanical to electrical*; i.e., the turbine is connected to a generator for the production of electricity.

The basic principles of thermodynamics are used in the design of boilers, internal combustion engines, gas turbines, air compressors, steam turbines, refrigeration, and air conditioning, and also include the study of the flow of fluids and heat transfer. This book concentrates on boilers and steam turbines as well as combined cycle systems.

The following are typical steam plant examples that help illustrate the preceding information.

Example: Heating Water   How much heat is required to increase the temperature of 120,000 lb/h of feedwater from 180 to 300°F?

Solution

$$h_f \text{ at } 300°F = 269.6 \text{ Btu/lb}$$
$$h_f \text{ at } 180°F = \underline{147.9 \text{ Btu/lb}}$$
$$\Delta h_f = 121.7 \text{ Btu/lb}$$

$$\text{Heat required} = 120,000 \text{ lb/h} \times 121.7 \text{ Btu/lb}$$
$$= 14.6 \times 10^6 \text{ Btu/h}$$

Example: Fuel Required to Heat Water   If a gas-fired hot-water boiler operates at 75 percent efficiency and the natural gas has a heating value of 1000 Btu/ft^3, how much gas is required in the preceding example?

Solution

$$\text{Quantity of gas} = \frac{14.6 \times 10^6 \text{ Btu/h}}{0.75 \times 1000 \text{ Btu/ft}^3}$$
$$= 19,467 \text{ ft}^3/\text{h}$$

Questions and Problems

3.1 Of the various codes that are used in a boiler design, which is the most widely used?

3.2 Define *stress* as experienced by a boiler, and describe its importance in the design of a boiler.

3.3 If the force on a 2-in-diameter support rod is 8000 lb, what is the cross-sectional area of the rod, and what is the stress on the rod in pounds per square inch?

3.4 Define the following material characteristics: elastic limit, ultimate strength, and ductility.

3.5 Carbon steel is used extensively in many boiler designs. What is the maximum metal temperature that limits the use of this material?

3.6 What is the area of the head on a boiler drum that is 60 in in diameter?

3.7 A water-tube boiler operates at a steam pressure of 750 psi, and its steam drum diameter is 60 in. What is the total pressure on the head of the drum?

3.8 A boiler tube has an outside diameter of 2 in and a wall thickness of $\frac{3}{16}$ in. What is the tube's inside diameter?

3.9 What is meant by the *tensile strength* of a material? What is *shearing strength?*

3.10 What is an *alloy?*

3.11 What is meant by *factor of safety?*

3.12 Name some of the most common materials found in a modern boiler design.

3.13 Why has the welding of joints become the method of manufacturing boilers as compared with riveting?

3.14 What is used on modern boilers to prevent water carry-over from the steam drum?

3.15 What procedures must be followed for the welding of boilers and pressure vessels?

3.16 How is stress relieving of a welded component accomplished?

3.17 The inspection and testing of welded components are critical to quality control. Describe the methods of destructive and nondestructive testing that are used and the advantages of each.

3.18 A steel rod is $\frac{3}{4}$ in in diameter, and it reaches its elastic limit when subjected to a force of 12,000 lb and it fails under a load of 25,000 lb. Calculate its elastic limit and ultimate strength in pounds per square inch.

3.19 A 600-psi pressure boiler has a drum that is 36 in in diameter. What will be the total pressure on the head?

3.20 If the drum in Problem 3.19 is 30 ft long, calculate the total pressure on the longitudinal projected area, as shown in Fig. 3.5.

3.21 A tube has a 6-in inside diameter and is $\frac{3}{8}$ in thick. Its ultimate tensile strength is 65,000 psi, and its factor of safety is 6. Calculate the estimated internal design pressure.

3.22 Calculate the minimum required wall thickness of a welded steam drum that has an inside diameter of 60 in. The selected material is SA-516 Grade 70, which has an allowable stress of 17,500 psi. The design pressure of the boiler is 750 psi, and $E = 1.0$. What would the commercial plate size have to be for the ordering of this plate?

3.23 How are steam drums and other pressure vessels normally supported? What types of loads must be considered in the design of the structural supports for these vessels? Provide examples of the types of loads that must be evaluated as part of the design.

3.24 In describing the capacity or size of a boiler, how is it best defined for modern boilers?

3.25 Define the heating surface of a boiler and its major heat-transfer components of superheater, economizer, and air heater.

3.26 Calculate the heating surface of a return tubular boiler that is 6 ft in diameter, 18 ft long, and contains seventy 3.5-in tubes. The boiler is bricked in one-third from the top, and the area of the heads is neglected.

3.27 What is meant by *boiler horsepower?* In Problem 3.26, what would be the boiler horsepower of this HRT boiler?

3.28 What is meant by the *saturated steam temperature?*

3.29 A boiler produces 220,000 lb/h of saturated steam at 735 psig when supplied with feedwater at 240°F. With a boiler efficiency of 81 percent, what is the heat input requirement for the boiler?

3.30 What is *wet steam,* and how is the enthalpy of wet steam calculated?

3.31 What is *superheated steam,* and what does *degrees of superheat* mean?

3.32 What are the advantages of using superheated steam?

3.33 Compare the enthalpy of saturated steam at 900 psia with the enthalpy of superheated steam at 900 psia and 950°F. What are the degrees of superheat?

3.34 A water-tube boiler has a capacity of 150,000 lb/h of steam at a pressure and temperature of 385 psig and 500°F. Feedwater enters the economizer at 250°F and leaves at 346°F. The efficiency of the boiler is 87 percent. The heating surfaces of the boiler are as follows: (*a*) boiler, 6500 ft^2; (*b*) furnace waterwall, 1125 ft^2; (*c*) superheater, 440 ft^2; and (*d*) economizer, 4500 ft^2. The furnace volume is 2000 ft^3. Calculate the heat transfer in Btu/h/ft^2 for (i) combined boiler and waterwall areas, (ii) superheater, and (iii) economizer. Also calculate the heat release per furnace volume in Btu/h/ft^3.

3.35 What is the science of thermodynamics? What forms of energy transformation does it study?

4

Combustion of Fuels

Combustion is the rapid chemical combination of oxygen with the combustible elements of a fuel, resulting in the production of heat. Combustion is accomplished by mixing fuel and air at elevated temperatures. The air supplies oxygen, which unites chemically with the carbon, hydrogen, and a few minor elements in the fuel to produce heat.

Steam has been generated from the burning of a variety of fuels. In addition to the common fuels of coal, oil, and natural gas, today an increased amount and varied supply of waste and by-product fuels are used, such as municipal solid waste (MSW), coal mine tailings, and biomass wastes such as vine clippings and bagasse, a sugar cane by-product. MSW also can have a large percentage of biomass because it contains yard waste. These fuels must be burned and their combustion products properly handled. They create a unique challenge in their use because the fuel quality is significantly reduced owing to the fuel's lower heating value and poor combustion characteristics. In addition, such fuels often present more restrictive emission limitations. The designs of the boilers are therefore unique to the combustion of each of these fuels.

4.1 The Combustion Process

The combustion process follows fundamental principles that must be understood by the designers and operators of boilers and associated equipment to ensure reliable service and high efficiency.

1. *Control of air supply.* The amount of air required depends on the fuel, the equipment used for combustion, and the operating conditions and is determined from manufacturers' recommendations,

which are based on actual performance tests and operating experience. Too much air results in an excessive release of hot gases from the stack with a correspondingly high heat loss and reduction in efficiency. A deficiency of air permits some of the fuel, unburned or only partially burned, to pass through the furnace, which also results in a reduction in efficiency. It is therefore important that the best proportion of air to fuel be determined and maintained in order to obtain the highest efficiency possible.

2. *Mixing of air and fuel.*   Air and fuel must be mixed thoroughly, since each combustible particle must come into intimate contact with the oxygen contained in the air before combustion can take place. If the air distribution and mixing are poor, there will be an excess of air in some portions of the fuel bed or combustion chamber and a deficiency in others. Combustion equipment is designed with this principle in mind in an attempt to obtain the best possible mixing of fuel and air.

3. *Temperature required for combustion.*   All around us we see combustible material in intimate contact with air, and still it is not burning. Actually, a chemical reaction is taking place, but it is so slow that it is referred to not as *combustion* but as *oxidation.* The corrosion (rusting) of steel when exposed to the atmosphere is an example of this oxidation.

When the combustible material reaches its ignition temperature, oxidation is accelerated and the process is called *combustion.* It is the rapid chemical combination of oxygen with the combustible elements of a fuel. Therefore, it is evident that it is important to maintain the fuel and air mixture at a temperature sufficiently high to promote combustion.

When the flame comes into contact with the relatively cool boiler tubes, the carbon particles are deposited in the form of soot. When boilers are operated at a very low capacity, the temperatures are lower, which can result in incomplete combustion and excessive smoke if combustion controls are not set properly.

4. *Time required for combustion.*   Air supply, mixing, and temperature determine the rate at which combustion progresses. In all cases an appreciable amount of time is required to complete the process. When the equipment is operated at an excessively high capacity, the time may be insufficient to permit complete combustion. As a result, considerable unburned fuel is discharged from the furnace. The rejected material may be in the form of solid fuel or combustible gases. The resulting loss may be appreciable and therefore must be checked and controlled.

These principles involving the process of burning (combustion) may be understood by reference to Figs. 4.1 and 4.2. Here the principles of

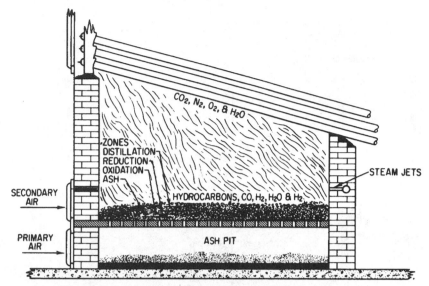

Figure 4.1   The combustion process as applied to hand firing.

combustion are applied to solid fuels burned on grates and to pulverized coal, gas, and oil burned in suspension.

Figure 4.1 illustrates a hand-fired stationary grate installed under a water-tube boiler. The coal is supplied by hand through the fire door. Air for combustion enters through both the ashpit and the fire door. Primary air comes through the stoker grate, and secondary air enters through the fire door in this illustration. For the purpose of illustration, the fuel bed may be considered as having four zones. Coal is added to the top or distillation zone; next are the reduction and oxidation zones and, finally, the layer of ash on the grates. The primary air that enters the ashpit door flows up through the grates and ash into the oxidation zone, where the oxygen comes into contact with the hot coal and is converted into carbon dioxide (CO_2). As the gases continue to travel upward through the hot-coal bed, this carbon dioxide is reduced to carbon monoxide (CO). The exposure of the coal in the upper zone to the high temperature results in distillation of hydrocarbons (chemical compounds of hydrogen and carbon), which are carried into the furnace by the upward flow of gases. Therefore, the gases entering the furnace through the fuel bed contain combustible materials in the form of carbon monoxide and hydrocarbons. The oxygen in the secondary air that enters the furnace through the fire door must combine with these combustibles to complete the combustion process before they enter the boiler tube bank and become cooled.

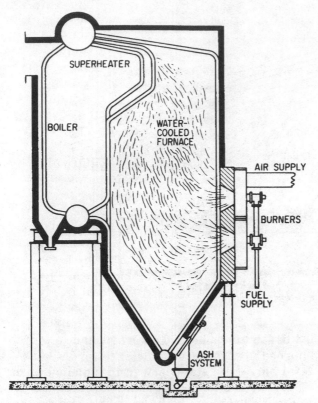

Figure 4.2 The combustion process as applied to suspension firing.

From this discussion of the process it is evident that with hand firing a number of variables are involved in obtaining the required rate of combustion and complete utilization of the fuel with a minimum amount of excess air. The fuel must be supplied at the rate required by the steam demand. Not only must the air be supplied in proportion to the fuel, but the amount entering through the furnace doors and the ashpit must be in correct proportions. The air that enters through the ashpit door and passes up through the fuel bed determines the rate of combustion. The secondary air that enters directly into the furnace is used to burn the combustible gases. Thorough mixing of the combustible gases and air in the furnace is necessary because of the short time required for these gases to travel from the fuel bed to the boiler tubes. Steam or high-pressure air jets are used to assist in producing turbulence in the furnace and mixing of the gases and air (see Sec. 5.11). A failure to distribute the coal evenly on the grates, variation in the size of the coal, and the formation of clinkers result in unequal resistance of the fuel bed to the flow of gases. (A *clinker* is a hard, compact, congealed mass of fuel matter that has fused in the furnace. It is often called *slag*.) Areas of low resistance in the fuel bed permit high velocity of gases and accelerated rates of combustion, which deplete the fuel and further reduce the resistance. These areas of low resistance have been called *holes in the fire*.

As a result of these inherent shortcomings in hand firing and the physical labor involved, mechanical methods of introducing solid fuel into the furnace and automatically controlling the air supply have been developed. These are explained in Chap. 5.

Suspension firing of fuel in a water-cooled furnace is illustrated in Fig. 4.2. This method may be utilized in the combustion of gaseous fuels without special preparations, of fuel oil by providing for atomization, and of solid fuels by pulverization. The fuel particles and air in the correct proportions are introduced into the furnace, which is at an elevated temperature. The fine particles of fuel expose a large surface to the oxygen present in the combustion air and to the high furnace temperature. The air and fuel particles are mixed either in the burner or directly after they enter the furnace. When coal is burned by this method, the volatile matter—hydrocarbons and carbon monoxide—is distilled off when the coal enters the furnace. These combustible gases and the residual carbon particles burn during the short interval of time required for them to pass through the furnace. The period of time required to complete combustion of fuel particles in suspension depends on the particle size of the fuel, control of the flow of combustion air, mixing of air and fuel, and furnace temperatures. Relatively large furnace volumes are required to ensure complete combustion. The equipment used in the suspension burning of solid, liquid, and gaseous fuels is discussed in Chap. 5.

From these illustrations we note that the requirements for good combustion are sufficient time of contact between the fuel and air, elevated temperature during this time, and turbulence to provide thorough mixing of fuel and air. These are referred to as the three T's of combustion—time, temperature, and turbulence.

4.2   The Theory of Combustion

Combustion is a chemical process that takes place in accordance with natural laws. By applying these laws, the theoretical quantity of air required to burn a given fuel can be determined when the fuel analysis is known. The air quantity used in a furnace, expressed as percentage of excess above the theoretical requirements (excess air), can be determined from the flue gas analysis.

In the study of combustion we encounter matter in three forms: solid, liquid, and gas. *Melting* is the change of phase from solid to liquid. Heat must be added to cause melting. The change in the reverse direction, liquid to solid, is *freezing* or *solidifying*. The change of phase from liquid to gas is called *vaporization,* and the liquid is said to *vaporize* or *boil*. The change from the gaseous or vapor phase to liquid is *condensation,* and during this process the vapor is said to be *condensing*. Matter in the form of a *solid* has both volume and shape. A *liquid* has a definite volume, in that it is not readily compressible, but its shape conforms to that of the container. A *gas* has neither a definite volume nor shape, since both conform to that of the container.

When liquids are heated, a temperature is reached at which *vapor* will form above the surface. This vapor is only slightly above the liquid state. When vapor is removed from the presence of the liquid and heated, a gas will be formed. There is no exact point at which a substance changes from a gas to a vapor or from a vapor to a gas. It is simply a question of degree as to how nearly the vapor approaches a gas. Steam produced by boiling water at atmospheric pressure is vapor because it is just above the liquid state. On the other hand, air may be considered a gas because under normal conditions it is far removed from the liquid state (liquid air). Gases follow definite laws of behavior when subjected to changes in pressure, volume, and temperature. The more nearly a vapor approaches a gas, the more closely it will follow the laws.

When considering gas laws and when making calculations in thermodynamics (i.e., the relationship between heat and other forms of energy), pressures and temperatures must be expressed in absolute units rather than in *gauge* values (read directly from gauges and thermometers).

Absolute pressures greater than atmospheric are found by adding the atmospheric pressure to the gauge reading. Both pressures must be expressed in the same units. The atmospheric pressure is accurately determined by means of a barometer, but for many calculations the approximate value of 14.7 psi is sufficiently accurate. For example, if a pressure gauge reads 150 psi, the absolute pressure is 164.7 psia (read as "pounds per square inch absolute"). Absolute pressures below zero gauge are found by subtracting the gauge reading from the atmospheric pressure. When a gauge reads −5 psi, the absolute pressure would be 14.7−5.0 = 9.7 psia. Pressures a few pounds above zero gauge and a few pounds below (in the vacuum range) are frequently measured by a U-tube containing mercury and are expressed in inches of mercury. Many of the pressures encountered in combustion work are nearly atmospheric (zero gauge) and can be measured by a U-tube containing water.

The zero on the Fahrenheit scale is arbitrarily chosen and has no scientific basis. Experiments have proved that the true or absolute zero is 460° below zero on the Fahrenheit thermometer. The absolute temperature on the Fahrenheit scale is found by adding 460°F to the thermometer reading. Absolute temperatures on the Fahrenheit scale are called *degrees Rankine* (°R). On the centigrade or Celsius scale, the absolute temperature is determined by adding 273°C to the centigrade thermometer reading. Absolute temperatures on the centigrade scale are measured in *kelvins* (K).

Expressed in absolute units of pressure and temperature, the three principal laws governing the behavior of gases may be stated as follows (where V_1 and V_2 are, respectively, the initial and final volumes, P_1 and P_2 are, respectively, the initial and final absolute pressures, and T_1 and T_2 are, respectively, the initial and final absolute temperatures):

Constant temperature. When the temperature of a given quantity of gas is maintained constant, the volume will vary inversely as the pressure. If the pressure is doubled, the volume will be reduced by one-half:

$$\frac{V_1}{V_2} = \frac{P_2}{P_1}$$

Constant volume. When the volume of a gas is maintained constant, the pressure will vary directly as the temperature. When the temperature is doubled, the pressure also will be doubled:

$$\frac{P_1}{P_2} = \frac{T_1}{T_2}$$

Constant pressure. When a gas is maintained at constant pressure, the volume will vary directly as the temperature. If the temperature of a given quantity of gas is doubled, the volume also will be doubled:

$$\frac{V_1}{V_2} = \frac{T_1}{T_2}$$

In combustion work the gas temperature varies over a wide range. Air enters the furnace or air heater, for example, at 70°F, is heated in some instances to over 3000°F in the furnace, and is finally discharged from the stack at between 300 and 400°F. During these temperature changes, the volume varies because the gases are maintained near atmospheric pressure. This is most important because fans, flues and ducts, boiler passes, etc., must be designed to accommodate these variations in volume.

In addition to these physical aspects of matter, we also must consider the chemical reactions that occur in the combustion process. All substances are composed of one or more of the chemical elements. The smallest particle into which an element may be divided is termed an *atom.* Atoms combine in various combinations to form molecules, which are the smallest particles of a compound or substance. The characteristics of a substance are determined by the atoms that make up its molecules. Combustion is a chemical process involving the reaction of carbon, hydrogen, and sulfur with oxygen.

The trading about and changing of atoms from one substance to another constitute an exacting procedure. Substances always combine in the same definite proportions. The atoms of each of the elements have a weight number referred to as the *atomic weight.* These weights are relative and refer to oxygen, which has an atomic weight of 16. Thus, for example, carbon, which is three-quarters as heavy as oxygen, has an atomic weight of 12. The chemical and physical properties of substances involved in the combustion process are given in Table 4.1.

When oxygen and the combustible elements or compounds are mixed in definite proportions at an elevated temperature under ideal conditions, they will combine completely. [The theoretical proportions (no deficiency and no excess) of elements or compounds in a chemical reaction are referred to as the *stoichiometric ratio.*] This shows that a given combustible element requires a definite amount of oxygen to complete combustion. If additional oxygen is supplied (more than necessary for complete combustion), the excess will not enter into the

TABLE 4.1   Properties of Substances in the Combustion Process

Name	Chemical			Physical			
	Molecular formula	Atomic weight*	Molecular weight*	Specific weight, lb/ft³†	Specific volume, ft³/lb†	Heating value, Btu/lb	State
Air	—	—	29‡	0.075	13.28	—	Gas
Carbon	C	12	—	—	—	14,540	Solid
Carbon dioxide	CO_2	—	44	0.114	8.75	—	Gas
Carbon monoxide	CO	—	28	0.073	13.75	4,355	Gas
Hydrogen	H_2	1	2	0.005	192.52	62,000	Gas
Nitrogen	N_2	14	28	0.073	13.75	—	Gas
Oxygen	O_2	16	32	0.083	12.03	—	Gas
Sulfur	S_2	32	64	—	—	4,050	Solid
Sulfur dioxide	SO_2	—	64	0.166	6.02	—	Gas
Water vapor§	H_2O	—	18	0.037	26.80	—	Vapor

*Approximate.
†At 14.7 lb/in² and 68° F.
‡Air is a mixture of oxygen and nitrogen. This figure is the average accepted molecular weight.
§At atmospheric pressure and 212°F.

reaction but will pass through the furnace unchanged. On the other hand, if there is a deficiency of oxygen, the combustible material will remain unburned. The law of combining weights states that the elements and compounds combine in definite proportions that are in simple ratio to their atomic or molecular weights.

Example The atomic and molecular weights of elements and compounds are useful in determining the weights and volume of gases. It has been determined that at the same temperature and pressure a given volume of all perfect gases will contain the same number of molecules. To test this theory, calculate the volume in cubic feet of 32 lb of oxygen and 28 lb of nitrogen (weights equivalent to their respective molecular weights).

Solution The specific volume in cubic feet per pound of oxygen and nitrogen as given in Table 4.1 are respectively 12.03 and 13.75 at a pressure of 14.7 lb/in^2 and at 68°F. Therefore,

$$32 \text{ lb} \times 12.03 \text{ ft}^3/\text{lb} = 385 \text{ ft}^3 \text{ of oxygen}$$

$$28 \text{ lb} \times 13.75 \text{ ft}^3/\text{lb} = 385 \text{ ft}^3 \text{ of nitrogen}$$

The weight of pounds of any substance, equal to its molecular weight, is known as a *pound mole*.

Therefore, a pound mole of oxygen equals 32 lb and a pound mole of nitrogen equals 28 lb.

The following is an explanation of some of the chemical reactions involved in combustion:

The volume of carbon dioxide (CO_2) produced is equal to the volume of oxygen (O_2) used. The carbon dioxide gas is, however, heavier than the oxygen. The combining weights are 12 lb of carbon (1×12) and 32 lb of oxygen (2×16), uniting to form 44 lb of carbon dioxide, or 1 lb of carbon requires 2.67 lb of oxygen ($32/12 = 2.67$) and produces 3.67 lb of carbon dioxide ($1 + 2.67 = 3.67$). (See Table 4.2.) The combustion of 1 lb of carbon produces 14,540 Btu.

TABLE 4.2   Carbon Burned to Carbon Dioxide

Substance	carbon	+	oxygen	→	carbon dioxide
Kind of matter	solid	+	gas	→	gas
Volume and chemical equation	C	+	O_2	→	CO_2
	C	+	O_2	→	CO_2
Atomic or molecular weight	12	+	32	=	44
Weight in pounds	1	+	2.67	=	3.67

TABLE 4.3   Carbon Burned to Carbon Monoxide

Substance	carbon	+	oxygen	→	carbon monoxide
Kind of matter	solid	+	gas	→	gas

Volume and chemical equation

$$2C + O_2 \longrightarrow CO \quad CO$$

or

$$2C + O_2 \longrightarrow 2CO$$

Molecular weight	$\{$	24	+	32	=		56
		2×12	+	32	=	$2 \times 28 =$	56
Weight in pounds		1	+	1.333	=		2.333

When carbon is burned to carbon monoxide (CO) [Table 4.3], which is incomplete combustion, the volume of oxygen used is only one-half of that required for completely burning the carbon to carbon dioxide; the volume of carbon monoxide produced is two times that of the oxygen supplied. The heat released is only 4355 Btu/lb, but it is 14,540 Btu when 1 lb of carbon is completely burned. The net loss is, therefore, 10,185 Btu/lb of carbon and shows the importance of completely burning the combustible gases before they are allowed to escape from the furnace.

In the reaction shown in Table 4.4, the two molecules of carbon monoxide (CO) previously produced, by the incomplete combustion of two molecules of carbon, are combined with the necessary one molecule of oxygen to produce two molecules of carbon dioxide. The 1 lb of carbon produced 2.333 lb of carbon monoxide. Finally, however, the 1 lb of carbon produces 3.67 lb of carbon dioxide regardless of whether the reaction is in one or two steps. The total amount of oxygen required, as well as the heat liberated per pound of carbon, is the same for complete combustion in both cases.

TABLE 4.4   Carbon Monoxide Burned to Carbon Dioxide

Substance	carbon monoxide	+	oxygen	→	carbon dioxide
Kind of matter	gas	+	gas	→	gas

Volume and chemical equation

$$CO \quad CO + O_2 \longrightarrow CO_2 \quad CO_2$$

or

$$2CO + O_2 \longrightarrow 2CO_2$$

Molecular weight	$\{$	2×28	+	32	=	2×44
		56	+	32	=	88
Weight in pounds		2.333	+	1.333	=	3.67

TABLE 4.5   Combustion of Hydrogen

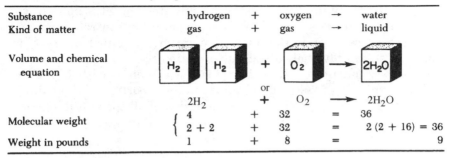

Substance	hydrogen	+	oxygen	→	water
Kind of matter	gas	+	gas	→	liquid

	$2H_2$	+	O_2		$2H_2O$
Molecular weight	4	+	32	=	36
	2 + 2	+	32	=	2 (2 + 16) = 36
Weight in pounds	1	+	8	=	9

Hydrogen is a very light gas with a high heat value. The combustion of 1 lb of hydrogen gas liberates 62,000 Btu (Table 4.1). To develop this heat, two molecules of hydrogen combine with one molecule of oxygen to form two molecules of water (Table 4.5). One volume of oxygen is required for two volumes of hydrogen. The weight relations are 1 lb of hydrogen and 8 lb of oxygen, producing 9 lb of water. This water appears as water vapor in the flue gases.

The equations for these and some of the other reactions involved in combustion are as follows:

$$C + O_2 \rightarrow CO_2$$

$$2C + O_2 \rightarrow 2CO$$

$$2CO + O_2 \rightarrow 2CO_2$$

$$2H_2 + O_2 \rightarrow 2H_2O$$

$$S + O_2 \rightarrow SO_2$$

$$2S + 3O_2 \rightarrow 2SO_3$$

$$\underset{\text{Methane}}{CH_4} + 2O_2 \rightarrow CO_2 + 2H_2O$$

$$\underset{\text{Acetylene}}{2C_2H_2} + 5O_2 \rightarrow 4CO_2 + 2H_2O$$

$$\underset{\text{Ethylene}}{C_2H_4} + 3O_2 \rightarrow 2CO_2 + 2H_2O$$

$$\underset{\text{Ethane}}{2C_2H_6} + 7O_2 \rightarrow 4CO_2 + 6H_2O$$

Sulfur is an undesirable constituent in fuels. It has a heating value of only 4050 Btu/lb (Table 4.1), contaminates the atmosphere with sulfur dioxide unless controlled with air pollution control equipment, and causes corrosion in the flues, economizers, and air heaters. Some forms of sulfur adversely affect pulverization.

In the combustion process, 1 lb of sulfur combines with 1 lb of oxygen to form 2 lb of sulfur dioxide. (Actually, a portion of the sulfur is converted to sulfur trioxide. The summation of all the sulfur oxides in the flue gases is referred to as SO_x.) The sulfur dioxide in the flue gases can be approximated as follows:

$$SO_2 \text{ lb/h} = K \times \text{lb/h fuel burned} \times 2 \times S/100$$

where K = the ratio of SO_2 in flue gases to theoretical amount resulting from the combustion of the sulfur in the fuel (frequently assumed to be 0.95) and S = the percentage of sulfur in the fuel. It is customary to express sulfur oxide emission in pounds per million Btus of fuel burned.

$$SO_2 \text{ lb/million Btu} = \frac{SO_2 \text{ lb/h} \times 1{,}000{,}000}{\text{lb fuel/h} \times \text{Btu/lb in fuel}}$$

Example Coal containing 1.5 percent sulfur with a Btu content of 11,500 Btu/lb burns at the rate of 3 tons per hour. What is the sulfur emission in pounds per million Btus input to furnace?

Solution

$$SO_2 \text{ lb/h} = 0.95 \times 3 \times 2000 \times 2 \times 1.5/100 = 171$$

$$SO_2 \text{ lb/million Btu} = \frac{171 \times 1{,}000{,}000}{3 \times 2000 \times 11{,}500} = 2.48$$

In practice, the oxygen supplied for combustion is obtained from the atmosphere. The atmosphere is a mixture of gases that for practical purposes may be considered as being composed of the following:

Element	Volume, %	Weight, %
Oxygen	20.91	23.15
Nitrogen	79.09	76.85

Only the oxygen enters into chemical combination with the fuel. The nitrogen combines in small amounts with the oxygen to form nitrogen oxides or commonly called NO_x. These are an atmospheric pollutant. The amount of NO_x produced depends on the combustion process. The higher the temperature in the furnace, the greater the amount of nitrogen oxides. The remainder of the nitrogen passes through the combustion chamber without chemical change. It does, however, absorb heat and reduces the maximum temperature attained by the products of combustion.

In order to supply 1 lb of oxygen to a furnace it is necessary to introduce

$$\frac{1}{0.2315} = 4.32 \text{ lb of air}$$

Since 1 lb of carbon requires 2.67 lb of oxygen, we must supply

$$4.32 \times 2.67 = 11.53 \text{ lb of air per pound of carbon}$$

The 11.53 lb of air is composed of 2.67 lb of oxygen and 8.86 lb of nitrogen.

$$11.53 - 2.67 = 8.86$$

$$\text{lb air} \quad \text{lb O}_2 \quad \text{lb N}_2$$

By referring to the equation for the chemical reaction of carbon and oxygen, we find that 1 lb of carbon produces 3.67 lb of carbon dioxide. Therefore, the total products of combustion formed by burning 1 lb of carbon with the theoretical amount of air are 8.86 lb of nitrogen and 3.67 lb of carbon dioxide.

In a similar manner it can be shown that 1 lb of hydrogen requires 34.56 lb of air for complete combustion. The resulting products of combustion are 9 lb of water and 26.56 lb of nitrogen.

Also, 1 lb of sulfur requires 4.32 lb of air, and therefore, the products of combustion are 3.32 lb of nitrogen and 2 lb of sulfur dioxide.

The condition under which, or degree to which, combustion takes place is expressed as *perfect, complete,* or *incomplete. Perfect combustion,* which we have been discussing, consists of burning all the fuel and using only the calculated or theoretical amount of air. *Complete combustion* also denotes the complete burning of the fuel but by supplying more than the theoretical amount of air. The additional air does not enter into the chemical reaction. *Incomplete combustion* occurs when a portion of the fuel remains unburned because of insufficient air, improper mixing, or other reasons. See Sec. 4.4 for a discussion of coal analysis.

The values given in Table 4.6 will now be used in determining the amount of air required and the resulting products involved in the perfect combustion of 1 lb of coal having the following analysis:

Constituent	Weight per lb
Carbon	0.75
Hydrogen	0.05
Nitrogen	0.02
Oxygen	0.09
Sulfur	0.01
Ash	0.08
Total	1.00

In the case of carbon, the values given in Table 4.7 are found as follows:

$$0.75 \times 2.67 = 2.00$$
lb C/lb coal   lb O_2 req'd/lb C   lb O_2 req'd/lb coal

$$2.00 \times 4.32 = 8.64$$
lb O_2/lb coal   lb air req'd/lb O_2 lb air req'd/lb coal

$$0.75 \times 3.67 = 2.75$$
lb C/lb coal   lb CO_2/lb C   lb CO_2/lb coal

$$8.64 - 2.00 = 6.64$$
lb air req'd/lb coal   lb O_2 req'd/lb coal   lb N_2/lb coal

The values for hydrogen and sulfur are found in a similar manner. Note in Table 4.7 that the weight of fuel and air supplied is equal to the weight of the resulting quantities.

$$0.92 \text{ lb} + 10.022 \text{ lb} = 10.942 \text{ lb}$$
fuel less ash   air req'd   total input

$$2.75 \text{ lb} + 7.722 \text{ lb} + 0.45 \text{ lb} + 0.02 \text{ lb} = 10.942 \text{ lb}$$
CO_2 N_2 water vapor SO_2 total output

In practice, it is necessary and economical to supply more air than the theoretical amount in order to obtain complete combustion. The air supplied to a combustion process in an amount above that theoretically required is known as *excess air*.

TABLE 4.6   Theoretical Quantities Involved in Combustion of Fuel

Constituent	Required		Resulting quantities			
	O_2	Air	CO_2	N_2	H_2O	SO_2
Carbon (C)	2.67	11.53	3.67	8.86	—	—
Hydrogen (H)	8.00	34.56	—	26.56	9.00	—
Sulfur (S)	1.00	4.32	—	3.32	—	2.00

NOTE: All values are expressed in pounds per pound of fuel.

TABLE 4.7 Theoretical Quantities Involved in Combustion of Coal

Constituent	Weight per lb	Required		Resulting Quantities			
		O_2	Air	CO_2	N_2	H_2O	SO_2
Carbon	0.75	2.00	8.64	2.75	6.64	—	—
Hydrogen	0.05	0.40	1.728	—	1.328	0.45	—
Sulfur	0.01	0.01	0.043	—	0.033	—	0.02
Total	0.81	2.41	10.411	2.75	8.001	0.45	0.02
Correction for							
N_2 in coal	0.02	—	—	—	+ 0.02	—	—
O_2 in coal	0.09	−0.09	−0.389	—	−0.299	—	—
Corrected total	—	2.32	10.022	2.75	7.722	0.45	0.02

NOTE: All values are expressed in pounds per pound of coal.

The flue gas analysis is effective in determining the amount of air supplied for combustion, as indicated by Fig. 4.3. This graph shows how the amount of excess air used in the combustion process can be calculated by the percentage of either carbon dioxide or oxygen in the flue gases. When a single fuel is burned, the carbon dioxide content of the flue gases provides a satisfactory index of the amount of excess air being used. This can be explained by the fact that with the complete combustion of 1 lb of carbon, 3.67 lb of carbon dioxide is produced.

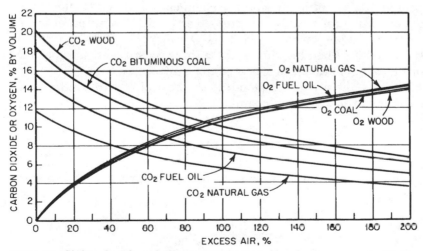

Figure 4.3 Carbon dioxide and oxygen in percentage by volume compared with the excess air used when various fuels are burned.

Therefore, the amount of carbon dioxide formed depends on the amount of carbon burned. When a relatively large amount of air is used, the fixed amount of carbon dioxide gas will be diluted and the percentage correspondingly lowered. Conversely, if only a small amount of excess air is used, there will be less dilution, and the percentage of carbon dioxide will be relatively high. For a given percentage of excess air, fuels with higher carbon-hydrogen ratio will have a higher percentage of carbon dioxide in the flue gases than fuels with lower carbon-hydrogen ratio. For a given percentage of excess air, the flue gases from a coal-fired furnace will have a higher percentage of carbon dioxide than when fuel oil is burned. For example, flue gas will contain 12 percent carbon dioxide when 54 percent excess air is used with bituminous coal and only 27 percent excess air with fuel oil (see Fig. 4.3).

The percentage of oxygen in the flue gases provides an adequate measurement of excess air when either single or multiple fuels are being used. The oxygen in the flue gases represents that portion which entered but did not combine with the combustible elements in the fuel. This oxygen in the flue gases and the nitrogen with which it was mixed are the excess air. The theoretical oxygen and therefore air requirement is approximately proportional to the heat content of the fuel even with variations in the carbon-hydrogen ratio. For a given percentage of oxygen, the excess air is approximately the same for either coal or fuel oil. For example, a flue gas will contain 6 percent oxygen when 40 percent excess air is used with bituminous coal and 38.8 percent excess air with fuel oil.

The flue gas analysis is obtained by use of the Orsat, as explained in Sec. 4.8, which is an apparatus where gaseous constituents are measured by absorption in separate chemical solutions. Modern facilities use *continuous emission-monitoring* (CEM) equipment. The analysis includes carbon dioxide (CO_2), oxygen (O_2), and carbon monoxide (CO). When the sum of these three constituents is subtracted from 100, the remainder is assumed to be nitrogen (N_2).

The excess air in percentage of the theoretical requirements can be calculated by the following formula:

$$\text{Percentage of excess air} = \frac{O_2 - \frac{1}{2}CO}{0.263N_2 + \frac{1}{2}CO - O_2} \times 100$$

However, when there is no carbon monoxide present, the formula becomes

$$\text{Percentage of excess air} = \frac{O_2}{0.263N_2 - O_2} \times 100$$

The use of these formulas will be shown by application to the following flue gas analyses.

Analysis	CO_2	O_2	CO	N_2
A	13.7	3.5	1.8	81.0
B	13.5	5.5	0.0	81.0

Analysis A:

$$\text{Percentage of excess air} = \frac{3.5 - \frac{1}{2} \times 1.8}{0.263 \times 81.0 + \frac{1}{2} \times 1.8 - 3.5} \times 100 = 13.9$$

Analysis B:

$$\text{Percentage of excess air} = \frac{5.5}{0.263 \times 81.0 - 5.5} \, 5 \times 100 = 34.8$$

For analysis A, the excess air is 13.9 percent; for B, it is 34.8 percent. Free oxygen is present in both analyses, but the presence of carbon monoxide in A indicates incomplete combustion.

When 1 lb of carbon is burned to carbon dioxide, 14,540 Btus are released, but when it is burned to carbon monoxide only 4355 Btus are released. Therefore, it is important that the carbon monoxide content of the products be at a minimum. It also means that the carbon monoxide content of the products of combustion can be used to control the amount of excess air. When there is a trace of carbon monoxide in the flue gases, the air supply is increased.

When the percentages of carbon, hydrogen, and sulfur in solid fuels are known, the heating value can be approximated by using Dulong's formula:

Heating value, Btu/lb of fuel = $14{,}540C + 62{,}000(H - \frac{1}{8}O) + 4050S$

Theoretical air required, lb/lb fuel = $11.53C + 34.56(H - \frac{1}{8}O) + 4.32S$

It is preferable to determine the heating value of a fuel by actually developing and measuring the heat. This is accomplished by completely burning a carefully weighed sample of the fuel in a calorimeter. The heat produced causes a temperature rise in a known quantity of water. The temperature rise is indicative of the heating value of the fuel.

Physical characteristics of fuel are, in many cases, more important in practical application than chemical constituents. Means have been devised for determining some of these characteristics by simple control tests that can be made by the plant operators.

TABLE 4.8 Ranges of Excess Air Requirements for Various Fuels and Methods of Firing

Fuel	Excess air, % by weight
Pulverized coal	15–20
Coal	
Fluidized bed combustion	15–20
Spreader stoker	25–35
Water-cooled vibrating grate stoker	25–35
Chain and traveling grate stoker	25–35
Underfeed stoker	25–40
Fuel oil	3–15
Natural gas	3–15
Coke oven gas	3–15
Blast furnace gas	15–30
Wood/bark	20–25
Refuse-derived fuel (RDF)	40–60
Municipal solid waste (MSW)	80–100

As noted previously, fuel and air mixing to ensure complete combustion is not perfect, and therefore excess air is required. However, because the excess air that is not used for combustion leaves the unit as part of the flue gas at stack exit temperatures, the amount of excess air should be minimized. The energy required to heat the air from ambient to stack temperature is lost heat. Each design of combustion equipment has its excess air requirements, and Table 4.8 shows typical ranges for various fuels and methods of firing.

4.3 The Air Supply

Supplying oxygen, as contained in air, is an important consideration in the combustion process. In a typical case in which 15 lb of air is required per pound of coal (assuming approximately 50 percent excess air), it is necessary to deliver 15 tons of air to the furnace and to pass almost 16 tons of gases through the boiler for each ton of coal burned. (See Table 4.7, where 10 lb of air is required per pound of coal and, with 50 percent excess air, 15 lb of air is required.) In many installations the ability to supply air is the limiting factor in the rate of combustion. The number of pounds of fuel that can be burned per square foot of stoker grate area depends on the amount of air that can be circulated through the fuel bed. A means must be provided to supply the required amount of air to the furnace. The products of combustion must be removed from the furnace and circulated over the heat-absorbing surfaces. Finally, the excess pollutants must be removed from the resulting flue gases before they are discharged through the stack to the atmosphere.

This circulation of gases is caused by a difference in pressure, referred to in boiler practice as *draft*. Draft is the differential in pressure between two points of measurement, usually the atmosphere and the inside of the boiler setting.

A differential in draft is required to cause the gases to flow through a boiler setting. This required differential varies directly as the square of the rate of flow of gases. For example, when the flow is doubled, the difference in draft between two points in the setting will increase four times. For a given amount of fuel burned, the quantity of gases passing through the boiler depends on the amount of excess air being used. The draft differential across the boiler tube bank and the differential created by an orifice in the steam line are used to actuate a flowmeter.

Recording flowmeters are calibrated under actual operating conditions by use of the flue gas analysis so that when the steam flow output and the gas flow coincide, the optimal amount of air is being supplied.

A draft gauge in the form of a U-tube is partly filled with water. A sampling tube inserted in the boiler setting is then connected to one end of the U-tube, while the other end remains open to the atmosphere. The difference in the height of the two columns of water is a measure of the draft.

The scales on these gauges are calibrated in inches, and the difference in height of the columns is read in inches of water. The draft gauges in common use are mechanical or dry-type gauges. However, the scales are calibrated in inches of water.

A pressure in the furnace slightly lower than that of the atmosphere (draft) causes the air to enter, thus supplying the oxygen required for combustion. A draft at the boiler outlet greater than that in the furnace causes the products of combustion to circulate through the unit. The rate of flow or quantity of air supplied can be regulated by varying the draft differential.

The principle of draft and air regulation is explained by reference to the hand-fired boiler, using a natural draft, shown in Fig. 4.4 and the graph of Fig. 4.5. The stack produces a draft of 1.0 in of water, which is regulated by the stack damper to give the required furnace draft. As the capacity increases, more air is required to burn the additional fuel, and the stack damper must be opened to compensate for the draft loss caused by the increased flow of gases. Draft loss occurs across the fuel bed, the boiler, the damper, and the breeching (flue). Finally, at 100 percent capacity, the stack damper is wide open and no more air can be supplied, thus limiting the ability of the furnace to burn additional coal efficiently.

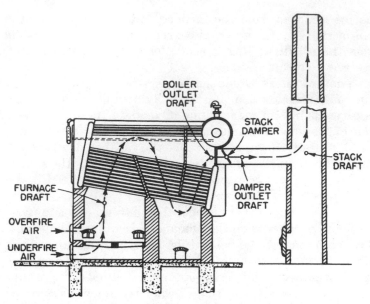

Figure 4.4 Flow of air and gases through a hand-fired boiler.

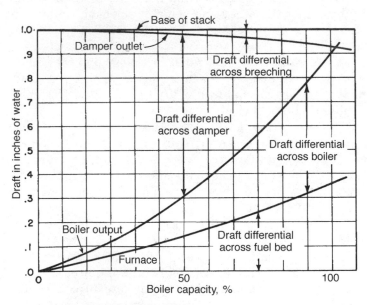

Figure 4.5 Draft in a hand-fired boiler.

The type of fuel determines the amount of draft differential required to produce a given airflow through the fuel bed. This is one of the reasons that more load can be carried with some fuels than with others. Figure 4.6 shows the draft normally required to burn several different types of fuels at varying rates.

Early boiler designs met their total draft requirements with natural draft that was supplied by the height of the stack. As units became larger and included additional heat traps, such as superheaters, economizers, and air heaters (and thereby having a higher draft loss), it was not practical to draft the entire unit from the stack. These units required fans in addition to the stack, using a *forced-draft* (FD) fan alone or in combination with an *induced-draft* (ID) fan.

Most combustion equipment uses forced-draft fans for supplying air to the furnace either through the burners or through the grates. The fan supplies the air at a pressure above that of the atmosphere and forces it into the furnace. Figures 4.7 and 4.8 show the application of forced draft to a stoker and the draft and wind-box pressure

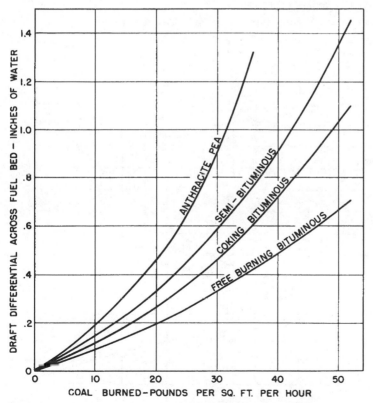

Figure 4.6 Fuel bed draft differential required for various fields.

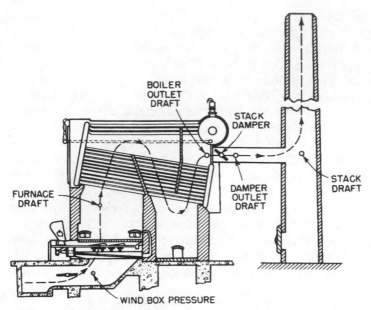

Figure 4.7 Flow of air and gases through an underfeed stoker-fired boiler.

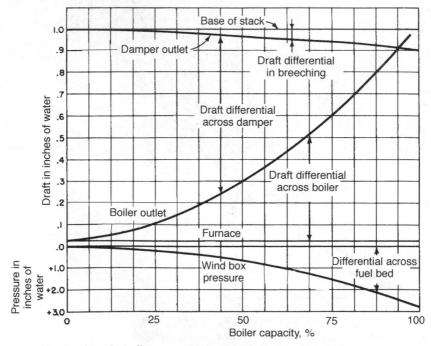

Figure 4.8 Draft and windbox pressure in an underfeed stoker-fired boiler.

at various boiler capacities. This is the same unit as that shown in Fig. 4.4 except that an underfeed stoker has been substituted for the hand-fired grates. The forced-draft fan produces a pressure under the stoker and causes the necessary air to flow up through the fuel bed. An increase in capacity requires an increase in pressure under the stoker to cause the additional air to flow through the fuel bed. The stack produces the draft necessary to circulate the gases through the boiler and breeching (flue). By automatic regulation of the stack damper, the furnace draft is maintained constant at 0.05 in of water. Operating a furnace at a constant draft slightly below atmospheric pressure is referred to as *balanced draft*. When the combustion rate increases, more air is added to the furnace, and a corresponding increased amount of flue gas is removed. This results in a greater flow but maintains a constant pressure in the furnace. Since the total effect of the stack is now available in overcoming the resistance through the boiler, a higher capacity can be obtained than with the same unit operating without a forced-draft fan and with hand-fired grates.

The stack provided an adequate means of circulating the air and gases through hand-fired grates or burners and boilers. However, the application of stokers necessitated the use of forced-draft fans, and induced-draft fans are required when economizers, air heaters, or flue gas cleaning equipment is applied to balanced-draft boilers. Modern boilers that burn solid fuels are nearly all balanced-draft units incorporating a forced-draft (FD) fan and an induced-draft (ID) fan.

The pressurized furnaces used in connection with package oil- and gas-fired boilers require only forced-draft fans. Figure 4.9 shows how the pressure developed by the forced-draft fan is used to produce the flow of air and flue gases through the entire unit. Combustion air flows from the forced-draft fan through the supply duct to the wind box, into the furnace, and through the boiler, economizer, and interconnecting flues to the stack. All the energy required is supplied by the forced-draft fan. At the maximum rating of 200,000 lb of steam per hour, the static pressure at the forced-draft fan outlet must be 19.6 in of water (gauge). The pressure drop across the wind box is maintained nearly constant at all ratings by adjustable louvers. This creates a high velocity at the burners and promotes thorough mixing of air and fuel, thereby maintaining good combustion efficiency at lower steam loads. Note the draft loss differential for the boiler and economizer and how it increases as capacity increases.

The pressure-furnace principle had been applied to large pulverized-coal-fired boilers. The necessary air and gas flow was developed with the forced-draft fan, but a number of design and operating problems were introduced such as flue gas leaks from the furnace into operating areas of the plant. In addition, the requirement for flue gas

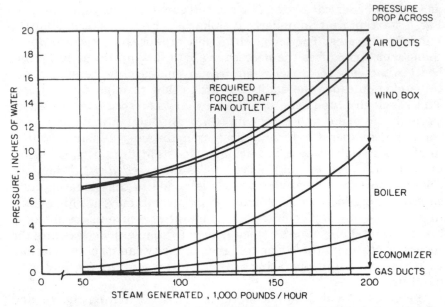

Figure 4.9 Air and flue gas pressure in a 200,000 lb/h pressurized fuel oil– and gas-fired package boiler.

cleaning equipment, such as SO_2 scrubbers, increased the draft loss and made the pressurized unit unattractive. Therefore, these types of units are designed today as balanced-draft boilers.

4.4 Coal

Coal was formed by the decomposition of vegetation that grew in prehistoric forests. At that time the climate was favorable for very rapid growth. Layer upon layer of fallen trees was covered with sediment, and after long periods of aging, the chemical and physical properties of the now ancient vegetation deposits were changed, through various intermediate processes, into coal. The process of coal formation can be observed in the various stages on the earth today. However, present-day formations are insignificant when compared with the magnitude of the great coal deposits.

It is estimated that 100 years is required to deposit 1 ft of vegetation in the form known as *peat,* and 4 ft of peat is necessary for the formation of 1 ft of coal. Therefore, it requires 400 years to accumulate enough vegetable matter for a 1-ft layer of coal. The conversion from peat to coal requires ages of time. In some areas where other fuel is scarce, the peat is collected, dried, and burned.

The characteristics of coal depend on the type of vegetation from which it was formed, the impurities that became intermixed with the vegetable matter at the time the peat bog was forming, and the aging, time, temperature, and pressure. It is apparent that the characteristics of coal vary widely.

For example, peat often contains partially decomposed stems, twigs, and bark. Peat is progressively transformed to lignite, which eventually can become anthracite when provided with the proper progression of geologic changes. However, this transformation takes hundreds of years to complete.

Coal is a heterogeneous material that varies in chemical composition according to location. In addition to the major organic ingredients of C, H_2 and O_2, coal also contains impurities. The impurities that are of major concern are ash and sulfur.

The ash results from mineral or inorganic material that was introduced during formation of the coal. Ash sources include inorganic substances, such as silica, which are part of the chemical structure of the plants. Dissolved mineral grains that are found in swamp water are also captured by the organic matter during the formation of coal. Mud, shale, and pyrite are deposited in pores and cracks of the coal seams.

Sulfur occurs in coal in three forms:

1. Organic sulfur, which is part of the coal's molecular structure

2. Pyritic sulfur, which occurs as the mineral pyrite

3. Sulfate sulfur, which is primarily from iron sulfate

The highest sulfur source is sulfate iron, which is found in water. Fresh water has a low sulfate concentration, whereas salt water is high in sulfate. Bituminous coal is found deposited in the interior of the United States, where oceans once covered the region, and this coal has a high sulfur content.

Coal can be classified as follows, and typical analyses are shown in Table 4.9.

Peat. Peat is the first product in the formation of coal and consists of partially decomposed plant and mineral matter. Peat has a moisture content of up to 70 percent and a heating value as low as 3000 Btu/lb. Although it is not an official coal classification, it is used as a fuel in some parts of the world.

Lignite. Lignite is the lowest ranking of coal with a heating value of less than 8300 Btu/lb and a moisture content as high as 35 percent. The volatile content is also high, and therefore lignite ignites easily.

Subbituminous.   These coals are noncoking; i.e., they have minimal swelling on heating and have a relatively high moisture content of 15 to 30 percent. They are high in volatile matter and thus ignite easily. They also have less ash and burn cleaner than lignite. They have a low sulfur content, generally less than 1 percent, and a heating value between 8300 and 11,500 Btu/lb. Because of the low sulfur content, many power plants have changed to subbituminous coal in order to limit SO_2 emissions.

Bituminous.   This coal is the one most commonly burned in electric utility boilers, and it has a heating value between 10,500 and 14,000 Btu/lb. As compared with lignite and subbituminous coals, the heating value is higher and the moisture content and volatile matter are lower. The high heating value and its relatively high volatile matter enable these coals to burn easily when fired as pulverized coal. Some types of bituminous coal, when heated in the absence of air, soften and release volatiles and then form coke, a porous, hard, black product. Coke is used as fuel in blast furnaces to make iron.

Anthracite.   This is the highest ranked coal. It has the highest content of fixed carbon, ranging from 86 to 98 percent. It has a low volatile content, which makes it a slow-burning fuel. Its moisture content is low at about 3 percent, and its heating value can be as high as 15,000 Btu/lb. Anthracite is low in sulfur and volatiles and burns with a hot, clean flame. It is used mostly for domestic heating as well as some metallurgical processes.

Preparation plants are capable of upgrading coal quality. In this process, foreign materials, including slate and pyrites, are separated from the coal. The coal may be washed, sized, and blended to meet the most exacting power plant demands. However, this processing increases the cost of coal, and an economic evaluation is required to determine whether the cost can be justified. If the raw coal available in the area is unsatisfactory, the minimum required upgrading must be determined. Utility plants obtain the lowest steam cost by selecting their combustion equipment to use the raw coal available in the area. However, in order to meet sulfur dioxide emission limits, many utilities use low-sulfur coals that often have to be transported significant distances. For example, many plants located in the Midwest of the United States use coals shipped from Montana and Wyoming.

All the many factors involved in obtaining the lowest-cost steam production must be taken into consideration. For a new facility, this would involve the selection of combustion equipment for the coal that is available in the region or for the coal that may be imported to the plant. For an existing facility, a selection of the type of coal to best match existing combustion equipment may be necessary.

Because of the varied and complex nature of coal, a number of methods for evaluating and specifying have been developed. One of the simplest methods of controlling coal quality is the proximate analysis, which includes moisture, volatile matter, fixed carbon, ash, heating value, and sometimes sulfur.

The procedure for obtaining a proximate analysis must be in accordance with ASTM laboratory procedures, which are generally as follows:

Moisture. A 1-g sample of coal is placed in an oven where the temperature is maintained at 220°F for 1 h. The difference between the weight before and after drying is the amount of moisture removed.

Volatile matter. The sample is next placed in a furnace in a covered crucible, where the temperature is maintained at 1700°F for a period of 7 min. The gaseous substance driven off is called *volatile matter.*

Fixed carbon. The lid is now removed from the crucible, the furnace temperature is increased, and the crucible is allowed to remain in the furnace until the combustible has been completely burned. The loss in weight as a result of this burning is the amount of fixed carbon.

Ash. The residual material in the crucible is ash.

Heating value. The heating value of coal may be determined by use of a bomb calorimeter. The bomb calorimeter provides a means of burning a small sample of coal under controlled conditions and measuring the resulting temperature rise in a given quantity of water. A 1-g sample (a gram is 1/453.6 lb) of pulverized and dried coal is placed in a tray. The tray is then placed in a steel bomb with a fuse wire arranged to extend into the tray of coal. The bomb is then closed, connected to an oxygen tank, and pressurized. A measured amount of water is poured into the calorimeter bucket. The bucket is placed in the calorimeter, and the bomb is carefully submerged in the water. A stirring device agitates the water to maintain uniform temperature. A calibrated thermometer permits the operator to observe the temperature. The coal is now ignited by means of the fuse wire and the temperature rise is noted. The heating value of the coal in Btu/lb is found by multiplying the temperature rise by a constant for the specific calorimeter.

Coal analyses are expressed in three different ways depending on the constituents included. The designation *as received* or *as fired* refers to an analysis in which the actual moisture is included. When the expression *moisture-free* or *dry coal* is used, the analysis considers the moisture as having been removed. Since neither moisture nor ash adds to the heating value of coal, analyses are calculated to exclude these constituents and in this way to give a true indication of the nature of the combustible material. When the moisture and ash are not included, the analysis is referred to as *moisture-* and *ash-free* or *combustible*. Sulfur has some heating value but is nevertheless an objectionable constituent of coal because of its corrosiveness and because of its contribution to air pollution, which is commonly referred to as *acid rain*.

The ultimate analysis of coal includes carbon (C), hydrogen (H), sulfur (S), oxygen (O), and nitrogen (N). This analysis is performed in a chemical laboratory. It is required for heat-balance calculations and the determination of the theoretical air requirements. Table 4.9 gives a typical representation of the proximate and ultimate analysis of fuels on an as-received basis, classified by rank according to the progressive changes from wood to anthracite. Analyses of coal types vary based on their location throughout the world.

In this ultimate analysis, the moisture content is included in the hydrogen and oxygen values. The calculations in Table 4.10 show how the ultimate analysis for bituminous coal can be expressed with the moisture as a separate item. The calculations also show how the analysis can be converted to a dry and ash-free basis.

Ash is an inert material, but its characteristics frequently determine the desirability of a coal for a given installation. Because of the importance of the fusion or melting temperature of the ash, tests have been devised to determine this property.[1] The ash to be tested is molded into a small pyramid, placed in a test furnace, and exposed to a steadily increasing temperature. The atmosphere surrounding the sample is controlled, and the temperature is measured while the pyramids are observed through a peephole in the side of the furnace (Fig. 4.10). The temperature of the pyramids is noted and recorded at three stages of melting: *initial deformation* (Fig. 4.10a), when the tip of the pyramid first shows a change; *ash-softening temperature* (Fig. 4.10b), when the pyramid forms in a sphere; and *melting point* (Fig. 4.10c), when the ash becomes fluid and the sphere flattens. These three temperatures are reported in reference to ash fusion. Fusion temperatures provide ash melting characteristics and are used to determine the potential for slagging under normal operating conditions. The degree of slagging can determine the firing method for the boiler.

When coal is heated in the absence of air or with a large deficiency of air, the lighter constituents are volatilized and the heavier hydrocarbons crack, liberating hydrogen and leaving a residue of carbon. This carbon residue that contains the ash and a part of the sulfur of the original coal is called *coke*. The principal uses for coke are in the

---

[1] Such tests are used by the American Society of testing Materials (ASTM).

TABLE 4.9 Representative Analyses of Wood, Peat, and Coal on an "As Received" Basis

Kind of fuel	Proximate analysis					Ultimate analysis						Calorific value, Btu/lb
	Moisture	Volatile matter	Fixed carbon	Ash	Sulfur	Hydrogen	Carbon	Nitrogen	Oxygen			
Wood	—	—	—	—	—	6.25	49.50	1.10	43.15			5800
Peat	56.70	26.14	11.17	5.99	0.64	8.33	21.03	1.10	62.91			3586
Lignite	34.55	35.34	22.91	7.20	1.10	6.60	42.40	0.57	42.13			7090
Subbituminous	24.28	27.63	44.84	3.25	0.36	6.14	55.28	1.07	33.90			9376
Bituminous	3.24	27.13	62.52	7.11	0.95	5.24	78.00	1.23	7.47			13,919
Semibituminous	2.03	14.47	75.31	8.19	2.26	4.14	79.97	1.26	4.18			14,081
Semianthracite	3.38	8.47	76.65	11.50	0.63	3.58	78.43	1.00	4.86			13,156
Anthracite	2.80	1.16	88.21	7.83	0.89	1.89	84.36	0.63	4.40			13,298

SOURCE: Adapted from E. S. Moore, *Coal*, Wiley, New York.

TABLE 4.10 Bituminous Coal Ultimate Analysis: Calculation of Moisture and Ash-Free Basis

	From tables	Separate moisture		Dry		Combustible
S	0.95	0.95		0.98		1.06
H	$5.24-3.24\times\frac{1}{9}*$ =	4.88		5.04		5.44
C	78.00	78.00		80.62		87.01
N	1.23	1.23	$\div(1-0.0324)$ =	1.27	$\div(1-0.0735)$ =	1.37
O	$7.47-3.24\times\frac{8}{9}*$ =	4.59		4.74		5.12
Ash	7.11	7.11		7.35		—
Moist	—	3.24		—		—
	100.00	100.00		100.00		100.00

$*1 \text{ lb } H_2 + 8 \text{ lb } O_2 \rightarrow 9 \text{ lb } H_2O.$

Figure 4.10 Ash fusion temperature determination.

production of pig iron in blast furnaces and in iron foundries. Because it is smokeless in combustion, it has been used for space heating.

The coking tendency of coal is expressed by the *free-swelling index* (FSI). This test is made by grinding a sample of coal to pass a no. 60 sieve and then heating 1 g under specified conditions that also have been described by ASTM. The profile obtained by heating the sample is compared with a standard set of profiles to determine the FSI of the sample. The standard profiles are expressed in one-half units from 1 to 9. Coals having an FSI below 5 are referred to as *free-burning,* since particles do not tend to stick together and form large lumps of coke when heat is applied but remain separate during the combustion process. Coals having an FSI above 5 are referred to as *caking* or *coking,* since the particles swell and tend to stick together when heated.

The caking, or coking, characteristics of coal affect its behavior when it is burned on hand-fired grates or stokers. When a coal cakes, the smaller particles adhere to one another, and large masses of fuel are formed on the grates. This action reduces the surface area that is exposed to oxygen and therefore retards the burning. Since these large pieces of coke do not burn, a portion is discharged to the ashpit as unburned carbon. For efficient combustion, a coking coal requires some agitation of the fuel bed to break up the coke masses in order to maintain uniform air distribution. Free-burning coals, on the other hand, may be burned successfully without fuel bed agitation. These characteristics (coking and free-burning) need not be considered when coal is burned in the pulverized form.

Coal that has a relatively high percentage of volatile matter is termed *soft*; that which has a lower percentage of volatile matter is termed *hard*. Many bituminous coals are considered soft coals, whereas anthracite coal is considered a hard coal. When coal is heated, the volatile matter has a tendency to be distilled off in the form of combustible gases known as *hydrocarbons*. These volatile gases liberated from coal must be burned in the combustion space above the fuel bed. A large combustion space must be provided to burn these gases and thereby eliminate fuel loss and smoke. Because hard coal has a relatively lower percentage of volatile matter, it burns with a short flame, and most of the combustion takes place in the fuel bed.

When soft coal is burned in pulverized form, the volatile material is distilled off and burns as a gas. This makes it relatively easy to maintain ignition and complete combustion with a minimum flame travel. Hard coal is also burned in the pulverized form, and in this case, each particle is a small portion of carbon that must be burned by contact with oxygen. The combustion of these carbon particles requires an appreciable amount of time, resulting in a long flame travel and a tendency for the fire to puff out at low loads and when starting up. Other conditions being equal, it is necessary to resort to finer pulverization when hard coal is burned than when soft coal is burned. Thus consideration must be given to the volatile content of the coal both in designing equipment and in operation.

Coal as it is removed from the mine contains some moisture, and the amount may be increased by exposure to the weather before it reaches the plant. Moisture represents an impurity in that it adds to the weight but not to the heating value of coal. It enters the furnace in the form of water and leaves as steam in the flue gas. Heat generated by the fuel actually must be expended to accomplish this conversion. Normally it is to the operator's advantage to burn coal with a low moisture content to prevent the loss of heat that results from converting the water into vapor or steam. However, when coal is burned on grates or stokers, there are conditions that make it advantageous to have a small percentage of moisture present. This moisture tends to accelerate the combustion process, keep the fuel bed even, and promote uniform burning. The advantages gained by the presence of the moisture may then balance the loss resulting from the heat required for its evaporation. Coals having 7 to 12 percent moisture content are recommended for use on chain- and traveling-grate stokers. The addition of moisture to promote combustion of coal is referred to as *tempering*.

Coal transportation costs are a significant part of the overall fuel cost. Nearly 60 percent of all coal deliveries is made by rail, and slightly less than 20 percent is transported by barge. The remaining coal delivery is equally divided between truck shipment and a continuous belt system when the power plant is located at the coal mine.

Coal with a high moisture content presents some difficult handling problems. During the winter season, this moisture freezes while the coal is in transit, making it very difficult to remove the coal from the railroad cars.

The methods used to thaw coal cars are gas burners under the cars, thawing sheds equipped with radiant electric heaters, and thawing sheds with steam coils for heating air, which is circulated around the cars. Car shakers are helpful, but they will not remove frozen coal.

Coal removed from the storage pile during snowy or rainy weather also may contain a high percentage of moisture. The wet coal adheres to the chutes, causing problems in the flow of the coal to the stoker or pulverizer mill. In many plants this has become a serious operating problem. Improvised methods of rapping the pipe, using an air lance, etc., have proved marginally effective in getting the coal to flow. In the design of the plant, the coal feed pipe from the bunkers to the stoker or mill should be as nearly perpendicular as possible with no bends or offsets. Access openings should be provided so that when stoppages do occur they can be relieved quickly.

Electric vibrators have proved beneficial when the pitch of the chute is insufficient to promote flow. In some installations, hot air is passed through the coal pipe with the coal. The larger the coal size, the less water it will retain. Therefore, some relief from freezing in cars and the stoppage of chutes can be obtained by using a coarser grade of coal. This, however, usually means an increase in cost.

The practice of applying antifreeze solutions to the coal when it is loaded at the mines has proved effective in reducing moisture pickup in transit. The added coal cost can be justified by the reduced handling costs at the plant. In fact, for hot, dry weather, where winds could create coal dust clouds, the coal is often sprayed with oil or an antifreeze solution to settle the fines.

Particle size is an important consideration in the selection of coal. The size requirement for different equipment varies widely. Coal burned on grates must have a certain size composition to regulate the passage of air through the fuel bed, while for a pulverized-coal burner the coal must be reduced in size to small, powderlike particles to promote rapid and complete combustion.

There is a tendency for coal particles to fracture when mechanical force is applied. Dense, hard coals resist fracture and retain their size during handling. Soft coals shatter easily, break up into small particles when handled, and therefore are said to be *friable*. Size degradation during shipment depends on the coal friability, which identifies its ease of crumbling, and the techniques, methods, and number of transfers that should be used. This size degradation is not a concern when the coal is to be pulverized.

When coal containing a range of sizes is dropped through a chute or other coal-conveying equipment, the fine and coarse particles segregate. If this condition occurs in the coal supply to a stoker, the fine is admitted to one section and the coarse to another. The resistance of the fuel bed to airflow varies, resulting in different rates of airflow through the grates and subsequent variations in the rates of combustion. Coal-conveying systems for supplying stokers should be designed to prevent this segregation.

Fine coal particles have a greater tendency to retain moisture, and wet fine coal will not flow readily through chutes and spouts. The clogging of coal-conveying equipment by fine wet coal becomes so troublesome that coal with coarser particles sometimes must be selected. Once the fine wet coal is in the furnace, there is a tendency for the combustion gases to carry the small particles of coal. This results in a loss of heat due to the unburned carbon.

The type of combustion equipment dictates the size of coal that is required. Therefore, it is necessary to designate coal size by stating the largest and smallest pieces and the percentage of the various-sized particles, such as $1\frac{1}{4}$ in by $\frac{1}{4}$ in with not more than 15 percent minus $\frac{1}{4}$. For stoker-fired units, coal sizing is specified to meet the stoker requirements, and therefore no additional sizing equipment is required at the plant. For pulverized-coal-fired boilers, a maximum delivered top size is specified with no limitation on the percentage of fines.

A characteristic known as *grindability* is considered when selecting coal for pulverizer plants. Some coals are harder and therefore more difficult to pulverize than others. The grindability of coal is tested, and the results are reported in accordance with the Hardgrove standard, which has an index of 100. A weighed, screened sample is placed in a laboratory test mill, and a given preestablished amount of energy is applied. The ratio of the fineness produced in the test sample to that produced when the same amount of energy was expended on a sample of standard coal is the Hardgrove value for the coal tested.

The index is relative, since large values, such as 100, represent coals that are more easily pulverized than those having smaller values, such as 40, which are difficult to pulverize. A knowledge of the grindability of coal is used in the selection of pulverizer mills and in the procuring of satisfactory coal for a pulverized-coal-fired plant.

When coal for an installation is being selected, a study should be made of the fusing or melting temperature of the ash that it contains. The exposure of low-fusing-temperature ash to the high-temperature zones of a fuel bed causes hard clinkers to form. These clinkers interfere with the movement of fuel in a stoker and make it difficult to clean a hand-fired grate. This results in an increased amount of carbon being removed with the ash, uneven fuel bed with increased excess air, and burned-out grates or stoker castings. In pulverized-coal-fired furnaces and in some cases even with stokers, the molten ash is carried by the gases to the walls and into the boiler passes, where it accumulates in large quantities. This slagging of ash causes deterioration of any refractory furnace lining, and its accumulation on the boiler tubes progresses until it bridges across from one tube to the other. This reduces the effective area of the gas passage, thereby restricting the flow. When the melting temperature of the ash is low, it is sometimes necessary to operate the boiler at reduced capacity or with increased excess air in order to lower the furnace temperature to prevent slag formation.

The study of fusion temperatures originally was developed to evaluate the clinkering tendency of coal ash that was produced by combustion on a stoker grate. In a pulverized-coal-fired unit, however, rather than slow heating and gradual melting of ash, the process is reversed. Ash particles are heated rapidly and then cooled at a relatively slow rate as they pass through the furnace and convection passes of the boiler. In a pulverized-coal-fired unit during combustion, the coal particles are heated nearly instantaneously to approximately 3000°F. As the flue gas cools in the furnace, the ash is cooled in approximately 2 seconds to a temperature of about 2000°F at the furnace exit.

Fusion temperatures provide an indication of the temperature range over which portions of the ash will be in a molten state or a plastic state. High fusion temperatures indicate that ash released in the furnace will cool quickly to a nonsticky state, which results in minimum slagging. However, low fusion temperatures indicate that ash will remain molten or sticky longer, and therefore slagging in the furnace or on boiler convection passes has a higher probability. The boiler must be designed for the coals that are to be used.

When temperatures in the furnace are below the initial deformation temperature, a large proportion of the ash particles would be in a dry state and would not adhere to the heating surfaces. Some settlement of this dust could appear on heating surfaces, but this could be removed by sootblowers. At temperatures above initial deformation, the ash becomes more sticky and can adhere readily to heating surfaces. This slagging potential can seriously affect boiler performance when this situation occurs.

The amount of ash in coal also should be given consideration, since it is an impurity that is purchased. It produces no heat and must be removed from the furnace and hauled from the plant. Since ash is an inert material, the heating value of coal can be expected to be low when the ash content is high.

A minimum of 4 to 6 percent ash is required in coal burned on hand-fired grates and on some stokers. The layer of ash forms an insulator that prevents an excessive amount of heat from reaching the grates. When the ash content of the coal is too low and the grates are unprotected, the grates become overheated, and maintenance costs are increased. Combustion equipment is designed to use high-ash coal and to operate without difficulty and with a minimum of carbon loss in the ash. Therefore, when coals are selected, the minimum ash requirement to protect the grates must be given first consideration if this is a factor. Then the economics of coals with various ash contents must be considered. Coals having a low ash content can be obtained either by selection or by processing. However, this selection of low-ash coal must be justified on an overall economic basis. With the use of adequate combustion equipment, many plants find high-ash coal more economical. In fact, as noted in Chap. 2, fluidized bed boilers are designed to burn waste coals that have a very high ash content and a low heating value. For these facilities, fuel costs are very low.

When selecting coal and the equipment used in the burning of coal, the sulfur content must be considered. The corrosive effects of sulfur necessitate the use of special materials in the construction of conveyors and bunkers. The dew point of the flue gases is lowered by the presence of the sulfur oxides. The flue gas temperature reduction in economizers and air heaters must be limited to prevent the metal temperatures from being reduced to or below the dew point. This precaution is necessary to prevent serious corrosion. Furthermore, the sulfur oxides discharged with the flue gases pollute the atmosphere unless adequate air pollution control equipment is utilized, such as flue gas scrubber systems, as described in Chap. 12.

Since it is seldom possible to schedule coal deliveries to meet plant needs, it is necessary to maintain a storage. This storage of coal frequently presents a problem in that the coal has the potential for overheating and starting to burn. This self-excited burning is known as *spontaneous combustion*. Actually, it is not entirely self-excited because the oxygen is obtained from the air. At first the oxidation proceeds slowly at atmospheric temperature, but if the heat produced is not carried away, the temperature increases. This higher temperature accelerates the process, and the heat is generated more rapidly. When the ignition temperature is reached, the coal starts to burn. Many tons of coal have been wasted as the result of spontaneous combustion.

Some of the precautions against spontaneous combustion are as follows:

1. Take the temperature of the coal by means of a thermocouple inserted in a pointed pipe.

2. In outside areas, use a means of storage that will prevent segregation; tightly pack the coal to exclude air. Isolate any hot region from the remainder of the pile.

3. Prevent the direct radiation of heat from the boiler, etc., against the side of bunkers by providing an insulating air space between the inner and outer bunker walls.

4. Do not allow a storage of coal to remain in excess of 1 month in an overhead bunker, and monitor the temperature of the coal with thermocouples.

5. Should the temperature exceed 150°F, inject carbon dioxide gas at once, by use of a perforated pipe driven into the coal. The introduction of carbon dioxide should not, however, be considered a final answer because the coal must be moved as soon as possible.

6. Provide a nitrogen "blanket" over the coal in the bunker to reduce contact with the oxygen in the air.

4.5   Fuel Oil

The origin of petroleum has not been definitely established, but several theories have been offered. The combination of carbon and hydrogen as they exist in petroleum could have been generated from either animal or vegetable matter. One theory holds that large amounts of some form of animal or vegetable matter, and perhaps both, decayed and aged under pressure and in the presence of salt water. Those who offer this theory point to the fact that petroleum deposits are usually found in the presence of salt water. Another theory holds that petroleum was formed by inorganic substances, i.e., without the presence of either animal or vegetable life. This theory is based on the fact that material containing hydrogen and carbon has been produced experimentally by the action of certain carbides and water. Petroleum is believed, by the advocates of this theory, to have been formed in the interior of the earth by the action of minerals, water, and gases.

Deposits of petroleum are found floating on subterranean lakes of salt water. A dome-shaped layer of nonporous rock holds the petroleum in place. The dome-shaped rock formation usually also entraps a quantity of natural gas. The deposit consists of natural gas, petroleum, and water, which are separated from the rock down in the order named by virtue of their specific gravity. Oil is obtained by drilling through the layer of rock that covers the oil and gas. This procedure releases the natural gas, which is stored underground or transmitted through pipelines for use in distant cities. The oil is removed from the ground by a pumping operation that frequently fails to remove all the oil deposit.

When compared with coal, fuel oils are relatively easy to handle and burn, and there is significantly less ash to dispose of. In most oil burners, the oil is atomized and mixed with combustion air. In this atomized state, the characteristics of oil approximate those of natural gas.

Fuel oils include virtually all petroleum products that are less volatile than gasoline. They range from light oils, which are used in internal combustion or turbine engines, to heavy oils, which require heating. The heavier fuel oils are those used for steam generation in boilers.

The specific gravity of fuel oil is used as a general index of its classification and quality in much the same manner as the proximate analysis is used to specify the characteristics of coal. In determining the specific gravity of oil or any substance, the weight of a given volume of the substance is divided by the weight of an equal volume of water when both are measured at the same temperature.

Special laboratory determinations of fuel oil are made in terms of specific gravity, but in practical fieldwork the gravity is measured by a hydrometer and read in degrees Baumé or API. The API scale as adopted by the American Petroleum Institute is now generally accepted to determine the relative density of oil. It differs slightly from the Baumé scale. The specific gravity as expressed in relation to water may be converted into degrees API by the use of the following formula:

$$°API = \frac{141.5}{\text{sp. gr. at } 60/60°F} - 131.5$$

Note: Sp. gr. at 60/60°F means that oil at 60°F has been referred to water at 60°F. Since water has a specific gravity of 1, we find from this formula that it has an API gravity of 10. The API gravity of commercial fuel oil varies from 10 to 40.

Example A sample of fuel oil has a specific gravity of 0.91 at 60°F. What is its gravity in degrees API?

Solution

$$°API = \frac{141.5}{0.91} - 131.5 = 24.0$$

Specific gravity, therefore, is the ratio of the density of oil to the density of water, and it is important because oil is purchased by volume, either by the gallon or by the barrel.

Heavier, liquid fuels have a lower API gravity, as noted in Table 4.11. The viscosity or resistance to flow of an oil is important because it affects the ease with which it can be handled and broken up into a fine mist by the burner. An increase in temperature lowers the viscosity of an oil and causes it to flow more readily. It is therefore necessary to heat heavy oils in order to handle them effectively in pumps and burners.

Light oils contain a larger proportion of hydrogen than heavy oils. These light oils ignite easily and are said to have low flash and fire points. The *flash point* of an oil is the temperature at which the gases given off will give a flash when ignited. The *fire point* is the temperature at which the gases given off may be ignited and will continue to burn. An oil that has a low flash point will burn more readily than one with a high flash point.

The heating value of fuel oil is expressed in either Btu/lb or Btu/gal. The commercial heating value of fuel oil varies from approximately 18,000 to nearly 20,000 Btu/lb. The calorimeter provides the best means of determining the heating value of fuel oil. When the gravity in API degrees is known, the heating value may be estimated by use of the following formula:

Btu/lb of oil = 17,687 + 57.7 × API gravity at 60°F

Example If a sample of fuel oil has an API gravity of 25.2 at 60°F, what is the approximate Btu/lb?

Solution

Btu/lb of oil = 17,687 + 57.7 × 25.2 = 17,687 + 1454 = 19,141

The calculation of the heating value of fuel oil is generally not required because ASTM test methods have established the specifications for fuel oils, and these are used when fuel oil is purchased.

Some of the impurities in fuel oil that affect its application and should be determined by analysis are ash, sulfur, moisture, and sediment. These are established by ASTM test procedures.

There is a wide variation in the composition and characteristics of the fuel oil used for the generation of heat. Practically any liquid petroleum product may be used if it is feasible economically to provide the necessary equipment for its combustion.

Fuel oils may be classified according to their source as follows:

1. *Residual oils.* These are the products that remain after the more volatile hydrocarbons have been extracted. The removal of these hydrocarbons lowers the flash point and makes the oil safe for handling and burning. This residual oil is usually free from moisture and sediment except for that which is introduced by handling and in transit from refinery to consumer. These oils are high-viscosity fluids and require heating for proper handling and combustion. Fuel oils in grades 4 and 5 (see Table 4.11) are less viscous and therefore more easily handled and burned than no. 6 oil. Number 6 usually requires heating for handling and burning.

2. *Crude petroleum.* This is the material as it comes from the oil well without subsequent processing. Sometimes, because of exposure to the weather or for other reasons, the crude petroleum is too low in quality to justify refining, and it is then used as fuel under power boilers. Since it may contain some volatile gases, it must be handled with care. Present-day refinery practices make it possible to recover some high-quality product from almost all crude petroleum. Therefore, only a small quantity of this material is now available for use as fuel oil. The processing of crude oil yields a wide range of more valuable products. The average product yield of a modern refinery is

Gasoline	44.4%
Lube oil	16.4
Jet fuel	6.2
Kerosene	2.9
Distillates	22.5
Residual fuel	7.6
Total	100.0%

3. *Distillate oils.* These are obtained by fractional distillation and are of a consistency between kerosene and lubrication oils. Fuel oil produced in this manner does not contain the heavy tar residue found in others. The light grades of fuel oils are produced in this manner. These oils are typically clean and essentially free of sediment and ash and are relatively low in viscosity. These fuel oils are generally in grades 1 and 2 (see Table 4.11). Although no. 2 oil is sometimes used

TABLE 4.11   Commercial Classification of Fuel Oil

Fuel oil no.	Designation	API	Btu/gal
1	Light domestic	38–40	136,000
2	Medium domestic	34–36	138,500
3	Heavy domestic	28–32	141,000
4	Light industrial	24–26	145,000
5	Medium industrial	18–22	146,500
6	Heavy industrial	14–16	148,000

as a fuel for steam generation, it is used primarily for home heating and industrial applications where there is a very high importance on low ash and low sulfur. For steam-generation applications, these fuel oils are used as a startup or supplemental fuel to a solid-fuel combustion process.

4. *Blended oils.*   These are mixtures of two or all of the above, in proportions to meet the desired specifications.

For commercial purposes, fuel oils are divided into six classes according to their gravity (see Table 4.11). The no. 5 oil is sometimes referred to as "bunker B" and the no. 6 as "bunker C," although bunker C also frequently refers to oil heavier than 14°API.

Fuel oil is used for the generation of heat in preference to other fuels when the price warrants the expense or when there are other advantages that outweigh an unfavorable price difference. In some instances, the cleanliness, reduced stack emissions, ease of handling, small space requirements, low cost of installation, and other advantages derived from the use of fuel oil compensate for a considerable price difference when compared with solid fuels. The fact that oil can be fired automatically and that there is a minimum amount of ash to be removed makes its use desirable. The location of the plant with respect to the source of fuel supply and the consequent freight rates are determining factors in the economic choice of fuels. Many plants are equipped to burn two or more kinds of fuel in order that they may take advantage of the fluctuating costs over the life of the plant.

In burning fuel oil, it is essential that it be finely atomized to ensure mixing of the oil particles with the air. To accomplish this, it is necessary to have pumping equipment for supplying the oil and burners suitable for introducing the fuel into the furnace.

The heavy grades of fuel oil cannot be pumped or atomized properly by the burner until the viscosity has been lowered by heating. The storage tanks that contain heavy oil must be equipped with heating coils and the temperature maintained at 100 to 120°F to facilitate pumping. This oil must then be passed through another heater before it goes to the burners. With intermediate grades, the tank heater is not necessary, since the required temperature for atomization is obtained by the heater in the oil lines. The temperature to which the fuel oil should be heated before it enters the burners varies with different equipment, but the generally accepted practice is as follows:

Fuel oil no.	Temperature at burner, °F
4	135
5	185
6	220

Fuel oil provides a satisfactory source of heat for many types of services when the correct equipment is used and good operating practices are followed. The ash content is usually low, but even a small amount may react with the refractory in a furnace and cause rapid deterioration. Completely water-cooled furnace designs have solved this problem. The flame should be adjusted to prevent impingement on the furnace walls. Moisture, emulsified oil, abrasive particles, and other foreign matter, referred to collectively as *sludge,* settle to the bottom of the storage tank. When this sludge has accumulated in sufficient quantity to be picked up by the oil pump, strainer and burner nozzles become stopped up and abrasive material causes pump and burner wear.

Because of its relatively low cost, no. 6 fuel oil is the oil most often used for steam generation. Its ash content is very low when compared with coal, since it ranges between 0.01 and 0.5 percent. Seldom does it exceed 0.2 percent. Yet operating problems do exist because the fuel oil contains vanadium, sodium, and sulfur in the ash, causing deposits and corrosion in the boilers. The deposition of oil ash constituents on furnace walls and on superheater surfaces can be a serious problem, and the design and operation of the boiler must take this into account.

Sulfur is another objectionable constituent of fuel oil, and it can range from a low of 0.01 percent for no. 1 fuel oil to a high of 3.5 percent for no. 6 fuel oil. Some of the sulfur compounds formed during the combustion process raise the dew point of the flue gases, mix with the moisture present in the gases, and corrode metal parts. This corrosive action results in plugging of gas passages and rapid deterioration of boiler tubes, casings, and especially economizers and air heaters. Furthermore, the sulfur compounds discharged with the flue gases contaminate the atmosphere. The reduction of this source of contamination is the subject of much debate among air pollution control authorities.

The corrosive action is reduced by maintaining the flue gases at a temperature higher than the dew point. The gases must be permitted to discharge at a higher temperature, thus carrying away more heat and reducing the efficiency. If uncontrolled emissions from an oil-fired facility exceed established limits, then SO_2 scrubbers are required.

Burner and accessory equipment capable of operating with very low percentages of excess air provide another method of combating the problems encountered in burning fuel oil having a high sulfur content. For a fuel oil of given sulfur content, the dew point of flue gases can be lowered by operating with a low percentage of excess air. This method also increases the efficiency, and the contaminants discharged to the atmosphere are less objectionable. However, established limits must be met.

4.6   Gas

Natural gas usually occurs in the same region as petroleum. The gas, because of its low specific gravity, is found above the petroleum and trapped by a layer of nonporous rock. Since natural gas and petroleum are usually found together, it is believed that they have a similar origin. The gas in its natural state is under pressure, which causes it to be discharged from the well.

Electric production by utilities has in the past consumed relatively small amounts of natural gas. However, the use of gas turbine cycles is increasing this consumption significantly. For industrial steam generation, natural gas is used widely because of its ease in distribution, the requirement for a smaller boiler space, and its relative simplicity in overall plant design.

Methane (CH_4) is the largest component of natural gas, with the remainder primarily ethane (C_2H_6). Natural gas is the most desirable fuel for steam generation because it can be piped directly to the boiler, which eliminates the need for storage.

Natural gas is free of ash and mixes easily with air, providing complete combustion without smoke. However, because of the high content of hydrogen in natural gas as compared with coal or oil, more water vapor is produced in the combustion process. This results in a lower boiler efficiency as compared with burning coal or oil. In addition, the fuel cost and availability are directly related, and high fuel costs can result when demand is high or the supply is limited.

Manufactured substitutes for natural gas may be produced from solid or liquid fuels. These gases are produced for special industrial applications and for domestic use. Because they cost more than other fuels, they are seldom used under power boilers.

Producer gas is manufactured from coke or coal. Steam and air are blown up through the incandescent fuel bed. Both carbon dioxide and carbon monoxide are generated here and pass up through what is termed a *reduction zone*. In this zone the shortage of oxygen causes the oxygen and carbon to combine to form carbon monoxide. As the gases pass through this zone, they encounter a heavy layer of coal. The volatile gases are distilled from this coal and, together with the carbon monoxide obtained from the region below, pass from the top. The combustibles are carbon monoxide, hydrogen, and a small amount of volatile gases. The heating value of this gas is slightly above that of blast-furnace gas.

Water gas and oil gas are manufactured for industrial applications where a gaseous fuel is required and for domestic supply, sometimes being used to augment the natural gas supply. These gases are too expensive to use under power boilers unless the unit is small, and if they are used, full automatic control is necessary.

In most cases, gaseous fuels that are derived from coal have currently been replaced by natural gas and fuel oil because of their greater availability and less cost.

An analysis of gases consists of expressing the chemical compounds of which they are composed in percentage by either volume or weight. The combustible constituents are composed of various combinations of hydrogen and carbon known as *hydrocarbons*. Also, there are various inert gases such as carbon dioxide and nitrogen. The chemical analysis of a gas is useful in determining the air required for combustion and in calculating the products of combustion, the heating value, the type of burners required, etc.

The heating value of gas is determined by burning a sample in a gas calorimeter. This is an instrument in which a given quantity of gas is burned and the resulting heat transferred to a measured quantity of water. Unlike a solid- or liquid-fuel calorimeter, the gas calorimeter utilizes a constant measured flow of gas and water. The heat generated by the combustion of gas is absorbed by the cooling water. The temperature rise of the water is an index of the heat produced and, by use of suitable constants for a given calorimeter, may be converted into Btu/ft^3 of gas.

Gas used for fuel is classified according to its source or method of manufacture. Table 4.12 gives typical analyses of natural and more generally used manufactured gases.

In order for gas to be burned efficiently, it must be mixed with the correct proportion of air. Since the fuel is already in the gaseous state, it is unnecessary to break it up, as is the case with solid and liquid fuels. Correct proportioning and mixing are, however, essential. This process is accomplished by the burner and by admission of secondary air. Combustion must be complete before the mixture of gases reaches the boiler heating surfaces.

TABLE 4.12   Analysis of Fuel Gases (by Volume)

Gas	Natural	Manufactured		
		Producer	Coke oven	Blast furnace
CH_4	83.5	2.5	29.1	—
C_2H_6	14.4	—	—	—
C_2H_4	—	0.5	3.1	—
CO_2	—	5.5	1.7	12.7
CO	—	22.3	4.9	26.7
N	2.1	59.0	4.4	57.1
H	—	10.2	56.8	3.5
Btu/ft^3 at 60°F and atmospheric pressure	1080	140	550	98.1

With optimal conditions, it is possible to obtain complete combustion of gas with a very low percentage of excess air. These conditions include furnace and burner design for correct mixing of fuel and air, sensitive controls, and scheduled maintenance and calibration of all equipment.

Gas is a desirable fuel because it is easy to control, requires no handling equipment, and leaves no ash to remove from the furnace.

4.7   By-product Fuels

Industrial processes frequently produce materials that may be burned in industrial furnaces. The utilization of these by-products not only reduces the cost of fuel and thus lowers operating costs but also frequently solves a disposal problem at the same time. However, the economic advantage to be derived from the use of these fuels must be evaluated in terms of added investment for equipment and of increased operating costs. By-product fuel is often not available in quantities sufficient to meet the total steam demand. In these situations it is necessary to provide for a supplementary fuel.

Wood is available as a by-product fuel in sawmills, paper mills, and factories that manufacture articles made of wood. The sawmill waste products consist of slabs, edgings, trimmings, bark, sawdust, and shavings in varying percentages. This material is passed through a hogging machine, which reduces it to shavings and chips in preparation for burning. A hogging machine consists of knives or hammers mounted on an element that rotates within a rigid casing. The knives shred the wood waste and force it through a series of spacer bars mounted in a section of the casing. Dry wood waste has a heating value of 8000 to 9000 Btu/lb. This shredded material (hog fuel) contains approximately 50 percent moisture and, therefore, has a low heating value. Woodworking plants that manufacture furniture and similar articles use seasoned wood. The moisture content of the waste material is from 20 to 25 percent. This waste material consists of sawdust and shavings and only a small amount of trimmings. Precautions are necessary in its handling and storing to prevent fires and explosions.

Bark is removed from logs before they are ground into pulp in paper mills. This bark may contain 80 percent moisture, a part of which is often pressed out before it can be utilized for fuel. Even these moisture contents are often above 65 percent, and a large part of the heat from combustion is required to evaporate the inherent moisture.

A by-product solution (black liquor) containing wood fibers and residual chemicals is produced in some paper mills. The wood fibers are burned, and the chemicals are recovered as ash in the furnace. In this operation a savings is realized both by the reclamation of heat and by the recovery of the chemicals used in the pulping process. The black liquor is concentrated before being introduced into the furnace, where it is atomized and burned in a manner similar to fuel oil.

Bagasse is a by-product material produced when the juice is removed from the cane in a sugar mill. The moisture content is 40 to 55 percent, and the heating value, 4000 to 5200 Btu/lb. The utilization of this by-product material is an important factor in the economic operation of a sugar mill. Special furnaces have been developed to provide a means of utilizing the heat from the furnace to evaporate some of the moisture before attempting to burn the fuel. Bagasse is frequently supplemented with auxiliary fuels.

The disposal of municipal solid waste (MSW) has become a worldwide problem and requires special attention. The methods available for its disposal including recycling and waste-to-energy facilities are described in Chap. 13.

Coke-oven gas is the volatile material removed when coal is heated to produce coke. These volatile gases are collected and burned as a by-product of the coke production. Coke-oven gas has a heating value of 460 to 650 Btu/ft^3. This gas burns very well because of its high concentration of hydrogen, and it presents few problems when used as a fuel to produce steam.

Blast-furnace gas is produced as a by-product of steel mill furnaces. The air is blown up through the hot fuel bed of the blast furnace, where it combines with carbon to form carbon monoxide. The resulting gases contain carbon monoxide and small quantities of hydrogen as combustibles in addition to inert gases consisting of nitrogen and carbon dioxide. The heating value varies from 90 to 110 Btu/ft^3.

Because the production of this gas is irregular and the steam demand does not coincide with the gas production, gas storage is necessary and supplementary fuel is required. When used, the deposits from the combustion of blast-furnace gas adhere readily to boiler heating surfaces, and provisions for sootblowers must be made for cleaning.

Refinery gas is a by-product of crude oil refining operations. The heating value of this gas is high because of the large amount of heavy hydrocarbons present and will vary from 1000 to 1800 Btu/ft^3 depending on the refining process. The gas produced is frequently adequate to generate the steam required by the refinery. Refineries also produce petroleum coke, which has a high carbon but low hydrogen and ash content. This fuel is used by the refinery or sold for use in neighboring plants.

Although not truly defined as a by-product fuel similar to those described previously, various sludges are produced in industrial processes, and these can be used in a boiler as a fuel to produce steam. In the papermaking process, a number of sludges are generated. The now common recycling of old newspapers, magazines, and corrugated paper containers has increased the quantity of sludge because of the deinking process. This has created a disposal problem that generally has been solved by depositing the sludge in a landfill.

Landfills, however, are becoming filled, as well as costly (see Chap. 13), and new landfills are becoming more difficult to locate because of permitting and environmental problems. Therefore, the use of sludge as a fuel has become a viable alternative. The available combustion technologies all have certain limitations for the burning of sludge, and the characteristics of the sludge must be carefully evaluated for such things as quantity, heating value, and moisture content in order to select the proper technology.

Typical sludges from pulp mills are very good fuels because they have heating values between 8000 and 10,000 Btu/lb (dry) and ash contents of approximately 10 percent. However, sludges from the deinking process are less suitable fuels, having heating values between 4000 and 6000 Btu/lb (dry) and ash contents from 20 to 50 percent. Therefore, for a deinking sludge having a moisture content of 50 to 60 percent, the heating value is only about 2800 Btu/lb. The combination of high moisture content and low heating value makes it difficult to burn.

The burning of this high-moisture sludge on a stoker presents some significant problems because the removal of the moisture in the fuel results in a lower furnace temperature, which, in turn, affects combustion. High air temperatures are limited on both a traveling-grate stoker and a water-cooled hydrograte stoker because of limitations on stoker grate materials. Therefore, on stokers, sludge is generally fired in combination with another fuel such as bark.

This has led to the successful use of fluidized bed combustion technology as a viable solution and in particular the bubbling bed technology (see Sec. 2.15). Not only can the bubbling bed boiler handle this high-moisture, low-heating value fuel, it also can burn sludge alone, which eliminates the need for high-cost dewatering equipment.

Sludge is also a product of treating sewage. Rather than disposal into a landfill, this high-moisture product also can be burned in a boiler using the fluidized bed boiler technology. As with sludge from the pulping process, not only does this save on fuel costs, but it also lowers disposal costs, and reduced landfill requirements result, which provide significant benefits to the overall process.

4.8 Control of the Combustion Process

Each unit of fuel contains a given amount of heat in the form of chemical energy. The amount of this energy is readily determined in the laboratory and is expressed in Btus per unit (i.e., Btu/lb or Btu/ft^3). The burning of fuel and the subsequent operation of the boiler make it necessary to convert the chemical energy into heat, apply it to the water in the boiler, and thereby generate steam.

Fuel supplied to the furnace of a power boiler may be rejected unburned, lost as heat, or absorbed by the boiler. The effectiveness of the process can be determined by noting the losses. When the losses have been reduced to a practical minimum, the highest efficiency is being maintained.

When solid fuel is burned, a certain portion of the combustible carbon becomes mixed with the ash and is removed without being burned. It is called *unburned combustible,* and its efficiency loss is called *unburned carbon loss.* The mixture of carbon and ash discharged from a furnace is referred to as *residue.* This term has now replaced *refuse* as the term for solid waste products resulting from combustion. (Today, *refuse* most commonly refers to municipal solid waste, as discussed in Chap. 13.) The magnitude of this unburned carbon loss is determined by noting the reduction in weight when a sample of residue is burned completely in a laboratory muffle furnace.

There are two possible sources of unburned carbon loss. The first, and most important, is the one due to the carbon in a solid fuel being trapped in the ash found in the ash hoppers and in the collected fly ash. This loss varies with different coals because of the differences in the volatile matter and ash content of coals. It also can vary as a result of the design of the burning equipment and the furnace. To have optimal efficiency, the designer must minimize this unburned carbon loss.

The second source of unburned carbon loss is found in the incomplete combustion of the fuel, as determined by the presence of CO in the flue gas. In well-designed boilers, this loss is negligible, and it should be assumed as zero. In design calculations, liquid and gaseous fuels generally are presumed to burn without any unburned carbon loss.

When high carbon loss is indicated by a boiler efficiency test, a few areas can be concentrated on to correct the problem. High levels of unburned combustibles in the ash indicate that adjustments must be made in the fuel preparation and burning equipment. For example, improperly sized fuel particles on a pulverized-coal-fired unit or improperly atomized oil in an oil-fired unit will result in high carbon loss. Improperly adjusted fuel-air mixing equipment will result in high carbon loss, and improper fuel-air ratios also will contribute to unburned carbon loss.

By the continuous monitoring of flue gas temperatures and flue gas constituents, and by periodic analysis of ash samples, both bottom and fly ash, an operator can determine whether the expected boiler efficiency is being met. For a modern pulverized-coal-fired boiler, unburned carbon loss (UCL) is approximately 0.5 percent.

In order to illustrate the importance of minimizing unburned carbon loss, assume a coal-fired boiler produces 250,000 lb/h of steam at 450 psig and 650°F when supplied with feedwater at 220°F. Assuming an efficiency of 86.9 percent, the boiler input is 328.6×10^6 Btu/h. Assume the coal heating value is 13,000 Btu/lb.

$$\text{Additional fuel required with 0.5 percent UCL} = \frac{328.6 \times 10^6 \times 0.005}{13,000}$$

$$= 126 \text{ lb/h or } 3033 \text{ lb/day}$$
$$\text{or } 1.5 \text{ tons/day}$$

If the UCL were to increase to 2.0 percent, then the additional fuel would increase to 6 tons/day, with significantly increased fuel costs. A utility boiler with, say, 10 times the steam output and boiler input requirements would increase its coal input proportionately.

It has been shown that water is formed during the combustion of hydrogen. In the conventional boiler, the gases are discharged to the stack at a temperature in excess of the boiling point of water (212°F), and the water therefore leaves in the form of vapor or steam. A part of the heat produced by the combustion of the fuel is utilized in vaporizing the water and is therefore lost with the stack gases. This heat loss is high when fuels (such as natural gas) containing large amounts of hydrogen are burned. A similar loss occurs in the evaporation of moisture contained in solid fuels. The operator cannot do much to reduce the loss except to have the fuel as dry as possible.

By far the largest controllable or preventable loss occurs in the heat rejected from the stack in the dry products of combustion. These losses occur in three different ways as follows:

1. Combustible gases, mainly carbon monoxide, contain undeveloped heat. This loss can be determined by the following formula:

$$h_2 = C \times \frac{10,160 \; CO}{CO_2 + CO}$$

where h_2 = heat loss (Btu/lb of fuel burned as a result of carbon monoxide in flue gases)
  CO = percentage by volume of carbon monoxide in flue gases
  CO_2 = percentage by volume of carbon dioxide in the flue gases
  C = weight of carbon consumed per pound of fuel

2. The temperature of gases discharged from the boiler is higher than that of the incoming air and fuel. The heat required to produce this increased temperature is lost. The magnitude of this loss can be decreased by maintaining the lowest possible boiler exit gas temperature consistent with economics and with the potential for corrosion.

3. The amount of gases per unit of fuel discharged from the boiler depends on the percentage of excess air used. The more gas discharged, the greater will be the heat loss. In the interest of economy, the excess air should be kept as low as practicable. The relation of the carbon dioxide content of flue gases to heat loss for several flue-gas temperatures is given in Fig. 4.11.

The heat carried to waste by the dry gases may be calculated by using the following formula:

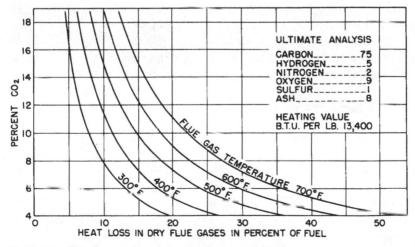

Figure 4.11 Dry flue gas loss for various flue gas temperatures and percentages of carbon dioxide.

$$h_3 = w(t_s - t)0.24$$

where h_3 = heat loss (Btu/lb of fuel)
$\quad w$ = weight of dry gases (lb per pound of fuel)
$\quad t_s$ = temperature of boiler exit gases (°F)
$\quad t$ = temperature of air entering furnace (°F)
$\quad 0.24$ = specific heat of flue gases

This heat loss therefore can be reduced by lowering the boiler exit gas temperature or by decreasing the quantity of gases discharged, thus lowering the percentage of excess air used. (Note that it is important to use the boiler exit gas temperature. With many units requiring flue gas scrubbers to reduce the acid gases, the stack temperatures will be lower.)

Conventional oil and gas burners controlled by standard systems require 3 to 15 percent excess air for complete combustion. However, *low-excess-air* burners and accessory control equipment are available for reducing the air necessary for combustion to nearly the theoretical requirement. Advanced methods of atomizing oil are used. The temperature of the oil supplied to the burners is carefully controlled. Both the fuel and air supplies to the burners are metered. This low-excess-air operation can be obtained with a clear stack to meet stringent air pollution ordinances and with not more than a trace of carbon monoxide.

The necessary equipment for low-excess-air operation is more costly to install and requires more precise adjustment and care in operation than standard equipment. The reduction in excess air results in an improvement in efficiency, a reduction in the corrosive effect of the products of combustions on the heating surfaces, and a reduction in objectionable contaminants (including nitrogen oxides).

The measure of stack losses is based on analysis of the products of combustion and a determination of the exit temperature of the gases. A commercial portable gas analyzer is shown in Fig. 4.12. This Orsat gas analyzer has been used for a long time, and it remains a trusted standard for the verification of sophisticated electronic equipment.

The Orsat uses chemicals to absorb the CO_2, SO_2, CO, and O_2, and the amount of each is determined by the reduction in volume from the original flue gas sample. A sample of gas is obtained by inserting a sampling tube into the flue gas stream in the boiler, economizer, or air heater, depending on the information required. A $\frac{1}{4}$-in standard pipe is satisfactory if the temperature of the gases in the pass does not exceed 1000°F. Care must be exercised to locate the end of the sampling pipe to obtain a representative sample of gases. It is advisable to take samples with the pipe located at various places in the pass to make certain that the analysis is representative.

Figure 4.12   Portable Orsat gas
analyzer.

The procedure for operating the gas analyzer in Fig. 4.12 is shown
diagrammatically by steps in Fig. 4.13.

Step 1.   The inlet tube is carefully connected to the sampling tube
to prevent air leakage from diluting the sample. The gas is then
drawn into the measuring burette by means of the rubber bulb. The
excess is expelled through the level bottle. The aspirator bulb is
operated long enough to ensure a fresh sample of gas in the burette.

Step 2.   The handle of the three-way cock is then moved to a vertical position, closing off the sampling line and opening the burette to
the atmosphere. The sample is measured by bringing the water
level in the measuring chamber exactly to zero. The excess gas is
discharged through the three-way valve to the atmosphere, and the
amount remaining is measured at atmospheric pressure to ensure
accuracy.

Step 3.   The handle of the three-way cock is next moved away from
the operator to a horizontal position, entrapping the sample of gas
in the measuring chamber. The valve above the absorption chamber
is now opened and the leveling bottle raised to force the gas sample
into contact with the chemical that absorbs the carbon dioxide.
Care must be exercised in this operation to prevent the chemical
and water from becoming mixed.

It is necessary to raise and lower the leveling bottle and expose
the gas to the chemical from two to four times to ensure complete
absorption of the carbon dioxide.

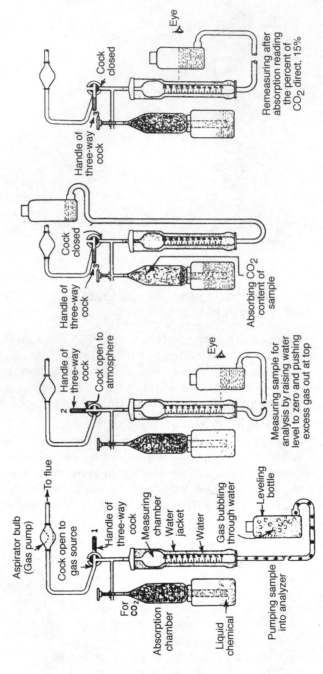

Figure 4.13 Operation of an Orsat gas analyzer.

Aspirator bulb (Gas pump)

To flue

Cock open to gas source

Handle of three-way cock

1

For CO_2

Measuring chamber

Water jacket

Water

Gas bubbling through water

Leveling bottle

Absorption chamber

Liquid chemical

Pumping sample into analyzer

Handle of three-way cock

Cock open to atmosphere

2

Eye

Measuring sample for analysis by raising water level to zero and pushing excess gas out at top

Cock closed

Handle of three-way cock

3

Absorbing CO_2 content of sample

Cock closed

Handle of three-way cock

3

Eye

Remeasuring after absorption reading the percent of CO_2 direct, 15%

Step 4. The remainder of the gas sample is drawn back into the measuring chamber, and when the chemical level in the absorption chamber has been restored to zero, the valve closes. The leveling bottle is held with the water at the same level as that in the measuring chamber. The percentage of carbon dioxide in the sample is then read directly on the etched wall of the measuring chamber.

To obtain the percentage of oxygen and carbon monoxide, respectively, the same sample of gas is exposed to the chemicals in the second and third absorption chambers of the gas analyzer. See Fig. 4.13 for the step-by-step procedure.

To determine the percentage of excess air, as explained in Sec. 4.2, and for heat-balance calculations, it is necessary to make a complete flue gas analysis. For a routine check on operation, when there is no change in the character of the fuel, carbon dioxide determination is sufficient for comparison with established satisfactory readings. The carbon monoxide absorption chamber has been omitted from the gas analyzer shown in Fig. 4.12. The solutions have a limited gas absorption ability and must be replaced after continued use or long periods of storage. They may be purchased mixed ready for use or compounded using standard chemicals.

There are disadvantages with the use of an Orsat. It lacks the accuracy of more refined devices, it requires an experienced operator, and the results do not lend themselves to electronic recording unless input separately by hand. As stated before, however, it is used often as a trusted verification of electronic equipment or when specialized equipment is not available at a particular plant.

The use of automatic recording equipment for such areas as steam and airflow is recommended, as well as automatic analyzers to record the carbon dioxide, oxygen, sulfur oxides, and other substances that are present in the flue gases. They are valuable aids to the operator in maintaining the optimal combustion conditions at all times and ensuring compliance with any permit requirements. Not only does this equipment provide continuous emission monitoring, but also the data can be recorded for immediate access and evaluation.

Continuous emission monitoring (CEM) and associated reporting systems are absolutely essential to the compliance tracking of many power plants. New pollution regulations have forced plants to evaluate their use of various fuels to produce electricity.

A CEM analyzer can continuously measure up to six gases simultaneously such as HCl, NO, NO_2, N_2O, NH_3, SO_2, HC, CO, CO_2, etc., as well as opacity, dust loading, and flue gas flow. Applications for operating permits include environmental compliance monitoring plus combustion and process control for industrial and utility boilers.

Questions and Problems

4.1 Describe the four major conditions to ensure good combustion.

4.2 What are the typical gases that are released from combustion of a solid fuel?

4.3 Describe the purpose of primary and secondary air.

4.4 Discuss the causes of smoke.

4.5 What are the three T's of combustion, and describe their importance?

4.6 What are the forms of matter? What is condensation?

4.7 What is absolute pressure, and how is it determined?

4.8 What is the absolute temperature on the Fahrenheit scale?

4.9 State the three principal laws that govern the behavior of gases.

4.10 The ultimate analysis of coal is as follows: carbon, 60 percent; H_2, 7 percent; N_2, 3 percent; O_2, 12 percent; S, 5 percent; and ash, 13 percent. How many pounds of O_2 would be required to burn this fuel? How many pounds of air (theoretically) would be required?

4.11 What is incomplete combustion? What is the approximate heat loss that results from incomplete combustion?

4.12 Why is sulfur in a fuel considered undesirable?

4.13 Define *perfect, complete,* and *incomplete combustion.*

4.14 What is meant by *excess air?*

4.15 If the flue gas analysis shows CO_2, 15.2 percent, O_2, 4.7 percent, and N_2, 80.1 percent, what is the percentage of excess air?

4.16 What two methods are used to analyze flue gas and the results used to calculate the percentage of excess air?

4.17 How many pounds of air would be required to burn a pound of coal with the analysis given in Problem 4.10 if 50 percent excess air were used?

4.18 Using the analysis in Problem 4.10 and Dulong's formula, determine the heating value of the fuel.

4.19 Excess air is important to ensure complete combustion. Why, then, should it be minimized?

4.20 In a boiler design, what is *draft,* and what is its purpose?

4.21 What is *draft loss?* In what units is it normally defined?

4.22 Define boilers that are balanced draft as compared with pressurized. Provide the advantages and disadvantages of each.

4.23 Define and describe the various classifications of coal.

4.24 Describe the proximate and ultimate analyses of coal.

4.25 Name the three ways that can express coal analyses.

4.26 Why are the fusion characteristics of ash so important to a boiler design?

4.27 What are *soft* and *hard* coals? What types of coal fall into these categories?

4.28 Discuss the characteristics to be considered when buying coal. Is coal size a consideration? Why?

4.29 Define the *grindability* of coal. Why is this important?

4.30 A fuel oil has a specific gravity of 0.92. What is its specific gravity on the API scale?

4.31 Define the various classifications of fuel oils. Provide examples of those oils which are used as a fuel oil.

4.32 What are important characteristics to be considered when selecting fuel oil for a particular plant?

4.33 When must fuel oils be heated?

4.34 What fuel oil is used most often for the production of steam? Why?

4.35 Why is natural gas the most desirable gas for steam generation? What are the disadvantages for its use?

4.36 Name some of the major by-product fuels that are burned in boilers. Provide the major advantages for the burning of these fuels.

4.37 Sludges are a common by-product. What combustion technology is often used to burn this fuel, and what advantages does this provide?

4.38   What is *unburned carbon loss?* Why should it be minimized?

4.39   What is the purpose of an Orsat gas analyzer? Name its advantages and disadvantages.

4.40   What is a *continuous-emission monitoring (CEM) system?*

5

Boiler Settings, Combustion Systems, and Auxiliary Equipment

In order to apply the principle of combustion, it is necessary to design, construct, and operate equipment. A wide variety of equipment is required to burn solid, liquid, and gaseous fuels satisfactorily. The functions that must be performed are supplying and mixing fuel and air, confining the high temperature during combustion, directing the heat to the heat-absorbing surfaces of the boiler and heat-recovery equipment, and discharging the ash and products of combustion from the unit at an emission rate within the limits imposed by environmental requirements.

5.1 Boiler Settings

As defined by today's modern boiler designs, the boiler setting includes all the water-cooled walls, casing, insulation, lagging, and reinforcement steel that form the outside envelope of the boiler and furnace enclosure and which safely contains the high-temperature flue gases that result from the combustion process. Fire-tube boilers have been supported by the brickwork setting or independently suspended from overhead steel beams. Figure 2.5 shows an older design of a horizontal-return tubular boiler supported by brickwork. Note that the boiler rests on rollers so that it can expand independently of the setting. The overhead suspension was considered best because it relieved the brickwork of extra load, kept the boiler at the proper level, and permitted repairs to the brickwork without disturbing the boiler.

Water-tube boilers may be either bottom or top supported. On a two-drum bottom-supported boiler, the lower drum and tube sections are held rigid by the structural steel. The steam drum and tube sections are free to expand upward. For top-supported boilers, the steam drum and tube sections are held in place by structural steel. The lower drum and tube are free to expand downward without interfering with the setting and combustion equipment such as a stoker. Seals are required to permit this movement and at the same time prevent air leakage into the furnace.

At least 50 percent of the combustion of soft coal (i.e., those coals having high volatile matter) consists of the burning of gases distilled from the coal, and it takes place in the space above the fuel bed. It is therefore necessary that the furnace space above the grates be large enough to ensure complete combustion before the gases reach the heating surface. The combustible gases rising from the fuel bed are composed of hydrocarbons and carbon monoxide. Air must be admitted above the fuel bed to burn these gases, and this is called *secondary air.* If the flue gas reaches the heating surface before combustion has been completed, it becomes cooled. The result is smoke, soot, poor efficiency, and higher fuel costs. The setting must provide not only the proper furnace volume but also bridge walls, baffles, arches, etc., for properly directing the flow of gases with respect to the heating surfaces.

The *bridge wall* is located behind the grates. It keeps the fuel bed in place on the grates and deflects the gases against the heating surface.

Baffles direct the flow of gases and ensure maximum contact with the heating surface.

Arches are in reality roofs over parts of the furnace. They are made of refractory material or water-cooled walls and are used to direct the flue gas and to protect parts of the boiler from the direct heat of the furnace.

In the *flush-front-type* HRT boiler setting, the front tube sheet sits back from the front of the boiler casing (Fig. 2.5). The gases leave the tubes and flow through this space to the breeching and stack. A small arch is required at the front of the furnace to seal off this space and prevent the furnace gases from short-circuiting the boiler and going directly from the furnace to the stack.

Other horizontal-return tubular boilers (Fig. 2.6), known as the *extended-front type,* have their front tube sheets set in line with the front of the boiler setting. The bottom half of the shell extends beyond the tube sheet and forms part of the breeching or flue. This part of the shell that extends past the tube sheet has hot gases on one side but no water on the other and is called the *dry sheet.*

In some older designs, the combustion space was enclosed within the boiler and no setting was required. The combustion space was surrounded by two sheets of boilerplate, held together by means of stay bolts, which were bolts threaded through or welded at each end into two spaced sheets of a firebox for the purpose of supporting flat surfaces against internal pressure. The space between the plates was filled with water and was part of the boiler. Figures 2.3 and 2.4 show a vertical boiler with the firebox enclosed within the boiler. This self-contained feature made the unit semiportable.

In early designs, boilers were set very close to the grates, and the combustion space was limited. The result was smoke and poor efficiency. As more was learned about the combustion of fuels, furnaces were made larger by setting the boilers higher above the grates. A hand-fired furnace with large combustion space would burn a wide variety of coal and operate over a wider capacity range at higher efficiencies than one with a small combustion space.

Large-capacity stokers operate at high burning rates and heat release. This high-capacity operation requires furnaces having large volume to ensure complete combustion before the gases reach the convection heating surfaces and to prevent excessive ash deposits from forming on the lower boiler tubes. Because of the high rates of combustion on current designs, water-cooled furnace walls are required rather than refractory furnace wall construction, which results in effective heat transfer and higher-capacity units.

Arches are installed in furnaces to protect parts of the boiler from the direct heat of the furnace and to deflect the flow of gases in the desired path. Ignition arches provide surfaces for deflecting the heat of the furnace to ignite the incoming coal and other solid fuels such as refuse-derived fuel (RDF). Advanced furnace design has reduced the requirement for arches. However, some spreader stokers and chain-grate stokers use ignition arches in connection with over-fire air jets to assist in igniting the coal and to improve the efficiency of combustion. These arches usually are constructed of boiler tubes with refractory covering on the furnace side. The refractory surface permits the desired high surface temperature required to ignite the coal, while the water-cooled construction results in low maintenance.

In hand-fired furnaces and those equipped with underfeed or chain-grate stokers, a considerable portion of the coal is burned on the grates. However, the proportion depends on the percentage of volatiles in the coal. When coal is introduced into a high-temperature furnace, the volatiles are separated to form a gas containing hydrocarbons that must be burned in suspension. Coals that contain a low percentage of volatile matter burn on the grates. The furnace volume therefore can be less than when low-volatile coals are used, but a larger grate area must be provided. When fuels are burned in suspension, furnace volumes must be larger than when a portion of the fuel is burned on the grates.

Pulverized coal, oil, and gas are examples of fuels burned in suspension. The spreader stoker uses a combination of suspension and on-grate burning. Fine coal particles and high volatiles increase the percentage burned in suspension, while large particles and low volatiles reduce the amount of fuel burned in suspension and therefore increase the amount that must be burned on the grates.

Furnaces may be designed to maintain the ash below or above the ash-fusion temperature. When the ash is below the fusion temperature, it is removed in dry or granular form. When low-fusion-ash coal is burned, it is difficult to maintain a furnace temperature low enough to remove the ash in the granular state. Cyclone and some pulverized-coal-fired furnaces are operated at temperatures high enough to maintain ash in a liquid state until it is discharged from the furnace. These units are referred to as *intermittent* or *continuous slag-tapped furnaces* depending on the procedure used in removing the fluid ash from the furnace. Since this process permits high furnace temperatures, the excess air can be reduced and the efficiency increased.

High rates of combustion, suspension burning, large furnace volume, and liquid-state ash discharge are all factors that increase the severity of service on the furnace lining. Refractory linings were adversely affected by high temperature and the chemical action of the ash. Therefore, walls and arches are constructed of water-cooled tubes that form a part of the boiler. A large percentage of the heat required to generate the steam is absorbed in these tubes, thus reducing the amount of heating surface required in the convection banks of the boiler.

5.2   Hand Firing

Although hand firing is seldom used, it is included here to explain a simple application of combustion fundamentals. The grates (Fig. 5.1) serve the twofold purpose of supporting the fuel bed and admitting primary air. The front end of the grate bars is supported from the dead plate and the rear end by the bridge wall. The grate bars are not set level but are inclined toward the bridge wall. This is done to aid the operator in getting the fuel to the rear of the furnace and to carry a slightly heavier fire near the bridge wall, where the effect of the draft is greatest.

Hand-fired grates are made of cast iron. They vary in design to suit the type of fuel to be burned (Fig. 5.1). The air openings in the grates vary from $\frac{1}{8}$ to $\frac{1}{2}$ in in width. Larger air openings can be used for

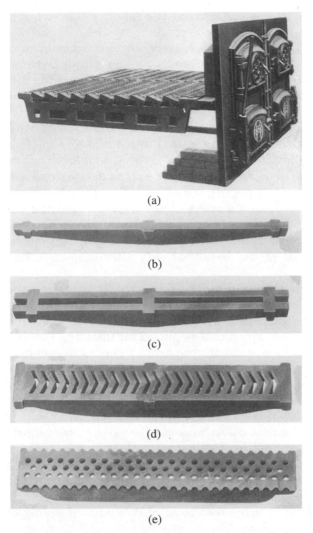

(a)

(b)

(c)

(d)

(e)

Figure 5.1 Hand-fired grates and furnace parts. (*a*) Shaking grate. (*b*) No. 9 grate, 1 in wide. (*c*) No. 8 grate, 3 in wide. (*d*) Tupper grate for coal, 6 in wide. (*e*) Sawdust grate, 6 in wide. (*ABB Combustion Engineering Systems.*)

burning bituminous than for burning anthracite coal. The bridging action of bituminous coal prevents it from sifting through the grates. Grates designed to burn bituminous coal have air openings equal to 40 percent of the total grate area, while those intended for anthracite have air openings equal to about 20 percent.

The width of the boiler limits the width of the furnace. The length of the furnace is limited to 6 or 7 ft by the physical ability of the operator to handle the fire. The actual grate area required for a given installation depends on its heating surface, the capacity at which it is to be operated, and the kind of fuel to be used.

Shaking grates, as shown in Fig. 5.1, were an improvement over stationary ones. The ash was discharged into the ashpit without greatly disturbing the operation of the boiler. The agitation of the fuel bed by the grates helped to keep it even and to prevent holes in the fuel bed.

To start a fire on hand-fired grates, the grates are covered with coarse coal and then with fine coal to a total depth of 3 to 4 in. This coal is covered with a layer of wood and shavings, and oily rags or waste are placed on the wood and then lit. The stack damper and ash pit door are partly opened to permit the air to circulate through the fuel. By this procedure, the coal starts to burn from the top, and the volatile gases that are distilled off must pass through the hot fuel bed. They ignite and burn, thus reducing to a minimum the amount of smoke produced. Coal should be added only after the entire fuel bed has been ignited.

The *coking method* of maintaining the fire consists of placing the fresh coal on the dead plate and allowing it to be exposed to the heat of the furnace. (A *dead plate* is a plate adjacent to a grate through which no air passes and on which coal is placed for distilling the coal volatiles before the coal is moved onto the grate.) The volatile matter is distilled off, mixed with the secondary air entering the furnace through the door, and burned. After coking in this manner, the coal is pushed back over the active area of the grates. This method is satisfactory if the coal is not of the caking kind and if sufficient coking time is permitted. If the coal cakes and runs together, the volatile matter in the interior will not be distilled off, and excessive smoking will result when the coal is pushed over the hot fuel bed. The coking process is slow because only one side of the coal pile is exposed to the heat of the furnace.

An alternative method of firing was to have coal applied evenly to one side of the furnace at a time. The other side was not disturbed, so the heat generated by it helped to burn the coal. However, just after the coal had been applied, the volatile gases often were distilled off so rapidly that they could not be burned properly. If this occurred, smoke was produced as a result of incomplete combustion.

Another method was to *spread* the coal over the fuel bed evenly or in such a manner that holes were eliminated. The coal came into direct contact with the hot fuel bed, and the volatile gases were quickly distilled off. If smoke was to be prevented, the coal had to be fired in relatively small quantities. Care had to be exercised to admit sufficient air over the fire to burn these volatile gases. Any tendency of coal to clinker was likely to be increased by this method of firing because the ash was exposed to the direct heat of the burning fuel. Caking coals would form a surface crust and cause unequal distribution of air. This was especially true if too much coal was added at one time.

Hand firing has been almost completely replaced by mechanical devices for introducing the fuel into the furnace. These include underfeed and overfeed stokers and pulverized-coal and fluidized bed combustors.

5.3 Stokers

All stokers are designed to feed solid fuel onto a grate where the fuel burns as primary air is passed up through it and over-fire air is also introduced to enhance combustion. The stoker is located in the furnace, and it is also designed to remove the residue of ash that remains after combustion.

Stoker firing consists of various systems having the following major features:

1. A fuel feed system

2. A grate assembly, either stationary or moving, that supports the burning fuel and admits the majority of combustion air to the fuel

3. An over-fire air system for the completion of combustion and reduction of emissions such as NO_x

4. An ash discharge system

Stokers can handle a vast variety of solid fuels. They handle all forms of coal, wood, bark, bagasse, rice hulls, peach pits, vineyard trimmings, coffee grounds, and municipal solid wastes (MSW). There are two general types of stoker systems:

1. *Underfeed stokers:* Both fuel and air supply are from under the grate

2. *Overfeed stokers:* The fuel is supplied from above the grate and air is supplied from below the grate.

For overfeed stokers, there are two types:

1. *Mass feed stoker.* Fuel is fed continuously to one end of the stoker grate, and it travels horizontally (or sometimes inclined for fuels such as MSW) across the grate as it burns. Ash is removed from the opposite end. Combustion air is introduced from below the grate and moves up through the burning bed of fuel. Figures 5.9 and 5.11 are examples of this type of stoker.

2. *Spreader stoker.* Fuel is spread uniformly over the grate area as it is thrown into the furnace. The combustion air enters from below through the grate and the fuel on the grate. Since the fuel has some fines, these burn in suspension as they fall against the airflow moving upward through the grate. The heavier fuel burns on the grate and the ash is removed from the discharge end of the grate. Figures 2.28 and 5.16 are examples of this system.

In today's market, as a result of high cost and the difficulty in meeting environmental requirements, there is little demand for underfeed and small overfeed coal-fired stoker units. The steam demand of these size boilers is now being met with package boilers that burn oil and natural gas. However, there are stoker units of this type that continue to operate and meet their operating requirements.

The burning rate on a stoker depends on the type of fuel, and this rate varies based on the stoker system being used. The resulting steam capacity of the boiler depends on this stoker heat release. Table 5.1 lists typical heat releases and the approximate steam capacity for various stoker types.

TABLE 5.1   Heat Releases and Steam Capacities for Various Stokers

Stoker type	Fuel	Heat release, $Btu/ft^2/h$	Steam capacity, lb/h
Underfeed			
Single retort	Coal	425,000	25,000
Double retort	Coal	425,000	30,000
Multiple retort	Coal	600,000	500,000
Overfeed			
Mass			
Vibrating, water-cooled	Coal	400,000	125,000
Traveling chain	Coal	500,000	80,000
Reciprocating	Refuse	300,000	350,000
Spreader			
Vibrating			
Air-cooled	Coal	650,000	150,000
Air-cooled	Wood	1,100,000	700,000
Water-cooled	Wood	1,100,000	700,000
Traveling	Coal	750,000	390,000
	Wood	1,100,000	550,000
	RDF*	750,000	400,000

*Refuse-derived fuel.

5.3.1   Underfeed stoker

As the name implies, underfeed stokers force the coal up underneath the burning fuel bed. Small boilers (approximately 30,000 lb of steam per hour) are supplied with single- or, in some cases, twin-retort stokers. Larger boilers (up to 500,000 lb of steam per hour) are equipped with multiple-retort stokers. It is these smaller underfeed stokers that have a limited market today as new installations.

The front exterior of a single-retort underfeed stoker is shown in Fig. 5.2. The motor drives a reduction gear, which in turn operates the main or feed ram, secondary ram or pusher plates, and grate movement. The two horizontal cylinders located on the sides of the stoker front are actuated by either steam or compressed air to operate the dumping grates. The doors above the dump-grate cylinders are for observing the fuel bed. The doors below these cylinders are for removing the ash after it has been dumped into the ashpits.

Figure 5.3 shows a longitudinal sectional view of this stoker with the various external and internal components. Figure 5.4 is a cross section of the stoker showing the retort, grates, dumping grates, and ashpits. The illustration also shows how the grates are moved up and down to break up the coke formation and to promote air distribution through the burning fuel bed.

The feed ram forces the coal from the hopper into the retort. During normal operation, the retort contains coal that is continually pushed out over the air-admitting grates, which are called *tuyères*. The heat

Figure 5.2   Front exterior of a single-retort underfeed stoker. (*Detroit Stoker Company, subsidiary of United Industrial Corp.*)

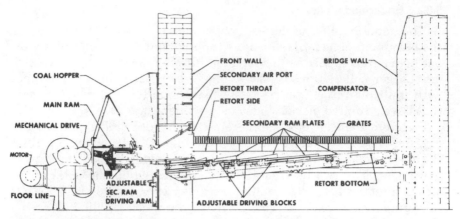

Figure 5.3 Side view of a single-retort underfeed stoker. (*Detroit Stoker Company, subsidiary of United Industrial Corp.*)

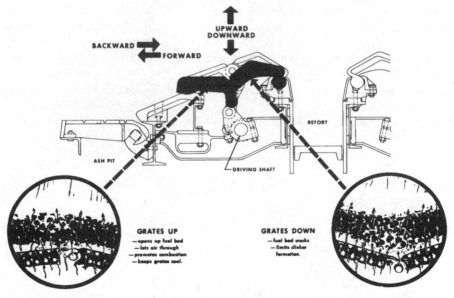

Figure 5.4 Cross section of a single-retort underfeed stoker. (*Detroit Stoker Company, subsidiary of United Industrial Corp.*)

absorbed from the fuel bed above and the action of the air being admitted through the grates cause the volatile gases to be distilled off and burned as they pass up through the fuel bed. The burning fuel slowly moves from the retort toward the sides of the furnace over the grates. As the fuel moves down the grates, the flame becomes shorter, since the volatile gases have been consumed and only coke remains. A portion of the coke finds its way to the dump grates, and a damper is provided to admit primary air under the grates to further complete combustion before the ashes are dumped. The combustion control automatically regulates the rate of coal supplied. The secondary-ram driving arm is adjustable so that the movement can be varied to obtain optimal fuel-bed conditions.

Underfeed stokers are supplied with forced draft for maintaining sufficient airflow through the fuel bed. The air pressure in the wind box under the stoker is varied to meet load and fuel-bed conditions. Means are also provided for varying the air pressure under the different sections of the stoker to correct for irregular fuel-bed conditions. When too much coke is being carried to the dumping grates, more air is supplied to the grate section of the stoker. The use of forced draft causes rapid combustion, and when high-volatile coals are used, it becomes necessary to introduce *over-fire air* that produces high turbulence and burns the resulting volatile gases with less smoke. Figure 5.3 shows how air from the wind box was admitted above the fuel bed as secondary air. A further improvement in combustion and reduction in smoke was obtained by use of steam or high-pressure air jets for over-fire air, as explained in Sec. 5.11.

All single-retort stokers feature the principles explained above, but there are numerous variations in details to comply with specific requirements. The feed ram may be driven directly by a steam or hydraulic piston. A screw may be used in place of the reciprocating ram to feed the coal into the retort. Stationary grates may be used in place of the moving grates shown in Fig. 5.4. When stationary grates are used, the action of the coal as it is pushed out of the retort is utilized to obtain movement over the grates and for maintenance of the correct fuel bed. The grates at the sides of the stoker may be stationary, making it necessary to remove the ashes from the furnace rather than dumping them into the ashpit.

Single-retort stokers provided a satisfactory means of burning coal to produce steam, provided certain principles of operation were observed. It was advisable to have a thick fuel bed on the stoker to protect the grates and tuyères from the furnace heat and to prevent the formation of holes through which air could pass without coming in contact with the fuel. In addition, large amounts of fuel on the grates are an advantage when there is an increase in the demand for steam. First, the steam pressure will drop, and the controller, sensing the drop, will increase the airflow to the stoker. The increase in air will cause the coal on the stoker to burn more rapidly, generating the added heat required. The converse is true when there is a decrease in the steam demand and the pressure increases. Low-fusion-ash coal is to be avoided because of its tendency to form clinkers. The tuyères depend on a covering of ash to protect them from the furnace heat, and low-ash coal may not provide adequate protection, resulting in outages and excessive maintenance.

Multiple-retort stokers have essentially the same operating principle as that of single- and twin-retort stokers. They are used under large boilers to obtain high rates of combustion. A sufficient number of retort and tuyère sections are arranged side by side to make the required stoker width (Fig. 5.5). Each retort is supplied with coal by means of a ram (Fig. 5.6). These stokers are inclined from the rams toward the ash-discharge end. They are also equipped with secondary rams, which, together with the effect of gravity produced by the inclination of the stoker, cause the fuel to move toward the rear to the ash discharge. The rate of fuel movement and hence the shape of the fuel bed can be regulated by adjusting the length of the stroke of the secondary rams. Some stokers produce agitation of the fuel bed by imparting movement to the entire tuyère sections. After the fuel leaves the retorts and tuyères, it passes over a portion of the stoker known as the *extension grates* or *overfeed section* (Fig. 5.5). By this time the volatile gases have been consumed, and only a part of the carbon remains to be burned.

These large stokers have mechanical devices for discharging the ash from the furnace. Dumping grates receive the ash as it comes from the extension grates. When a sufficient amount has collected, the grates are lowered and the ash is dumped into the ashpit. These dumping grates, even when operated carefully, frequently permit large quantities of unburned carbon to be discharged with the ash.

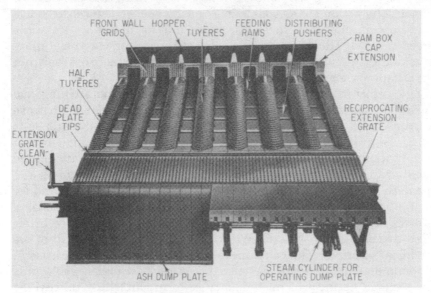

Figure 5.5 Multiple-retort stoker showing steam-operated ash dumping plates and coal and air distribution mechanisms. (*Detroit Stoker Company, subsidiary of United Industrial Corp.*)

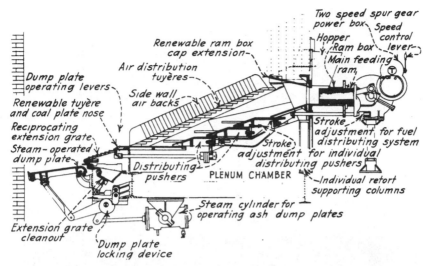

Figure 5.6 Cross section of a multiple-retort stoker with dumping grates. (*Detroit Stoker Company, subsidiary of United Industrial Corp.*)

Furnace conditions are disturbed during the dumping period, and efficiency is lowered. Power-operated dumps have been used to shorten the dumping period (Fig. 5.6). The combustible in the ash discharged from an underfeed stoker with dumping grates will vary from 20 to 40 percent. It is because of this high unburned carbon loss, and thus low boiler efficiency and higher fuel costs, that underfeed stokers have been generally replaced by spreader stokers and vibrating-grate stokers.

Rotary ash discharge also has been used to regulate the rate of ash discharged from multiple-retort underfeed stokers (Fig. 5.7). Stokers using this type of ash discharge may be referred to as *clinker grinders*. They consist of two rollers with protruding lugs installed in place of the dumping grates. These rollers are operated at a variable speed to discharge the ash continuously and thereby maintain a constant level in the stoker ashpit. The ash is removed without either loss of capacity or upsets in boiler operating conditions. The combustible in the ash from these clinker grinders will vary from 5 to 15 percent, still high, but an improvement over the dump grate.

Multiple-retort stokers generally are driven by motors. Provisions are made for varying the stoker speed either by changing the speed of the driving unit or by use of a variable-speed device between the driving unit and the stoker. Automatic controls vary the stoker speed to supply coal as required to meet the steam demand.

Irregularities in the fuel bed are corrected by controlling the distribution of air under the stoker. The wind box is divided into sections, and the air supply to each section is controlled by a hand or power

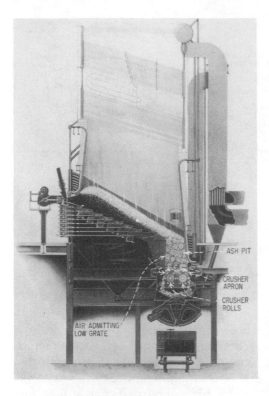

ASH PIT

CRUSHER APRON

CRUSHER ROLLS

AIR ADMITTING LOW GRATE

Figure 5.7 Multiple-retort stoker with rotary ash discharge. (*Detroit Stoker Company, subsidiary of United Industrial Corp.*)

operated damper. This arrangement enables the operator to correct for thin or heavy places in the fuel bed by reducing or increasing the wind-box pressure at the required spot.

The setting of an underfeed stoker is relatively simple in that no special arches are required. Mixing of fuel and combustible gases with air is accomplished by means of the high-pressure air supply and by the addition of over-fire air. Since there are no arches, the stoker is directly under the boiler, and a large quantity of heat is transmitted by radiation, resulting in a decrease in the temperature of stack gases and improved efficiency. Multiple-retort stokers are provided with sidewall tuyères or air blocks, which improve combustion and prevent the formation of clinkers (Fig. 5.6). The application of waterwall tubes to the furnace of a stoker-fired boiler as a means of preventing slagging difficulties and reducing maintenance is shown in Fig. 5.7.

With certain limitations as to the type of coal used, the underfeed stokers are well suited for continuous operation at their rated capacity. They are especially adapted to the burning of high-volatile coals, which can be burned economically by the underfeed principle. It has been found by operating experience that the use of over-fire air improves combustion efficiency. The fuel bed attains a very high temperature, to which the ash is exposed as it is moved to the surface. Therefore, clinkers are formed when coal having a low-fusion-temperature ash is burned on these stokers. Clinkers shut off the air openings in the tuyères and interfere with the movement of fuel on the stoker. This results in increased operating labor, reduced economy, and high maintenance. In general, these stokers will burn coking coal successfully; however, those which provide for agitation of the fuel bed are best suited for this type of coal.

Underfeed stokers have been built in varying sizes and applied to a wide variety of conditions. They have varied in size from the small domestic type to the large industrial multiple-retort unit. The use of forced draft and the relatively large quantity of fuel on the stoker make them responsive to rapid changes in load. These stokers have been developed to a high state of perfection and when applied and operated correctly give satisfactory results.

Before a stoker is operated, it should be lubricated in accordance with the manufacturer's instructions. Reduction gears run in lubricants, and the gear housing must be filled to the required level with the grade of lubricant specified. The oil level should be checked daily and the lubricant changed on schedule. The crankshaft bearings and other slow-moving parts are lubricated by medium grease applied with a high-pressure grease gun. These bearings should receive daily attention.

When preparing to put a stoker in service, operate the mechanism and check to see that it is in satisfactory condition. Feed coal into the furnace and spread it manually to a uniform thickness of 3 in over the tuyères. Place oily waste, shavings, and wood on the bed of coal. Then open the stack damper or, if mechanical draft is used, start the induced-draft fan to ventilate the setting and furnace, consequently eliminating the possibility of an explosion that might result from the presence of combustible gases. At this stage the wind-box door may be opened and the fire lighted. The drafts are then regulated to keep the fire burning.

When the coal is burning freely, the wind-box door is closed, the forced-draft fan is started, and the air supply is regulated to maintain the desired rate of combustion. The rate at which the unit can be brought "on line" depends on the design of the boiler. During this period, enough coal must be fed into the stoker to replace that which is being burned and to establish the fuel bed.

Underfeed stokers operate with furnace pressures at or only slightly below atmosphere; the condition is referred to as a *balanced draft*. This procedure prevents excessive air leakage into the setting and does not cause excessive heat penetration into the furnace setting.

The balanced draft is maintained by utilizing the wind-box pressure, as created by the forced-draft fan, to produce a flow of air through the fuel bed. Similarly, the induced-draft fan is used to remove the gases from the furnace by overcoming the resistance of the boiler, economizer, etc. Actually, the draft in the furnace should be maintained between 0.05 and 0.10 in of water during both operating and banked periods. The wind-box pressure varies from 1 to 7 in of water depending on the stoker design, fuel, and capacity. Control equipment will maintain the desired draft conditions automatically. As a safety precaution, the furnace draft should be increased when operating sootblowers or dumping ashes.

The following analysis on an "as fired" basis will serve as a preliminary guide to the selection of coal for use on underfeed stokers: moisture 0 to 10 percent, volatile matter 30 to 40 percent, ash 5 to 10 percent, heating value 12,500 Btu/lb minimum, ash softening temperature 2500°F minimum, top size 1 in by $\frac{1}{4}$ in, and nut and slack with not more than 20 percent less than $\frac{1}{4}$ in. The free-swelling index of coal for use on stationary-grate stokers should be limited to 5, maximum. For use on moving-grate and multiple-retort stokers, the free-swelling index may be 7, maximum. When the analysis of the available coal is known, these stokers can be designed and selected to burn a wider range of coal with some increase in initial cost and decrease in efficiency.

Coarse coal tends to segregate in the coal-conveying equipment, causing uneven feeding and burning. Wet coal may clog in the chutes and restrict feeding to the stoker. Stokers are supplied with shear pins that prevent serious damage to the mechanism as a result of foreign material in the coal, but the breaking of a shear pin and the removal of the foreign material cause interruption to service; consequently, every effort should be made to have the coal supply correctly sized and free from foreign material. Never allow the coal hopper to run empty or the supply chute to clog while the stoker is in operation, since this will cause the coal to burn in the retort and the tuyères to become overheated.

When operating the underfeed stoker, the airflow must be adjusted in proportion to the coal feed and regulate the supply to the various sections of the grates in order to maintain the correct contour of the fuel bed (Fig. 5.8). For example, insufficient air to the tuyère section or too great a travel of the secondary rams will cause the fuel to pass over the tuyères before it is burned. This will result in an excessive amount of unburned carbon reaching the dumping grates, where the combustion activity must be accelerated to prevent an excessive amount of carbon from being rejected with the ash. The travel of the secondary rams must be adjusted by observing the fuel bed and the

Figure 5.8 Contour of fuel bed on a multiple-retort underfeed stoker. (*Detroit Stoker Company, subsidiary of United Industrial Corp.*)

flue gas analysis until satisfactory results have been obtained. The fuel-bed depth 3 ft from the front wall should be 16 to 24 in. The correct depth depends on a number of factors, including the coking tendency of the coal and the amount of coarse and fine particles that it contains. The operator must take into account the fact that on entering the furnace, coking coals tend to swell and produce a deep fuel bed. If compensation is not made for this tendency, the fuel bed will become too heavy, and excessive amounts of coke will be discharged with the ash. The designer of the stoker should have experience on various coals with operation in many plants. Therefore, the operator should obtain operating guidelines from the designer and adjust the operating procedures as required by the specific situation.

Regardless of whether stationary grates, dumping grates, or clinker grinders are used to remove the ash from the furnace, the operator must exercise judgment in "burning down" before discharging the ash. Insufficient burning down will cause high carbon content in the ash; on the other hand, the indiscriminate admission of air will result in a high percentage of excess air (low carbon dioxide) and consequent high stack loss. The air chambers under the stoker should be inspected daily and the accumulation of siftings removed before they become excessive and interfere with the working parts of the stoker or form a fire hazard.

Banking of a boiler is a term that defines the operation of a boiler at a combustion rate that is just sufficient to maintain normal operating pressure at zero steam capacity. In order to bank the fire on an underfeed stoker, the fans are stopped, but the stack damper is not completely closed. There is danger that the coal will cake in the retort and block the movement of the rams. If a fire is to be banked for several hours, the stoker should be run one revolution at a time at regular intervals during the banked period.

Stoker maintenance may be divided into three classes: (1) coal feeders, speed reducers, and driving mechanisms; (2) internal parts, including tuyères and grates; and (3) retorts and supporting structures.

Satisfactory and long maintenance-free operation of the driving and external coal-feeding mechanism of a stoker depends on the correct lubrication and the elimination of foreign matter from the coal. If the driving mechanism should fail to start, examine the stoker carefully to determine whether foreign material is lodged in the ram or screw. Do not allow coal siftings and dirt to accumulate to a point where they impede the operation. Repairs to these external parts can be made without waiting for the furnace to cool.

The tuyères and grates have to be replaced occasionally. The frequency of replacement depends on the fuel used, the type of service, and the care used in the operation of the stoker. Coal with a low-fusion-temperature ash results in clinkers that adhere to the grates, restrict the airflow, and cause the stoker castings to become overheated. When the ash content of the coal is low, there is insufficient protection, and maintenance will be high. Even when suitable coal is used, the stoker castings may be overheated if the fuel bed is allowed to become so thin that parts of the grates are exposed. These castings are easily replaceable, but it is necessary to have an outage of equipment long enough to allow the furnace to cool sufficiently for someone to enter and perform the maintenance.

Unless the stoker is improperly operated, the retorts and structural parts should not have to be replaced for a long period of time. The replacement of these parts is expensive and requires an extended outage.

5.3.2 Overfeed stokers

As noted earlier, there are two types of overfeed stokers, mass feed and spreader, and these are described below.

Mass-feed stoker. Two types of mass-feed stokers are used for coal firing:

1. The moving-grate stokers, either chain or traveling grate
2. The water-cooled vibrating-grate stoker

The character of these stokers is that fuel is fed by gravity onto an adjustable grate that controls the height of the fuel bed. This firing method involves a fuel bed that moves along a grate with air being admitted under the grate. As it enters the furnace, the coal is heated by furnace radiation, which drives off the volatiles and promotes ignition. The coal burns as it is moved along the depth of the furnace. The fuel bed decreases in thickness until all the fuel has burned and the relatively cool ash is discharged into the ash hopper.

The *chain-grate stoker* uses an endless chain that is constructed to form a support for the fuel bed. The *traveling-grate stoker* also uses an endless chain but differs in that it carries small grate bars that actually support the fuel bed (Figs. 5.9 and 5.10). In either case, the

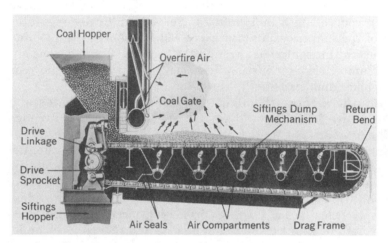

Figure 5.9 Traveling-grate stoker.

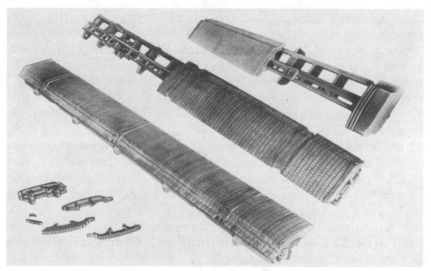

Figure 5.10 Typical grate-bar assembly in traveling grates. (*ABB Combustion Engineering Systems.*)

chain travels over two sprockets, one at the front and one at the rear of the furnace. These sprockets are equal in length to the width of the furnace. The front sprocket is connected to a variable-speed driving mechanism. The air openings in the grates depend on the kind of coal burned and vary from 20 to 40 percent of the total area. The steam capacity of these units is approximately 80,000 lb/h.

Coal is fed by gravity from a hopper located in the front of the stoker. The depth of fuel on the grate is regulated by a hand-adjusted gate (Fig. 5.9). The speed of the grate varies the rate at which the coal is fed to the furnace. The combustion control automatically regulates the speed of the grate to maintain the steam pressure. The burning progresses as the grate travels through the furnace. The ash, containing a small amount of combustible material, is carried over the rear end of the stoker and deposited in the ashpit.

Air enters the furnace through the openings in the grates and through over-fire jets. These over-fire air jets enter the furnace through the front arch or front wall above the arch. The intense heat at the front of the stoker results in distillation of gases from the fuel bed. The combination of over-fire air and the air passing through the fuel bed provides turbulence required for rapid combustion. In some designs, over-fire air jets are also located at the rear wall to provide a counterflow of gases in the furnace, promoting increased turbulence and further improving the combustion process and reducing the emission of smoke.

The grates are entirely enclosed by a suitable casing to prevent air infiltration at undesirable points and to promote combustion. Air leakage along the sides of the moving grates is reduced by the use of adjustable ledge plates. Some difficulty is encountered with excessive air leakage at the point where the ash is discharged into the ashpit. This leakage is reduced or the effect minimized by using a damper to seal off the ashpit, by using close clearance between the stoker and the setting, and by a rear arch that forces the air to pass over the fuel bed, mix with the combustible gases, and be utilized in the combustion process.

When the clearance between the rear of the furnace and the stoker is reduced to control air leakage, the hot ash comes into contact with the wall. Prior to the advent of water-cooled furnaces, ash adhered to the refractory-lined furnace and formed clinkers. This condition is now prevented by the use of a water-cooled furnace.

Chain- and traveling-grate stokers were operated originally with natural draft. However, in order to increase the capacity, they were later designed to utilize forced draft. In order to accomplish this, the entire stoker was enclosed. Forced-draft combustion air entered from the sides of the stoker between the upper and lower sections. It then flowed through the upper grate section and fuel bed. The stoker was divided into zones from front to back. Hand or power operated dampers were provided to regulate the flow of air to the zones to compensate for irregularity in the fuel bed depth.

The vibrating-grate stoker (Fig. 5.11) operates in a manner similar to that of the chain- or traveling-grate stoker, except that the fuel feed and fuel-bed movement are accomplished by vibration. The grates consist of cast-iron blocks attached to water-cooled tubes. These tubes are equally spaced between headers that are connected to the boiler as part of the boiler circulation system. The connecting tubes between these headers and the boiler circulating system have long bends to provide the flexibility required to permit vibration of the grates. The movement of the grates is extremely short, and no significant strains are introduced in this system. The space beneath the stoker is divided into compartments by means of flexible plates. Individual supply ducts with dampers permit regulation of the air distribution through the fuel bed. The grates are actuated by a specially designed vibration generator that is driven by a constant-speed motor. This consists essentially of two unbalanced weights that rotate in opposite directions to impart the desired vibration to the grates. The steam capacity of the vibrating-grate design is approximately 125,000 lb/h.

The depth of the coal feed to the grates is regulated by adjustment of the gate at the hopper. However, the rate of fuel feed is automatically controlled by varying the off-and-on cycle of the vibrating mechanism. The vibration and inclination of the grates cause the fuel bed to move through the furnace toward the ashpit. The compactness

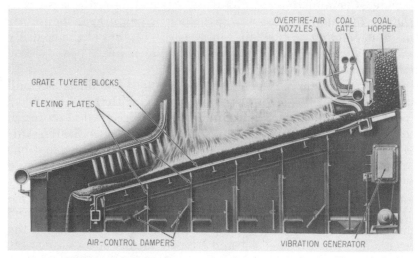

Figure 5.11 Vibrating-grate stoker.

resulting from the vibration conditions and fuel bed and promotes uniform air distribution. This action tends to permit the use of a wider range of coals. The water cooling decreases the minimum ash requirement for grate protection and permits the use of preheated air for combustion, which improves the combustion efficiency. The water cooling of the grates makes this stoker more adaptable when an alternative fuel such as natural gas or fuel oil is used because special grate protection is not required other than the normal bed of ash left from coal firing.

Chain- and traveling-grate stokers can burn a wide variety of solid fuels, including peat, lignite, bituminous, anthracite, and coke breeze, which is a fine coke generally less than $\frac{3}{4}$ in in size. These stokers often use furnace arches, which are not shown in Fig. 5.9. These front and rear arches improve combustion by reradiating heat to the fuel bed. Vibrating-grate stokers are suitable for medium- and high-volatile bituminous coals and for low-volatile bituminous coals and lignites at reduced burning rates. Coking coal requires agitation and is therefore not suited for use on these stokers.

The most satisfactory coal for use on chain- and traveling-grate stokers has a minimum ash-softening temperature of 2100°F, moisture of 0 to 20 percent, an ash content of 6 to 20 percent, a free-swelling index of 5 maximum, and a size range of 1 in by 0 with not more than 60 percent less than $\frac{1}{4}$ in.

Water-cooled vibrating-grate stokers operate best with coals having a moisture content of 0 to 10 percent, ash content of 5 to 10 percent, an ash-softening temperature less than 2300°F, and a coal size range of 1 or $\frac{3}{4}$ in by 0 with not more than 40 percent less than $\frac{1}{4}$ in.

The rate of combustion for chain- and traveling-grate stokers is 425,000 to 500,000 Btu per square foot of grate area per hour (Btu/ft²/h) depending on the ash and moisture content of the coal. By comparison, a water-cooled vibrating-grate stoker has a heat release of approximately 400,000 Btu/ft²/h.

To start a fire on a chain-grate stoker, the feed should be set and the stoker started. The grates should be covered with 3 to 4 in of coal for two-thirds of the stoker's length and a wood fire started on top of the coal. If a gas or fuel oil burner is part of the boiler design, this would be used. The stoker should not be restarted until the coal has been thoroughly ignited and the ignition arch is hot. Then it should be operated at its lowest speed until the furnace becomes heated.

There are two ways of regulating the rate of coal supplied by a chain-grate stoker: by controlling the depth of coal on the grates and by adjusting the rate of travel of the grates. The depth of coal on the grates is maintained between 4 and 6 in by means of an adjustable gate (Fig. 5.9). The average operating speed of the chain or traveling grate is 3 to 5 in/min. Variations in capacity are obtained by changing either or both of these fuel-feed adjustments.

Dampers are provided in stokers to permit the operator to control the air supply to the various zones. These air-supply adjustments enable the operator to control the rate of burning in the various zones and thereby reduce to a minimum the unburned carbon carry-over into the ashpit. If satisfactory operation cannot be accomplished by adjusting these dampers, the next step is to adjust the fuel-bed depth; finally, one can experiment by tempering the coal, which is the addition of water to certain coals that have insufficient moisture content for proper combustion on stokers. Excessive unburned carbon in the ash results not only in a fuel loss but also in burned stoker-grate links. Moreover, the residue continues to burn in the ashpit, and with low-fusion-ash coal, this causes a clinkering mass of ash that is difficult to remove.

Improvements in operation can be obtained by tempering (adding water) when the coal supply is dry, contains a high percentage of fine particles, or has a tendency to coke. Once coke has formed on the grates, the flow of air is restricted, making it very difficult to complete combustion and prevent an excessive amount of the coke from being carried over the end of the grate into the ashpit. Best results are obtained when the moisture content of the coal is 8 to 10 percent. The water should be added 10 to 24 h before the coal is burned. The addition of water at the stoker hopper results in excessive amounts of surface moisture.

The approximate amount of fuel consumed by a chain- or traveling-grate stoker can be calculated when the grate speed and thickness of coal on the grate are known. The grate speed can be determined by timing with a watch or by a revolution counter on the driving mechanism. When the revolution counter is used, a factor must be used to convert the reading into grate speed. This factor may be determined by comparing the revolutions of the driving mechanism with the grate speed. This may be done when the unit is out of service and at room temperature. In converting from volume to weight, assume bituminous coal as 48 lb/ft^3, anthracite as 60 lb/ft^3, and coke breeze as 40 lb/ft^3.

Example   A chain-grate stoker 8 ft wide is burning bituminous coal. The grate speed is 4 in/min and the coal is 6 in deep. At what rate is coal being consumed, in tons per hour?

Solution

8	$\times$	$\frac{4}{12}$	$\times$	$\frac{6}{12}$	$=$	$\frac{4}{3}$
width stoker, ft		rate of feed, ft/min		depth of coal, ft		coal burned, ft^3/min
$\frac{4}{3}$	$\times$	60	$\times$	48	$\div$ 2000	$=$ 1.92
coal burned, ft^3/min		min/h		lb coal/ft^3	lb/ton	coal burned, ton/h

Banking of a boiler burning solid fuel on a grate is defined as burning the fuel at a rate to maintain ignition only. To bank the fire on a chain- or traveling-grate stoker, close off the air supply under the stoker a little at a time. Allow the fire to burn down before closing off the air supply to prevent damage from overheating. Fill the hopper and close off the coal supply. When the furnace is relatively cool, raise the gate and run all the coal from the hopper into the furnace. In order to guard against possible overheating, do not leave the coal in contact with the gate. Admit a small amount of air under the stoker at the end of the supply of coal. Open the stack damper sufficiently to maintain a slight draft in the furnace or start the induced-draft (ID) fan. As the coal burns and the fire moves from the rear to the front of the furnace, the point of admission of air also must be varied by adjusting the zone dampers. Before the fire reaches the gate, proceed to fill the hopper and introduce another charge of coal. In starting from bank, it is frequently necessary to use a bar to break up the coal that has caked near the gate. The coal feed should not be started until the air has blown through the grates for a period of 3 to 5 min in order to increase the furnace temperature. Operating procedures vary as experience is gained. Operators should consult with equipment designers for their recommended procedures.

The chain should be tight enough to keep it from dipping down appreciably when it comes over the front sprocket. The chains are adjusted by a take-up screw that moves the sprocket. The plates between the sidewalls and the grates must be adjusted to prevent excessive air leakage. The clearance between the plate and grate is adjusted to $\frac{1}{8}$-in. Excessive burning of the link ends is an indication that combustible material is passing over the ash end of the stoker. Broken or burned links must be replaced at the earliest opportunity.

Spreader stoker. The spreader-stoker installation consists of a variable-feeding device, a mechanism for throwing the coal or other suitable fuel such as bark, wood, or refuse-derived fuel (RDF) into the furnace, and grates with suitable openings for admitting air. The coal feeding and distributing mechanism is located in the front wall above the grates. A portion of the volatile matter and of the fine particles in the coal burns in suspension, and the remainder falls to the grate, where combustion continues. As a result of the pressure created by the forced-draft fan, primary air enters the furnace through the openings in the grates. A portion of this air is used to burn the layer of fuel on the grates, and the remainder passes into the furnace, where it is utilized to burn the volatile matter and fine particles. The over-fire air fan supplies additional air through jets in the furnace wall. This adds to the air supply for suspension burning and produces turbulence. The modern spreader stoker is the most versatile and most commonly used stoker today. The spreader stoker can be used with boilers that are designed for steam capacities in the range of 75,000 to 700,000 lb/h. The stoker responds rapidly to changes in steam demand, has a good turn-down capability (i.e., can operate efficiently at part loads), and can use a wide variety of fuels. It is generally not used for low-volatile fuels such as anthracite.

Although coal size is specified when ordered, the actual size when received does vary from coarse to fine-sized particles. In order to have efficient operation and proper combustion, the mixture of this coarse and fine coal should be distributed across the stoker hopper inlet. If done properly, a uniform fuel bed, low unburned combustibles, higher boiler capacity, and lower maintenance will result. One means of achieving this even distribution is with the conical distributor, as shown in Fig. 5.12. Such a distributor has proven effective on new stoker-fired boiler designs and in replacing the original equipment.

The conventional mechanical spreader stoker feeding and spreading device is arranged to supply coal to the furnace in quantities required to meet the demand. The feeder delivers coal to the revolving rotor with protruding blades. These revolving blades throw the coal into the furnace. The spreader system must be designed to distribute the coal evenly over the entire grate area. This distribution is accomplished by the shape of the blades that throw the coal. Variations in performance of the spreader, due to changes in coal size, moisture content, etc., are corrected by means of an external adjustment of the mechanism.

The suspension burning causes a portion of the unburned carbon particles to be carried out of the furnace by the gases. Some of these particles collect in the hoppers beneath the convection passes of the boiler. However, a portion is retained in the exit boiler gases and must be captured by a mechanical dust collector. Since these particles contain carbon, the efficiency of steam generation can be improved by returning the particles to the furnace to complete the burning. Multiple injection ports are located across the width of the boiler and provide uniform mixing of the unburned carbon into the furnace for complete burnout. However, it is undesirable to reinject significant amounts of ash because this causes erosion within the boiler. This is minimized because a mechanical dust collector is effective in collecting the larger particles that contain the majority of the unburned carbon in the carry-over. Carbon reinjection improves coal fired boiler efficiency by about 2 to 4 percent. In order to meet particulate emission requirements, the system design also requires an electrostatic precipitator (ESP) or a baghouse prior to the release of flue gas to the stack.

Figure 5.12   Conical nonsegregating coal distributor across stoker inlet. (*Stock Equipment Company.*)

Spreader stokers can handle a wide range of fuels and can be used with different boiler sizes. A wide variety of coal can be burned on a spreader stoker, but the variation must be considered when the unit is specified. It is inadvisable and uneconomical to provide a stoker that will burn almost any coal in a location where a specific coal is available. In addition to coal, spreader stokers are used to burn bark, wood chips, and many other kinds of by-product fuels.

When burning coal, spreader stokers are used with high- and low-pressure boilers from the smallest size to those having a steam-generating capacity of 400,000 lb of steam per hour. At lower capacities, the installation cost of spreader stokers is less than that for pulverized coal. However, at between 300,000 and 400,000 lb of steam per hour, the cost tends to equalize, and at higher capacities pulverized-coal plants are less costly. Because of the relatively high percentage of fuel burned in suspension, spreader-stoker-fired boilers will respond quickly to changes in firing rate. This feature qualifies them for use where there is a fluctuating load. Because these stokers discharge fly ash with the flue gases, they were at one time subjected to severe criticism as contributors to air pollution, but with efficient dust-collecting equipment they meet the most stringent air-pollution requirements.

In the smaller sizes, spreader-stoker installations are comparatively costly because of the necessity for ashpits, fly-ash reinjection, and fly-ash-collection equipment required for efficient and satisfactory operation. Difficulty is also encountered in designing a unit that will operate satisfactorily when the lowest steam requirement is less than one-fifth the designed capacity (i.e., a unit having a turn-down ratio less than 5). The *turn-down ratio* is defined as the design capacity divided by the minimum satisfactory operating output.

The spreader-stoker feeder functions to vary the supply of coal to the furnace and to provide even distribution on the grates. The mechanisms that propel the coal into the furnace include steam and air injection as well as mechanical rotors. In the feeder and distribution system shown in Fig. 5.13, there is a reciprocating feed plate in the bottom of the hopper. The length of stroke of this plate determines the rate at which coal is fed into the furnace. The automatic combustion

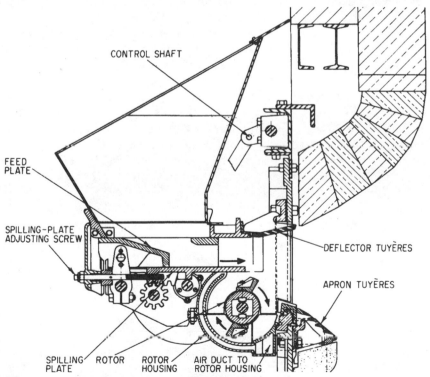

Figure 5.13 Spreader-stoker reciprocating-type coal feeder and rotary distribution mechanism. (*Detroit Stoker Company, subsidiary of United Industrial Corp.*)

regulator varies the length of the feed-plate stroke. The coal leaving the hopper drops from the end of the spilling plate into the path of the rotor blades, which distribute the coal on the grates. The distribution is regulated by adjustment of the spilling plate and by the speed of the rotor. The in-and-out adjustment of the spilling plate changes the point at which the coal comes in contact with the rotor blades. Moving the spilling plate back from the furnace allows the coal to fall on the rotor blades sooner. The blades impart more energy to the coal, and coal is thrown farther into the furnace. In a like manner, increasing the speed of the rotor imparts more force to the coal, throwing it farther into the furnace.

The ability of spreader stokers to utilize coal having a high moisture content is limited by the performance of the feeder and distribution mechanism. A maximum moisture content of 25 percent is considered a limit for a typical coal. Figure 5.14 is a drum-type feeder that uses a pocket wiper to prevent the recycling of wet coal. When there is a decided change in either the size or the moisture content of the coal, the feeder mechanism must be adjusted to maintain a uniform fuel bed on the grates. Air enters the furnace both above and below the feeders to keep the feeder cool and provide air for combustion. In addition to being air cooled, the rotor bearings are water cooled.

Spreader-stoker feeders are divided into two classes: overthrow, when the coal comes into contact with the upper part of the rotor blade assembly (Figs. 5.13 and 5.14), and underthrow, when it comes into contact with the lower part.

The simplest types of grates are stationary and are similar to those used for hand firing. The feeder automatically deposits the coal on the grates, and air for combustion enters the furnace through the holes in the grates. At least two feeders are used, and before the ash deposits become deep enough to seriously restrict the airflow, one of the feeders is taken out of service. The fuel on the grate is allowed to "burn down," and the ash is raked through the furnace door. The feeder is then started, and after combustion has been reestablished and stabilized, the remaining grate sections are cleaned in a similar manner.

Dumping grates (Fig. 5.15) provide a means for tipping each grate section and thereby dumping the ash into a pit. The tipping of the grates can be accomplished either by hand operation or by means of steam- or air-powered cylinders. The procedure of taking one feeder section out of service long enough to remove the ash is the same as when stationary grates are used, but the time required for the operation and the amount of labor involved are reduced. The intermittent discharge of ash from the furnace has the following disadvantages: Considerable labor is needed; skill is required to prevent a drop in steam pressure; efficiency is reduced because of the high excess air introduced during

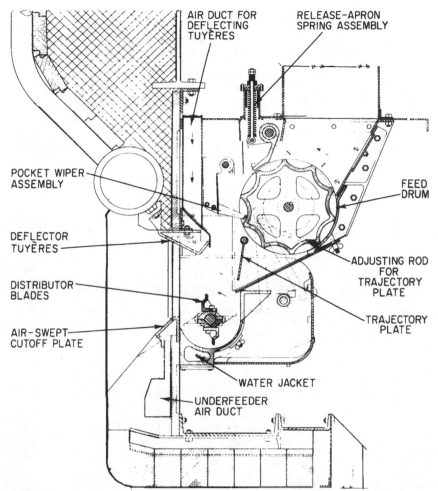

Figure 5.14 Spreader-stoker drum-type coal feeder and rotary distribution mechanism (*DB Riley, Inc.*)

the cleaning period; and there is a tendency to discharge considerable combustible with the ash, produce smoke, and increase the discharge of fly ash while the grates are being cleaned. The disadvantages of dumping grates have led to their general disuse. Traveling- and vibrating-grate stokers are now the most common designs used today.

A traveling-grate spreader stoker installed in a water-cooled furnace is shown in Fig. 5.16. The coal falls on the grate, and combustion is completed as it moves slowly through the furnace. The ash remains and falls into the pit when the grates pass over the sprocket. The stoker moves from the rear of the unit to the front, and therefore, the ashpit is located at the front of the stoker (beneath the feeder). The

Figure 5.15 Four-unit spreader stoker with power-operated dumping grates. (*Detroit Stoker Company, subsidiary of United Industrial Corp.*)

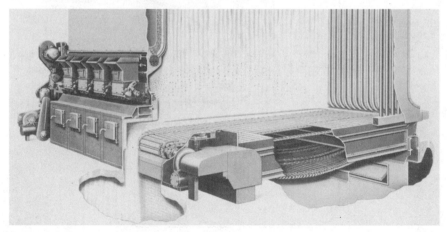

Figure 5.16 Traveling-grate continuous-ash-discharge spreader stoker. (*Detroit Stoker Company, subsidiary of United Industrial Corp.*)

ashpit is located in the basement to provide adequate storage for the ash produced on an 8-hour shift or for continuous ash removal with an ash-handling system. The rate of grate movement is varied to produce the required depth of ash at the discharge end.

The ash-removal method used in vibrating-grate stokers is similar to that used in removing ash from spreader stokers. The grates are mounted on a pivoted framework, a motor vibrates the assembly, and the ash is moved along the grates toward the ashpit. The motor that produces the vibration is run at intervals by a timer. The off-and-on cycles are varied to obtain the desired depth of ash at the discharge end of the grate.

Another method of obtaining continuous ash discharge consists of overlapping grates similar to shingles on the roof of a house. The grate bars are mechanically driven and move back and forth, alternately increasing and decreasing in the amount of overlap. This motion causes the ash to shift from one grate to the other and slowly move toward the ashpit. The rate of ash discharge is varied by changing the amount of travel of the grate bars.

The ash discharged from the furnace by these methods of continuous bottom-ash removal has an unburned combustible content of 4 to 8 percent. In contrast, the spreader stoker, with its carbon reinjection system, is capable of reducing the amount of combustible lost and therefore of improving combustion efficiency by 2 to 4 percent.

With a spreader stoker, the fuel burned in suspension comes from two sources, the fines in the fuel fed to the furnace and the carbon returned in the ash reinjection system from the boiler, economizer, air heater, and dust-collector hoppers. The amount of fuel burned in suspension depends on the percentage of volatiles, moisture, and fines in the fuel and the capacity at which the boiler is operated. Much of the ash produced by this suspension burning is carried by the combustion gases into the boiler convection passes. This ash contains from 40 to 60 percent combustibles and therefore represents an appreciable heat loss. Ash is collected in the boiler, economizer, air heater, mechanical dust collector, and baghouse or precipitator hoppers. Ash with the highest combustible content is deposited in the boiler hoppers and that with the lowest in the baghouse or precipitator hoppers. The fine particles removed by the baghouse or precipitator must be discarded to the ash-removal system. The baghouse or precipitator is required to reduce the particulate emission from the stack gases sufficiently to meet clean-air standards.

As the permissible particulate emission from the boiler stack gases was reduced by law, attempts were made to meet these more stringent requirements by reducing the amount of ash recycled to the furnace. More of the ash collected in the various hoppers was rejected to the ash-disposal system. This procedure sacrificed efficiency in an attempt to meet the emission standards.

Another attempt to reduce recycling and to lower stack emissions was by gravity return of some of the ash, which contains the majority of carbon, to the traveling grate. The ash from the various collection sources is collected in a hopper extending across the rear of the furnace and is permitted to flow by gravity, in a number of streams, onto the traveling grate.

Now baghouses or precipitators are used in a final cleanup of stack gases. Ash from the various hoppers may be either discarded to the ash-disposal system or returned to the furnace for reburning and partial recovery of the heat content in combustible particles. This increases the boiler efficiency. The advantages and disadvantages of reburning must be considered in order to determine the ratio of ash to be returned to the furnace and that to be discarded to the ash-conveying system.

Reinjection in theory increases boiler efficiency, but extensive recirculation of the ash can result in operating problems. Large quantities of ash handled by the reinjection system can result in accelerated wear and require replacement of parts. The inert portion of the reinjected ash retards combustion in the furnace and can result in clinker formation on the stoker grates and slagging of the furnace waterwall tubes. In boilers designed for high-velocity gases in the tube passes, the heavy concentration of abrasive ash particles can result in boiler-tube erosion. Nevertheless, the impact of reinjection must be seriously considered because it does improve boiler efficiency and reduce fuel cost. For example, by reinjecting 50 percent of the fly ash, the efficiency can be improved by over 2 percent, which leads to significant savings in fuel costs.

Suspension burning requires a supply of air directly to the furnace. This air is supplied by an over-fire air fan and enters the furnace through ports in the walls. Such over-fire air fans perform the triple function of providing combustion air, creating turbulence in the furnace, and reinjecting ash from the collection hoppers.

Approximately 35 percent of the total air is used as over-fire air, with the remaining 65 percent being under-grate air. The over-fire air is injected through a series of small nozzles that are arranged in the front and rear walls, often at different elevations, and the air through these nozzles creates high penetration and mixing with the fuel. With the combustion air split between under-grate and over-fire, spreader-stoker firing does provide a form of staged combustion that is effective in controlling NO_x.

The stoker shown in Fig. 5.17 has forced draft, power-operated dumping grates, and a combination over-fire air and a cinder (ash) return system. Cinders deposited in the hoppers beneath the convection passages of the boiler are transferred to the side of the unit by screw conveyors and then picked up by streams of air and returned to the furnace. This illustration shows how the cinders from the dust-collector hopper are returned to the furnace. Consideration should be given to discharging them to the ash-conveyor system. Ashes deposited on the grates are dumped into the shallow pit and transferred by hand to the ash conveyor located in front of the stoker.

A simple combustion control for a spreader stoker consists of a pressure-sensing element in the steam outlet that activates a controller on the fuel-feed rate and changes the flow of air from the forced-draft fan. The fuel-feed change is accomplished by adjusting the length of stroke of the feed plate (Fig. 5.13). The amount of air supplied under

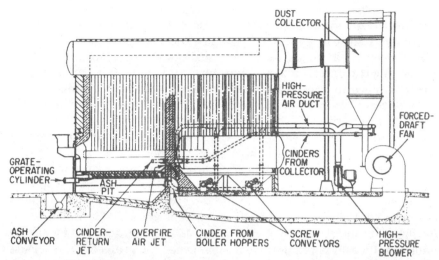

Figure 5.17 Spreader stoker with dumping grates and cinder return. (*Detroit Stoker Company, subsidiary of United Industrial Corp.*)

the grates is usually regulated by a damper or by inlet vanes on the fan, although sometimes the speed of the fan is varied. The induced-draft-fan damper or the fan speed is varied to maintain a constant furnace draft of 0.05 to 0.10 in of water. A recommended refinement in the control consists of a means by which the forced-draft-fan air supply is readjusted to maintain a definite ratio of airflow to fuel feed regardless of the depth of the fuel bed. Other refinements are the control of the rate of ash discharge from the grates and the over-fire air pressure. Adjustments can be made to vary the distribution of coal on the grates. On stokers equipped with stationary dumping grates, manipulation of the feeder and air supply, as well as the dumping of the grates, is made during the cleaning period.

Consideration must be given to the size when selecting coal for use on a spreader stoker. The larger particles are more easily thrown to the rear of the furnace than the smaller particles. In order to obtain a uniform depth of bed on the grates, the coal should have a top size of $1\frac{1}{4}$ by $\frac{3}{4}$ in with not more than 40 percent able to pass through a $\frac{1}{4}$-in mesh screen. The more fines, the more suspension burning and the higher the percentage of ash and combustible carried from the furnace by the gases. For best operation, coal on an "as fired" basis should contain at least 18 percent volatile, 5 to 10 percent ash (up to 15 percent on a traveling grate), a heating value of 12,000 Btu/lb minimum, and an ash-softening temperature of 2000°F on traveling grates but somewhat higher on other bottom ash-disposal systems. By special design, a wider range of coal can be burned on these stokers.

Because of the effect of size on distribution in the furnace, it is essential that segregation of the coarse and fine particles of coal be prevented. Coal conveyors and bunkers serving spreader-stoker-fired boilers must be arranged to reduce segregation to a minimum. Coal size segregation can be a problem on any stoker, but the spreader stoker is more tolerant because the feeder rate can be adjusted and many of the fines burn in suspension. The coal distributor shown in Fig. 5.12 also minimizes coal size segregation.

The spreader stoker will burn a wide range of fuels varying from lignite to bituminous coal. Fuels having volatile matter of less than 18 percent are generally not suitable. Clinkering difficulties are reduced, even with coals that have clinkering tendencies, by the nature of the spreading action. The fuel is supplied to the top of the fuel bed, and the ashes naturally work their way to the bottom, where they are chilled below the fusion temperature by contact with the incoming air. The coking tendency of a coal is reduced before it reaches the grates by the release of volatile gases that burn in suspension. The utilization of high-ash coal lowers the efficiency and at the same time increases the work of the operator because there is a greater volume of ash to be handled. The cleaning periods must be more frequent when the ash content of the coal being burned is high. In general, the spreader stoker will burn good coal with an efficiency comparable with that of other stokers and in addition will utilize poor coal successfully but at reduced efficiency.

The spreader-stoker principle can be applied successfully to burn waste-type materials. Both coal and waste may be burned in the same furnace. However, experience has shown that some waste materials, such as RDF from municipal solid waste (MSW), do not burn well in combination with coal. A system burning both fuels performs the twofold purpose of disposing of waste material and at the same time reducing the consumption of costly fossil fuel. In addition, the waste fuel frequently has a lower sulfur content, and its use reduces the sulfur dioxide content of the flue gases. Figure 5.18 shows wood waste being blown into a furnace. The initial burning occurs in suspension, and the final burning takes place on a traveling-grate stoker. The chain feeder regulates the flow of wood waste from the overhead bin to the burner. As the fuel falls on the distribution trays, it is blown into the furnace by streams of air. Fuel trajectory is controlled by adjusting the distribution trays. Figure 5.19 shows refuse (or waste) feeders installed in a furnace above the conventional coal feeders. This arrangement provides for the use of waste material and coal in varying proportions. It results in ultimate utilization of waste material and at the same time supplies a variable steam demand. Almost any kind of combustible waste can be used provided it is milled to a 4-in top size. The partial list of waste material suitable for use as fuel includes municipal solid waste as refuse-derived

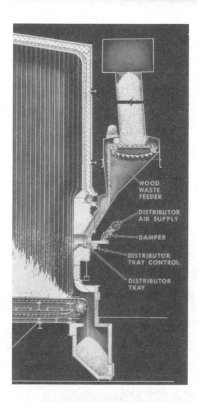

Figure 5.18 Spreader stoker for burning wood waste. (*DB Riley, Inc.*)

Figure 5.19 Conventional coal-burning spreader stoker with air-swept refuse distribution spouts. (*Detroit Stoker Company, subsidiary of United Industrial Corp.*)

fuel (RDF), bark, bagasse, wood chips, shavings, sawdust, and coffee grounds. Boilers can be designed for combination fuels or for dedicated firing of a waste material such as RDF. The moisture content of these waste materials may be as high as 50 percent, and the combustion air is heated to 550°F or to the recommended limit of the grate bars.

The burning of biomass and waste fuels on stokers not only reduces operating costs, due to low fuel cost, but also the reduction of these otherwise waste materials minimizes landfill requirements and related disposal costs. These systems burn a vast variety of fuels to produce low-cost electrical energy. These fuels include

Refuse-derived fuel (RDF) from municipal solid waste

Bark, wood waste, and sawdust

Agricultural waste such as bagasse (sugar cane residue) and coffee refuse

Chicken litter

Sunflower hulls from grain production

The spreader stoker is the most widely used technology for the burning of these fuels, and the hydrograte stoker shown in Fig. 5.20 is commonly used.

Sized fuel is metered to a series of distribution devices that spread it uniformly over the stoker-grate surface. The fine particles are burned rapidly in suspension by being assisted with the over-fire air system. The heavier fuel particles are spread evenly on the grate, forming a thin, fast-burning fuel bed. The combination of suspension firing and a fast-burning fuel bed results in a system that is very responsive to load demand.

The hydrograte stoker combines advanced spreader-stoker technology with automatic ash discharge and water-cooled grates. It can handle a variety of high-moisture, low-ash fuels over a broad load range. Because the stoker is water cooled, its firing can be based on combustion conditions instead of limitations on cooling air temperature requirements in non-water-cooled grate designs. This makes it possible to maintain the higher combustion air temperatures necessary for burning high-moisture biomass or waste fuels without damaging the grates. The hydrograte is also ideal for burning low-ash waste fuels because no ash cover is needed to protect the grate. This stoker also minimizes the need for under-grate airflow. This ensures the most effective use of over-fire air that optimizes emission control and combustion efficiency. The higher percentage of over-fire air also results in effective staged combustion, which results in controlling NO_x emissions. Automatic ash discharge results from the intermittent

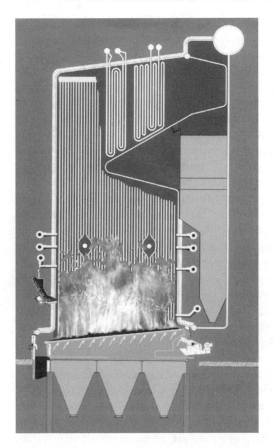

Figure 5.20 Hydrograte stoker operates with higher than normal combustion air temperatures because of the water-cooled grate that permits the burning of high-moisture biomass fuels without grate damage. (*Detroit Stoker Company, subsidiary of United Industrial Corp.*)

grate vibration that moves the fuel bed forward through the furnace for discharge of ash at the front of the boiler.

Because there is a minimum of moving parts in the design of a hydrograte stoker, down time and maintenance costs are significantly reduced. In addition, because of the water cooling of the grate, replacement of the grate is minimized.

Before placing a spreader stoker in operation, it should be checked carefully to see that it is in good condition. This precaution may prevent a forced outage when the unit is urgently needed. The oil level in the reduction-gear case should be checked, and the bearings of all moving parts should be greased and oiled. The flow of cooling water should be adjusted through the bearings, the dampers and automatic control mechanism should be checked, and the fans should be test operated. The stoker should be started and operated without sticking or binding. The over-fire draft controller should be set and adjusted to maintain 0.1 in of water draft in the furnace.

When a stoker is found to be in satisfactory condition, the feeder is operated until approximately a $\frac{1}{2}$-in layer of coal has been deposited on the grates. The coal is covered with a generous supply of kindling wood and paper. (Many units are designed with auxiliary natural gas or oil burners that are used during startup.) The fans are operated and the dampers adjusted to create a draft in the furnace. The fire is lit and the air and draft regulated to produce a bright, smokeless flame. Then the stoker (if of the moving-grate type) is started, and the coal is fed slowly to sustain ignition. The unit is brought on-line slowly to avoid rapid temperature changes. The over-fire air fan is started to avoid smoke and to keep the ash-reinjection system from overheating. If smoke becomes excessive during this startup period, the coal feed should be reduced and the over-fire air adjusted. Specific startup instructions for the unit should be followed to speed up the stoker, increase the fuel feed, and adjust the airflow and draft.

Even when spreader stokers are provided with controls by which both the fuel feed and the air supply are automatically regulated to meet the steam demand, there are numerous details to which the operator must attend to obtain good operation. The spreader mechanism must be adjusted to distribute the coal evenly over the grates. Irregularities in the fuel bed caused by segregation of coal, wet coal, clinker formation, etc., must be corrected. The fuel bed should be maintained at sufficient thickness to sustain combustion for 2 to 3 minutes with the fuel feed shut off. The fuel-bed thickness may be changed by adjusting the ratio of fuel to air. An increase in the air pressure under the grates with the same fuel feed will decrease the thickness of the fuel bed; a decrease in air pressure will allow the fuel-bed thickness to increase. The application of over-fire air must be regulated to obtain the lowest amount of excess air (highest carbon dioxide) without permitting temperatures detrimental to the furnace lining, clinkers in the fuel bed, and excessive smoke. The over-fire air jets should be regulated to supply additional air for burning the fuel in suspension as well as to increase turbulence and consequently to improve combustion. The furnace draft must be maintained at between 0.05 and 0.10 in of water to prevent overheating of the furnace lining, the coal-feeding mechanism, and the furnace doors. The bearing-cooling water flow must be adjusted to maintain an outlet temperature not exceeding 150°F or the manufacturer's recommended temperature.

Ash is removed from the spreader stokers equipped with a stationary grate by using a hand hoe. The ash should be removed when it is approximately 3 in deep on the grates. The cleaning should be performed as quickly as possible to minimize disturbance to boiler operation. These stationary-grate stokers should have at least two sections with separate feeders and coal distributors and divided wind boxes so that each section can be cleaned separately. This gives the operator an opportunity to burn up the carbon before removing the ash and makes it possible to maintain the steam pressure during cleaning. The cleaning operation is performed by first stopping the coal feed to the unit to be cleaned. When the fuel on the grates has been burned down, close the hand-operated damper to this unit. If the fuel bed is of the correct thickness, the grates will be ready for cleaning in 2 to 3 minutes. After the fuel has been consumed, quickly hoe the ash from the grates and remove any clinker that has formed on the side of the furnace or bridge wall. Immediately start the feeder, and spread a thin layer of coal on the grates. After the coal has been ignited, slowly open the hand-operated forced-draft damper. After the fire has become fully established and the steam pressure has been restored, clean the other grate units in a similar manner.

The removal of ash from hand- or power-operated dumping grates is similar to the procedure with stationary grates except that less labor is involved and the operation can be performed more quickly. After the fire has burned down and the forced-draft damper has been closed, operate the dumping grates. Some operators open and close the grates several times to make sure they are clean. Be certain that the grates are back in position and perfectly flat before applying coal to reestablish the fire. When the stoker is composed of more than two units, it is advisable to clean the units alternately. After cleaning one unit, permit steam pressure to recover before dumping the next section of grates.

When traveling grates are provided, the ash is continuously discharged from the furnace into the ashpit. When properly sized and operated, the ash discharged from a spreader stoker with a traveling grate is relatively cool and clinker free. The discharged ash in the hopper may be removed by a conventional ash-handling system, generally without clinker grinders. The coal is thrown to the rear of the furnace, and the speed of the traveling grates is adjusted to give sufficient time for combustion to be completed before the ash is discharged at the front. When the grates are correctly adjusted, the process is continuous, and there is no interference with boiler operation.

The furnace and ash hoppers must be emptied frequently enough to prevent the accumulation of ash from interfering with the operation. Excessive ash in the furnace ashpit will block the air and cause the grates to be overheated. When too much material accumulates in the ash hopper, its effectiveness is reduced. When means are provided for returning the ash from the hopper to the furnace, frequent checks must be made to see that the nozzles are not clogged. If these transfer nozzles stop operating, the ash will accumulate in the hopper.

At times of no steam demand, it is normally advisable to allow the fire to go out and then restart it when there is need for steam. For short periods and in anticipation of sudden increased demands for steam, however, it may be advisable to resort to banking. During banking periods, care must be exercised to prevent injury to the grates and to the coal-distribution rotor.

A spreader stoker, equipped with stationary or dumping grates, should be prepared for banking by allowing a layer of ashes to accumulate on the grates. The air supply to the grates is reduced, and the feeder and speed of the rotor are adjusted to deposit a layer of coal on the front end of the grates. The coal feed and forced-draft fan are stopped, but the rotor continues to operate to prevent its overheating. A slight furnace draft (about 0.1 in of water) is maintained during the banking period by manipulating the dampers. The feeder is operated at intervals to restore the fuel on the grates, and the draft is regulated to control the rate of combustion in order to prevent the fire from going out and the steam pressure from decreasing below the level necessary for the anticipated standby requirements. The operators of small spreader stokers sometimes prefer to resort to hand firing when banking. When ready to return the boiler to service, the remaining fuel is distributed over the grates, and then the fans are started to ignite the coal quickly as it is fed into the furnace. Controls are provided to apply coal automatically to the grates in sufficient quantities to keep the fire from going out and to maintain the steam pressure required for quickly returning the boiler to service.

To bank a traveling-grate spreader stoker, the grate is stopped, and the coal feed is decreased slowly until the steam output is nearly zero and the fuel bed appears nearly black. Both the forced- and induced-draft-fan dampers should be closed and the furnace draft reduced to nearly zero. The fuel bed should be allowed to darken and get "spotty" with live coals. Then coal is fed to the front section of the grates. This coal will smother the live coals. Waiting until the fuel bed has cooled sufficiently before introducing the coal reduces the tendency to smoke and increases the time interval between feedings. The forced-draft fan should be stopped, and a slight draft in the furnace should be maintained either by operating the induced-draft fan or by manipulating the damper. When the steam pressure has dropped to 50 percent of the line pressure, the fans should be run to restore the pressure, and then coal should be applied as before.

The coal-feeder bearing-cooling water flow should be checked several times during each shift. The bearings should be lubricated daily with a solid lubricant that will not thin when heated to 500°F. Dampers and accessory control and ash dumping equipment should be checked whenever the stoker is out of service. The maintenance of grates and furnace linings requires an outage of the boiler, but some repairs can be made to the feeding-and-spreader mechanism when the unit is in operation.

Careful attention to the following details reduces maintenance: Prevent excessive furnace temperature by avoiding overloading the unit and by supplying an adequate amount of air to the furnace; maintain a draft in the furnace at all times; and follow established procedures when dumping ashes, blowing down, blowing soot, banking fires, placing the unit in service, and taking it out of service. The ash-return system must be checked for possible stoppage resulting from slagging over in the furnace or from large pieces of slag becoming lodged in the ash-inlet nozzles. In every situation, the recommended operating instructions of the equipment supplier should be followed carefully.

5.4   Pulverized Coal

Coal is an abundant and low-cost fuel for boilers. It can be burned in a number of ways depending on the characteristics of the coal and the particular boiler application. Various firing methods for the burning of coal are used: cyclone furnaces, stokers, pulverized coal, and fluidized bed. Of these applications, pulverized-coal firing is the dominant method in use today in terms of the quantity of coal burned. Pulverized-coal (PC) firing is predominant on the large, highly efficient utility boilers that are used to provide the base-loaded electric capacity in utilities throughout the world.

Pulverized-coal firing uses much smaller particle sizes, which allows high combustion rates, and, therefore, it is significantly different from the other coal combustion technologies. The combustion rate of coal is controlled by the total particle surface area. By pulverizing coal, the coal can be burned completely in 1 or 2 seconds. This is similar to burning oil and natural gas. By comparison, the other coal-firing technologies use crushed coal of various sizes and provide longer combustion times, up to 1 minute.

The pulverization of coal exposes a large surface area to the action of oxygen and consequently accelerates combustion. The increase in surface area, for a given volume, can be expressed by reference to an inch cube of coal. The cube has six faces, each having an area of 1 in^2—a total of 6 in^2 of surface. Now, if this 1-in cube of coal is cut into two equal parts, each part will have two sides with an area of 1 in^2 and four with an area of $\frac{1}{2}$ in^2, making a total of 4 in^2 of surface. The two pieces together have a total surface of 8 in^2 as compared with the original 6 in^2. Imagine the increase in surface when the process proceeds until the coal particles will pass through a 200-mesh sieve. The improved combustion and flexibility of the unit must, of course, justify the cost of power and equipment required for pulverization.

Electric-generating steam plants were among the first to use pulverized coal with boilers. Originally, many plants used the *central system,* which consisted of a separate plant for drying and pulverizing the coal. This system provided a high degree of flexibility in the utilization of equipment and in boiler operation. However, it required additional building space to house the coal preparation equipment and presented fire and explosion hazards. Both the initial installation and operating costs were high. A separate crew of operators was required in the coal preparation plants. For these reasons, the central system has long been replaced by the *unit system.*

When the coal particle enters the furnace, its surface temperature increases because of the heat transfer from the furnace gases and other burning particles. As the coal particle's temperature increases, the contained moisture is vaporized, and the volatile matter in it is released. This volatile matter ignites and burns almost immediately, and this further raises the temperature of the particle, which now consists primarily of carbon and mineral matter (ash). The particle is then consumed at high temperature, leaving the ash and a small amount of unburned carbon.

In the combustion of pulverized coal, the following are important influences on the process:

1. *Volatile matter.* This is critical for maintaining flame stability and accelerating the particle burnout. Coals with low volatile matter, such as anthracite and low-volatile bituminous, are difficult to ignite and require specially designed combustion systems.

2. *Coal particles.* The rate of combustion of the coal particle depends on its size and its porosity.

3. *Moisture content.* Moisture content in the coal influences combustion behavior. Pulverized-coal-fired systems convey all the moisture to the burners, unless preheating is used in the fuel-preparation (pulverization) phase. The moisture presents a burden to coal ignition because the water in the coal must be vaporized as the volatile matter in the coal particles is burned. It is because of this problem that coal drying is done in the pulverization process with the use of preheated air. Not all the moisture in the coal is eliminated, however.

4. *Mineral matter (ash).* The mineral matter or resulting ash in the coal is inert, and it dilutes the heating value. Consequently, with coals of higher ash content, more fuel is required to meet the heat input that is required in the furnace. The ash absorbs heat and interferes with the heat transfer to the coal particles, thus deterring the combustion process with high-ash coals. The type of ash varies in coal, and this reflects the tendency for slagging, which must be accounted for in the boiler design.

Figure 5.21 MPS-type coal pulverizers at a modern power plant. (*Babcock & Wilcox, a McDermott company.*)

With the unit system of pulverization, each boiler is equipped with one or more pulverizing mills through which the coal passes on its way to the burners. A typical arrangement of pulverizers for a large modern utility is shown in Fig. 5.21. The coal is fed to these pulverizing mills by automatic control to meet the steam demand. No separate drying is necessary because heated air from an air preheater is supplied to the pulverizer mill, where drying takes place. This stream of primary air carries the fine coal from the pulverizer mill through the burners and into the furnace. Combustion starts as the fuel and primary air leave the burner tip. The secondary air is introduced around the burner, where it mixes with the coal and primary air. The velocity of the primary and secondary air creates the necessary turbulence, and combustion takes place with the fuel in suspension.

Modern pulverizer systems for boilers are direct-fired systems and consist of the following major features:

1. A raw coal feeder that regulates the coal flow from the coal bunker to the pulverizer

2. A heat source that preheats the primary air for coal drying, either the boiler air heater or a steam-coil air heater

3. A primary air fan that typically is located ahead of the pulverizer (pressurized mill) or after the pulverizer (suction mill)

4. A pulverizer

5. Piping that directs the coal and primary air from the pulverizer to the burners

6. Burners that mix the coal and combustion air, both primary and secondary

7. Controls

The usual pulverizer exit temperature is 150°F, but for various types of coal ranging from lignite to anthracite, exit temperatures vary from 125 to 210°F.

The satisfactory performance of the pulverized-coal system depends to a large extent on mill operation and reliability. The pulverizer mill should deliver the rated tonnage of coal, having a nominal rate of power consumption, produce a pulverized coal of satisfactory fineness over a wide range of capacities, be quiet in operation, give dependable service with a minimum of outage time, and operate with low maintenance cost.

There is considerable variation in the pulverized-coal fineness requirement. Coals with low-volatile content must be pulverized to a higher degree of fineness than those with higher-volatile content. For normal conditions bituminous coal will burn satisfactorily when 70 percent will pass through a 200-mesh sieve. In order to minimize unburned carbon loss, which is contributed to by coarse particles, the amount passing through a 200-mesh sieve is often increased to 80 percent. It is a waste of energy to pulverize coal finer than required to obtain satisfactory combustion.

The three fundamental designs used in the construction of pulverizer mills are contact, ball, and impact mills. Each of these designs attempts to meet certain requirements that are necessary for a good pulverizer design:

1. Optimal fineness for design coals over the operating load range of the pulverizer

2. Rapid response to load changes

3. Safe operation over load range

4. Reliable operation

5. Low maintenance, particularly in grinding elements

6. Capable of handling a variety of fuels

7. Ease of maintenance

8. Minimum space requirements

5.4.1 Contact mills

Contact mills contain stationary and power-driven elements, which are arranged to have a rolling action with respect to each other. Coal passes between the elements again and again, until it has been pulverized to the desired fineness. The grinding elements may consist of balls rolling in a race or rollers running over a surface. A stream of air is circulated through the grinding compartment of the mill (Fig. 5.22a). The rotating classifier allows the fine particles of coal to pass in the airstream but rejects the oversize particles, which are returned for regrinding. The airstream with the pulverized coal flows directly to the burner or burners. (Refer to Figs. 2.25, 2.33, 2.40, and 2.41 for typical arrangements of pulverizer systems, each with a different pulverizer design.)

In the sectional views of Fig. 5.22, the primary air fan supplies a mixture of room air and heated air for drying the coal and for transporting it from the mill and then through the burners and into the furnace. The proportion of room air to heated air is varied, depending on the moisture content of the coal, to provide sufficient drying. Coals having a moisture content of up to 40 percent have been handled successfully in pulverizers. The fan is on the inlet and therefore is not subject to being worn by the air and pulverized-coal mixture. However, the mill is pressurized, and any casing leakage will cause pulverized coal to be blown into the room.

This mill uses steel balls and rings or races as grinding elements. The lower race is power driven, but the upper is stationary. Pressure is exerted on the upper race by means of springs. The force exerted by these springs can be adjusted by the screws that extend through the top of the mill casing. The coal is pulverized between the balls and the lower race. The capacity of these mills ranges from 1.5 to 20 tons per hour.

The grinding elements of these mills are protected from excessive wear and possible breakage that can result when heavy foreign objects are found in the coal. These heavy particles (pyrites) resist the upward thrust of the stream of primary air and collect in a compartment in the base, where they are removed periodically.

The coal supply to the burners is regulated automatically by the combustion control. When the control senses an increased demand for steam, more primary air is supplied to carry additional coal to the burners. In this way the residual coal in the mill is utilized to give quick response to a demand for more steam. The mill coal-feed controller senses a drop in differential pressure across the grinding

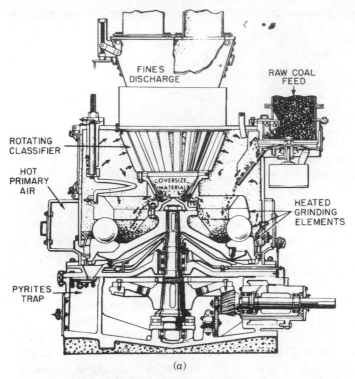

(a)

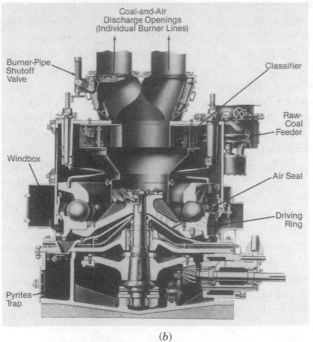

(b)

Figure 5.22 Ball-and-race-type pulverizer mill. (a) Cross section showing path of coal and air. (b) Type EL single-row pulverizer. (*Babcock & Wilcox, a McDermott company.*)

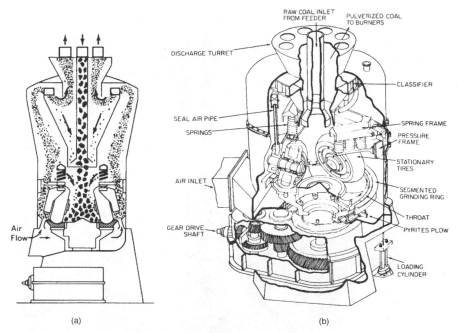

Figure 5.23 Roll-and-race-type pulverizer. (*a*) Cross section showing path of coal and air. (*b*) Type MPS pulverizer. (*Babcock & Wilcox, a McDermott company.*)

elements, due to the decrease in quantity of coal, and increases the coal feed to supply the deficiency.

The mill shown in Figs. 5.23*a* and 5.23*b* uses rolls and a ring as grinding elements. The ring is power driven, but the roll axles are stationary. The revolving ring causes the rolls to rotate, and the coal is ground between the two surfaces. Grinding pressure on the rolls is set at three external points. Coal drying takes place with the air sweeping around the track. The air is also used for transporting the pulverized coal through the classifier to the burners. The segmented grinding ring and replaceable "tires" on the rolls are provided to reduce and facilitate maintenance. These pulverizers can be designed for capacities ranging from 17 to 105 tons per hour.

In the ring-and-roller-type mill shown in Fig. 5.24, coal enters through the center tube and is deposited on the power-driven grinding table and ring. In this vertical spindle-type of pulverizer, pulverization is accomplished by this ring and three contacting rollers. Pressure between the rollers and the grinding ring is maintained by independent grinding roller assemblies and individual roller tensioners. The motion of the grinding ring is transmitted to the rollers, and they rotate at about one-half the speed of the ring. The pulverized coal is carried by primary air up through the adjustable classifier and

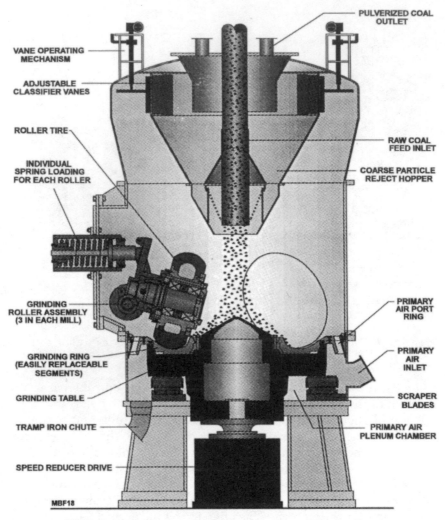

PULVERIZED COAL
OUTLET

VANE OPERATING
MECHANISM

ADJUSTABLE
CLASSIFIER VANES

ROLLER TIRE

INDIVIDUAL
SPRING LOADING
FOR EACH ROLLER

RAW COAL
FEED INLET

COARSE PARTICLE
REJECT HOPPER

GRINDING
ROLLER ASSEMBLY
(3 IN EACH MILL)

GRINDING RING
(EASILY REPLACEABLE
SEGMENTS)

GRINDING TABLE

TRAMP IRON CHUTE

SPEED REDUCER DRIVE

PRIMARY
AIR PORT
RING

PRIMARY
AIR
INLET

SCRAPER
BLADES

PRIMARY AIR
PLENUM CHAMBER

MBF18

Figure 5.24   MBF ring-and-roller-type coal pulverizer. (*Foster Wheeler Corp.*)

out the angular opening at the top of the mill and into the burner
lines.

The pulverizer shown in Fig. 5.25 has grinding elements consisting
of stationary rollers and a power-driven bowl in which pulverization
takes place as the coal passes between the sides of the bowl and the
rollers. A primary air fan draws a stream of heated air though the
mill, carrying the fine particles of coal into a two-stage classifier
located on the top of the pulverizer. The vanes of the classifier return
the coarse particles of coal through the center cone to the bowl for further grinding, while the coal that has been pulverized to the desired

Figure 5.25 Bowl-type pulverizer with primary air fan. (*ABB Combustion Engineering Systems.*)

fineness passes out of the mill, through the fan, and into the burner lines. The automatic control changes the supply of coal to the bowl of the mill by adjusting the feeder speed and the flow of primary air by regulating a damper in the line from the pulverizer to the fan. This arrangement of primary air fan has the advantage of having a cleaner facility because the mill is under suction rather than being pressurized, thereby eliminating coal leakage. However, the fan now must handle a coal-air mixture and, therefore, its design must incorporate additional preventive-wear features to offset the erosion caused by the coal-air mixture.

These mills have several mechanical features. The roller assembly bearings in Fig. 5.26 are pressure lubricated to exclude coal dust. This figure also shows how the roller-support bearing and spring may be adjusted to hold the rollers in place relative to the grinding-ring surface of the revolving bowl. The roller does not come into contact with the grinding ring but allows some space for the coal to pass. The classifier located in the top of the mill may be adjusted to change the coal fineness while the mill is operating. As stated previously, the primary air fan is located on the outlet, which results in the mill's oper-

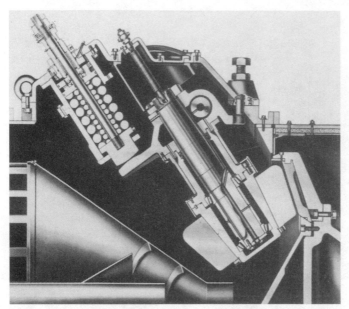

Figure 5.26   Roller assembly and spring pressure mechanism of bowl-type pulverizer. (*ABB Combustion Engineering Systems.*)

ating under a draft or negative pressure, and this eliminates the leakage of coal dust from the mill casing. The impurities in coal compose the heavy particles, and when these enter the rotating bowl, they are thrown over the side by centrifugal force. These heavy particles then fall into the space below the bowl and are discharged from the mill through a specially provided spout.

Pulverized coal may be readily transported through a pipe by means of a high-velocity stream of air, but the coal will not be equally distributed through the cross section of the pipe. This causes difficulty when two burners are supplied from one pulverizer mill. The distributor (Fig. 5.27) divides the cross section of the main pipe into several narrow strips. The mixture of pulverized coal and air that enters these strips is directed into the two burner pipes in order to obtain equal amounts of coal to the burners.

5.4.2   Ball mills

The typical ball mill, shown in Fig. 5.28, consists of a large cylinder or drum partly filled with various-sized steel balls filling approximately 30 percent of the cylinder volume. The cylinder is slowly rotated while coal is fed into the drum. The coal mixes with the balls and is pulverized. Hot air enters the drum, dries the coal during pulveriza-

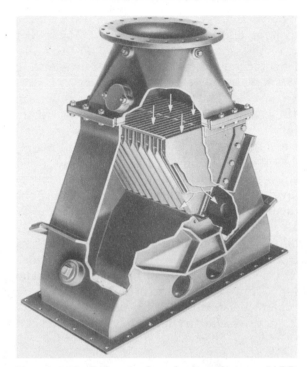

Figure 5.27 Pulverized-coal distributor. (*ABB Combustion Engineering Systems.*)

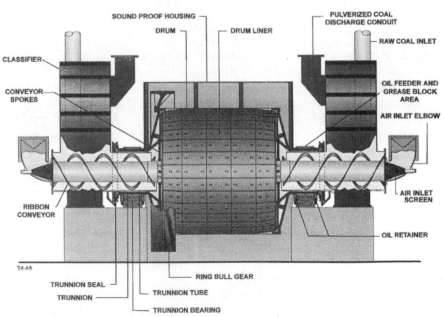

Figure 5.28 Ball-mill-type coal pulverizer. (*Foster Wheeler Corp.*)

tion, and then carries the pulverized fuel out of the drum through the classifiers and to the burners. The classifiers return the oversized particles to the drum for further grinding. In addition to the mills, coal feeders, a source of heated air, primary air fans, burners, and controls are required for a complete pulverized-coal system.

These pulverizers have several features. Unlike the grinders in most other pulverizers, the grinding elements in these units are not seriously affected by metal and other foreign matter in the coal. The pulverizers contain enough coal for several minutes of operation. This feature prevents the fire from going out when there is a slight interruption in fuel feed caused by coal clogging in bunkers or spouts. These mills have been used successfully for a wide range of fuels including anthracite and bituminous coal, which are difficult to pulverize.

For new boilers, ball mills have in most cases been replaced by vertical air-swept pulverizers. Ball mills require more building volume and consume more power. They are also more difficult to control, and erosion is more prominent. However, they are highly efficient in grinding very abrasive and low-moisture coals, and because they have a long coal residence time, they are effective for fine grinding.

5.4.3 Impact mills

In a pulverizer mill that uses the impact principle, the coal remains in suspension during the entire pulverizing process (Fig. 5.29). All the grinding elements and the primary air fan are mounted on a single shaft. The primary air fan induces a flow of air through the pulverizer, which carries the coal to the primary stage, where it is reduced to a fine granular state by impact with a series of hammers, and then into the final stage, consisting of pegs carried on a rotating disk and traveling between stationary pegs, where pulverization is completed by attrition. The now finely pulverized coal is carried to the center of the pulverizer, where it encounters the rapidly rotating scoop-shaped rejector arms, which throw the large particles back into the grinding section, while those of the desired fineness are passed through the fan and discharged into the burner lines.

The principle of operation of the impact mills results in several interesting characteristics. These pulverizers may be connected directly to the motor drive and operated at high speed. There is only a few seconds' coal supply in the mill, and any variation in the rate of feed to the mill is almost immediately reflected in the output. This means that the power required to drive the pulverizer is nearly proportional to the coal pulverized over a wide load range. Because the mill operates at high speed and the fan is integral with the pulverizer, a minimum of floor space is required.

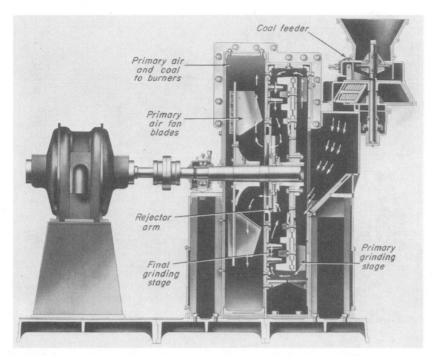

Figure 5.29   Cross section of Atrita coal pulverizer. (*DB Riley, Inc.*)

Control of pulverizer output is accomplished by varying the coal feed and the flow of primary air by automatic control.

5.4.4   Mill feeders

Care must be exercised to prevent scrap metal from entering pulverizer mills. Figure 5.30 shows a feeder with a magnetic separator used in connection with an impact pulverizer. The rotating drum (A) has eight coal pockets. The spring-loaded apron (B) levels off the coal and prevents flooding past the feed, yet it will release if there is a large lump of coal in one of the pockets. The wiper (C) is geared to the main feeder and rotates eight times as fast as the main feeder shaft. This wiper completely removes the coal from each pocket, thus helping to maintain uniform feed even when the coal is so wet that it has a tendency to stick in the feeder. From the feeder the coal drops onto a belt that runs over the magnetic pulley (D). The coal drops vertically from the belt, while magnetic material is carried to the left and is deposited in the box, from which it may be removed periodically. The raw coal to all pulverizers should be passed over a magnetic separator.

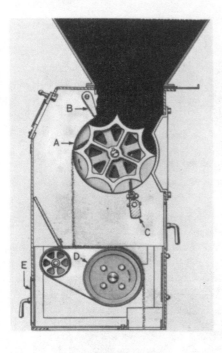

Figure 5.30 Pulverizer feeder with magnetic separator. A = Rotating drum. B = spring-loaded apron. C = wiper. D = magnet pulley. E = inspection door. (*DB Riley, Inc.*)

5.4.5 Gravimetric feeders

In any combustion process, there are two primary control requirements:

1. Feed just enough fuel to meet the demand for heat input.

2. Provide just enough air to burn the fuel.

In a boiler, the first requirement is relatively easy to control by measuring steam pressure. If too much fuel is being burned, steam pressure rises, and if too little fuel, steam pressure falls. Thus a boiler control uses steam pressure as a major input signal.

The second requirement is critical for the efficiency of the boiler. Too much air wastes heat because the additional air is heated, and the heat is lost to the atmosphere from the stack. Too little air results in incomplete combustion, and the unburned fuel is lost with the ash. Insufficient air also creates a dangerous condition in that an explosive potential exists because of the fuel-rich mixture.

Coal is not a uniform material; it varies in heating value, ash and moisture content, and density even when it comes from the same mine. It is because of these variations that a gravimetric feeder is used for coal feed as compared with a volumetric feeder.

When handling a heterogeneous material such as coal that consists of coarse and fine particles, the density changes significantly with any change in surface moisture. When firing coal, there is no effective way to measure heating value (or Btu) flow, so this value must be determined from either volume or weight flow. The volume measurement will be in error by the variation in both density and heating value. However, a weight flow signal will be in error only by the variation in the heating value.

The combustion-control system handles these variations in heating value and other measurement inaccuracies by feedback control using signals from the boiler such as steam pressure, air-fuel ratio, etc. The control system allows sufficient excess air to prevent a dangerous fuel-rich mixture. The gravimetric feeder provides a more accurate fuel input measurement to the combustion-control system, which results in higher boiler efficiencies and potentially lower maintenance. A schematic of a gravimetric feeder system is shown in Fig. 5.31, and an illustration of the feeder is shown in Fig. 5.32.

The gravimetric feeder weighs material on a length of belt between two fixed rollers located in the feeder body. A third roller, located midway in the span and supported at each end by precision load cells, supports half the weight on the span. As material passes over the span, the load cell generates an electric signal directly proportional to the weight supported by the center roller. The microprocessor takes samples of the amplified output of each load cell several times a second. This output is compared with parameters that are stored in the system memory. If the load cell outputs are within the expected range, each is converted into a signal equivalent to weight per unit of belt length. They are then added to obtain total weight per unit of belt length.

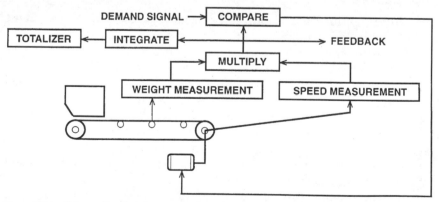

Figure 5.31 Schematic showing operation of a gravimetric feeder. (*Stock Equipment Company.*)

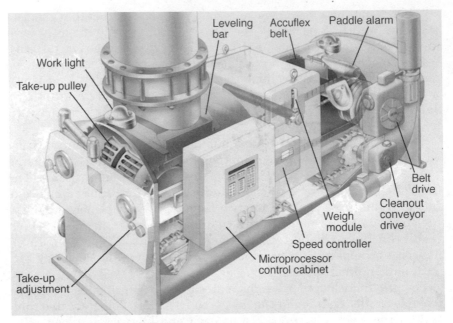

Figure 5.32 Gravimetric feeder. (*Stock Equipment Company.*)

The microprocessor matches the feeder output to the demand output by adjusting the motor speed on the feeder. This corrects for changes in either demand or material density. In Fig. 5.33, the location of a gravimetric feeder is shown in a typical coal bunker-to-pulverizer inlet arrangement. Note the requirements for various coal valves in the system.

5.4.6 Burners

The efficient utilization of pulverized coal depends to a large extent on the ability of the burners to produce mixing of coal and air and to develop turbulence within the furnace. Emission requirements also dictate that efficient combustion takes place with low NO_x emissions.

The pulverized-coal-fired burner shown in Fig. 5.34 is a modern burner that produces a less turbulent, slower mixed flame that reflects gradual coal-air mixing and low NO_x. It is unlike the short, highly turbulent, actively mixed coal flame of burners that have been used in the past and resulted in high NO_x emissions.

A venturi section at the end of the coal nozzle concentrates fuel and air in the center of the burner nozzle. As this fuel-rich mixture passes over a coal spreader, the blades divide the coal stream into a small number of individual streams. This results in a gradual mixing of the

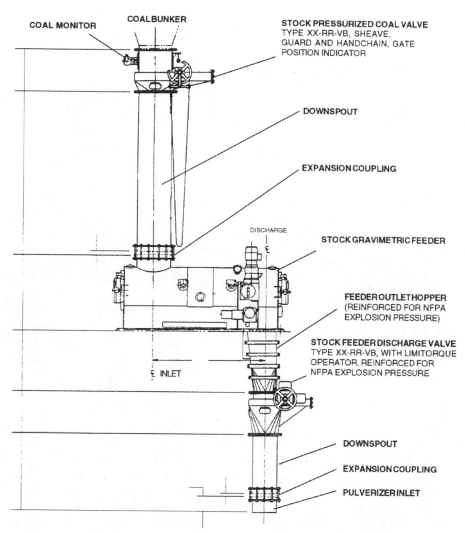

COAL MONITOR   COAL BUNKER   STOCK PRESSURIZED COAL VALVE
TYPE XX-RR-VB, SHEAVE,
GUARD AND HANDCHAIN, GATE
POSITION INDICATOR

DOWNSPOUT

EXPANSION COUPLING

DISCHARGE

STOCK GRAVIMETRIC FEEDER

FEEDER OUTLET HOPPER
(REINFORCED FOR NFPA
EXPLOSION PRESSURE)

STOCK FEEDER DISCHARGE VALVE
TYPE XX-RR-VB, WITH LIMITORQUE
OPERATOR, REINFORCED FOR
NFPA EXPLOSION PRESSURE

℄ INLET

DOWNSPOUT

EXPANSION COUPLING

PULVERIZER INLET

Figure 5.33 Arrangement of equipment from coal-bunker outlet to pulverizer inlet.
(*Stock Equipment Company.*)

fuel and air with a subsequent reduction in peak flame temperature,
and these conditions minimize the formation of NO_x.

The burner uses curved, overlapping air register blades to provide
swirl control and low burner pressure drop. Secondary air flow is controlled by a movable shroud that slides over the vanes. These separa-
rate adjustments allow the controlling of combustion and of NO_x.

Figure 5.35 shows a circular single-register-type burner designed
for firing pulverized coal only. Many pulverized coal burners have oil

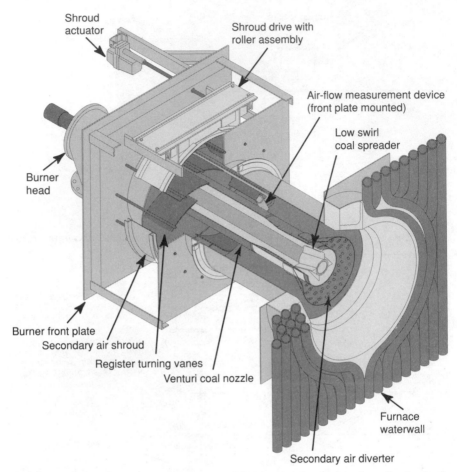

Figure 5.34 Controlled combustion venturi (CCV) burner for the burning of pulverized coal and low NO$_x$ production. (*D. B. Riley, Inc.*)

burners that extend through the coal-carrying tube (see Fig. 5.43). These multifuel burners are effective during emergency conditions, when fuel availability dictates their use, or during certain economic situations.

Adequate combustion volume must be provided for satisfactory combustion of pulverized coal. Flame temperatures are high, and refractory-lined furnaces were proven inadequate many years ago. The use of waterwalls permits higher rates of heat release with a consequent reduction in furnace size. The furnace volume must be sufficient to complete combustion before the gases reach the relatively cool heating surface. However, the use of burners that produce turbulence in the furnace makes it possible to complete combustion with flame travel of minimum length.

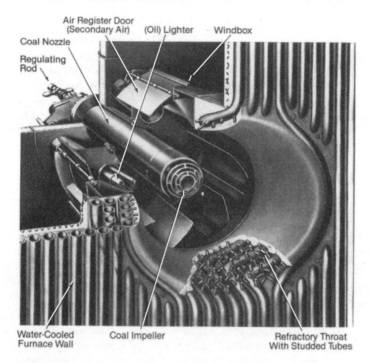

Air Register Door
(Secondary Air)   (Oil) Lighter   Windbox

Coal Nozzle

Regulating
Rod

Water-Cooled   Coal Impeller   Refractory Throat
Furnace Wall   With Studded Tubes

Figure 5.35 Circular register burner for pulverized-coal firing.
(*Babcock & Wilcox, a McDermott company.*)

Until the early 1970s, the primary requirement of combustion system development was to design compact boilers that were cost-effective. As a result, the burner system focused on maximizing the heat input per unit of volume, and this resulted in small furnace volumes, rapid mixing burners, and very high flame temperatures. Although these burners performed very well, this design produced an unintended side effect of high levels of nitrogen oxides (NO_x) that now are considered an emission pollutant. The circular burner shown in Fig. 5.35 has been used for many years and is considered a conventional pulverized-coal burner. For certain applications it is still used today.

Regulations have been implemented throughout the world on NO_x emissions. The most cost-effective means is the use of low-NO_x combustion technology either by itself or in combination with other techniques.

More than 75 percent of the NO_x formed during conventional pulverized-coal firing is from the fuel (fuel NO_x), with the remaining coming from dissociation of nitrogen and oxygen in the air during the combustion process and the recombination of these two elements to form nitrogen oxides (thermal NO_x). Therefore, the most effective combustion technique is to limit the formation of fuel NO_x. Fuel NO_x

is formed by oxidation of the nitrogen that is bound to the fuel during devolatilization and particle burnout. Coal contains nitrogen, generally ranging from 0.5 to 2 percent. The availability of large quantities of oxygen in combustion air together with high flame temperatures leads to the conversion of this fuel-bound nitrogen to NO_x.

The most effective means of reducing fuel-based NO_x formation is to reduce the availability of oxygen, i.e., combustion air, during fuel-air mixing (primary air). Additional air (secondary air) is added later in the process to complete burnout of the coal and to obtain high combustion efficiency.

The availability of oxygen can be reduced in several ways: A portion of the combustion air can be removed from the burner and introduced at another location in the furnace. This method is called *two-stage combustion.* Alternatively, the burner can be designed to supply all the combustion air but to limit the air to the flame. Only a portion of the required air is allowed to mix with the coal. The remaining air is mixed downstream in the flame to complete combustion. Air-fuel mixing is reduced, and the flame envelope is larger as compared with the rapid mixing of conventional burners.

When compared with uncontrolled levels, NO_x emissions are reduced from 30 to 70 percent when using low-NO_x burners. These burners also can be used in combination with air staging (two-stage combustion) through NO_x ports to further reduce NO_x and meet emission limits. There are designs of low-NO_x burners that can meet emission requirements without the use of NO_x ports.

A low-NO_x burner design is shown in Figs. 5.36 and 5.37. It is a dual-register burner providing two air zones, each of which is controlled by a separate register, where the air zones are around an axially positioned coal nozzle. A pitot tube is located in the burner barrel for the measurement of secondary airflow. With this information, secondary air can be uniformly distributed between all burners by adjusting the sliding air damper. Balanced air and fuel distribution among all burners is critical to good combustion efficiency. Downstream of the pitot tube are the inner and outer air zones.

Adjustable vanes in the inner zone stabilize ignition at the burner nozzle tip. The outer zone is the main air path, and it is equipped with two stages of vanes. The upstream set is fixed and improves peripheral air distribution within the burner. The downstream set is adjustable and provides proper mixing of this secondary air into the flame.

The burner nozzle incorporates a conical diffuser and flame-stabilizing ring. These combine to improve flame stability while further reducing NO_x by using fuel-staging technology. This burner reduces NO_x by 50 to 70 percent from uncontrolled levels without using NO_x

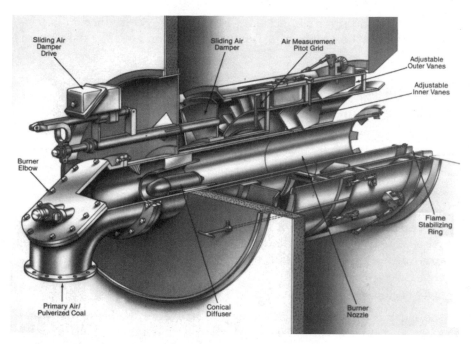

Figure 5.36 Low-NO$_x$ pulverized-coal-fired burner. (*Babcock & Wilcox, a McDermott company.*)

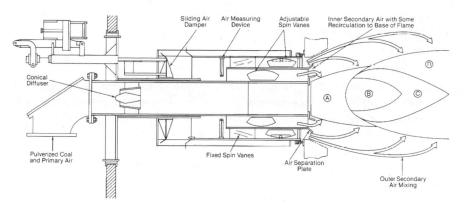

Figure 5.37 Combustion zones for low-NO$_x$ pulverized-coal-fired burner. A = high-temperature, fuel-rich devolatization zone. B = production of reducing species. C = NO$_x$ decomposition zone. D = char oxidizing zone. (*Babcock & Wilcox, a McDermott company.*)

ports for secondary air injection. This level of NO_x reduction is possible through the creation of flame combustion zones, as shown in Fig. 5.37.

5.4.7 Lighters

A lighter is required to initiate combustion in the following situations:

1. As pulverized coal is first introduced to the burners

2. As burners are being shut down

3. As required for flame stability

Lighters, similar to the one shown in Fig. 5.35, use an electrically generated spark to ignite a lighter fuel, either natural gas or no. 2 fuel oil. Modern units have automated controls that activate, operate, purge, and shut down the lighters.

5.4.8 Flame safety system

Modern pulverized-coal-fired boilers are equipped with a flame safety system. A scanner is used at each burner to electronically monitor the flame conditions. Flame scanners evaluate the performance of the lighters and the burners. The lack of satisfactory flame causes the system to automatically take corrective action, which is to shut down the burner. This prevents unburned fuel from entering the furnace, and, therefore, the system significantly reduces the risk of an explosion.

5.4.9 Operation

The ash in the coal presents a problem that must be given careful consideration in furnace design. Small and medium-sized units are arranged for the removal of ash from the furnace while it is in the dry state. In this case, care must be exercised to prevent the temperature of the ash from exceeding its fusion temperature. If this precaution is not taken, large masses of slag will form. In some large units the ash in the furnace is maintained at a temperature in excess of the melting point. The slag is "tapped" and allowed to flow from the furnace in the liquid state. After leaving the furnace, it is rapidly chilled by the application of water. This quick change in temperature causes slag to shatter into small pieces as it is collected in the ash-handling system.

The use of slag-tapped furnaces enables the operator to use a minimum of excess air because the furnace can be maintained at a high temperature without slagging difficulties. Since the usual slagging difficulties are overcome, it is possible to use a wide variety of coals, including those having a low-fusion-temperature ash.

The principle of pulverized-coal firing has long been widely accepted, first in large utility systems and then in industrial plants. The choice of pulverized coal in preference to other firing methods depends on the size of the boiler unit, the type of coal, the kind of load (constant or fluctuating), the load factor, the cost of fuel, and the training of operating personnel.

Pulverized-coal equipment has not been adapted generally to small units. The benefits in efficiency and flexibility do not warrant the complications in equipment, space requirements, initial costs, and operating technique.

Some of the characteristics of coal that are so important when it is burned on grates need less consideration in the pulverized-coal-fired plant. Coal in sizes less than 2 in can be pulverized without difficulty, and segregation usually does not affect operation. Coking and caking characteristics, so important in underfeed stoker operation, require less consideration. In general, high-volatile coals are harder to grind (have lower grindability) than low-volatile coals, requiring larger mills for the same capacity, more power per ton, and more maintenance. The disadvantages of high-volatile coal are partly compensated by the fact that the gases are more easily distilled and burned, making it unnecessary to pulverize the coal as finely as is required for low-volatile coal. For ease in maintaining ignition, it is recommended that the volatile content of coal used in a pulverized-coal-fired plant be over 30 percent. Lower-volatile coals may require pilot fuel testing and special burner design, and coals having a low volatile content have been pulverized effectively. These coals have included low-volatile bituminous and anthracite.

Because there is only a small amount of fuel at any one time in a pulverized-coal furnace, it is possible to change the rate of combustion quickly to meet load demands. This is frequently one of the factors worthy of consideration when combustion equipment is selected. The possible rate of change in output, from the banked condition to full load, is usually limited by the boiler rather than by the pulverizer and combustion equipment. Automatic control applied to pulverized-coal-fired boilers is effective in maintaining an almost constant steam pressure under wide load variations.

A pulverized-coal plant costs more than some of the other methods used in burning coal, and this difference in cost must be repaid in lower operating costs including decreased maintenance and increased efficiency resulting in lower fuel costs. To realize the greatest possible financial benefit from this increased efficiency, the equipment must be utilized at high steam flows and for a large part of the time, i.e., high capacity factors. A pulverized-coal plant can utilize a wide variety of coal, including the lower grades. The use of low-cost fuel frequently justifies the increased cost of pulverizing equipment. When the fuel cost is high, the installation refinements in combustion and heat-recovery equipment can be justified. The question is, Will the cost of equipment required for increased economy be justified by fuel saving?

The pulverized-coal plant requires trained operators who are familiar not only with the dangers involved but also with safe operating practices. To control the rapid combustion of the pulverized coal, as it is burned in suspension, requires an alert operator.

In order to have an explosion with pulverized coal there must be the correct mixture of coal dust and air, as well as a source of ignition. It follows that for safe operation the mixture of coal and air must be either too "lean" (too much air) or too "rich" (too much coal) to explode, except in the furnace. This principle is applied both in the boiler room and in the pulverizer. In the boiler room all areas are kept clean so that there is no dust to be agitated by a draft or minor explosion so as to produce sufficient concentration in the air to be exploded by a chance spark or an open flame. Inside the pulverizer a high ratio of coal to air (rich mixture) is maintained so that there will not be enough air to cause an explosion even if a spark is produced in the pulverizer. Once a combustible mixture has been established, the application of a spark or an open flame will cause an explosion. The pulverized coal must be burned as fast as it is introduced into the furnace or an explosive mixture can be produced.

Operators of pulverized-coal-fired equipment should carefully review all operating instructions for the equipment as provided by the supplier. These instructions should provide specific information on the design, calibration, adjustment, startup, operation, and shutdown procedures for the various pieces of equipment and the system in its entirety. In addition, recommended maintenance and ideas for troubleshooting should be understood by the operators.

The advantages and disadvantages of a pulverized-coal-fired plant are

Advantages

Can change rate of combustion quickly to meet varying load

Requires low percentage of excess air

Reduces or eliminates banking losses

Can be repaired without cooling the unit down, since equipment is located outside the furnace

Can be easily adapted to automatic combustion control

Can utilize highly preheated air successfully

Can burn a wide variety of coal with a given installation

Limits use of fine wet coal only through the ability of the conveying and feeder equipment to deliver the coal to the pulverizers

Increases in thermal efficiency

Can produce high steam capacity

Disadvantages

Is costly to install

Requires skilled operating personnel because of explosion potential

Requires multiple mills and burners to obtain a satisfactory operating range

Requires extra power to pulverize the fuel

Requires greater housekeeping to ensure a clean plant

Before attempting to operate a pulverized-coal-fired boiler, the following should be thoroughly inspected: fuel-burning equipment, controls, fans, and safety interlock equipment. The failure of any one might result in an explosion. If, for example, the induced-draft fan should fail and the pulverizer mill and primary air fan continue to operate and supply coal to the furnace, an explosion might occur. In order to reduce such a possibility, the boiler auxiliaries are equipped with interlocks, by means of which when one auxiliary fails, the others stop automatically. When the induced-draft fan fails, the interlocks are arranged to stop the mill feeder, the mill, and the forced-draft fan in the order listed. Electric-eye flame detectors are used to indicate flame failure and to disrupt the fuel supply when flame failures occur. Television receivers are installed in the control panel to permit constant observation of the flame. Study the safety equipment on the boiler, check it at regular intervals, and make sure that it is in good operating condition at all times. Remember, it was placed there for your protection, very probably as the result of difficulty encountered with similar equipment.

The exact procedure for lighting a pulverized-coal-fired furnace depends to some extent on the types of pulverizer, burners, and other equipment. There are, however, certain general practices that are common to all. First, operate the induced-draft fan and set the dampers to produce a furnace draft sufficient to prevent a pressure from being developed when the fire is lighted. Then start the igniter and make sure that it is producing sufficient flame in the path of the pulverized coal. Now start the pulverizer and feeder, and adjust the flow of primary air and the mill-coal level to supply a rich mixture to the burner. Regulate the coal and air supply to the burner, and maintain as rich a mixture as possible without smoke until the furnace has been heated. Do not depend on hot walls or other fire from an adjacent burner to light a pulverized-coal burner. Use an igniter for each burner. If the coal as it is discharged from a burner fails to ignite after even a few seconds, the furnace will be filled with an explosive mixture. When this occurs, it is necessary to operate the induced-draft fan until the explosive mixture has been removed. The flame safety system will safely shut the system down.

Care and judgment on the part of the operators are necessary during the warming-up period. There is a low limit of output below which a given pulverizer mill cannot be operated satisfactorily. When the primary air drops below a minimum velocity, it fails to remove sufficient coal from the mill, and a lean mixture results. This minimum mill output usually results in too high a firing rate during the warming-up period. The only solution is to fire the boiler at the lowest rating, intermittently and frequently enough to bring the unit on line in the specified time. Safety precautions and procedures must be carefully followed during each lighting.

Since the rate of change in firing may be high and the flames may be extinguished by momentary interruptions of coal feed, more precautions are necessary with pulverized coal than when coal is burned on grates. Pulverized-coal boilers are equipped with automatic controls, which relieve the operators of tedious adjustments of fuel and air. These controls do not, however, lessen an operator's responsibility for the equipment. In case of failure or emergency, the unit must be operated safely by hand control. The control regulates the rate of coal feed to produce the required amount of steam in order to maintain the established pressure. The air is regulated to burn the coal with the amount of excess air considered most satisfactory for the given installation. The steamflow-airflow meter and the carbon dioxide and oxygen recorders are used to indicate the amount of excess air. Burners, louvers, and primary air pressure must be adjusted to produce a flame that will not strike the furnace walls yet will result in almost complete combustion before the gases enter the boiler passes.

When the mixture of coal and primary air becomes too lean, the flame will have a tendency to leave the burner tip and then return. This condition may become so severe that the combustion consists of a series of small explosions with danger of loss of ignition. This pulsating condition usually can be corrected by decreasing the flow of primary air and thereby producing a richer mixture at the burner tip.

Slag deposits in the furnace are a guide to the operator in adjusting the flame shape. Excessive temperatures in the furnace are caused by a deficiency of air, unequal distribution of fuel, or excessive ratings, and they result in slag formation and high furnace maintenance. The heated and tempering air to the pulverizer mill must be proportional to produce an air and pulverized-coal outlet-mill temperature of 150 to 180°F for most bituminous coals. For anthracite coals having a low volatile content, exit temperatures are generally 200 to 210°F. The moisture content of the coal entering the mill determines the amount of preheated air that must be mixed with the room tempering air to produce the desired outlet temperature.

Pulverized-coal-fired boilers are banked by closing off the coal supply and thus extinguishing the fire. Insert the igniter and stop the coal feed to the mill. Allow the pulverizer to operate until all the coal has been removed. When the pulverizer is nearly empty, the coal feed will be irregular, and so it is necessary to have the igniter in place to prevent the flame from being extinguished. If it is necessary to have the boiler ready for service on short notice, light the burners at intervals to maintain the steam pressure.

Practically all the pulverized-coal system is located outside the furnace and therefore may be serviced, as has been said, without waiting for the boiler to cool down. Boilers are equipped with multiple mills to permit part-load operation with one mill out of service. This makes it possible to adjust and maintain the pulverizer mills without taking the boiler out of service. Some boilers are arranged so that they can be supplemented with fuel oil if the pulverizer equipment fails.

The principal parts of a unit pulverizer system that require adjustment and maintenance are the grinding elements, mill liners, primary air fan rotors and liners, piping, burners, and classifiers.

Maintenance costs for all types of pulverizers depend on wear life of the components, which varies with the abrasiveness of the coal. Abrasiveness is related to the quantity and size distribution of the quartz and pyrite that are found in the coal. Pyrite is a compound of iron and sulfur that naturally occurs in coal.

Maintenance and power costs also depend on operating practices. An experienced operator, for example, may notice an increase in pulverizer pressure differential, which may mean a need for spring adjustment. As wear increases on a mill, major decisions have to be made on whether to incur maintenance repair or replacement costs on worn parts or whether to absorb the higher power costs that occur during the end of wear life. Planning maintenance during scheduled outages ensures that optimal availability of the plant is obtained.

In the contact mill shown in Fig. 5.22, the spring tension must be adjusted to compensate for wear of balls and races. This adjustment is made by means of the bolts that extend through the top of the mill. These bolts are adjusted to maintain a specified spring length. Once arranged to give the correct fineness, the classifier does not require further adjustment. The primary air fan handles the clean air to the mill and therefore requires very little maintenance. The satisfactory operation of the mill depends, to a considerable extent, on the quantity of coal maintained in the grinding compartment, as determined by the mill-feeder-controller setting. When there is insufficient coal in the mill, the mixture supplied to the burner is too lean for satisfactory combustion. If there is too much coal in the mill, a considerable amount will be discharged to the impurities-reject compartment or pyrites trap (see Fig. 5.22), causing the operator unnecessary work.

The roller-type contact pulverizer requires occasional adjustment of the grinding elements. During operation, the bowl is partly filled with coal. The lower end of the rollers and the bottom of the bowl liner ring are subjected to more wear than the other parts of these grinding elements. As this wear increases, the space between the rollers and the ring increases, and the coal level in the bowl must be higher for a given output. It then becomes necessary to adjust grinding surfaces so that they will be parallel and $\frac{3}{8}$ to $\frac{1}{4}$ in apart. Failure to maintain this adjustment will result in overloading the driving motor and in spillage of coal from the impurities-discharge spout, especially at high ratings. The roller-spring tension is established by the manufacturer to satisfy specific load and coal conditions. This tension is determined by the spring length and should not be altered without thorough investigation. When it is impossible to adjust the rollers to maintain the relation between the grinding surfaces, the rollers and grinding ring must be replaced. Some operators have found it practical to replace the rollers twice for every one replacement of the grinding ring.

The primary air fan in the roller-type contact pulverizer shown in Fig. 5.25 handles the mixture of air and coal as it comes from the mill and is therefore subject to wear. This fact has been taken into consideration in the design of both the fan blades and the casing in that they have replaceable liners. The stationary classifier by which the fineness of the coal can be regulated may be readily adjusted while the mill is in operation.

The ball mill shown in Fig. 5.28 is supplied with the necessary assorted sizes of steel balls when first placed in service. Wear reduces the size of the balls, and the grinding capacity is maintained by adding a sufficient number of the large balls. Normal wear reduces the size of the balls and thus provides the necessary variation in size. The liners in the cylinder are wear resistant and seldom require replacement. The coal and air mixture passes through the primary air fan, which is therefore subject to some wear, but the parts subject to wear are easily replaceable.

The impact pulverizer shown in Fig. 5.29 operates at the same speed as the motor and therefore eliminates the use of reducing gears. Wear occurs in the pins that perform the secondary or final pulverization. These parts are designed, however, so that considerable wear can take place before the fineness is appreciably affected. All parts, including the primary air fan, are readily accessible for inspection and replacement by removal of the upper casing in much the same manner that a centrifugal pump is inspected.

Whatever the type of mill, every effort should be made to keep the pulverizer system tight, since the leakage of pulverized coal is dangerous; many explosions have resulted from the careless practice of allowing pulverized-coal dust to settle on the equipment. When pulverized-coal dust leaks out of the system, it should be cleaned up immediately and the leaks repaired.

All portions of the pulverized-coal system should be inspected and maintained with the goal of meeting optimal operating conditions. Coal feeders, burner lines, burners, fans, and control dampers all must be kept in good operating condition to ensure good performance.

In general, pulverized-coal-fired systems should meet the following operating requirements:

1. The coal and air feed rates must comply with the load demand over the operating range. For modern systems with high-volatile bituminous coal, flames should be stable without the use of lighters from about 30 percent full load to full load.

2. Unburned combustible loss (UCL) should reduce efficiency by less than 1 percent and for many coals by less than 0.2 percent.

3. The burner should maintain performance without continual adjustment.

4. Automatic flame safety and combustion-control systems should be used to ensure the safest operation under all conditions.

5.5 Fuel Oil

The burning of fuel oil in a furnace requires consideration in the selection of equipment and grade of oil as well as attention to operating details. The necessary amount of plant storage capacity depends on the amount of oil consumed, availability of the supply, and transportation facilities. The oil-heating equipment must be adequate to heat the heaviest oil that is to be burned at the maximum rate of consumption in the coldest weather. The pumps should deliver the maximum quantity at the pressure specified for the type of burner used. The pumps should be installed in duplicate, each having sufficient capacity for maximum requirement. The burners must be selected to deliver the required quantity of oil to the furnace in a fine mist to ensure quick and thorough mixing with the air.

Recommendations and requirements for safe storage and transport of fuel oil are explained by the National Fire Protection Association. In addition, the installation also must comply with the local codes and insurance company requirements. Meeting these requirements is the responsibility of the designer, but the operator also must be familiar with the requirements to make certain that codes are not violated by maintenance changes, operating practice, or neglect.

The use of supertankers for the transport of crude oil has reduced the cost of transportation significantly, and refineries are generally located in areas of consumption rather than adjacent to oil fields.

Extensive piping and valves as well as pumping and heating equipment are necessary for the transportation and handling of fuel oil from storage tanks to the oil burners. Storage tanks, piping, and heaters for heavy oils all must be cleaned at scheduled intervals to reduce the potential for fouling and the accumulation of sludge.

Underground steel tanks afford a generally accepted method of storing fuel oil, and the added safety justifies the added cost. However, there are situations where underground storage is too costly and otherwise not feasible. Aboveground tanks are used with restrictions on how close they can be placed to other structures. Moreover, dikes or another type of secondary containment must be provided to prevent the spread of oil if the tank should rupture or leak.

Fuel-oil tanks require connections for the following: filling, venting, pump suction, oil return, sludge removal, and level measurement. When the heavier grades of oil are to be used, the tank must have steam coils for heating the oil, thereby lowering its viscosity, to facilitate pumping. This is accomplished by use of a tank suction heater. The heating coils, suction lines, and return lines terminate within an enclosure in the tank. A small portion of oil is heated to operating temperature while the remainder is at a lower temperature. The oil to the pumps is maintained at optimum temperature.

When the tank is located above the pump level, special precautions should be taken to prevent the oil from being siphoned from the tank if a leak should develop. One method is to provide an antisiphon valve that will open under the suction action of the pump but resist siphoning. In the event of a fire, oil could be siphoned from a tank located above grade. To guard against this possibility, a shutoff valve is installed in the discharge line. This valve is held open by a system containing a fusible link. A fire would open the fusible link, closing the valve and so preventing the flow of oil.

When the tank is below the suction of the pumps, a check valve must be installed in the suction line above the tank to prevent oil from draining back into the storage tank. The suction line must be free from leaks to ensure satisfactory operation of the pumps. This discharge line must be tight to prevent a fire hazard and unsightly appearance caused by leakage.

Supply and return lines intended for use with heavy oil must be insulated and traced to prevent the oil from congealing. Tracing consists of applying heat beneath the insulation, either by a tube supplied with steam or by passing electricity through a resistance wire.

Fuel oils frequently contain sludge that settles in the bottom of storage tanks, causing irregular pump and burner operation. Additives are available that hold this sludge in suspension and reduce operating difficulties. A manhole must be provided to permit cleaning and inspection of the interior of the tank. Since the sludge and water collect in the bottom of the tank, it is advantageous to have a connection to permit its removal without draining the tank.

A typical fuel-oil-supply system is shown in Fig. 5.38. Oil flows from the storage tank through the suction strainers, steam- or motor-

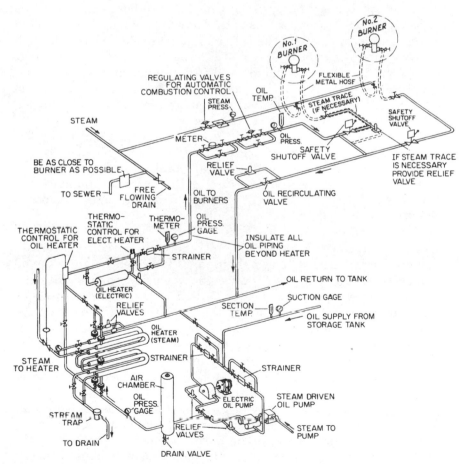

Figure 5.38 Typical fuel oil pumps, heaters, and piping arrangement. (*Factory Mutual System.*)

driven pump, steam or electric heater, discharge strainer, meter, regulating valve, and safety shutoff valve to the burners or to the return line to the tank. Atomizing steam is supplied from the main steam header through a regulating valve to the burners. The combustion-control system positions the regulating valves in both oil and steam lines to proportion the fuel to boiler output steam requirements.

Strainers are provided to prevent foreign material from being delivered to the burners. It is good practice to use a coarse screen, 16- to 20-mesh, in the suction line, and a finer screen, 40- to 60-mesh, in the discharge line. The discharge screen is located in the line after the oil heater. These screens are of the twin type, which enables the operator to clean one screen basket while the oil is passing through the other.

For the delivery of oil from the storage tank to the burners, rotary positive-displacement pumps (see Chap. 8) are best suited for this purpose. A major reason for this is that at constant speed and with oil having a uniform viscosity, the rate of flow on a volume basis through a rotary pump is relatively constant regardless of the pressure. Centrifugal pumps are seldom used in oil pump and heating sets. Reciprocating pumps generally are not desirable either, because of their inherent pressure pulsation that is not compatible with burner, boiler safety, and combustion-control equipment.

Pumps deliver a quantity of oil in excess of that required by the burners. The excess is discharged through a spring-loaded bypass valve to the pump suction or the storage tank. This bypass valve is set to maintain the desired pressure at the burners, and since the pump has a capacity in excess of the burner requirements, it is possible to supply the demand automatically.

Oil is heated by means of steam or electric heaters to reduce the viscosity to a range of 100 to 150 SSU (Seconds Saybolt Universal). This involves oil grades heaver than no. 2. Tank heaters raise the oil temperature sufficiently to reduce the viscosity, thus facilitating straining and pumping. The heater located on the high-pressure side of the pump lowers the viscosity of the oil so that it can be atomized effectively by the burners. Oil heaters are supplied with thermostats that maintain a constant outlet-oil temperature. Thermometers in the lines permit the operator to check the temperature of the oil and hence the performance of the regulator. (Recommended temperatures for various grades of fuel oil are given in Chap. 4.)

The efficient and satisfactory utilization of fuel oil depends primarily on the ability of the burners to atomize the oil and mix it with the air in the correct proportions. The burner admits fuel and air to the furnace in a manner that ensures safe and efficient combustion while meeting the capacity requirements of the boiler.

Oil for combustion must be atomized into the furnace as a fine mist and properly mixed with the combustion air. Proper atomization is the important factor to efficient combustion.

Fuel oil is atomized by either mechanical or dual fluid atomizers that use either steam or air. The choice of design depends on the boiler design and operating requirements. Generally, steam-assisted atomizers produce a higher-quality spray and are more appropriate for low-NO_x applications. However, mechanical atomization is often used where the economics dictate that steam cannot be used for atomization but for producing required power.

Mechanical-type atomizers are used in burners, generally defined as rotary- or pressure-type burners, while steam- or air-atomizing burners are another type of burner. A general description of these types of burners follows. The general preference of oil burners is for steam atomizing because of better operating and safety characteristics, which include lower NO_x production.

5.5.1 Rotary burners

The rotary burner shown in Fig. 5.39 is a self-contained unit in which

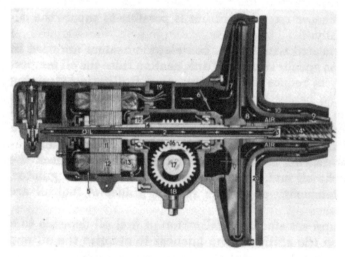

Figure 5.39 Rotary oil burner with integral pump. 1 = oil-metering valve. 2 = fuel-oil tube. 3 = oil-distribution head. 4 = rotary atomizing cup. 5 = cooling-air passages around motor. 6 = adjustable primary air shutter. 7 = primary air fan rotor. 8 = primary air passage to burner. 9 = angular-vaned air nozzle. 10 = passage for induced air to cool burner front plate and sleeve. 11 = motor rotor. 12 = motor stator. 13 = motor-stator windings. 14 = motor shaft for driving oil pump and rotating cup. 15 = ball bearings. 16,17 = oil pump driving gears. 18 = splash-feed lubricating-oil reservoir. 19 = motor-leads junction box. 20 = sleeve to provide clearance for water legs, refractory walls, or dead plates. (*Petro Oil Burner Company.*)

the motor drives the oil pump, the primary air fan, and the atomizing cup. The pump supplies oil to the atomizing cup (4), which is rotated at 3450 rpm. The centrifugal force created by this rotation causes the oil to leave the edge of the cup in a fine spray. Primary air produced by the fan (7) is introduced through the passage (8) into the fine spray of oil. This high-velocity air introduces a secondary flow through the passage (10), which serves the dual purpose of cooling the burner and aiding combustion.

5.5.2   Steam- or air-atomizing burners

The spray-type burner uses air or steam as the atomizing agent. In the inside-mixing-type burner, the atomizing agent and oil mingle before they leave the tip, whereas in the outside-mixing type they come together after leaving the tip. A steam-atomizing inside-mixing type burner is shown in Fig. 5.40. Atomization is accomplished by projecting steam tangentially across the jets of oil. This results in the formation of a conical spray of finely divided oil after the mixture has left the orifice plate. This atomizer may be used in connection with a multiple-fuel burner similar to that shown in Fig. 5.43. The vanes control the flow of secondary air. The burning rate can be varied from 500 to 3500 lb of fuel per hour without changing the spray plate and with pressure variation of 20 to 70 psi in the steam supply and 100 to 140 psi in the oil supplied. With careful operation, the steam required for atomization does not exceed 1 percent of the quantity generated by the boiler.

A proportioning inside-mixing-type oil burner using low-pressure air as an atomizing agent is shown in Fig. 5.41. The oil enters at the rear of the burner and flows through a central tube. Upon reaching the end of this tube, the oil combines with primary and secondary atomizing air. The primary atomizing air passes through tangential openings, which impart a swirling motion to this air as it passes around the stream of oil. The mixture next meets the secondary air and leaves the burner in a converging cone to complete atomization. The control lever at the side of the burner regulates both the oil flow and the airflow. Once adjusted, the correct ratio of air is maintained over the entire range of operation. The effective turn-down operating range of this burner is approximately 5 to 1.

There are many designs for oil burners, each with unique features. For atomizing burners, dry saturated steam at the specified burner pressure is used for fuel-oil atomization, or moisture-free

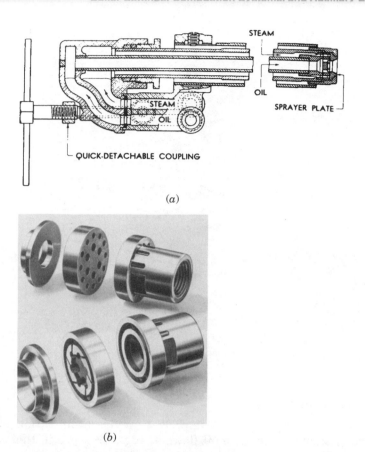

(a)

(b)

Figure 5.40 (a) Cross section of a steam-atomizing oil burner assembly. (b) Sprayer plate and orifice assembly of a steam-atomizing oil burner. (*Babcock and Wilcox, a McDermott company.*)

compressed air is used. The obvious disadvantage of a steam atomizer is its consumption of steam. For a large boiler, this amounts to a large quantity of steam and consequent heat loss. When a boiler supplies a substantial amount of steam for a process where condensate recovery is small or none at all, the additional makeup is negligible. However, for a large utility boiler where makeup is kept to a minimum, the use of atomizing steam is a significant factor. Steam consumption rates as high as 0.2 lb of steam per pound of oil are possible, but some designs have steam consumption rates less than 0.1 lb of steam per pound of oil. This lower rate still results in significant steam consumption.

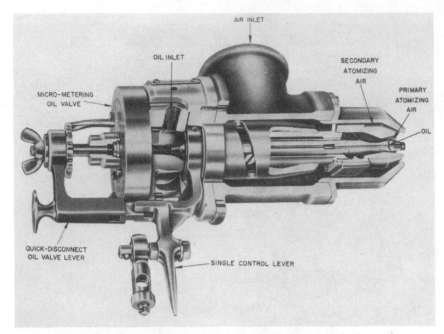

Figure 5.41 Proportioning air-atomizing oil burner. (*Hauck Mfg. Co.*)

5.5.3 Pressure-type burners

The pressure-type burner shown in Fig. 5.42 provides atomization without the use of steam or air. The oil flows at high pressure in the center tube and is discharged through tangential slots in the sprayer-plate swirling chamber. The swirling oil then passes with undiminished energy through the sprayer-plate orifice and into the space between the two plates. The centrifugal action forces some of the oil into the openings that lead to the return line. The amount of oil thus returned is determined by the position of the return-line control valve. The oil that is not returned passes through the orifice plate and into the furnace in the form of a hollow cone-shaped spray. The same amount of oil flows in the burner at all ratings; therefore, the whirling action within the burner is maintained constant at all rates of oil flow. The pump supplies an excess of oil, and the amount that passes through the outer orifice to be burned depends on the opening of the return valve. This type of atomizer can be used with the burners shown in Figs. 5.43 and 5.44. For this type of burner design, the required oil pressure at the atomizer must be 600 to 1000 psig.

Fuel oil is used in the generation of steam and in general industrial applications. Compared with solid fuel, it has various advantages but also some disadvantages.

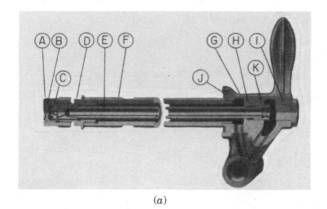

(a)

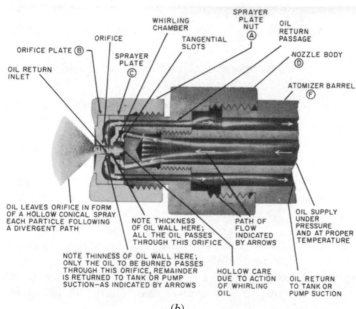

(b)

Figure 5.42 Mechanical pressure atomizing fuel oil burner. (a) Atomizer assembly. A = sprayer plate nut. B = orifice plate. C = sprayer plate. D = nozzle body. E = inlet tube. F = atomizer barrel. G = atomizer handle. H = ferrule. I = plug. J = gasket. K = packing. (b) Detail of atomizer tip. (*Todd Combustion, Inc.*)

The advantages and disadvantages of fuel oil as compared with solid fuel are

Advantages

Oil can be stored without deterioration or danger of spontaneous combustion.

Plant can be operated with less labor (ease of operation and no ashes to be removed).

Combustion process can be controlled automatically.

Initial cost of the plant is less because coal- and ash-handling equipment is unnecessary.

Plant can be kept clean.

Less air pollution control equipment is required.

Disadvantages

Price of oil is generally higher than that of solid fuel.

Availability of oil may be a problem depending on world supply and demand.

Storage tanks are more expensive than storage for solid fuel.

Cleaning of oil tanks and disposal of sludge are costly and trouble-some.

Impurities in oil result in operating difficulties caused by plugged strainers and burners.

Sulfur in the fuel oil causes corrosion of boiler casings and heating surfaces and pollution of the atmosphere.

Oil unsuited for other purposes may be used successfully as fuel under power boilers. The equipment selected, however, must be suitable for handling this inferior grade of waste oil, which usually contains impurities in the form of sludge and grit. Impurities clog the strainers, cause wear of the pumps, and must be removed periodically from the storage tank.

Oil is used extensively as a standby fuel for emergencies when there is an outage of the other combustion equipment. It is also used in lighting pulverized-coal burners. The oil burner is lighted with an electric spark, and the pulverized coal is safely ignited by the oil burner.

Fuel oil may be delivered to the plant in barges, tankers, tank cars, or tank trucks. Except for very high rates of consumption, tank trucks are used. The haul from supplier to customer is usually comparatively short, and enough heat is retained in the oil for it to flow into the storage tank without the use of further application of heat. When the heavy grades of oil are shipped in railroad tank cars, it is usually necessary to supply steam to the coils in the tank before the oil will flow or can be pumped from the car.

Before attempting to operate a fuel-oil-burning plant, check the piping, pumps, strainers, heaters, and valve arrangement. The burners should be inspected and cleaned if necessary. The conventional mechanical steam- and air-atomizing burners can be removed readily and the tip disassembled for cleaning. The rotating-cup-type unit is mounted on a hinge and may be swung out of the furnace position for inspection and service.

When no. 1, 2, or 3 oil is used, no heat is required, but for heavier oil it is necessary to admit steam to the storage tank heating coil. The pump is started, and oil is circulated through the system, and the pressure and temperature must be closely observed.

There is a difference of opinion regarding the correct temperature to which oil must be heated because of the various types of burners. In general, satisfactory results are obtained with burner-oil temperatures of 135°F for no. 4 oil, 185°F for no. 5, and 220°F for no. 6. Another rule is to heat the oil to within 10°F of the flash point of the oil being used. Mechanical-atomizing-type burners require temperatures near the flash point. Steam-atomizing burners require lower oil-admission temperatures because some heat is available in the steam. Rotary-type burners will utilize unheated no. 4 oil and nos. 5 and 6 at a lower temperature than the spray-type burner.

An ignition flame must be placed in front of the burner before the oil is admitted. A gas flame provides a satisfactory means of igniting an oil burner, and spark ignitors are commonly used. If an ignitor is not part of the burner design, a torch may be used, made from a ¼-in pipe by wrapping rope around one end for about 10 in. Wire should be used to hold the rope tightly in place. This torch should be kept in a convenient place in a container of kerosene or light oil. A suitable container can be made by welding a plate on the end of a short section of 3-in pipe. Hand torches have become less common. Lighters are either stationary or designed with a retractable mechanism that allows them to be removed during burner operation and protected from the furnace-radiated heat.

With the fuel supply in readiness, operate the draft equipment to free the furnace and setting from all combustible gases. Insert the torch (or ignite the lighter) and make certain that the flame is in front of the burner. Admit oil to the burner tip and watch to make certain that it ignites immediately. If the fire should go out, turn off the oil and allow the draft to clear the unit of combustible gases before attempting to relight. Modern boilers have flame safety systems that perform these functions automatically.

As soon as ignition has taken place, the oil and air mixture must be adjusted to produce a stable flame. At this point it is best to have the flame slightly yellow, indicating a deficiency of air. If steam is being used, the supply should be reduced to the minimum required for atomization. Insufficient steam or improper functioning of the burner may result in oil dripping on the furnace floor, creating a dangerous condition that must be prevented. The operator must learn how to adjust the fuel and air supply to obtain the lemon-colored flame required during the starting-up period. After the correct flame conditions have been obtained, the boiler is brought up to steam temperature and pressure slowly as with other fuels.

It is essential when burning fuel oil that the operator know the equipment and attend to certain details to ensure satisfactory operation. The burners must be regulated to prevent the flames from striking the boiler heating surface or furnace walls. If this precaution is not taken, localized heating will cause a burned-out tube and rapid deterioration of the furnace wall. The burner tips should be cleaned frequently enough to ensure good atomization. When retrieving-type burners are used, a spare set should be available at all times. Even with automatically controlled heaters, the oil temperature should be observed frequently. If the oil temperature is too low, atomization will be poor, and if it is allowed to get too high, it could exceed the flash point. Remember that an appreciable amount of steam is required in burners that use this atomizing agent. Economies may be optimized by keeping the amount of steam used for atomization to a minimum.

The flow of air and gases in a modern oil-fired furnace may be produced by (1) a forced-draft fan to introduce the air and an induced-draft fan to remove the gases or (2) only a forced-draft fan having enough pressure to supply the air and to discharge the products of combustion. In either case the air supply must be regulated to correspond to the rate of fuel feed. This is accomplished by movable dampers in the air supply duct or wind box. Normally the supply of both fuel and air is automatically controlled.

The operator of a fuel-oil-fired unit is always confronted with the possibility of foreign material interrupting the flow of oil. Sediment may collect in the bottom of the tank and enter the suction line. The strainers either before or after the pumps may become clogged and restrict or stop the flow of oil. Then, too, the sediment may reach the burner tip and cause trouble. An interrupted and self-restored flow of oil from a burner is very dangerous. When the oil flow stops, the fire goes out, and then if the flow is restored, it may not reignite until the furnace has been filled with oil vapor. If this happens to be a combustible mixture, a disastrous explosion could result. Safety devices are used that have a flame detector so that when the fire goes out, the absence of light on the sensitive element of this device trips a relay that closes off the oil supply. This prevents the restoration of the oil flow until the operator has purged the furnace, applied the torch or ignitor, and manually tripped out or reset the safety device.

A summary of the difficulties that may be encountered in operating an oil-fired plant are as follows:

1. Failure of the oil to flow to the pumps due to ineffective venting of the tank
2. Clogged strainers caused by sludge deposits in the tank
3. Failure of pumps to operate due to air leaks in the suction line
4. Vapor formation in the pump suction line as a result of too high a temperature in the tank
5. Oil too heavy to flow to the pumps because of insufficient heating
6. Faulty atomization caused by
   a. Too low an oil temperature
   b. Carbon formation on burner tips
   c. Worn burner tips
   d. Too low an oil pressure
   e Insufficient steam or air

7. Smoke, poor combustion, and flame striking furnace walls caused by
   a. Faulty atomization of oil
   b. Incorrect adjustment of air supply
   c. Automatic control not functioning properly

8. Fire hazard and unsightly appearance caused by leaks in the pump discharge lines

Oil-fired boilers are banked by closing off the oil and air supply. The fans are shut down and the stack damper is closed to prevent natural circulation of air and thus retain the heat in the unit. If it is necessary to have the boiler available for service on short notice, the fire must be relighted often enough to maintain the steam pressure at a predetermined minimum pressure.

When sufficient sediment collects in the fuel-oil supply tank, it will frequently clog the strainers and perhaps even pass on to the burners, causing irregular firing. The real solution to this situation is to drain and thoroughly clean the tank.

Some oil contains abrasive materials that cause wear to the pumping equipment and result in low oil pressure. It is not economical to run with low oil pressure, especially when mechanical atomization is used. When wear in the pumps causes a reduction in pressure sufficient to affect operation adversely, the worn parts of the pumps should be replaced. This condition should be anticipated by selecting pumps that have wearing parts that can be replaced quickly and inexpensively. Redundancy in pumps also ensures high availability because it allows pump maintenance while the other pump maintains the oil flow requirement.

The importance of maintaining the oil system free from leaks cannot be overemphasized. Socket welding of oil-piping joints is an accepted method of reducing leakage. Many an oil system failure has been traced to a small leak in the suction pipe. The flexible connection at the burners should be replaced or repaired before serious leaks develop.

In most plants the cleaning of burner tips is considered an operating function. In time, with cleaning and wear from impurities in the oil, it becomes necessary to replace the tips.

Air registers and automatic control equipment must be checked frequently to ensure satisfactory operation.

As with coal fire burning, the formation of NO_x and its control are major operating concerns on an oil-fired unit. As explained previously, NO_x is formed during combustion by two methods, thermal NO_x and fuel NO_x. Thermal NO_x is the primary source of NO_x formation from natural gas and fuel oils because these fuels are generally low or absent of nitrogen. Thermal NO_x reactions occur rapidly at combustion temperatures in excess of 2800°F, which is the operating furnace temperature for oil- and gas-fired units. There are various methods of controlling NO_x, and these include the following.

Low excess air. By lowering the excess air, not only is NO_x reduced, but also thermal efficiency improves. If burner and combustion efficiency are maintained at acceptable levels, NO_x formation can be reduced by 10 to 20 percent. However, the success of this method depends on the careful control of the distribution of fuel and air to the burners, and it may require more sophisticated means to measure and regulate the fuel and air to all burners.

Two-stage combustion. This is a common method where air is directed to the combustion zone in quantities that are less than that required to burn the fuel. The remainder of the air is introduced through overfire air ports. This method limits the oxidation of nitrogen that is chemically bound in the fuel, and by introducing the combustion air in two stages, the flame temperature is also reduced. This method can reduce NO_x emissions by as much as 50 percent.

5.6   Gas

The gaseous fuels available are natural, manufactured, water, and oil gas and by-product gases such as coke-oven, blast-furnace, and refinery gas. Natural gas is widely distributed by means of high-pressure pipeline systems. Nevertheless, cost and availability vary widely and depend on the distance the gas must be transported. In some instances, gas at a reduced cost is available for steam generation during periods when primary customer demand is low. By-product gas usually is used in the same plant in which it is produced or in a neighboring plant.

Natural gas from high-pressure pipelines is reduced in pressure and distributed to local customers through a low-pressure network of piping. When the pressure in the supply mains is above that required by the burners, pressure-regulator valves are used to maintain constant pressure at the burners. However, when the supply main pressure is less than that required at the burners, it becomes necessary to install booster compressors. These boosters are either multistage centrifugal blowers or constant-volume units similar in principle to a gear pump. The outlet pressure of constant-volume units must be controlled by spring-loaded bypass valves. In some instances it is necessary to install coolers in these bypass lines to remove the heat developed in the compressor. Gas burners are designed and regulated for a given pressure, and this pressure must be maintained for efficient and satisfactory operation. In like manner, provision must be made to supply by-product gas to the burner at a controlled pressure.

Gas is a high-grade fuel and is used extensively for domestic heating and in industrial plants for heating and limited process applications. Industries that require large quantities of steam usually find other fuels more economical. Power boilers that burn gas generally have combustion equipment suitable for two or more kinds of fuel. The use of multiple fuels makes it possible to take advantage of market conditions, provide for shortages in supply, and guard against possible transportation difficulties. It is difficult to operate a pulverized-coal-fired boiler safely at widely variable capacities. When low capacities are necessary, it is advisable to use oil or gas as a substitute for pulverized coal. The burner shown in Fig. 5.43 is designed to use pulverized coal, oil, or gas. The gas is introduced through many small holes in a ring that is located in front of the burner proper. The gas mixes with a stream of air passing through the ring.

The recent availability and relatively low cost of natural gas have made cogeneration power facilities popular for utilities in the production of electricity. These facilities utilize the natural gas to power a gas turbine for electric production, and an additional steam cycle is incorporated to use the heat contained in the gas turbine exhaust gas and produce steam that is used in a conventional steam turbine cycle. Not only does this cogeneration cycle use relatively low-cost fuel, but also the plant's initial cost is lower than that of a coal-fired facility. The schedule for installation of a cogeneration plant is also significantly shorter. However, the acceptability of this system will remain

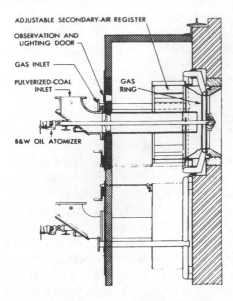

ADJUSTABLE SECONDARY-AIR REGISTER

OBSERVATION AND
LIGHTING DOOR

GAS INLET

PULVERIZED-COAL
INLET

GAS
RING

B&W OIL ATOMIZER

Figure 5.43 Multifuel burner.
(*Babcock & Wilcox, a McDermott
company.*)

only as long as the availability of low-cost natural gas is maintained
and the reliability and low maintenance costs on gas turbines also
remain acceptable.

Industries that have a supply of by-product gas utilize it to advantage as a fuel in their power boilers. They use the by-product fuel
first, and when the supply is insufficient to meet the demand, they
use a secondary fuel which they must buy. This method of operation is
most effective in reducing the cost when there is a source of waste or
by-product fuel. Combustion controls admit the proper mixture of by-
product and secondary fuels to the furnace to meet the necessary
combustion requirements. A burner arrangement for either oil or
blast-furnace gas is shown in Fig. 5.44. Because of the low heating
value of this gas, a correspondingly large volume must be handled.
The gas is discharged into the burner through a circular ring formed
by the burner tube and throat. The air enters through the burner
tube with a whirling motion and mixes with the gas.

Sufficient furnace volume is essential to the proper operation of
gas-fired furnaces. If the fuel has not been entirely burned before the
gases reach the heating surface, serious smoking and loss of efficiency
will result. Thorough mixing at the burners generally reduces the
length of the flame. The design should be such that the entire furnace
volume is utilized for combustion.

Many of the procedures and precautions that are required for oil
firing also apply to gas. Before attempting to light the burners, make

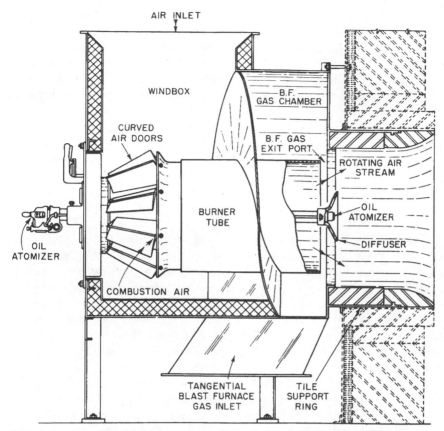

Figure 5.44   Combined blast-furnace gas and oil burner. (*Peabody Engineering Corp.*)

certain that the gas supply valves are closed, and then operate the draft equipment to remove all possible accumulations of gas. Adjust the shutters to admit a small amount of air. Then insert the lighted torch or start the igniter and turn on the burner valve. If the fire should go out, never relight until you are sure the draft has removed all combustible gases from the unit. Flame safety systems are used extensively on all modern boiler designs, and older units have been retrofitted to incorporate this critical safety feature.

The appearance of the flame is an unsatisfactory method of adjusting the air going to the burner and the furnace draft. The best procedure is to analyze the flue gases with an Orsat or continuous-monitoring equipment and regulate the air accordingly. With conventional burner design and careful adjustment, gas-fired boilers can be operated successfully with less than 15 percent excess air. Improper mixing of air and gas results in long flame travel. If the flames strike the heating surface, soot will be deposited and smoke will be produced. Gas flames frequently have a tendency to pulsate. This may be caused by poor distribution of the flame in the furnace. If the burner and furnace designs are correct, these pulsations can be eliminated by increasing the furnace draft. When the furnace is started up, the burners should be adjusted to produce a yellow flame that will not be extinguished easily.

The accurate control of the flow of gases to the burners depends on functioning of the pressure regulator. In some installations the pressure of the gas entering the burners is changed to vary the rate of firing, while in others the number of burners in service is varied. Automatic control is effectively used not only to regulate the fuel and maintain the steam pressure but also to maintain the correct ratio of air to fuel.

The maintenance of gas burners and accessories is less than that for other types of combustion equipment, but there are a few points that must receive attention. The air register operating mechanisms and gas regulators must be kept in good operating condition, the burners must be kept clean so that the flow of gas will not be obstructed, and the boiler setting must be checked to reduce air leakage to a minimum. The use of gas as a fuel eliminates stokers or pulverizers and coal- and ash-handling equipment, thus decreasing plant maintenance and operating costs.

The control methods for NO_x when firing natural gas are the same as those described previously for the burning of fuel oil. Thermal NO_x is the primary source of NO_x formation because natural gas and other gaseous fuels generally are low in or absent of nitrogen.

This control of NO_x also has led to the need for completely new burner designs. One such design is shown in Fig. 5.45, and this is a distributed airflow design that uses multistaging to reduce NO_x with minimum flame volume and minimum CO levels to achieve low NO_x and low CO. This burner is designed for high heat release where limited boiler furnace dimensions exist.

The burner consists of two separate air zones. The inner zone establishes a recirculating air zone that is downstream from the spinner, and the outer zone uses an adjustable swirl to shape the burner flame, which optimizes flame conditions and emission levels. This burner can be designed to burn natural gas, refinery gas, coke-oven gas, landfill gas, light and heavy fuel oils, and coal-water slurries.

The use of flue gas recirculation to the burners has been an effective way to reduce NO_x emissions when firing natural gas and fuel oils. Flue gas from the economizer outlet is introduced into the combustion air, and this results in the flame temperatures of the burner being lowered, and thus NO_x emissions are reduced significantly.

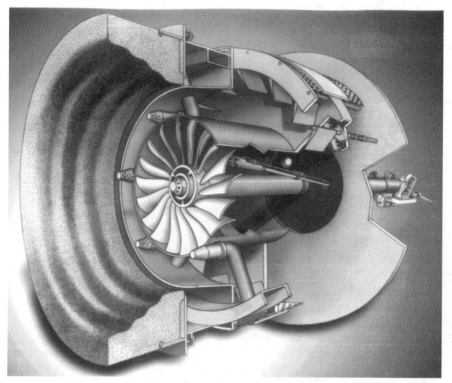

Figure 5.45 Multistage low-NO$_x$ burner for the burning of gaseous fuels and fuel oil. (*Coen Company, Inc.*)

Although this method is effective, the flue gas recirculation increases both the initial costs and the operating costs.

Burner designs continue to be developed, and Fig. 5.46 shows a burner capable of meeting the current NO$_x$ emission limits without the use of flue gas recirculation for certain applications. By using a combination of an air-fuel lean premix and staged combustion, peak flame temperatures are reduced without using flue gas recirculation, with a resulting reduction in the production of NO$_x$. While the use of flue gas recirculation in burner design is an effective technique to reduce NO$_x$ emissions, this unique burner design can be used on many boiler applications and effectively meet current NO$_x$ regulations. If this technology is applicable, there are obvious cost savings with elimination of the flue gas recirculation system.

Burner design development and cost-effective applications are continually being studied to meet the ever-changing lower limits for NO$_x$ production. New boilers as well as those operating boilers requiring retrofitting will benefit from these successful developments.

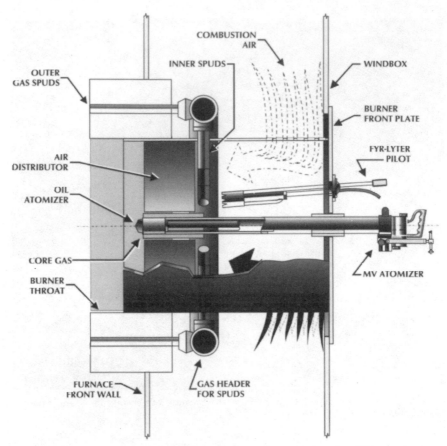

Figure 5.46 QLN burner designed for natural gas firing to meet low NO_x requirements without the use of flue gas recirculation. (*Coen Company, Inc.*)

5.7 Oil- and Gas-Firing Safety Precautions

The National Fire Protection Association (NFPA) has developed a set of procedures and recommended interlocks and trips for the safe and reliable operation of oil- and gas-fired boilers. These procedures are used in the design of burners and combustion-control systems and include the following:

1. The accumulation of oil and gas is unacceptable anywhere other than in the fuel-delivery system. If a gas odor is present, ventilation of the area must be done immediately and the source of the leak determined and remedied.

2. A minimum airflow purge rate of 25 percent of full-load airflow should be maintained. This includes prepurging the setting and

lighting of the lighters and burners until the air requirement exceeds the purge rate airflow.

3. A spark-producing lighter or a lighted torch must be in operation before any fuel is admitted into the furnace.

4. An airflow through the burners into the furnace and out the stack must be maintained at all times.

5. Adequate fuel pressure must be maintained at all times. For oil firing, both required pressure and temperature must be available, and for a dual fluid atomizer, steam or air pressure must be adequate at the atomizer.

By using proper procedures and operating skills, all fuels can be burned in a safe and efficient manner. Problems result when carelessness and misoperation of equipment occur and the fuel is burned in an unsafe manner.

5.8 Automatic Operation of Boilers

Combustion control regulates the fuel to meet the demands for steam or hot water and maintains the ratio of air to fuel required for the best economy. The operator performs the remaining functions, which include purging the furnace, lighting the burners, adjusting dampers, opening and closing fuel valves, starting and stopping pumps, and turning burners on and off.

The entire combustion process is made automatic by having the unit come into service and shut down, as required, without hand manipulation. The degree of automatic control is made possible by the use of safety monitoring that checks every step of the operation. This is accomplished by means of sensing devices that detect abnormal conditions, including flameout, high pressure, low water level, high temperature, and low fuel pressure. The control either corrects the situation or warns that trouble exists and shuts the unit down safely.

Automatic controls of this type are applied to units of all sizes. However, extensive economic justification has been found in the small and medium-sized package boilers. The package boiler (Fig. 5.47) is equipped with automatic controls capable of starting up and maintaining the firing rate as required for varying demands of steam. The control panel includes pressure and draft gauges, buttons for initiating the startup, lights for indicating the sequence of the cycle, and an annunciator to indicate specific causes of trouble. When two or more boilers are installed, the controls can be arranged to start additional units if a unit in service fails or the load increases above the capabilities of the units on the line.

Figure 5.47 Burner control package including burner, windbox, fan, fuel piping, and controls. (*Coen Company, Inc.*)

The design shown is a modular system that incorporates the burner, windbox, forced-draft fan, fuel piping, burner management, and controls in a complete package and attached to a package boiler.

Burner control systems of various types are applied to almost all boilers to assist the operator in the starting and stopping of burners and fuel equipment and thereby prevent hazardous furnace conditions. The most important burner control function is to prevent explosions in the furnace and in equipment such as pulverizers, where such explosions would threaten the safety of personnel and cause major damage to the boiler and associated equipment and buildings.

The various types of burner controls are manual control, remote manual sequence control, and automatic sequence control and these are generally described as follows.

Manual control. With manual control it is necessary to operate the burner equipment at the burner platform. The determination of conditions depends on the observation, evaluation, and experience of the operator. Good communication is essential between the local operator and the control room operator to coordinate the safe startup and shutdown of burner equipment, and this procedure is much more complex depending on the type of fuel, e.g., natural gas versus pulverized-coal firing. Manual control is generally found only on older boilers, since some form of automatic control is always installed on newer units.

Remote manual sequence control. This system is a major improvement over manual control. The operator performs remote manual functions by using plant instrumentation and position switches in the control room as information to determine the operational steps. With this system, the operator participates in the operation of the fuel equipment by controlling each sequence of the burner operating procedure from the control room, and no action takes place without the operator's command.

Automatic sequence control. Automating the sequence control allows the startup of burner equipment from a single pushbutton or switch control. This automation replaces the operator in the control of the operating sequence. The operator monitors the operating sequence as indicated by signal lights and instrumentation. This automatic control system is used on all modern natural gas–, oil- and coal-fired boilers.

The schematic drawing in Fig. 5.48 shows the control arrangement of an oil- and gas-fired boiler. The steam pressure from the boiler actuates the submaster (1). This impulse moves a shaft having linkages for controlling the forced-draft-fan damper through the airflow regulator (2) and the power unit and relays (3) and other linkages for controlling the burner register louvers; the oil supply to the burners, by adjusting the orifice-control valve (5); and the gas supply to the burners, by means of the adjustable-orifice gas valve (8). The gas supply to the pilot is turned on and off by the solenoid valve (16).

Automatic burners of this type are equipped with safety controls that sequence the startup and shutdown and override the operating controls to close the fuel valves if any of the components malfunction. These safety controls will shut the unit down when there is trouble, including low water, flame failure, low gas or oil pressure, and overpressure. They must be selected to meet the insurance underwriters' specifications.

These controls provide an option of recycle or nonrecycle for restarting the unit once it has tripped out as a result of a failure. A recycling control automatically attempts to restart the burner, while a nonrecycle control allows the burner to remain out of service until the start switch has been closed by hand. There are some insurance restrictions on the use of recycle control.

The automatic startup sequence and timing depend on the type of burners, specific conditions, and the use of oil, gas, or a combination. In general, however, the forced-draft fan will start immediately when

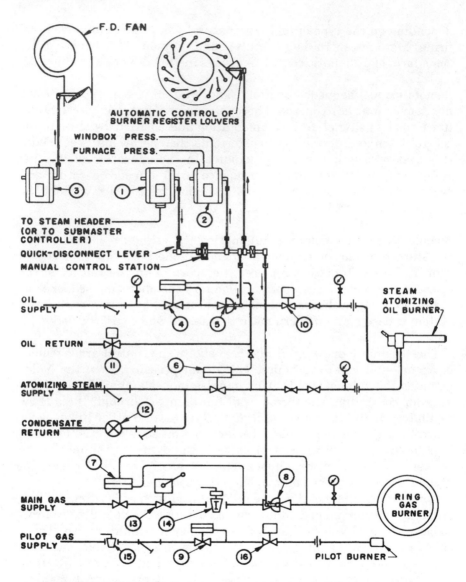

Figure 5.48 Schematic control diagram—automatic gas- and oil-fired unit. 1 = master actuator. 2 = airflow regulator. 3 = power unit and relays. 4 = differential oil control valve. 5 = adjustable-orifice oil-control valve. 6 = differential steam-control valve. 7 = differential gas-pressure-control valve. 8 = adjustable-orifice gas-control valve. 9 = pilot gas-pressure regulator. 10 = oil solenoid valve. 11 = oil-return solenoid. 12 = steam trap. 13 = main-gas solenoid valve. 14 = supervisory cock. 15 = pilot shutoff cock. 16 = pilot-gas solenoid valve. (*Coen Company, Inc.*)

the start switch is turned on. The air pressure produced by the forced-draft fan will close the air interlock. The flame safety control will be engaged for a preset purge period with the fuel control valve at high fire. Then the firing-rate controller will be adjusted to minimum fire in preparation for pilot ignition. The ignition transformer will be energized to light the pilot. After the pilot-proving period has elapsed, the main fuel valve will open and the burner will be placed in operation. After the burner has been "proved," the ignition transformer will be deenergized. The firing-rate controller will then be released from the minimum position to respond to the demand placed on the unit. In the event that the pilot or main valve has not been proved, the fuel valve or valves will close and an alarm will sound.

The safe operation of automatic burners depends on the reliability of the flame-detecting device. The type and location of flame detectors are an integral part of the burner and the control system design and are determined by the control system manufacturer.

Packaged automatic boilers require the same careful consideration to design limitations as apply to field-erected units. Satisfactory operation and low maintenance require limitations on fuel-burning rates, heat-absorption rates, and flue gas velocities. However, some decrease in the size of units for a given output can be obtained as a result of improvement in combustion equipment, extended use of water cooling in the furnace, and better circulation of water in the boiler. The operation of automatic boilers depends on carefully designed and located sensing devices, electric circuits, and electronic controls.

Electronic controls once considered unreliable by plant operators have been improved by the use of analog and digital microprocessor control systems. Replacements are made possible by preassembled "plug in" components that can be stocked and quickly installed by plant personnel. This automatic equipment should be checked for performance of the safeguards on a regular schedule. In addition to these checks by plant personnel, it is advisable to have semiannual or annual inspections by qualified control service personnel.

5.9 Stacks

An adequate flow of air and combustion gases is required for the complete and effective combustion of fossil fuels. The term *draft* is defined as the difference between atmospheric pressure and the static pressure of the flue gases in the furnace, a convection pass, a flue, or a stack. *Draft loss* is the reduction in the static pressure of a flue gas caused by friction and other pressure losses, which are not recoverable, as the flue gas flows through a boiler system.

The flow of gases through a boiler can be achieved by creating draft by four methods: (1) natural draft, (2) forced draft, (3) induced draft, and (4) balanced draft. These are defined as follows:

1. *Natural draft.*   The required flow of air and flue gas through a boiler can be obtained by the stack alone when the system draft loss is low and the stack is tall. In today's modern units, this is not a common method, since mechanical draft is required on boiler systems.

2. *Forced draft.*   Operates with air and flue gas maintained above atmospheric pressure. A fan (forced-draft, or FD, fan) at the inlet of the boiler system provides pressure to force air and flue gas through the system. Any openings in the boiler setting such as doors and penetrations allow air or flue gas to be released unless that opening is pressurized and thus has air being forced into the opening, thus preventing any release.

3. *Induced draft.*   Operates with an air and flue gas static pressure below atmospheric pressure. Any openings in the boiler setting allow air to be drawn into the unit. A fan (induced-draft, or ID, fan) is located near the outlet of the boiler system prior to the stack and draws the air and flue gas through the system. Using an ID fan alone is not common on modern boiler designs.

4. *Balanced draft.*   Operates with a forced-draft (FD) fan at the inlet and an induced-draft (ID) fan near the outlet prior to the stack. The static pressure is above atmospheric at the FD fan outlet and decreases to atmospheric pressure at some point within the system, usually in the lower furnace. The static pressure is below atmospheric and progressively decreases as the flue gas flows from the balance point to the ID fan. This arrangement prevents the flue gas leaking from the unit and is the system predominantly used in modern units other than perhaps package boiler designs, where a forced draft, or pressurized system, is used.

Stack effect is the difference in pressure caused by the difference in elevation between two locations that convey heated gases. It is caused by the difference in density between air and heated gases.

Stacks serve several purposes. In a natural-draft boiler, it produces the draft necessary to cause the air to flow into the furnace and discharge the products of combustion to the atmosphere. In all applications, it delivers the products of combustion to a high altitude. As noted above, either the stack alone or a combination of stack and forced-draft (FD) and/or induced-draft (ID) fans must produce the required pressure differential for the flow. The gases within the stack are at a higher temperature than that of the surrounding air. The weight of a column of hot gases in the stack is less than that of a column of air at outside temperature. The intensity of draft produced by the stack depends on the height and the difference between the outside air and inside gas temperatures.

The following formula may be used for calculating the approximate static, or theoretical, draft produced by a stack:

$$D = 0.52H \times P \left(\frac{1}{T_a} - \frac{1}{T_s} \right)$$

where D = draft at base of stack (inches of water)
H = height (ft) of stack above point where draft measurement is taken
P = atmospheric pressure, psia (can be taken as 14.7 up to elevation of 1500 ft above sea level)
T_a = outside temperature (absolute °F)
T_s = gas temperature in stack (absolute °F)

Example A stack is 175 ft high, and the exit-gas temperature is 610°F. The outside temperature is 80°F, and the plant is located at an elevation of 550 ft. What static, or theoretical, draft in inches of water will it produce?

Solution

$$D = 0.52H \times P \left(\frac{1}{T_a} - \frac{1}{T_s} \right)$$

$$H = 175 \, ft \qquad P = 14.7 \qquad T_a = 80 + 460 = 540$$

$$T_s = 610 + 460 = 1070$$

Substituting these values in the equation:

$$D = 0.52 \times 175 \times 14.7 \left(\frac{1}{540} - \frac{1}{1070} \right)$$

$$D = 1.23 \text{ in of water}$$

This represents the static, or theoretical, draft; in actuality, when gases flow, it will be reduced, due to friction. The amount of reduction will depend on the size of the stack, on the quantity of gases, and, to some extent, on the material used in the construction of the stack. The actual available draft is approximately 80 percent of the calculated static draft.

Whereas the height of a stack determines the amount of draft it will produce, the cross-sectional area limits the volume of gases it can discharge successfully. The quantity of gases that a stack must expel depends on the amount of fuel burned and the amount of air used for combustion. Air leaks in the boiler setting, flues, etc., dilute the stack gases, increase the volume, lower the temperature, and thus lower the effective draft.

It frequently becomes necessary to study the performance of an existing stack installation. Questions arise as to whether the stack has sufficient capacity for the existing boiler. Will it be adequate if the existing boilers are rebuilt to operate at a higher capacity? Can more boilers be added without building another stack? In these instances, draft- and gas-temperature readings should be taken at the base of the stack while the boilers are operating at full capacity. These results should be compared with the expected draft as calculated from the height of the stack, the temperature of gases, and the temperature of the outside air. If the results show the draft as read to be less than 80 percent of the calculated value, the stack is overloaded.

The adjustment of the stack damper is a good indication of the available draft. Must the damper be wide open when the boilers are operating at full load? A stack may become so badly overloaded that it is impossible to maintain a draft in the furnace when operating at maximum output.

Sometimes difficulties develop, and a plant that has formerly had sufficient draft finds that the furnaces develop a pressure before the boilers reach maximum capacity. This condition, although a draft problem, might be caused not by a defective stack, but by any one or a combination of the following conditions:

1. Stoppage of boiler passes, flues, etc., with soot, slag, or fly ash, resulting in abnormal draft loss

2. Baffles defective or shifted so that they restrict the flow of gases

3. The use of more excess air than was formerly used

4. Operating at higher capacities than indicated because the flow-meter is out of calibration

5. The damper out of adjustment so that it will not open wide even though the outside arm indicates that it is open

6. Air infiltration through leaks in the setting

There are limitations to the application of the natural-draft stack. The design of modern units is toward high rates of heat transfer, which result in increased draft loss. High fuel costs justify the installation of heat-recovery equipment, such as economizers and air heaters to reduce the heat loss by lowering the exit gas temperature. The limits imposed on the emission of fly ash and acid gases by local ordinances and operating permits necessitate the installation of flue gas cleaning equipment. This equipment increases the draft loss, and when combined with the lowering of the gas temperature, a higher stack is required. These conditions demand a stack higher than is physically or economically practical, and mechanical-draft equipment must be used.

Stacks are used in connection with both balanced-draft and pressurized-furnace units. Fans are used to produce the necessary draft to overcome the pressure drops (air resistance and draft loss) that are developed in the combustion equipment, boiler, flue gas cleaning equipment, flues and ducts, and the stack. The products of combustion are discharged at high elevations in order to disperse the combustion gases.

The stack disperses the combustion gases. By increasing the stack height, this increases the area of dispersion. In valleys, for example, increased stack height may be necessary for proper dispersion. However, there are areas where stack height is limited, and in these cases, the stack outlet diameter must be reduced to form a venturi and thus increase the discharge velocity of the flue gases to achieve the required height dispersion. This increases the flow resistance (draft loss) that must be accounted for in the fan designs, and additional power requirements must be absorbed.

Stacks are subject to the erosive action of particulates and the corrosive action from acid flue gases such as SO_2. The stack design should incorporate materials that minimize the effect of erosion and corrosion. These materials can be steel alloys, special brick, refractory lining, or a combination of these features.

5.10 Mechanical Draft: Fans

There are essentially three methods of applying fans to boiler units:

1. *Balanced draft.* A forced-draft fan delivers air to the furnace and an induced-draft fan or a stack produces the draft to remove the gases from the unit. The furnace is maintained at 0.05 to 0.10 in of water (gauge) below atmospheric pressure.

2. *Induced draft.* A fan or stack is used to produce sufficient draft to cause the air to flow into the furnace and the products of combustion to be discharged to the atmosphere. The furnace is maintained at a pressure sufficiently below that of the atmosphere to induce the flow of combustion air. This type of design is not used on modern designs.

3. *Pressurized furnaces.* A forced-draft fan is used to deliver the air to the furnace and causes the products of combustion to flow through the unit and out the stack. The furnace is maintained at a

pressure sufficiently above that of the atmosphere to discharge the products of combustion.

Forced-draft (FD) fans handle cold (ambient temperature), clean air and provide the most economical source of energy to produce flow through boilers and their auxiliary systems. Induced-draft (ID) fans handle hot flue gas and thus a larger flue gas volume. Because of this higher temperature and greater volume of gases, they require higher electric power and are also subject to a higher probability of erosion from the fly ash that is contained in the flue gas. With the ID fan located after a particulate collector, such as a precipitator or a bag-house, this erosion potential is minimized.

In addition to forced- and induced-draft applications, fans are used to supply over-fire air to the furnace and in the case of pulverized-coal units to deliver the coal to the furnace. These fans are generally called over-fire air fans and primary air fans, respectively.

Most fans consist of a rigid wheel designed to rotate at a constant speed in a fixed housing. The output characteristics of each fan are determined by the width, depth, curvature, and pitch of the blades; by the speed and diameter of the wheel; and by the shape of the housing. The different characteristic curves determine each fan's suitability for a particular application.

Manufacturers provide fans with a range of characteristic curves. These curves are plots of static pressure (in inches of water) against output (in cubic feet per minute, cfm). In addition, the fan's power requirements and efficiency are plotted against output. A typical characteristic curve is shown in Fig. 5.49. Note that as the output of a fan changes, the horsepower required to drive it also changes, as does the efficiency.

Centrifugal fans are widely used to handle both combustion air and flue gases. Pressure is developed by the rotation of the rotor blades within the casing. There are three principal types of fan blades: radial, forward-curved, and backward-curved. These principal types are modified to meet specific conditions.

General characteristics of these fan types are

1. *Radial blades* do not produce a smooth airflow, and, therefore, any particles in the flue gas stream are deflected away from the blade surface. This type of fan provides maximum resistance to abrasion.

2. *Forward-curved blades* can easily accumulate deposits from dirty flue gas or air streams and therefore can be used only for handling clean air. These types of blades are seldom found in power plants, but they are used extensively in small air-handling equipment applications.

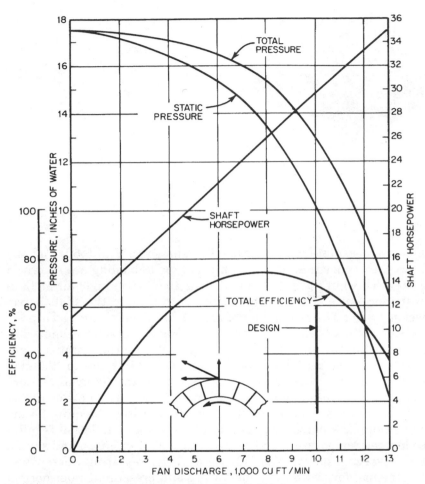

Figure 5.49 Characteristic curve for fan with radial blades.

3. *Backward-curved blades* produce a smooth air or flue gas flow and therefore are used where particles are entrained in the air or flue gas stream. A variation of this type blade that is used commonly is a radial-tip blade, sometimes called a *backward-inclined, forward-curved blade*. This fan is ideal for handling flue gas streams with moderate dust loadings and therefore is often used as an induced-draft fan.

The pressure developed by a fan is measured in inches of water as compared with pounds per square inch. For example, a forced-draft fan that supplies secondary air to a pulverized-coal-fired boiler provides this air at about 30 in of water, which is equivalent to about 1 psia.

Figure 5.50 Straight-radial-blade fan wheel.

A straight-radial-blade fan wheel is shown in Fig. 5.50, and the characteristic curves for a fan with this type of blading are shown in Fig. 5.49. This fan was selected to deliver 10,000 ft³/min of air at 13 in of water pressure. The efficiency is 69 percent, and the shaft horsepower is 29.60. The requirements are 10,000 ft³/min (cfm) at a pressure of 13 in of water, but consideration must be given to the performance of this fan under varying conditions. If the damper were completely closed (zero flow), the pressure would increase to 17.5 in of water. If, on the other hand, the resistance in the system were reduced (if a bypass damper were opened) to permit a flow of 13,000 ft³/min at a pressure of 6.7 in of water, the fan would require 35 hp. Furthermore, the horsepower of a fan in a given installation is influenced by the temperature and pressure of the air or gases, and since a fan is a volume device, the volume and the resulting horsepower requirements vary with the temperature and pressure. Under normal operating temperatures, an induced-draft fan may not overload its driving motor with the damper in the full open position. However, if this same fan is used to handle air at room temperature to provide ventilation during maintenance, the motor may be overloaded. The horsepower requirements of this radial-blade fan increase in an almost straight-line ratio with the volume of air delivered. This applies to constant-speed operation. When the speed of the fan changes, capacity varies directly as the speed, pressure varies as the square of the speed, and horsepower varies as the cube of the speed.

The radial-blade fan runs slower than the backward-inclined fan, and because of its design, any particles in the gas stream are deflected away from the blade surface, which provides significant resistance to vibration. This made it ideal for induced-draft fan applications. However, the efficiency of this fan is low (approximately 70 percent),

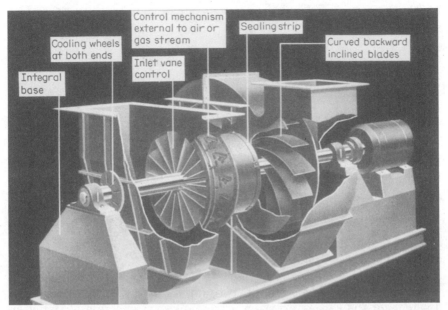

Figure 5.51 Centrifugal fan with backward-curved inclined blades and inlet-vane control. (*TLT-Babcock, Inc.*)

and with increasing energy costs, it has been virtually eliminated from consideration for ID fan applications. They still have some use, however, where the gas stream has a high particulate loading such as for flue gas recirculation and primary air fans.

The backward-curved blade fan, as shown in Fig. 5.51, is a design that has a low resistance to erosion from particles in the gas or air stream. It also has a relatively high efficiency (approximately 80 percent), which reduces energy costs. This type of blade design minimizes fan deterioration and the resulting maintenance and outages and is used in applications where dirty gas, ash particulates, heat, and high pressure all can cause fan deterioration.

Centrifugal fans are basically constant-output fans, and, therefore, the output flow must be throttled by inlet vanes or outlet dampers. The fan in Fig. 5.51 runs at constant speed, and adjustable inlet vanes control the output. This design prespins the flue gas, which means less work for the fan and thus lower power requirements. Inlet vanes also account for improved control stability over the entire operating range.

Power is required to operate fans, and the cost of this power is charged to plant operation. An effort must be made in both the design and operation of a plant to reduce the cost of power to operate the fans. Fan shaft horsepower is a function of the quantity of air or gases, the total differential pressure (static plus velocity pressure), and the efficiency of the fan.

$$\text{Fan shaft horsepower} = \frac{5.193 \times Q \times H}{33,000 \times E}$$

where Q = quantity of air in actual cubic feet per minute (ACFM)[1]
H = total pressure (inches of water)
E = efficiency of the fan (%/100)

In the sizing of a fan, it is customary to specify test block static head, temperature, and capacity requirements of the fan in excess of those calculated to allow for any departure from ideal conditions and to provide a satisfactory margin of reserve in the design.

The design of fans must provide not only sufficient capacity to obtain maximum capacity but also regulation when the unit is operated at partial capacity. There are three principal ways of controlling the output of a fan: by the use of dampers, by speed variation, and by the use of inlet vanes.

Damper control consists of operating the fan at constant speed and dissipating the excess pressure by means of a variable obstruction (damper) in the flow of gases or air. This provides a low-cost installation but is wasteful of operating power. To reduce this waste, two-speed motors are sometimes used in connection with dampers.

Variable-inlet vanes provide a method of controlling a fan that is less wasteful of power than a damper and less costly than variable-speed drives. This method of control has been applied widely to forced-draft fans, but fly ash introduces difficulties when it is used in connection with induced-draft fans. However, present-day units require particulate control equipment, and, therefore, since ID fans are located downstream of such equipment, significant amounts of particulates are removed from the gases. Thus ID fans also use inlet-vane control.

The output of a fan can be varied by changing the speed to meet the draft requirements, but variable-speed drives increase the installation cost. Variable speed can be obtained with constant-speed motor drives by use of magnetic or hydraulic couplings. The motor runs at constant speed, and the coupling is controlled to vary the fan speed. The speed of a steam-turbine-driven fan is varied by regulating the steam flow to the turbine. It is customary to use dampers in addition to these variable-speed devices.

Fan outputs are invariably regulated by the combustion controls. The combustion controls must be compatible with the type of fan con-

---

[1]ACFM—actual cubic feet per minute of gases at the existing temperature, pressure, and relative humidity. This is in contrast to SCFM, standard cubic feet per minute of gases at 29.92 in of mercury pressure (14.7 lb/in^2) and 68°F, 50 percent saturated with moisture, and weighing 0.075 lb/ft^3.

trol selected. For example, a variable-speed fan and damper arrangement requires more complex combustion control than when only a damper is used.

For those plant designs which operate frequently at different loads, the use of adjustable-pitch axial fans has become common because such fans can provide high efficiency at these different loads, thus providing lower operating costs. The adjustable-pitch axial fan consists of a large-diameter hub mounted on a bearing assembly. On the periphery of the hub are a number of blade shafts on which the fan blades are mounted. The inner end of the blade shaft carries a short crank and guide shoe that is located between two regulating rings. A hydraulic cylinder, mounted axially in the bearing assembly, moves the guide rings back and forth to alter the fan pitch.

Pitch can be controlled by a standard pneumatic signal, which controls the hydraulic positioning of the guide rings. Blade pitch varies continuously during boiler operation to accommodate changes in boiler load as well as any boiler upset conditions.

In designing a fan it is necessary to consider not only the correct characteristics but also the mechanical features. Fans are subjected to adverse conditions, such as heat, dust, and abrasive material carried by the air or gases.

The bearings are a most important part of a fan. Fans are designed with a variety of bearings, depending on their use. Sleeve-ring oiled bearings and antifriction oil- and grease-lubricated bearings are used.

The sealed grease-packed antifriction bearings require the least attention and are recommended for out-of-the-way places where periodic lubrication would not be feasible. For high-speed operation, oil lubrication is desirable. Antifriction bearings are subject to quick and complete failure when trouble develops but can be replaced easily and quickly.

Sleeve-ring oiled bearings, shown in Fig, 5.52, are used extensively in fans. Figure 5.52a shows a sleeve bearing with water cooling. This type of bearing would be used in an induced-draft fan. The heat in the flue gases is conducted to the bearing by the shaft and must be carried away by the cooling water to prevent overheating. Figure 5.52b shows a fixed-sleeve bearing provided with a thrust collar and setscrews for attaching to the shaft and thus restricting the axial movement of the shaft. Oil should be drained from the bearings at regular intervals and the sediment flushed out. After they have been cleaned, the bearings must be filled with a light grade of oil. When in operation, the oil level and operation of the oil ring must be checked regularly.

As with all rotating machinery, the alignment and balance must be correct. The fan must be aligned with the motor or turbine. Consideration must be given to expansion. Flexible couplings should

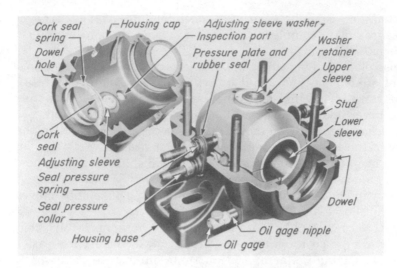

(a)

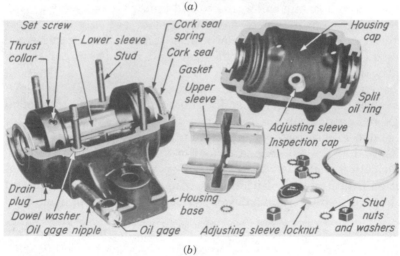

(b)

Figure 5.52 (a) Self-aligning water-cooled sleeve bearing. (b) Ring-lubricated sleeve bearing with thrust collar. (*American Standard, Inc.*)

not be depended on to compensate for misalignment. Even a slight out-of-balance condition will produce vibration detrimental to the bearings and to the entire unit. Wear will sometimes cause a fan to get out of balance.

The erosion of fan blades and casing by fly ash and other abrasive particles in the air and gases can cause serious and costly maintenance. However, particulate-collection equipment has significantly reduced the level of erosion that has been experienced by fans.

Flue gas recirculation fans and fly-ash reinjection fans, however, must be designed to minimize the impact of fly-ash erosion. In order to withstand the erosive action of fly ash, wear plates are installed on the fan blades and in other areas of the fan where erosion is expected. These wear plates generally are installed as weld overlay onto the base material of the fan blade. Periodic inspections of the fans are essential so that replacements or repairs can be made before wear has progressed beyond the renewable sections.

Noise pollution has become a significant concern for power plant operators as local laws have established maximum noise levels that cannot be exceeded. The sources of fan noise primarily involve the following:

1. *Blade passing frequency.* The most common source of noise that is caused by pressure pulsation as the blades pass a stationary object such as a cutoff sheet of a centrifugal fan or a straightening vane on an axial fan. The fan design should attempt to minimize these areas of noise.

2. *Turbulent noise.* Caused by vortices breaking away from the tracking edge of a fan blade. Blade designs can be made to smooth the flow of air with a minimum of turbulence, such as a hollow airfoil blade, thus generating less noise.

3. *Mechanical vibration.* Caused most frequently by a fan wheel that is not balanced. Misalignment in the couplings as well as inadequate foundations also can cause vibrations.

Correcting the problems is obviously the best remedy for reducing noise; however, unavoidable noise can be reduced by insulation and by enclosure of the fan.

Silencers can be used on a fan system if correction of noise cannot be made by other means. However, silencers produce a pressure drop in the system, which results in increased power consumption. Nevertheless, in order to reduce noise to acceptable levels, inlet silencers are often used on forced-draft fans, and outlet silencers are frequently used on fans exhausting to the atmosphere.

A major source of fan vibration is unbalance of the fan wheel. Fans that are balanced in the manufacturer's shop can become unbalanced as a result of the accumulation of deposits on the fan blades or of the erosion of the fan material. As part of the preventive maintenance program, fans must be inspected periodically.

5.11 Steam and Air Jets

The energy derived from a jet of expanding air or steam may be utilized in producing a draft. Devices to accomplish this consist of a high-pressure jet of steam or air discharging into a larger pipe. The aspirating effect produced causes an additional amount of gases to be delivered.

Jets are seldom used as a primary means of producing draft in stationary practice. However, they have been used for many years to introduce over-fire air, create turbulence, and thereby promote more complete combustion.

The secret of smokeless combustion is in burning the hydrocarbons before they form soot particles. To this end, jets must supply the required over-fire air, distribute it advantageously, and at the same time create the required turbulence. As a result of research and operating experience, it is possible to design and successfully apply jets to specific furnace installations.

While over-fire air jets are not a cure-all for faulty design and operation, they are capable of performing a very important function in boiler operation. Early installations often were makeshift arrangements, and the results were disappointing. Consideration must be given both to the quantity of air to be supplied and to the distance it is to be carried into the furnace. The jets must be so located and spaced that their effects will overlap to prevent lanes of unburned gases from escaping. The air must be introduced close to the fuel bed to prevent the formation of smoke, because smoke is very difficult to burn, once it has formed. However, the jets should not be close enough to the fuel bed to cause interference with the solid mass of burning fuel. Care also must be exercised to see that the jets do not impinge on the brickwork or waterwall tubes.

The jet action must be modified to meet specific conditions. In some cases the actual amount of air present in the furnace may be sufficient, and only turbulence need be provided. This calls for the installation of steam jets without the air tubes. In this case the steam jets must be placed near the inner furnace wall to obtain maximum penetration. Air is usually supplied by a high-pressure over-fire air fan. Over-fire air requirements and over-fire air systems are based on the experience gained by power plant operators and by boiler and combustion equipment designers.

5.12 Flues and Ducts

Flues and ducts are usually designed based on economic concepts related to velocity. These velocities generally vary from 2000 to 3000 ft/min. The higher the velocity, the smaller is the flue or duct area; however, the draft loss or air resistance increases, which results in higher fan power costs. An evaluation must be made of the initial costs versus the operating costs, and this evaluation is part of an integrated boiler design.

Questions and Problems

5.1 What is a boiler setting?

5.2 How are water-tube boilers supported? Briefly describe each method.

5.3 Which portion of a boiler is critical to providing complete combustion?

5.4 Describe the purpose of a furnace arch and a baffle located within a boiler.

5.5 Why are water-cooled furnaces the preferred design on modern boilers as compared with refractory-lined furnaces?

5.6 What is the purpose of a stoker, and what are its primary features?

5.7 Describe the types of stoker systems. Provide examples of the types of fuels that stokers are designed to handle.

5.8 What is meant by furnace volume? By stoker grate surface? Why are these important to a boiler design?

5.9 What is the purpose of over-fire air?

5.10 What is a typical heat release of an underfeed stoker as compared with a spreader stoker?

5.11 Describe the operation of an underfeed stoker.

5.12 What are some of the limitations of an underfeed stoker?

5.13 How is a fire started on an underfeed stoker?

5.14 What are the general coal characteristics for use on an underfeed stoker?

5.15 What does *banking of a boiler* mean, and how is a boiler with an underfeed stoker banked?

5.16 Describe the types of mass-feed stokers used for coal firing, and describe their operation.

5.17 Define the coal characteristics for use on a traveling-grate stoker and on a water-cooled vibrating stoker.

5.18 How is a fire started on a chain-grate stoker?

5.19 What is the tempering of coal, and when is it done in the operation of stokers?

5.20   Describe the operation of a spreader stoker. What are its advantages, and what is the primary difference from an underfeed-stoker design?

5.21   What is the advantage gained by the reinjection of a portion of the fly ash into the furnace? What potential added problem can occur?

5.22   For spreader-stoker operation, approximately what percentage of the total air requirement is over-fire air and under-grate air? In addition to enhancing combustion, what environmental control is also improved?

5.23   What coal characteristics are suitable for use on a spreader stoker?

5.24   Explain how a spreader-stoker operation should be adjusted to obtain satisfactory fuel distribution on the grates. Under what conditions does a change in these adjustments become necessary?

5.25   Other than coal, what other materials are suitable for burning in a spreader-stoker-fired boiler?

5.26   As compared with a spreader stoker with a traveling grate, what advantages does a spreader stoker with a hydrograte have?

5.27   Provide a startup procedure for a spreader stoker.

5.28   Why does pulverized-coal firing allow high combustion rates? Why is it responsive to load changes?

5.29   For pulverized-coal firing, describe the four elements that influence the process in a significant manner.

5.30   In a direct-fired pulverized-coal system, what are the major features of the system? What is the normal pulverizer exit temperature?

5.31   What must a pulverizer design provide to consider it as performing satisfactorily?

5.32   What is the purpose of a classifier in a pulverizer?

5.33   Discuss the advantages and disadvantages of the primary air fan location on a pulverizer, either on the inlet or on the outlet of the pulverizer.

5.34   For a multiple-burner arrangement, how is the coal-air mixture properly distributed to each burner from a pulverizer?

5.35   In a combustion process, what are the two primary control arrangements?

5.36   Since coal can vary because of heating value, ash, moisture content, and density, how does a gravimetric feeder assist in combustion control?

5.37 Discuss the operation of a pulverized-coal burner.

5.38 In pulverized-coal firing, how is NO_x formed? For a low-NO_x burner design, what must be the objective, and how is this accomplished?

5.39 As compared with uncontrolled levels of NO_x production, what reduction of NO_x can a well-designed low-NO_x burner achieve?

5.40 Describe a flame safety system.

5.41 Provide the advantages and disadvantages of a pulverized-coal-fired system.

5.42 Describe a general procedure for the startup of a pulverized-coal-fired plant.

5.43 What are the major operating requirements for a pulverized-coal-fired plant?

5.44 Under what conditions is it necessary to heat fuel oil?

5.45 In a fuel-oil burner, what two methods are used to atomize the oil for combustion? Describe the oil burners that use these methods. Which type of burner is preferred, and why?

5.46 Name the advantages and disadvantages of using fuel oil for power generation.

5.47 What are some of the major operating problems that can occur in the operation of a fuel-oil-burning power plant?

5.48 In a fuel-oil-burning plant, what is the primary source of NO_x, and how is the NO_x controlled?

5.49 Under what conditions is natural gas generally used as a fuel for power boilers?

5.50 Natural gas is being used as a primary fuel to power gas turbines in a cogeneration system. What are the advantages and disadvantages of this system?

5.51 How is NO_x formed when firing natural gas, and how is NO_x controlled?

5.52 What are the major procedures to ensure safe and reliable operation when firing fuel oil or natural gas?

5.53 What is the most important function of a burner-control system? Describe the three types of burner-control systems.

5.54 What is *draft* and *draft loss?* In a boiler, how is draft created?

5.55 A stack is 150 ft high, and the exit gas temperature is 550°F. The outside temperature is 45°F. What draft can be expected at the base of the stack when the plant is operating at the designed capacity?

5.56 One of the stack's purposes is to disperse the flue gases. If a stack height is limited for some reason, e.g., a permit, how are flue gases properly dispersed, and how is any additional draft loss accounted for?

5.57 Describe a forced-draft (FD) fan and an induced-draft (ID) fan. How are these fans part of a pressurized furnace and a balanced-draft boiler design?

5.58 What is a characteristic curve for a fan, and how is it used?

5.59 What are the three ways of controlling the output of a fan?

5.60 What are the primary sources of fan noise, and what can be done to correct a noise problem?

5.61 Discuss the possible ways in which over-fire air jets may improve combustion.

6

Boiler Accessories

6.1 Water Columns

Maintaining the correct water level in a boiler at all times is the responsibility of the boiler operator. Gauge glasses are provided for assistance and are installed to indicate the level of water in the boiler or boiler drum.

Drum water level is one of the most important measurements for safe and reliable boiler operation. If the level is too high, water can flow into the superheater, and subsequently, waterdroplets could be carried into the turbine. These waterdroplets will leave deposits in the superheater and possibly cause tube failure, and any waterdroplets carried to the turbine will cause serious erosion problems on the blades, resulting in high maintenance costs as well as costly outages. The results from too low of a water level are more severe, since this would result in a reduction in the water circulation and this could cause the tubes to overheat and ultimately fail, also causing costly maintenance and repairs.

For the small low-pressure boiler (Figs. 6.1 and 6.2), the gauge glass is attached directly to the drum or shell by screwed connections, or a water column may be used. The water column is a vessel to which the gauge glass or other water-level-indicating devices are attached (Figs. 6.3 and 6.4). The water column permits the gauge glass to be located where it can be seen easily and makes the installation accessible for inspection and repairs. The location of the gauge glass and water column varies for different types of boilers, but wherever they are located, the water in the glass must be maintained at the level required to avoid overheating of boiler surfaces.

The try cock is used to check the level of the water in the boiler when the gauge glass is broken or out of service. Thus, in addition to the gauge glass, each boiler must have three or more gauge cocks (see

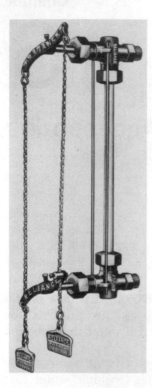

Figure 6.1 Water gauge valves. (*The Clark-Reliance Corp.*)

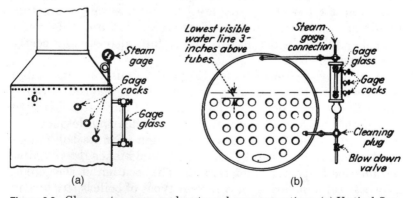

Figure 6.2 Glass water gauge and water column connections. (*a*) Vertical fire-tube boiler with gauge, cocks, and direct-connected glass water gauge. (*b*) Horizontal return tubular boiler showing water column, gauge, glass, gauge cocks, etc. (*The Clark-Reliance Corp.*)

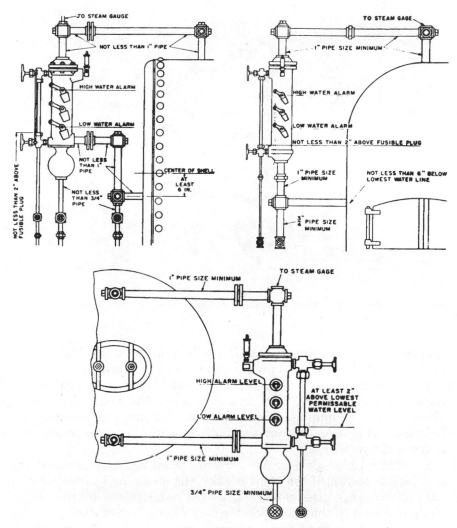

Figure 6.3 Installation of water columns. (*The Clark-Reliance Corp.*)

Figs. 6.2 and 6.3) located within the range of the visible length of the water glass when the maximum allowable pressure exceeds 15 psi, except when the boiler has two water glasses with independent connections to it located on the same horizontal line and not less than 2 ft apart. Boilers 36 in in diameter and under, in which the heating surface does not exceed 100 ft^2, need have but two gauge cocks. The lower cock is located at the lowest permissible water level, and the upper cock is located at the highest desired water level. If the cocks are located properly, water should always come from the lower cock

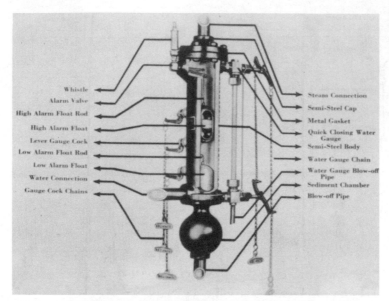

Figure 6.4 Safety water column. (*The Clark-Reliance Corp.*)

and steam from the upper cock. Depending on the water level in the boiler, either water or steam will come from the center cock.

For all boiler designs, drum-level instrumentation must be in accordance with the *ASME Boiler and Pressure Vessel Code*. For utility-sized boilers, the minimum drum water level instrumentation should be one water-gauge-glass and two indirect water-level indicators with alarm and trip points.

The water-gauge-glass connections are fitted with valves at the top and bottom so that if the glass breaks, the steam and water may be shut off for gauge-glass replacement. The hand valves are frequently chain operated so that the operator may remain out of danger. Many gauges have both hand and automatic shutoff valves. The automatic shutoff valves consist of check valves located in the upper and lower gauge-glass fittings. Should the glass break, the rush of steam and water would cause these valves to close.

The ball check valves are of the nonferrous type. They are designed to open by gravity. These automatic valves (balls) must be at least ½ in in diameter. It must be possible to remove the lower valve for inspection with the boiler in service. For added safety, the gauge glass is sometimes enclosed by wire-insert plate glass to protect the operator in the event that the glass breaks.

For higher pressures, water columns are made of flat glass (Fig. 6.5) having sheet mica to protect the water side of the glass from the etching action of the steam.

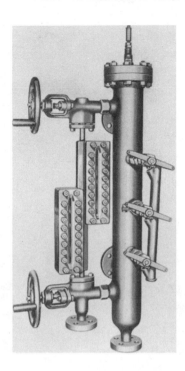

Figure 6.5 Water gauge with water column. (*Diamond Power, a McDermott company.*)

The water column (see Fig. 6.4) contains a high-low alarm that provides a signal (whistle) when the water rises or falls below the safe water level that is preset in the water column. Two floats are located in the water column, one above the other. To each float is attached a rod that opens a valve, thus actuating the whistle. Should the water rise high enough to raise the upper float, the rod would open the valve, admitting steam to create a whistling noise. The same thing would happen if the water level fell. Here the lower float would drop and repeat the warning. There are many different designs of high-low alarms that operate on the same principle. Whistles may be so designed that the tone denotes either a high- or a low-water alarm.

The Simpliport bicolor gauge (Fig. 6.6) is used for high-pressure service. It has the advantage of being sectionally constructed so that replacement of port assemblies is accomplished quickly without removing the gauge from the boiler. Normally, only one port need be serviced at a time. The gauge is illuminated to obtain maximum color contrast: Water is green and steam is red, and the water level is always where the colors meet.

The operation of the Simpliport gauge is based on the simple optical principle that the bending of a ray of light differs as the ray passes obliquely through different media. Therefore, when light passes through a column of steam, the amount of bending to which it is sub-

Figure 6.6 Simpliport bicolor gauge. (*The Clark-Reliance Corp.*)

jected is not the same as that when it passes through a similar column of water. If steam were to occupy the space between the windows ahead of the glass, the green light (water) would be bent out of the field of vision and the red light (steam) would appear in the glass. If water were to occupy the space between the windows, the red light would be bent out of the field of vision, and the green light would appear in the glass.

At times the water column is far removed from the operating-floor level. This problem is overcome by using a series of mirrors to bring the image of the gauge glass down to the operating level. The colors in the glass are directed into a hooded mirror, which in turn is reflected in a mirror at operating level. Distance is not a factor, and assurance of proper water-level indication is available at all times. On modern units, fiberoptics are often used to transmit drum-level indication to operator control rooms, which may be located a considerable distance from the boiler.

Other devices to indicate the water level are actuated by the height of the water in the drum. This differential pressure can be used to operate a pointer on a gauge (similar to a pressure gauge), or the instrument can be made to record the level. Devices of this type do not replace the gauge glass but provide an additional aid for the operator.

The gauge glass and water column are piped to the boiler so that the water level is the same in the glass as it is in the boiler. For the small vertical boiler, the gauge glass is attached directly to the shell. It must be so located that when water shows halfway in the glass, the boiler shell will be three-quarters full. For the HRT boiler (Fig. 6.2b), the lowest visible level in the water glass is 3 in above the top of the tubes. Water-column piping must be at least 1 in in diameter. Cross-connectors are used instead of ells (Fig. 6.3) so that the line may be easily cleaned. When the correct level is maintained in the boiler, the water column should show the water approximately in the center of the glass.

In the horizontal fire-tube boiler (Figs. 6.2b and 6.3), the water-column connection must enter the front top of the shell or as high as possible in the head. The lower water-column connection must be at least 6 in below the centerline of the boiler. Now, with the water column properly located, the lowest visible part of the water-column glass must be at least 3 in above the top row of tubes (Fig. 6.2b) and at least 2 in above the fusible plug (Fig. 6.3). A fusible plug is used on small low-pressure boilers and is a hollow threaded plug that has the hollow portion filled with a low-melting-point material that is usually located at the lowest permissible water level. The gauge glass must be at least ½ in in diameter, with the blowoff connecting pipe not less than ¾ in in diameter. The correct locations of the gauge glass and the water column are checked by filling the boiler with water to the normal operating level and measuring the height of water above the tubes. This reading can then be compared with the level as it appears in the glass.

For the firebox-type boiler (Fig. 6.3), the lower water-column connection must be taken at least 6 in below the lowest permissible water level. Boilers of the horizontal fire-tube type shall be set so that when the water is at the low level in the water-gauge glass, there shall be at least 3 in of water over the highest point of the tubes, flues, or crown sheet.

If shutoff valves are used between the boiler and the water column, they must be either outside screw-and-yoke-type gate valves or stopcocks. In either case the valves must be locked or sealed open. All water columns must be provided with a blowdown line of ¾ in minimum diameter. This line must be run to the ash sluice or other suitable drain.

No connection must be made to the water column except for the pressure gauge or feedwater regulators. Pipelines that are to supply steam or water must never be connected to the water column, since the flow of steam or water would cause the column to record a false level.

When replacing a broken glass, make sure no broken pieces remain in the gauge fittings. Prior to inserting a new glass, blow out the piping connections. Make certain that the glass is of the proper length. A glass that is too long will break because of its inability to expand; a short glass will continue to leak around the packing glands. Give leaks around the glands immediate attention, taking care first to close the valves before using the wrench. Dirty and discolored glasses should be replaced at the first opportunity. At the time of the annual inspection, check all component parts of the water column carefully, paying particular attention to floats and linkage, alarms, whistles, etc. Inspect all connecting pipes, removing scale and dirt that may have collected on them. For the high-pressure port and flat-type gauge glasses, replace leaky gaskets.

Reliable drum-water-level readings are essential for safe and efficient operation of a boiler. In cases where accurate drum-water-level signals were not transmitted to the operator, either locally or in the control room, serious boiler damage has resulted. Most problems related to drum water level are due to insufficient water, where the overheating of tubes results in their failure. Too much water results in water carry-over to the superheater and eventually to the turbine, causing tube failures and turbine blade erosion, respectively.

The ASME code states that any boiler operating above 400 psi should have two gauges, and any boiler operating over 900 psi should have two gauges or one gauge and two remote level indicators for reading the drum level.

Over the years, a number of methods have been developed to monitor the drum water level. The bicolor gauge is the most widely used because it is a direct means of viewing the drum water level. Most of these gauges have an illuminating source that incorporates red (for steam) and green (for water) colors to indicate water levels.

Although a direct-reading hood can be attached to the gauge for local viewing, it is desirable to transmit the light signals to the control room. A series of mirrors can accomplish this; however, various building or equipment interferences, distance, mirror alignment, and cleanliness can have an effect on this system.

Fiberoptic systems have become a widely accepted direct means of transmitting the water-level-gauge readings to the control room. A gauge-mounted hood accepts the red and green signals from the gauge, and the light signal is transmitted through the fiber to a control room display.

6.2 Fusible Plugs

Fusible plugs (Fig. 6.7) are used on small low-pressure boilers to provide protection against low water and consequent damage to the boiler. They are constructed of brass or bronze with a tapered hole drilled through the plug. In the ordinary direct-contact fire-actuated plug, this

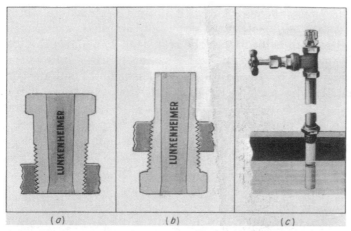

Figure 6.7 Fusible plugs and fuse alarms. (*a*) Inside type. (*b*) Outside type. (*c*) Fuse alarm. (*The Lukenheimer Company.*)

hole is filled with tin (or some other low-melting-point material), which has a melting point of 445 to 450°F. They are usually located at the lowest permissible water level. *Fire-side* plugs are those inserted from the fire side of the plate, flue, or tube to which they are attached. On the water side they are to project at least ¾ in on the other (gas) side, as little as possible but not more than 1 in. *Water-side* fusible plugs are those inserted from the water side of the plate or tube to which they are attached. Fusible plugs are installed with the large end of the core exposed to the water and made to blow through the narrow bore of the plug. For fire-actuated plugs, the smallest diameter of the tapered hole is not less than ½ in unless the maximum allowable pressure is over 175 psi or unless such a plug must be placed in a tube, in which case the smallest diameter of fusible metal is not less than ⅜ in. If a fire-actuated fuse plug is inserted in a tube, the tube wall shall not be less than 0.22 in thick or sufficiently thick to give four threads. For pressures over 225 psi, the fire-actuated plug is not used.

One side of the fuse plug is exposed to the hot furnace gases. Only water cooling on the other side of the plug permits the heat to be carried away fast enough to keep the plug from melting. Should the water drop below the fuse plug, the heat would melt the tin and be blown out of the core. The pressure, on being released, would sound an alarm and warn the operator of low water level. The fuse plug can be replaced only by taking the boiler out of service, cooling, and draining. For the *fuse-alarm*-type plug (Fig. 6.7c), the valve can be shut and another fuse plug inserted. Before inserting the plug and continuing with the operation, the extent of damage due to low water should be determined.

The fusible plug must be located at the lowest permissible water level and extend to make contact with the hot combustion gases. In the HRT boiler, the plug is located in the rear head 1 in above the top row of tubes, as measured from the upper surface of the tubes to the center of the plug. In a vertical fire-tube boiler with nonsubmerged tubes, it is located in an outside tube not less than one-third of the length of the tube above the lower tube sheet. In a vertical fire-tube boiler of a submerged-tube type, it is located in the upper tube sheet. In the horizontal-drum water-tube boiler, the plug is located not less than 6 in above the bottom of the drum, projecting through the sheet not less than 1 in and above the first pass of the products of combustion. For other bodies and fusible-plug locations, consult the boiler code and the boiler designer.

If fusible plugs are used, they should be inspected frequently, since scale and dirt on the water side and soot on the fire side may foul the plug. A plug in this condition will not function properly and will not provide the necessary protection.

With each boiler outage, the plug should be examined, cleaned, and scraped to a bright surface of the fusible metal. If it is not sound, or if its soundness is in question, replace it; do not fill it. All fusible plugs should be replaced at least once a year.

Fusible plugs are not used in today's boiler designs as they had been in the past, although many older boilers still have them installed. The plugs in these older designs are limited to operating pressures below 250 psig.

6.3 Pressure Measurement

The pressure gauge was probably the earliest instrument used in boiler operation. Even in today's modern power plants, with their complex control systems, a pressure gauge is still used to determine steam-drum pressure, over 100 years after the first water-tube boiler went into operation. The Bourdon-type tube pressure gauge is shown in Fig. 6.8. Although improvements have been made in construction and accuracy, its basic principle of operation remains unchanged. A closed-end oval tube in the shape of a semicircle tends to straighten with internal pressure. The movement of the closed end is converted to an indication by means of a needle position on a visible gauge face.

Pressure-measuring instruments are of various forms, and these depend on the magnitude of the pressure, the desired accuracy, and the application. Manometers are considered an accurate means of pressure or pressure-differential measurement. These instruments contain a variety of fluids depending on the pressure, and they are capable of a high degree of accuracy. The fluids used can vary from a

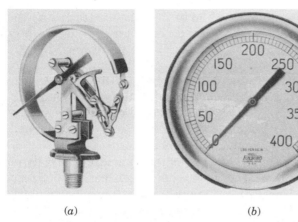

(a)   (b)

Figure 6.8 Pressure gauge. (a) Bourdon tube and linkage. (b) Exterior view. (*The Foxboro Company.*)

fluid that is lighter than water for use in low-pressure situations to mercury, which is used for relatively high pressures.

Incorporated into many pressure-measurement devices is the capability of producing an output signal. This output signal can be transmitted to a central measurement system and also to a control system. Pneumatically transmitted signals are often used in control systems, but the more modern designs use electric circuitry. These electric circuits are easily adaptable to computer-based systems.

All boilers must have at least one pressure gauge. It should have a range and dial graduation of at least 1½ times the maximum allowable working pressure. The connection may be made to the steam space or attached to the upper part of the water column. The gauge itself must be located so that it can be seen easily by the operator. Piping should be as direct as possible. If a valve is used in the gauge line connected to the boiler, it should be locked or sealed open. If a cock is used in place of a valve, it should be of the type that indicates by the position of its handle whether it is open or closed (open with the handle in line with the pipe). The piping to the gauge should be arranged so that it will always be full of water. A ¼-in branch connection and valve are provided so that a "test" gauge can be installed without removing the permanent gauge. If the temperature exceeds 406°F, brass or copper pipe or tubing should not be used. The connection to the boiler should be made with not less than a ¼-in standard pipe size, but when steel pipe or tubing is used, with not less than a ½-in inside-diameter size.

Steam gauges used in modern practice are usually of the Bourdon- or spring-tube type (Fig. 6.8). The Bourdon tube consists of a curved tube with an oval cross section. One end of the tube is attached to the frame and pressure connection; the other end is connected to a pointer by means of links and gears. Movement of the pointer is directly proportional to the distortion of the tube. An increase in pressure tends to straighten out the tube; as the tube position changes, its motion is transmitted to a rack and pinion through connecting linkage to position the pointer. Motion from the tube-connecting linkage is then transmitted to the pinion, to which the pointer is attached, moving the pointer over the range of the gauge.

The steam gauge for a small boiler is usually mounted directly on top of the water column, and, therefore, the gauge reads the correct pressure in the boiler. However, on many installations, particularly large boilers, the gauge is brought down to the operating level. At this level the gauge reads the steam pressure plus the hydraulic head of water in the line. Therefore, the gauge is inaccurate unless this head is compensated for. In this case we must measure the vertical distance between the point at which the connection is made (assume that it is the top of the water column) and the center of the gauge and correct for this water column. For each foot of vertical distance between the connection and the gauge, the gauge reading must be corrected by 0.433 psi per foot, and this correction is subtracted from the gauge reading.

Example   A gauge is located 45 ft below the point at which it is connected to the steam line or water column; the gauge reads 175 psi. What is the true gauge reading?

Solution   Pressure due to head of water:

$$45 \times 0.433 = 19.5 \text{ psi}$$

$$175 - 19.5 = 155.5 \text{ psi, actual pressure at point measured}$$

In this case the pointer on the steam gauge is reset by approximately 20 psi to indicate the true header pressure, or 155 psi (155.5 actual).

Pressure gauges frequently are mounted above the point of pressure measurements and piping connections made with ¼-in pipe or tubing. In many instances lines may be run horizontally before they proceed vertically to the gauge board and pressure gauge. These lines remain full of water, and the gauge will read inaccurately unless a correction is made for the water column as in the preceding example. For such installations measuring steam pressure, a siphon is often used at the takeoff point to make certain the line is full of water. Here again, the pressure in the connecting line (vertical distance) is measured as 0.433 psi per foot, and the pressure is added to the gauge reading. The pointer on the gauge must be reset accordingly.

6.4 Temperature Measurement

The Fahrenheit and Celsius (formerly centigrade) scales are the most common temperature scales. In addition to the commonly used thermometer, the instruments used most often are the optical pyrometer and the thermocouple.

Optical pyrometer. This device compares the brightness of an object to a reference source of radiation. It is used widely for the measurement of temperatures in furnaces at steel mills and iron foundries. It is generally not used for the measurement of flue gas temperatures.

Thermocouple. A thermocouple consists of two electrical conductors of dissimilar materials that are joined at the end to form a circuit. If one of the junctions is maintained at a temperature that is higher than the other, an electromagnetic force, called *emf,* is generated that produces a current flow through the circuit. The relationship between emf and the corresponding temperature difference has been established by laboratory tests for common thermocouple materials for various temperature ranges. The thermocouple is a low-cost, versatile, durable, simple device that provides fast response and accurate temperature measurement.

The temperature of a fluid (liquid, gas, or vapor) that flows under pressure through a pipe is usually measured by a glass thermometer, an electrical resistance thermometer, or a thermocouple. The thermometer is inserted into a well (thermowell) that projects into the fluid flowing through the pipe. The thermowell is the preferred method of temperature measurement; however, a thermocouple that is properly attached to the outside of a pipe wall also can provide accurate measurements.

It is often desirable to know the metal temperature of tubes in various sections of the boiler. These tubes include furnace wall or boiler bank tubes that are cooled by water and steam at saturated temperatures, economizer tubes that are cooled by water below the saturation temperature, and superheater and reheater tubes that are cooled by steam above the saturation temperature. Surface thermocouples are used to measure both metal and fluid temperatures.

6.5 Feedwater Regulators

A boiler feedwater regulator automatically controls the water supply so that the level in the boiler drum is maintained within desired limits. This automatic regulator adds to the safety and economy of operation and minimizes the danger of low or high water. Uniform feeding of water prevents the boiler from being subjected to the expansion strains that would result from temperature changes produced by irregular water feed. The danger in the use of a feedwater regulator lies in the fact that the operator may be entirely dependent on it. It is well to remember that the regulator, like any other mechanism, can fail; continued vigilance is necessary.

The first feedwater regulator (Fig. 6.9) was very simple, consisting of a float-operated valve riding the water to regulate the level. If the level dropped, the feed valve opened; if the level was too high, the valve closed; at intermediate positions of water level, the valve was throttled. A more modern float-type regulator (Fig. 6.10a) is designed with the float box attached directly to the drum.

For high-capacity boilers and those operating at high pressure, a pneumatic or electrically operated feedwater control system is used. There are basically three types of feedwater-control systems: (1) single element, (2) two element, and (3) three element.

1. *Single-element control.*   This uses a single control loop that provides regulation of feedwater flow in response to changes in the drum water level from its set point. The measured drum level is compared to its set point, and any error produces a signal that moves the feedwater-control valve in proper response. Single-element control will maintain a constant drum level for slow changes in load, steam pressure, or feedwater pressure. However, because the control signal satisfies the requirements of drum level only, wider drum-level variation results.

2. *Two-element control.*   This uses a control loop that provides regulation of feedwater flow in response to changes in steam flow, with a second control loop correcting the feedwater flow to ensure the correct drum water level.

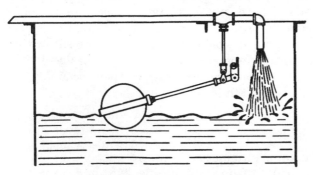

Figure 6.9   The first commercial feedwater regulator. (*Copes-Vulcan, Inc.*)

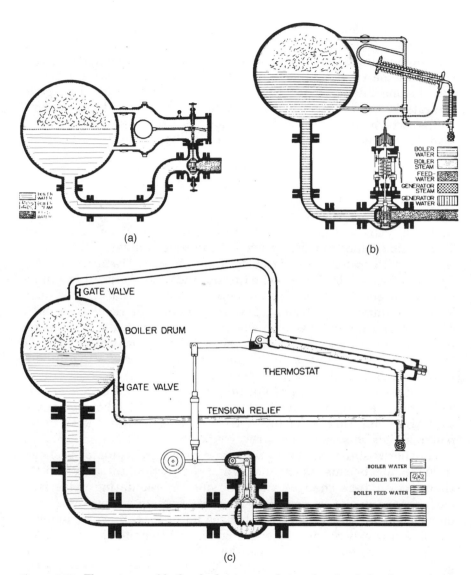

(a)

(b)

(c)

Figure 6.10   Three types of boiler feedwater regulators for simple level control. (*a*) Float-type regulator. (*b*) Thermohydraulic-type regulator. (*c*) Thermostatic expansion tube-type regulator. (*Copes-Vulcan, Inc.*)

The steam flow control signal anticipates load changes and begins control action in the proper direction before the drum-level control loop acts in response to the drum water level. The drum-level measurement corrects for any imbalance between the drum water level and its set point and provides the necessary adjustment to cope with the "swell and shrink" characteristics of the boiler.

3. *Three-element control.*   This uses a predetermined ratio of feedwater flow input to steam flow output to provide regulation of feedwater flow in direct response to boiler load. The three-element control regulates the ratio of feedwater flow input to steam flow output by establishing the set point for the drum-level controller. Any change in the ratio is used to modify the drum-level set point in the level controller, which regulates feedwater flow in direct response to boiler load. This is the most widely used feedwater-control system.

A thermohydraulic, or generator-diaphragm, type of boiler feedwater regulator is shown in Fig. 6.10*b*. Connected to the radiator is a small tube running to a diaphragm chamber. The diaphragm in turn operates a balanced valve in the feedwater line. The inner tube is connected directly to the water column and contains steam and water. The outside compartment, connecting the tube and valve diaphragm, is filled with water. This water does not circulate. Heat is radiated from it by means of fins attached to the radiator. Water in the inner tube of the regulator remains at the same level as that in the boiler. When the water in the boiler is lowered, more of the regulator tube is filled with steam and less with water. Since heat is transferred faster from steam to water than from water to water, extra heat is added to the confined water in the outer compartment. The radiating-fin surface is not sufficient to remove the heat as rapidly as it is generated, so the temperature and pressure of the confined water are raised. This pressure is transmitted to the balanced-valve diaphragm to open the valve, admitting water to the boiler. When the water level in the boiler is high, this operation is reversed.

The thermostatic expansion-tube-type feedwater regulator is shown in Fig. 6.10*c*. Because of expansion and contraction, the length of the thermostatic tube changes and positions the regulating valve with each change in the proportioned amount of steam and water. A two-element steam-flow-type feedwater regulator (Fig. 6.11) combines a thermostatic expansion tube operated from the change in water level in the drum as one element with the differential pressure across the superheater as the second element. The two combined operate the regulating valve. An air-operated three-element feedwater control (Fig. 6.12) combines three elements to control the water level. Water flow is proportioned to steam flow, with drum level as the compensating element; the control is set to be insensitive to the level. In operation, a change in position of the metering element positions a pilot valve to vary the air-loading pressure to a standatrol (self-standardizing relay). The resulting position assumed by the standatrol provides pressure to operate a pilot valve attached to the feedwater regulator. The impulse from the standatrol passes through a hand-automatic

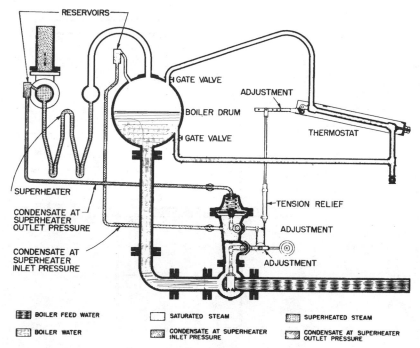

Figure 6.11 Two-element steam-flow-type feedwater regulator. (*Copes-Vulcan, Inc.*)

selector valve, permitting either manual or automatic operation. The hand-wheel jack permits manual adjustment of the feedwater valve if remote control is undesirable.

The simple float-operated regulator is satisfactory for small boilers with large water-reserve capacity. A more modern float-type regulator for the same purpose is shown in Fig. 6.10*a*. More accurate and dependable control is obtained with thermohydraulic, generator-diaphragm, or thermostatic expansion-tube-type regulators, and these are applied to water-tube boilers of moderate size and steam capacity. Such boilers have adequate water storage, and level fluctuations are not critical. The single-element control is affected only by the water level and is capable of varying the water level in accordance with the steaming rate.

Large boilers equipped with waterwalls, having relatively small water-storage space and subjected to fluctuating loads, use the two-element control, since feed characteristics are dependent on the rate of change rather than on the change in level. This change takes care of *swell* as well as *shrinkage* in boiler water level, and unless operating conditions are very severe, stability of water level can be maintained where load swings are wide and sudden, which is too difficult a

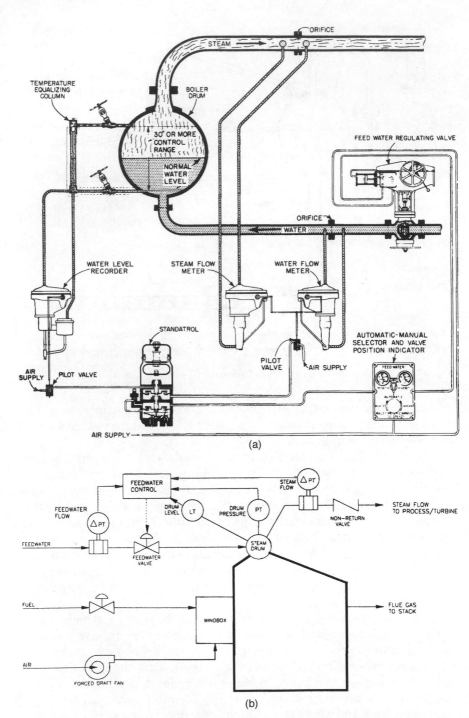

Figure 6.12 Three-element feedwater-control system. (*a*) Diagram layout of air-operated type. (*b*) Schematic of control system. (*Bailey Controls Company.*)

condition for a single-element regulator to control. In the two-element unit, steam flow predominates, and adjustment is provided from water level.

Three-element feedwater regulators (Fig. 6.12a) are used on large boilers subjected to wide and sudden load fluctuations and in installations where considerable variations in pressure drop across the feed valve are experienced. This type of control is recommended particularly for boilers equipped with *steaming* economizers and boilers that have small water-storage capacity. Three-element control is desirable with an increased rate of steaming because a much greater percentage of the volume below the surface of the water is occupied by steam. Only a small percentage of the total volume of water in a modern boiler is in the drum. Therefore, the drum water level will be seriously affected by changes in the steaming rate.

Feedwater control has evolved to provide the operator with improved boiler response to ensure the production of the required amount of steam. Figure 6.12b shows a schematic of a three-element feedwater control that is utilized on large modern boiler installations.

Effective steam production for power or process use depends on several factors. Matching fuel flow to boiler load is readily achievable because a boiler's efficiency does not vary greatly when it is operated at its designed output. However, the matching of boiler feedwater flow rate to steam flow rate is more difficult. In a smaller, less complicated boiler, the drum level can be monitored to adjust the boiler feedwater flow rate. With larger, more complex boilers, the drum-level response to load changes is the opposite of the expected response, especially if the load changes are rapid. To compensate for these factors, the mass balance around the boiler must be known, and this requires knowing the drum level, steam flow, and feedwater flow. As with the control design of Fig. 6.12a, these three inputs form the basics of the three-element feedwater control scheme.

To compensate for these changes in the steam drum, the drum pressure and level changes must be known. If both the pressure and level are increasing or decreasing, there is an imbalance in the relative rates of boiler feedwater flow and steam flow. Yet if the drum level is rising while the drum pressure is falling, there is a rapid load-demand increase. Drum-level information alone is not a sufficient indicator to determine the required feedwater flow rate. Figure 6.12b shows a three-element feedwater control system using a controller as the feedwater control component. The system requires four analog inputs: steam flow rate, boiler feedwater flow rate, drum level, and drum pressure. Transmitters provide the inputs.

At low flow rates (e.g., load demands less than 20 percent or at light-off), the system does not use these signals for control because of the reduction in flow-rate measurement accuracy. The reduced accuracy tends to offset the increased improvement in control. Therefore, at low loads the amount of drum inventory swell or shrink reduces, and the drum level alone becomes an adequate feedwater-control parameter. As load demand increases, the three inputs (drum pressure, steam flow, and feedwater flow) plus the drum level are used in feedwater flow control.

Maintenance is an important item in connection with control. All control lines should be checked for leakage at frequent intervals. Regulators equipped with a remote manual-automatic selector (Fig. 6.12a) can be checked for leakage by positioning the control knob on *reset*. In this position, the control is blanked off from either manual or automatic control and should remain in a fixed position. If the position of the pointer on the gauge varies, there is leakage between the selector and the control valve. Leakage along the control lines can be detected by the noise of escaping air (in the case of a large leak) or, if a soap solution is applied at points suspected of leaking, by the sight of the resulting bubbles.

The following are recommended procedures for the maintenance of these controls. The equipment suppliers' operating and maintenance (O&M) instructions should be followed carefully.

Semiweekly Blow down the water columns on the generator. Take care of water leaks around valves and fittings promptly.

Monthly. Lubricate control parts. Check meters and connections for leaks; check standatrol and automatic selector valves carefully; check flowmeters to zero to determine their accuracy, sensitivity, and response. Check automatic-control system for leakage.

Yearly. Disconnect the meter and all control lines; blow them out. Dismantle the meter; clean, inspect, overhaul, and calibrate by running a water-column test. Carefully inspect all control valves in the system, such as selector and standatrol valves. Also, dismantle and inspect the feedwater-regulating valves. If possible, dismantle and overhaul regulators semiannually or at least annually. At such times go over the entire control mechanism to eliminate wear in moving parts. Check valves for wear and replace parts where necessary. Give particular attention to all packing glands.

6.6 Safety Valves

Boilers are designed for a certain maximum operating pressure. If this pressure is exceeded, there is danger of an explosion unless this pressure is relieved. This danger is so great that it necessitates equipping all boilers with safety valves to maintain the boiler pressure within design limits.

Therefore, the most critical valve on a boiler is the safety valve. Its purpose is to limit the internal boiler pressure to a point below its safe operating level. One or more safety valves must be installed in an approved manner on the boiler pressure parts where they cannot be isolated from the steam. The valves must be set to activate at approved set-point pressure and then close when the pressure drops below the set-point level.

When open, the safety valves must be able to carry off all the steam that the boiler is capable of generating without exceeding the specified pressure. The *ASME Boiler and Pressure Vessel Code* specifies the minimum requirements for safety and safety relief valves that are applicable to boilers.

As defined by the ASME code, safety and relief valves are used as follows:

1. *Safety valve:* Used for gas or vapor service
2. *Relief valve:* Used primarily for liquid service
3. *Safety relief valve:* Used as either a safety or a relief valve

Power-actuated safety valves are used for some approved applications. These valves are fully opened at the set-point pressure by a controller with a source of power such as air, electricity, hydraulic fluid, or steam.

For drum boilers that have superheaters, the safety valves are set so that the superheater valves lift at all loads before those on the steam drum. This procedure maintains a flow of steam through the superheater and provides a means to prevent overheating in the superheater. This procedure also results in the lowest design pressure for the piping and valves downstream of the superheater. Again, the ASME code provides the specific requirements for each application.

Other rules governing safety valves, design, and installation are as follows: Each boiler shall have at least one safety valve, and if it has more than 500 ft^2 of water-heating surface, it shall have two or more safety valves. The safety-valve capacity for each boiler shall be such that the safety valve or valves will discharge all the steam that can be generated by the boiler without allowing the pressure to rise by more than 6 percent above the highest pressure at which any valve is set and in no case by more than 6 percent above the maximum allowable working pressure.

One or more safety valves on the boiler proper shall be set at or below the maximum allowable working pressure. If additional valves are used, the highest pressure setting shall not exceed the maximum allowable working pressure by more than 3 percent. The complete range of pressure setting of all the saturated-steam safety valves on a boiler shall not exceed 10 percent of the highest pressure to which any valve is set. The ASME code should be carefully reviewed for the required number and size of valves as well as their proper setting.

All safety valves shall be constructed so that failure of any part cannot obstruct the free and full discharge of steam from the valves. Safety valves shall be of the direct spring-loaded pop type. The maximum rated capacity of a safety valve shall be determined by actual steam flow at a pressure of 3 percent above the pressure at which the valve is set to blow.

If the safety-valve capacity cannot be computed (see the ASME code) or if it is desirable to prove the computations, the capacity may be checked in any one of the three following ways. If it is found insufficient, additional capacity shall be provided.

1. By making an accumulation test, shutting off all other steam-discharge outlets from the boiler, and forcing the fires to the maximum. The safety-valve equipment shall be sufficient to prevent a pressure in excess of 6 percent above the maximum allowable working pressure. This method should not be used on a boiler with a superheater or reheater.

2. By measuring the maximum amount of fuel that can be burned and computing the corresponding evaporative capacity upon the basis of the heating value of the fuel.

The approximate weight of steam generated per hour is found by the formula

$$W = \frac{C \times H \times 0.75}{1100}$$

where W = weight of steam generated (lb/h)
    C = total weight or volume of fuel burned (lb/h or ft^3/h)
    H = heating value of fuel (Btu/lb or Btu/ft^3)
    0.75 = assumed boiler efficiency of 75 percent
    1100 = assumed Btu/lb to produce steam

(*Note:* This is an approximation only.)

Example Assume that 6 tons of coal are burned each hour with a heating value of 12,000 Btu/lb. Approximately how much steam will be generated with a boiler efficiency of 75 percent and a heat requirement to produce 1 lb of steam at 1100 Btu/lb?

Solution

$$W = \frac{6 \times 2000 \text{ lb/ton} \times 12,000 \times 0.75}{1100} = 98,181 \text{ lb/h}$$

3. By determining the maximum evaporative capacity by measuring the feedwater. The sum of the safety-valve capacities marked on the valves shall be equal to or greater than the maximum evaporative capacity of the boiler.

When two or more safety valves are used on a boiler, they may be mounted separately or as twin valves made by placing individual valves on Y bases or duplex valves having two valves in the same body casing. Twin valves made by placing individual valves on Y bases or duplex valves having two valves in the same body shall be of equal size. When not more than two valves of different sizes are mounted singly, the relieving capacity of the smaller valve shall not be less than 50 percent of that of the larger valve.

The safety valve or valves shall be connected to the boiler independently of any other steam connection and attached as closely as possible to the boiler, without any unnecessary intervening pipe or fitting. Every safety valve shall be connected so that it stands in an upright position. The opening or connection between the boiler and the safety valve shall have at least the area of the valve inlet. The vents from the safety valves must be securely fastened to the building structure and not rigidly connected to the valves so that the safety valves and piping will not be subjected to mechanical strains resulting from expansion and contraction and the force due to the velocity of the steam.

No valve of any type shall be placed between the required safety valve or valves and the boiler or on the discharge pipe between the safety valve and the atmosphere. When a discharge pipe is used, the cross-sectional area shall not be less than the full area of the valve outlet or the total of the areas of the valve outlets that are being discharged into the pipe. The pipe shall be as short and straight as possible and shall be arranged to avoid undue stresses on the valve or valves.

Safety valves are intended to open and close within a narrow pressure range, and, therefore, safety valve installations require a careful and accurate design for both inlet and discharge piping. Safety valves always should be mounted in a vertical position directly on nozzles to provide unobstructed flow from the vessel to the valve. A safety valve should never be installed on a nozzle having an inside diameter smaller than the inlet connection to the valve or on excessively long nozzles. The safety valve (or valves) shall be connected to the boiler independent of any other connection and attached as close as possible to the boiler.

The discharge of a safety valve will impose a reactive load on the inlet of the valve, the mounting nozzle, and the adjacent vessel shell as a result of the reaction force of the flowing steam. These loads must be taken into account for the installation of the safety valve

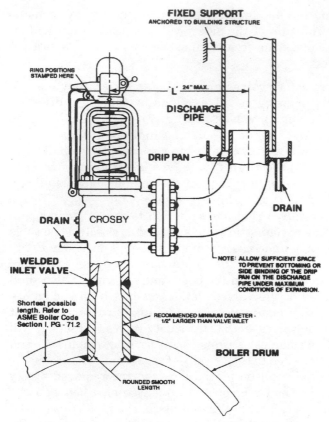

Figure 6.13 Typical arrangement of safety valve on steam drum showing steam-discharge piping and drains. (*Crosby Valves, Inc.*)

and associated piping. Figure 6.13 shows a typical installation arrangement.

The discharge piping from safety valves should be equal to or larger than the nominal valve outlet and should be as simple and direct as possible. The discharge pipe above the drip pan should be sized adequately to avoid blowback of steam from around the drip pan into the boiler room when the valve is discharging.

Provisions for drains are located in the valve bodies and should be piped to a drainage system to remove condensate from the valve bodies. Separate drains are recommended to drain the drip pan. All drains and piping in the discharge system must be piped to a safe disposal area to prevent possible injury to personnel when the valve discharges.

All safety-valve discharges shall have proper clearances from areas such as platforms. Ample provision for gravity drain shall be made in the discharge pipe at or near each safety valve and at locations where water or condensation may collect.

Safety valves shall operate without chattering, and shall be set and adjusted to close after blowing down not more than 4 percent of the set pressure but not less than 2 psi in any case. For spring-loaded pop safety valves at pressures between 100 and 300 psi, the blowdown shall not be less than 2 percent of the set pressure. Safety valves used on forced-circulation boilers of the once-through type may be set and adjusted to close after blowing down not more than 10 percent of the set pressure.

Each safety valve shall have a substantial lifting device by which the valve disk may be lifted from its seat when there is at least 75 percent of full working pressure on the boiler.

The spring in a safety valve in service for pressures up to and including 250 psi shall not be reset for any pressure more than 10 percent above or below that for which the valve is marked. For higher pressure, the spring shall not be reset for any pressure more than 5 percent above or below the safety valve's marked pressure.

Screwed openings can be used to attach the valve to the boiler when the proper number of threads is available. A safety valve over 3 in in size used for pressures greater than 15 psig shall have a flanged inlet connection or a welded-end inlet connection. On modern units, safety valves are attached to drums or headers by fusion welding, where the welding is done in accordance with the ASME code.

The safety valve shown in both Figs. 6.14 and 6.15 is designed specifically for saturated-steam service on boiler drums with design pressures over 1500 psig up to critical pressure. It incorporates an eductor control that permits the valve to attain full-capacity lift at a pressure 3 percent above popping pressure in accordance with Sec. I of the ASME code.

Springs on drum safety valves have very high preloads. As shown in Fig. 6.15, a thrust bearing (25) between the adjusting bolt (26) and top spring washer (21) makes set-point adjustments precise. The valve seats are protected from damage during set-point adjustment by lugs on the upper-spring washer (21). The lugs engage the bonnet (17) to prevent rotation of the spring (20), spindle (12), and disk insert (5).

The principal feature of the design is a dual-stage controlled-flow passage formed by the eductor (9A), disk holder (6), and adjustable guide ring (10). The nozzle ring (3) provides accurate and sharp pop action on opening.

Referring to Fig. 6.16, a typical valve operating cycle can be followed. As pressure in the boiler increases to the safety-valve set point, the valve will pop open. After the valve opens, steam passes through a series of annular flow passages (A) and (B) that control the

Figure 6.14 Boiler steam-drum safety valve designed for relief of saturated steam. (*Crosby Valves, Inc.*)

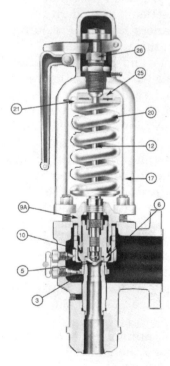

Figure 6.15 High-pressure, high-capacity safety valve designed for relief of saturated steam on steam drums. (*Crosby Valves, Inc.*)

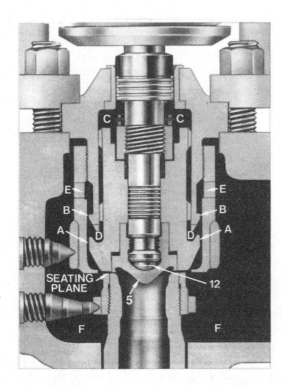

Figure 6.16 Safety-valve operating cycle. (*Crosby Valves, Inc.*)

pressure developed in chambers C and D, the excess steam exhausting through guide-ring openings (E) to the valve body bowl (F).

As the pressure in the boiler decreases, the forces on the lower face of the disk-holder assembly are reduced, and the safety valve disk begins to close. Assisted by pressure in chambers C and D, the valve at this point closes sharply and tightly. The seat-level loading of the spindle (12) on the disk insert (5) ensures uniform seat loading.

Every superheater has one or more safety valves near the outlet. If the superheater outlet header has a full and free steam passage from end to end and is so constructed that steam is supplied to it at practically equal intervals throughout its length, resulting in a uniform flow of steam through the superheater tubes and header, the safety valve or valves may be located anywhere in the length of the header.

The discharge capacity of the safety valve or valves attached to the superheater may be included in determining the number and size of the safety valves for the boiler, provided there are no intervening valves between the superheater safety valve and the boiler and provided the discharge capacity of the safety valve or valves on the boiler, as distinct from the superheater, is at least 75 percent of the aggregate valve capacity required. It is good practice to size the superheater safety valve to relieve approximately 20 percent of the total boiler capacity to protect the tubes against overheating.

Every independently fired superheater that may be shut off from the boiler and permitted to become a fired pressure vessel shall have one or more safety valves with a discharge capacity equal to 6 lb of steam per square foot of superheater surface measured on the side exposed to the hot gases. The total number of installed safety valves must meet capacity requirements.

The safety valve shown in both Figs. 6.17 and 6.18 is designed for saturated and superheated steam service with pressures and temperatures up to 900 psig and 750°F. The adjustable nozzle ring (3) and guide ring (9) utilize the reaction and expansive forces of the flowing steam to provide full lift. These valves shut off tight. The flat seat maintains continuous uniform seat contact at all times through a wide range of temperatures. A ball-bearing spindle point (11) ensures perfectly balanced transmission of spring loading to the disk insert (5).

Every reheater shall have one or more safety valves such that the total relieving capacity is at least equal to the maximum steam flow for which the reheater is designed. At least one valve shall be located on the reheater outlet. The relieving capacity of the valve on the reheater outlet shall be not less than 15 percent of the required total. The capacity of the reheater safety valves shall not be included in the required relieving capacity for the boiler and superheater.

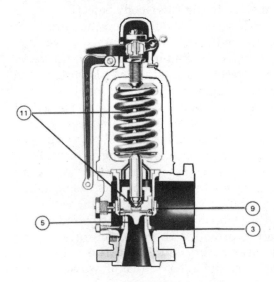

Figure 6.17 Safety valve for the relief of superheated steam showing tight shutoff design. (*Crosby Valves, Inc.*)

Figure 6.18 Superheater safety valve for steam temperatures to 1020°F. (*Crosby Valves, Inc.*)

Therefore, safety valves for boilers shall be sized in accordance with Sec. I of the ASME code. This is summarized as follows:

1. Boilers having more than 500 ft^2 of water-heating surface as well as a design steam-generating capacity exceeding 4000 lb/h must have two or more safety valves. If two valves are used, each valve must relieve approximately half the total boiler capacity.

2. Boilers having attached superheaters must have at least one valve on the superheater. The valves on the drum must be large enough to relieve at least 75 percent of the total boiler capacity. The superheater valve should relieve approximately 15 to 20 percent of the total steam generation to protect the tubes against overheating. The drum safety valves should then be sized to discharge the remainder of the boiler steam capacity.

3. Boilers having reheaters must have at least one safety valve on the reheater outlet capable of relieving a minimum of 15 percent of the flow through the reheater. The remainder of the flow through the reheater may be discharged by safety valves on the reheater inlet.

The design of safety valves varies between designers and manufacturers; therefore, the safety-valve capacities and any correction factors should be in accordance with the manufacturer's steam capacity and correction tables.

Every safety valve used on a superheater or reheater discharging superheated steam at a temperature over 450°F shall have a casing, including base, body, and spindle, of steel, of steel alloy, or of equivalent heat-resisting material. The valve shall have a flanged inlet connection or a welded-end inlet connection. The seat and disk shall be of suitable heat-erosion- and corrosion-resisting material, and the spring shall be fully exposed outside the valve casing so that it is protected from contact with the escaping steam.

When two or more boilers that are allowed different pressures are connected to a common steam line and all safety valves are not set at a pressure not exceeding the lowest pressure allowed, the boiler or boilers allowed the lower pressure shall each be protected by a safety valve or valves placed on the connecting pipe to the steam line. The area or the combined area of the safety valve or valves placed on the connecting pipe to the steam line shall be not less than the area of the connecting pipe, except when the steam line is smaller than the connecting pipe. Then the area or combined area of the safety valve or valves placed on the connecting pipe shall be not less than the area of the steam line. Each safety valve in the connecting pipe shall be set at a pressure not exceeding the pressure allowed on the boiler it protects.

The safety valve (Fig. 6.15) has the disk held on its seat by a spring. The tension on the spring can be adjusted to give some variation in popping pressure. This is accomplished by the compression screw, which forces the valve against its seat. The valve is correctly positioned by the valve extension fitting into the seat. An adjusting ring is used to regulate the blowback pressure and is provided to control the relieving pressure. This ring can be adjusted and fixed by a ring pin. A hand lever is furnished to permit popping the valve by hand.

To determine the capacity of a safety valve, refer to the valve manufacturer's *selection chart* for the pressure-temperature range in which the valve is to operate. Valve design varies to a considerable degree, so this is the most practical approach for determining capacities.

Problems with safety-valve leakage became increasingly severe as steam pressures increased, and so a high-capacity flat-seated reaction-type safety valve was developed (Fig. 6.19) to meet greater discharge capacity, shorter blowdown, etc., as required for high-pressure-temperature steam-generating equipment. Construction details and operation are as shown.

With reference to Fig. 6.20, in part (a) a 100 percent lift is attained by proper location of the upper adjusting ring (G). When full lift is attained in part (b), lift stop M rests against cover plate P to eliminate hunting, adding stability to the valve. When the valve dis-

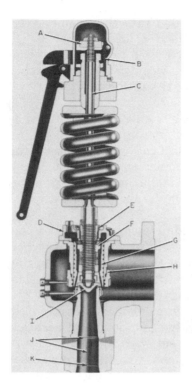

Figure 6.19 Consolidated maxi-flow safety valve. A = spring compression. B = lifting gear. C = spindle. D = backpressure. E = blowdown control. F = lift-stop adjustment. G = groove-disk holder. H = upper adjusting ring. I = thermdisk seat. J = inlet neck. K = inlet connection. (*Dresser Industries, Inc.*)

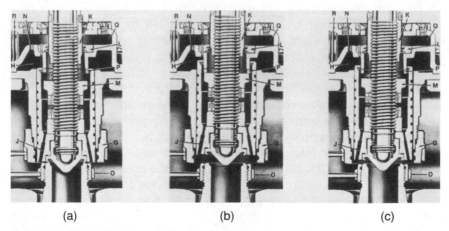

(a)   (b)   (c)

Figure 6.20 Operation of consolidated safety valve. (*Dresser Industries, Inc.*)

charges in an open position, steam is bled into the chamber (H) through two bleed holes (J) in the roof of the disk holder.

Similarly, the spindle overlap collar (K) rises to a fixed position above the floating washer (L). The area between the floating washer and the spindle is thereby increased by the difference in the two diameters on the overlap collar.

Under this condition, steam in chamber H enters into chamber Q through the secondary area formed by the floating washes L and the overlap collar (K) on the spindle, through orifice N, and escapes to the atmosphere through the pipe discharge connection (R). When closing in part (c), the spindle overlap collar (K) is adjusted so that it moves down into the floating washer (L), thereby reducing the escape of steam from chamber H.

The resulting momentary pressure buildup in chamber H, at a rate controlled by orifice N, produces a downward thrust in the direction of spring loading. The combined thrust of the pressure and spring loading results in positive and precise closing. Cushioning of the closing is controlled by the lower adjusting ring (O).

The valve includes several features such as (1) backpressure closing—lift and blowdown are separate valve functions and accurate control of each is possible; (2) thermodic seat—provides tight closure and compensates for temperature variations with thermal stresses minimized; (3) spherical-tip spindle with a small flat on the extreme end—provides a better point for pivoting than does a ball; (4) welded construction—forged neck and stainless steel nozzle; bypass leaks around a seal weld cannot occur with the three-piece construction.

The superjet safety valve (Fig. 6.21) is designed for steam pressures up to 3000 psi and temperatures to 1000°F. The spring here is protected from the heat by a shield and cradled by a special spring-saddle design to prevent distortion under compression. There are of course many other safety valves, all offering special features and advantages.

When boilers are equipped with superheaters and with safety valves on both the superheater outlet and the steam drum, the safety valve on the superheater outlet should open first. This produces a flow of steam through the superheater and prevents the superheater tubes from being damaged by high temperatures that would result if all the steam were discharged directly from the steam drum.

Safety-valve springs are designed for a given pressure but can be adjusted. However, if the change is greater than 10 percent above or below the design pressure, it becomes necessary to provide a new spring and in some cases a new valve. If a valve is adjusted for insufficient blowback, it is likely to leak and simmer after popping. Leakage after popping is also caused by dirt that gets under the seat

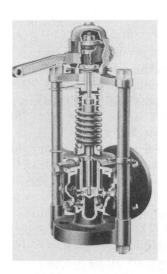

Figure 6.21 Superjet safety valve. (*Foster Engineering–Marine Industries, Inc.*)

and prevents proper closing of the valve. Safety valves should have the seat and disk "ground in" to prevent leakage. When valves are disassembled for grinding, the springs should be compressed by suitable clamps and held in place so that the adjustment of the spring will not be altered.

When a safety valve leaks at a pressure less than that at which it is set to close, the valves should be freed by operating the lifting lever. If this does not stop the valve from leaking, repair or replace the safety valve as soon as possible.

After changing the valve setting and adjusting the spring or the blowback ring, test the safety valve. This can be accomplished by slowly raising the steam pressure and noting the pressure-gauge reading when the popping pressure is reached. At the instant the valve pops, read the pressure gauge, after which the rate of steaming (or firing) should be reduced. Again read the pressure gauge when the valve closes, to note the blowback. Continued and repeated adjustments (if necessary) should be made to adjust the spring and blowback ring to obtain the desired popping pressure and blowback. Doing this will require that the pressure be raised or lowered until the correct setting has been obtained.

The set point of each safety valve is normally checked and adjusted immediately after reaching full operating pressure for the first time. Safety-valve seats are susceptible to damage from wet steam or grit, and therefore, cleaning the boiler and blowing out the superheater and steam line are essential prior to the testing of safety valves.

Safety valves on drum-type boilers are normally tested for both set-point pressure and closing pressure. This requires that the boiler pressure be raised until the valve opens and relieves sufficient pressure for the valves to close.

For high-pressure safety valves, in order to remain closed without leakage, they cannot tolerate any damage to the seats. They are not normally tested for closing pressure. These valves are checked without permitting them to open fully. Special gags are used to restrict valve lift and to close the valve as soon as it starts to simmer. The testing of safety valves always requires caution. Safety-valve exhaust piping and vent piping should not exert any excessive forces on the safety valve. If a hydrostatic test is made on the boiler, the safety valves should be removed and the openings blanked, or clamps should be applied to hold the valve closed.

In some power plant locations, noise is a major concern, and often it is limited by the facility's operating permit. These locations require that all safety valves include silencers that attenuate the safety-valve discharge noise to prescribed limits. Silencers are designed to break up the shock wave that occurs when the valve first opens and attenuate the steady-state noise that follows.

6.7 Blowdown Equipment

Water fed to the boiler contains impurities in the form of suspended and dissolved solids. A large portion of these impurities is left behind when the steam leaves the boiler. Some of these suspended impurities are of such a nature that they settle in the lowest part of the boiler. Others are light and float on the surface of the water. This condition frequently calls for the installation of both surface and bottom blowoff lines on some boilers.

In order to keep this concentration to a minimum, it is necessary to blow down the boiler periodically or even, at times, continuously. Blowdown therefore is the water that is bled from the boiler drum to control the concentration of total solids in the boiler water. In small power plants, this is done periodically by the operator by opening a blowoff valve for a few seconds and blowing out the water in the lowest part of the boiler, where the concentration is highest. In large plants, the amount of heat lost by such a blowdown practice would be high, so continuous blowdown systems are used. With these systems, a small amount of water is withdrawn continuously; however, before it is run to the waste discharge of the plant, it flows through a heat exchanger, where the heat from the blowdown water is transferred to the feedwater.

There are two types of surface blowoff arrangements. One consists of a pipe entering the drum approximately at the normal water level with the pipe fixed in location. The other arrangement is to have a swivel joint on the end of a short piece of pipe, the free end being held at the surface by a float. This floating-type surface blowoff is frequently referred to as a *skimmer*. Surface blowoff is advantageous in skimming or removing oil from the boiler water. A surface blowoff line shall not exceed 2½ in in diameter.

The surface blowdown is usually made on a continuous basis and after the feedwater has been tested. A flow-control valve (Fig. 6.22) is an orifice-type valve equipped with an indicator (for greater accuracy) and regulated by hand to control the quantity of water discharged, based on the water analysis. Blowing down at a slower rate and over a longer period of time reduces the concentration more effectively than is possible by opening wide the main blowoff valve. Therefore, closer control and more accurate regulation of the blowdown are achieved. Likewise, erosion and wear of valve parts are held to a minimum.

The continuous blowdown requires the use of a flash tank where the high-pressure water can be flashed into low-pressure steam and used for process or feedwater heating, or the continuous blowdown can be passed through a heat exchanger to preheat the makeup water. If required, a small portion of the higher-alkaline blowdown water may be introduced into the boiler feedwater line to raise the pH value of the water and eliminate feed-line or economizer corrosion.

To be effective, however, the continuous-blowdown takeoff must be placed at a point in the boiler where the water has the highest concentration of dissolved solids. This point is usually located where the greater part of the steam separates from the boiler water, ordinarily in the steam drum; hence the surface blowdown.

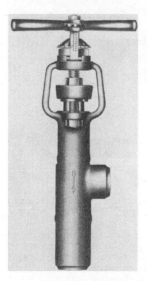

Figure 6.22 Continuous boiler flow-control blowdown valve. (*Dresser Industries, Inc.*)

All boilers must be equipped with a bottom blowoff pipe fitted with a valve directly connected to the lowest point in the water space so that the boiler can be drained completely. The blowoff lines must be extraheavy, not less than 1 in and not more than 2½ in in diameter, except that for boilers with 100 ft² of heating surface or less, the minimum size may be ¾ in. Too small a pipe might plug, and one too large would discharge the water too rapidly. Since there is no circulation of water in the pipe, scale and sludge frequently accumulate here. Unless this pipe is protected from the hot gases by suitable insulating materials, it may burn out.

On boilers operating at 100 psi or over, two blowoff valves are required. They may be two slow-opening valves or one slow-opening and one quick-opening valve. Over 125 psi, both valves and piping must be extraheavy. A slow-opening valve is one that requires at least five 360° turns of the operating device to change from fully closed to open, or vice versa. A quick-opening double-tightening valve is shown in Fig. 6.23. This valve is frequently used in tandem with a seatless valve. The quick-opening valve is installed next to the boiler and is opened *first* and closed *last*. In the tandem combination, the quick-opening valve becomes the sealing valve rather than the blowing valve. The valve shown is designed to operate at 320 psi maximum pressure. *Note:* For all other valves arranged in tandem, the sequence of operation is the reverse. Here the second valve from the boiler is opened first and closed last; blowing down takes place through the valve next to the boiler.

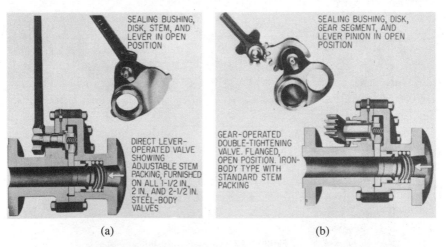

SEALING BUSHING, DISK, STEM, AND LEVER IN OPEN POSITION

SEALING BUSHING, DISK, GEAR SEGMENT, AND LEVER PINION IN OPEN POSITION

DIRECT LEVER-OPERATED VALVE SHOWING ADJUSTABLE STEM PACKING, FURNISHED ON ALL 1-1/2 IN, 2 IN, AND 2-1/2 IN. STEEL-BODY VALVES

GEAR-OPERATED DOUBLE-TIGHTENING VALVE. FLANGED, OPEN POSITION. IRON-BODY TYPE WITH STANDARD STEM PACKING

(a)   (b)

Figure 6.23 Quick-opening blowoff valve. (*a*) Sealing, bushing, disk, stem, and lever are in open position. Direct level-operated valve with adjustable stem packing, furnished on all 1½-in, 2-in, and 2½-in steel-body valves. (*b*) Gear-operated double-tightening valve, flanged open position, iron-body type with standard stem packing. Sealing bushing, disk, gear segment, and lever pinion are in open position. (*Yarnall-Waring Company.*)

 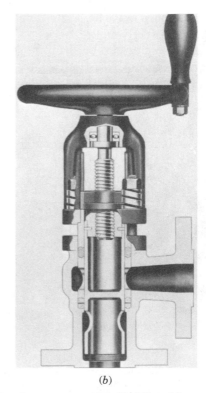

(a)   (b)

Figure 6.24 Blowoff valves. (a) Flanged angle valve—open position for full and free discharge. (b) Flanged angle valve—closed position for drop-tight shutoff. (*Yarnall-Waring Company.*)

For pressures to 450 psi, a seatless valve (Fig. 6.24) may be used. For pressures to 600 psi, a tandem arrangement (Fig. 6.25a) of a hard-seat blowing valve and a seatless sealing valve is used. With higher pressures (1500 to 2500 psi), the hard-head sealing valve is used. Here the blowing valve (nearest to the boiler) will have flow entering below the seat. The blowing valve (next to the boiler) should be opened *last* and closed *first*. The sealing valve (outside) should be opened first and closed last. For the hard-seat valve, the position of the handwheel above the yoke indicates the location of the disk in the valve, whereas for the seatless valve the position of the plunger indicates whether the valve is open or closed. While these valves are to be operated rapidly, they cannot be opened or closed quickly; waterhammer in the discharge line is thus avoided.

The tandem blowoff valve is a one-piece block serving as a common body for both the sealing and the blowing valves. The seatless valve is equipped with a sliding plunger, operated by the handwheel and non-rising steam. Leakage is prevented by packing rings above and below the inlet ports. As the seatless valve opens, blowdown is discharged

 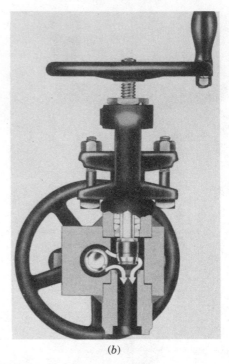

(a)   (b)

Figure 6.25 High-pressure tandem blowoff valves. (a) Open position—contains hard-seat blowing valve and seatless sealing valve. (b) Open position—contains hard-seat blowing and sealing valves. (*Yarnall-Waring Company.*)

through double ports. In the seatless-valve design, the annular space in the body permits pressure to surround the plunger, making the valve fully balanced and hence easy to operate at all pressures.

In operation, both valves are opened rapidly and fully to prevent erosion of the seat and disk faces and to increase the life of the packing and working parts. Blowdown should not be through a partly opened valve. If a hard-seat valve and a seatless blowoff valve are arranged in tandem, the hard-seat valve will be nearest to the boiler. When installing blowoff valves, take care that the piping is not restricted by the boiler setting but left free so that it can expand and contract.

In closing valves, they should not be forced to close, although firm pressure can be applied. If there appears to be some obstruction, the valves should be opened again before closing them finally. The blowdown valves are then tightly shut and remain that way when they are not blowing down. Leaky valves should be repaired as soon as possible. The valves should be dismantled at least once a year, and worn parts, such as scored plungers, packing rings, valve seats, etc., should be replaced if necessary. Prior to (and after) taking the boiler out of service for overhaul, it might be well to check for blowoff-valve leakage. Such leakage can be detected (with valves closed) by placing a hand on the discharge line to check the temperature, care being exercised not to get burned. If the line stays hot, leakage is evident. A rod held against the discharge line and used as a listening device also will detect leakage.

Blowoff connections cannot be run directly to a sewer or to the atmosphere. Steam and hot water might damage the sewer. Flashing steam might prove harmful to persons in the vicinity. Blowoff lines are run into a blowoff tank, entering at a point above the waterline maintained in the tank. The blowdown water and flashing steam are then discharged above the water level, where there is a vent in the top of the tank, which is necessary to avoid backpressure. A discharge line with an opening below the water level and near the tank bottom causes the cooler water in the tank to be discharged first. The discharge line leaves the tank at a point opposite the inlet. The line outside the tank is provided with a vent so that the water cannot be siphoned from the tank. The tank is provided with a drain outlet and valve at the bottom. The blowoff tank also acts as a seal to prevent sewer gas from backing up into a boiler that is out of service. A blowoff tank should never be relied on as a seal for boiler service. All valves must be closed tightly.

6.8 Nonreturn Valves

A *nonreturn valve* is a safeguard in steam power plants where more than one boiler is connected to the same header or a common main steam line. They must be installed between the boiler and the main steam line and should be attached directly to or adjacent to the nozzle outlet of the boiler. This prevents backflow of steam from the steam line into the boiler and also prevents steam from entering into a cold boiler. Pressure must be under the disk with the valve stem in a vertical position.

The valve will close instantly on loss of boiler pressure, such as caused by a tube failure, and it isolates the particular boiler to which it is attached when the pressure within that boiler drops below the pressure in the main steam line or main steam header. Likewise, it will open when the boiler to which it is attached reaches full pressure of the main steam line.

A nonreturn valve makes it possible to bank a boiler and return it to service without operating the steam shutoff valve. As the boiler is placed on the line, the valve opens automatically when the pressure in the boiler exceeds (slightly) the pressure in the steam header or main steam line on the discharge side of the nonreturn valve. The valve closes when the boiler pressure drops below the header pres-

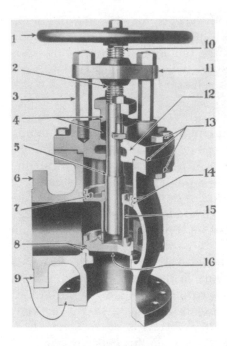

Figure 6.26 Nonreturn valve. 1 = handwheel. 2 = acme threads. 3 = hexagonal posts. 4 = stuffing box. 5 = packing under pressure. 6 = cast-iron flanges. 7 = thread, for removal and regrinding. 8 = seat and disk. 9 = center to face, to permit use for removal. 10 = steel stem. 11 = steel cross yoke. 12 = body and bonnet. 13 = bolted bonnet joint. 14 = piston rings. 15 = narrow neck piston to reduce pressure drop. 16 = full area-disk and piston guided and cushioned. (*Rockwell International, Flow Control Division.*)

sure. The use of the nonreturn valve provides additional safety in operation.

The automatic nonreturn valve (Fig. 6.26) is in reality a cushioned check valve. The cushioning is provided to keep the valve from chattering when it is being opened or closed. In some valves this cushioning is accomplished by using springs, but the usual method is to use a dashpot. The nonreturn valve has a stem not connected to the disk. When the stem is screwed down, it forces the disk shut and prevents it from opening. When the stem is screwed out, the disk can open or close automatically, depending on which has the higher pressure, the boiler or the main steam line.

The triple-acting nonreturn valve (Fig. 6.27a) has an additional feature in that it closes automatically in the event that the main steam line pressure decreases to a predetermined intensity below the boiler pressure. The cause of the decrease might be a break in the main steam line. The triple-acting nonreturn valve will (1) automatically open to cut a boiler into the main steam line when the pressures are approximately equal, (2) automatically close when the boiler pressure drops below the main steam line pressure, and (3) automatically close to isolate the boiler when the main steam line pressure drops by approximately 8 psi below the boiler pressure.

With reference to Fig. 6.27b, the main valve is installed with the

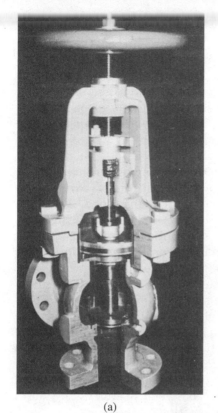

(a)

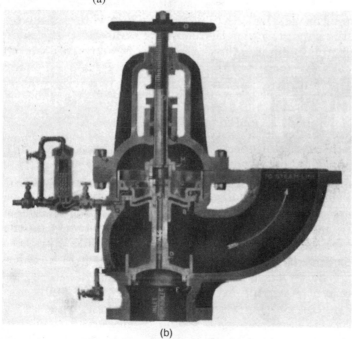

(b)

Figure 6.27 (a) Triple-acting nonreturn valve. (b) Interior view of triple-acting nonreturn valve, elbow pattern. (*G. A. Industries, Inc.*)

boiler pressure entering under the valve disk. The autopilot is piped so that pressure in the annular space between dashpots A and B of the main valve will enter under the automatic-pilot piston (H). The outlet of the autopilot is connected to the main steam line some 10 to 15 ft on the valve discharge side. Test line 4 is run to the operating-floor level for manual testing of the valve by the operator; it remains closed except during testing.

In operation, when the boiler pressure overcomes the main steam line pressure, the main valve (D) opens to admit steam. Steam from the boiler also passes through the bypass in the center of the disk, through the ball check valve (N), and then to the top of the piston (A). Small orifice holes (C) permit the passage of steam to the annular space between dashpots A and B, as well as to the autopilot (H) through valve 3.

In the event of a ruptured boiler tube, the pressure in the steam line forces the valve (D) to close. The steam located between the stationary disk (B) and the moving dashpot (A) tends to cushion the movement, preventing hammer or shock. If there is a break in the main steam line, the pressure on that side of the valve—and to the top of the pilot valve (H)—begins to drop. When it drops to about 8 psi below the boiler pressure, the boiler pressure raises the pilot valve (H), permitting steam to escape from between A and B; the main valve immediately moves to the closed position.

To test the main valve, the operator at the floor level opens valve 4 to accomplish the same thing as the foregoing (a break in the main steam line); opening valve 4 adjusts the valve to a closed position. Opening the test valve 4 creates an imbalance in the main valve and causes it to close, and the boiler-pressure gauge will immediately record an increase in pressure to prove that the main valve is closed. When the test valve is closed again, the main valve resumes its normal automatic position.

Several typical installations of nonreturn and stop valves are shown in Fig. 6.28. When boilers are set in series and are carrying more than 135 psi, they must be equipped with two steam valves, one of which can be a nonreturn valve. The nonreturn valve should be placed nearest the boiler as close to the boiler outlet as possible. It should be equipped with a drain or bleeder line for removing the water (condensation) before the valve is opened. In all cases, the valve design, arrangement, and requirements must be in accordance with the ASME code.

The nonreturn valve should be dismantled, inspected, and overhauled annually. The packing should be replaced and the corroded parts cleaned. The valves and seats should be checked for leakage, and the defective parts should be replaced.

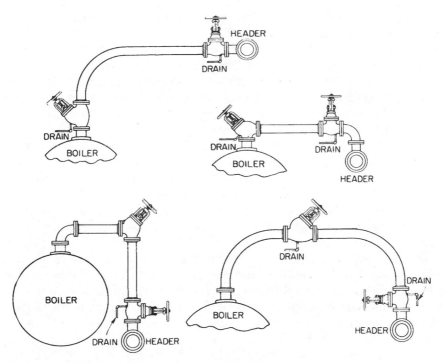

Figure 6.28 Typical installations of nonreturn valves. (*Crane Company, Valve Division.*)

6.9 Steam Piping

Pumps, turbines, auxiliaries, and process equipment that use steam are usually located at some distance from the boilers, and so it is necessary to transport the steam through piping. The main steam line picks up the steam lines from each boiler, and takeoffs are provided where necessary. This system of piping makes it possible to use one boiler or combination of boilers to supply steam for any equipment requiring steam.

Steam piping must be of sufficient thickness to withstand the internal pressure and shock due to the velocity of steam passing through it. Piping must be supported adequately to take care of strain due to expansion and contraction, and expansion joints or pipe loops (Chap. 11) must be located and installed properly. The piping is insulated and provided with drains and traps to remove the water of condensation, which prevents any water from entering the pumps, turbine, or other equipment.

Steam piping is designed for steam velocities of over 10,000 ft/min when supplying steam to turbines. The volume of steam necessary must be conveyed without excessive pressure drop, and a good design limits the pressure drop to a maximum of about 5 percent. Pressure drop is expressed as loss per 100 ft of lineal pipe; usually 2 to 3 psi is considered a permissible drop based on economic considerations.

Prior to operation, the steam lines must be blown to clean the line of dirt and any debris that remains after construction. This cleaning process generally requires the use of high-velocity steam (usually from a portable steam source as part of the initial steam cleaning equipment) to clean the superheater and main steam lines of any loose scale or foreign material before connecting the steam line to the turbine. Temporary piping to the atmosphere is required for all procedures. This temporary piping must be anchored properly in order to resist the nozzle reaction that is created during the period of high-velocity steam blowing. Subsequent cleaning of steam lines usually uses steam as produced by the boiler, and a portable steam source is generally not required.

For steam piping operating over 800°F, damage can occur due to creep, cycle fatigue, erosion, and corrosion, and the condition of the piping must be evaluated periodically during scheduled maintenance periods for the boiler. The most typical steam pipe failure is the cracking of attachment welds. These cracks are caused by thermal fatigue, improper support, or improper welding.

For steam piping operating at temperatures less than 800°F, damage by creep is generally not a problem. Failures of this piping typically are due to fatigue, erosion, or corrosion, and these low-temperature steam lines have much longer lives than high-temperature lines as long as proper inspection and maintenance are conducted.

6.10   Sootblowers

Boiler tubes and heating surfaces get dirty because of an accumulation of soot, slag deposits, and fly ash. These substances are excellent insulators and reduce the effectiveness of the heating surface. Therefore, they must be removed to ensure the continuation of optimal boiler performance. Removal can be accomplished by using a hand lance or a sootblower. Steam and compressed air are usually used for blowing, although water and shot are sometimes used to remove certain types of deposits that become baked hard and are difficult to remove with the conventional sootblower.

Sootblowers are mechanical devices that are used for on-line cleaning of gas-side boiler ash and slag deposits. They direct a cleaning medium through nozzles and against the ash that has accumulated on the heat-transfer surfaces of boilers in order to remove the ash deposits and maintain the effectiveness of the heat-transfer surfaces.

The type of sootblower required varies with the location in the boiler, the cleaning area required, and the severity of the accumulated ash. Sootblowers basically consist of

1. A tube element or lance that is inserted into the boiler and carries the cleaning medium

2. Nozzles in the tip of the lance to direct the cleaning medium and increase its velocity

3. A mechanical system for insertion and rotation of the lance

4. A control system

The cleaning medium can be saturated steam, superheated steam, compressed air, or water. In most cases, superheated steam is preferred because erosion of tube surfaces can occur with the use of saturated steam as a result of its moisture. Larger boilers often use compressed air furnished at high pressure from compressors.

Sootblower steam can be taken from intermediate superheater headers, reheat inlet or outlet headers, or secondary superheater outlet headers. The choice of air or steam as a cleaning medium is usually based on an economic analysis of operating costs and technical issues. Water is often used when either steam or air is ineffective, as in boilers that burn low-sulfur coals, where the ash deposits are often plastic in nature and strongly adhere to the tubes.

The types of sootblowers are as follows and are either fixed or retractable:

1. *Fixed-position blower.*   This is a nonretractable sootblower, either rotating or nonrotating, that is used to remove dusty ash from tube banks. It can only be used where lower flue gas temperatures are present.

2. *Short retractable furnace-wall blower.*   This is a short-travel retractable-type unit that is used primarily for cleaning furnace waterwall tubes.

3. *Long retractable blower.*   This is a long retractable blower that has a travel range of approximately 2 ft to over 50 ft.

Sootblowers are made of pipe and special alloys when required to withstand high temperatures. The element itself merely serves as a conduit or mechanical support for the nozzles, through which steam or air is transmitted at high velocity. Sootblowers are designed for many different applications, using various nozzle contours to meet specific needs under varying temperature conditions. The size, design, and location of sootblower nozzles are varied to meet the cleaning needs encountered in the generating tube bank and other heat-exchange equipment, including superheaters, air preheaters, and economizers. On bent-tube boilers, nozzles are set at a right angle to the element; on boilers with a staggered tube arrangement, the nozzles are set at an angle to clear the lanes between the tubes.

Figure 6.29 shows a sootblower installation for an HRT boiler that would burn a solid fuel. It consists of a revolving blow arm equipped with nozzles that are spaced to blow directly into each fire tube as the arm rotates. The arm can be operated from the front of the boiler by means of a chain and operating wheel while the boiler is in service. The spindle carrying the blow arm rotates and is supported by an adjustable bracket fastened to the inner door. Access to the boiler is gained by closing the steam valve and breaking the pipe union. Although this HRT boiler is an old design, it does illustrate how sootblowers can be designed to clean a boiler with a variable nozzle arrangement.

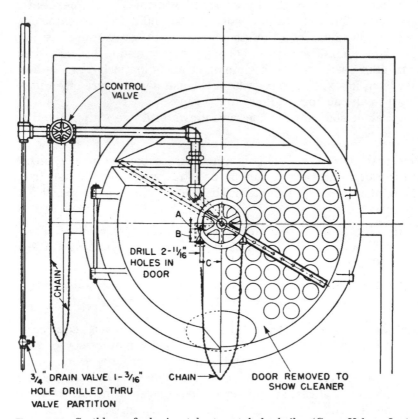

Figure 6.29 Sootblower for horizontal-return tubular boiler. (*Copes-Vulcan, Inc.*)

Automatic valve-head sootblowers control the admission of the blowing medium while rotating the element. The rotary element is attached to the sidewall or boiler casing and is supported on the tubes by bearings designed for that purpose. This type of sootblower permits blowing only while at the correct angle.

Hangers are of the intimate-contact type (Fig. 6.30), in which the bearing is provided with a smooth surface of large area for contact with a (comparatively) cool boiler tube, thus preventing excessive

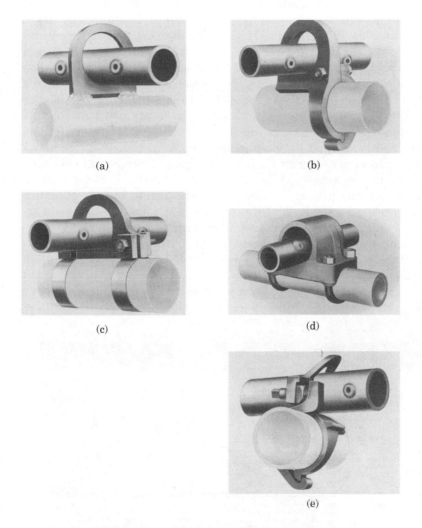

(a)

(b)

(c)

(d)

(e)

Figure 6.30 Sootblower bearings. (*a*) Welded bearing. (*b*) Intimate-contact bearing. (*c*) Crown bearing with straps. (*d*) Protective bearing. (*e*) Compression bearing. (*Copes-Vulcan, Inc.*)

temperature on blower and bearing. Details of bearings and element installations are shown for both the welded and the intimate-contact bearing. Bearings are machined to fit the tube snugly so as to give a heat-conducting bond or contact. With a protective bearing on each boiler tube and a jet between, only a small portion of the element is left exposed to the high temperature of the gases. A clamp-type bearing of the compression type is used where a welded-bearing installation is inconvenient.

For the conventional sootblower, only a small section of the element may not be in intimate contact with the tube or bearing (Fig. 6.30), yet such an element frequently overheats and becomes warped and thus inoperative. With higher flue gas temperatures, hangers and nozzles would be damaged by exposure to such temperatures. And so because it was costly to maintain the conventional multijet blower, a fixed-type rotary element was converted to a retractable sootblower (Fig. 6.31). The retractable blower element is located outside the furnace and is designed for traveling in excess of 50 ft. Paths requiring cleaning are traversed by blowing the jet as the element is being extended into or withdrawn from the furnace. Elements of this type are located in areas in which it was previously difficult to maintain a sootblower in continuous service. The retractable blower element, since it is located outside the furnace, requires no tube clamps, bearings, etc., which formerly necessitated continual maintenance.

Air, saturated or superheated steam, or water is used as the blowing medium without change in equipment. Outside adjustment of nozzle pressure is made possible by the mechanically operated head of the long retractable blower. Many varieties of nozzle arrangement may be used according to tube spacing. Nozzle inserts are of stainless

Figure 6.31 Long retractable sootblower. (*Copes-Vulcan, Inc.*)

Figure 6.32 Retractable-gun-type sootblower or wall deslagger. (*Copes-Vulcan, Inc.*)

steel, their shape and diameter depending on design and application. Either the rotating or the traversing speed of the element may be independently adjusted without affecting the other, merely by changing sprockets.

The gun-type deslagger (Fig. 6.32) is also of the retractable type. It is used for furnace walls and other inaccessible spots previously cleaned by hand lancing. In operation, one motor extends and retracts the swivel tube and another rotates the element. To save time, entrance and withdrawal are made quickly. The length of time can be altered to suit requirements. Single or dual nozzles can be used depending on the wall arrangement or slagging conditions. Wall deslagging is one sure way to minimize average slag thickness for maintaining constant high heat-transfer capability and uniform superheat and reheat temperature control.

For the operator, being aware of slagging and fouling conditions is very important in achieving reliability and availability on a coal-fired boiler. However, the cleanliness of heating surfaces within a boiler is one of the most difficult operating variables to determine. Indications of surface fouling are shown to the operator by indirect means such as increases in steam temperature, spray attemperator flows, and draft losses.

One indication of surface cleanliness is draft loss. By observing the draft loss across a tube bank, an operator can determine that sootblowing is required when the draft loss increases. However, it is often possible that by the time an increase in draft loss is noticed, deposits on the tubes have led to bridging, and removal by sootblowers may be too late.

On modern power plants, computer based boiler performance-monitoring systems are used to evaluate the cleanliness of the furnace and of the convection heating surfaces. Measurements of temperature, draft loss, flue gas flows, and flue gas analyses are used to perform heat-transfer analysis in the furnace and convection sections.

Potential slagging and fouling problems are recognized early, and selective sootblowing can be directed at the specific problem area. Sootblower sequencing can be optimized based on actual cleaning requirements rather than operating at certain time intervals that would waste the blowing medium of air or steam and possibly cause erosion by blowing clean tubes.

Automatic sootblowing systems are now installed in most power plants where sootblowing is done by remote control. Once the master button has been pushed, the entire system is in correct sequential operation. The operator can easily tell exactly which blower is being operated, and steam or air pressure is recorded or indicated. No longer is the hard-to-get-at sootblower or the one in the hot location neglected as in the past. Although such systems are expensive to install, they are justified because without automatic equipment this important operation might be neglected, with resulting loss in efficiency, boiler outage, and reduction in capacity.

The frequency of sootblowing is strongly dependent on the fuel characteristics and the operation of the installation. For example, the frequency of sootblowing would vary significantly from an installation burning coal with a high ash content to one using coal with a low ash content.

A modern boiler includes a comprehensive sootblower control system that ensures optimal cleaning and results in optimal boiler performance. Using microprocessor technology, the system is operated from a simplified set of controls with a single display screen. The frequency and sequencing of sootblower operation are controlled from a keyboard on the operator interface unit. Operating information is displayed on the monitor, and signals are transmitted to the individual sootblower units. Based on preprogramming, sootblower operation can be done based on actual cleaning requirements, as noted above, or at periodic intervals based on operating experience. The monitor (or CRT) has full color graphics that indicate the status of the boiler cleanliness and display information that is useful to the operator. The computer stores information for long-term evaluation. A backup system is readily available if the computer fails, since the operator can use the base sootblower panel with programmable sequences.

6.11   Valves

Valves are used to control the flow of water to the boiler and steam from the boiler to the main steam line and eventually to the equipment using the steam. They are also used in all auxiliary piping systems. They are attached to the piping in several ways: The body of the valve may be equipped with pipe threads so that the valve can be screwed into position, or the valve may be equipped with flanges and bolted in position, or, on high-pressure boilers, valves are welded into position.

The *globe* valve (Fig. 6.33) consists of a plug or disk that is forced into a tapered hole called a *seat*. The angle used on the taper of the seat and disk varies with the valve size and the kind of service to which the valve is applied. Globe valves are used when the flow is to be restricted or throttled. Whenever a globe valve is used on feed piping, the inlet shall be under the valve disk. The seats and disks on the globe valve are not cut by the throttling action as readily as with gate valves. Valve parts are easy to repair and replace. The disadvantages of the globe valve are (1) increased resistance to flow, i.e., high-pressure drop, (2) the fact that more force is required to close the valve because of the increased pressure under the disk, and (3) the possibility that foreign matter may cause plugging of the valve.

The *gate* valve (Fig. 6.34), as the name implies, consists of a gate that can be raised or lowered into a passageway. The gate is at right angles to the flow and moves up and down in slots that hold it in the correct vertical position. It is usually wedge shaped so that it will tighten against the sides of the slots when completely shut off. A gate

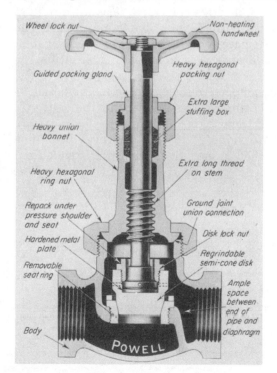

Figure 6.33   Globe valve. (*The Wm. Powell Company.*)

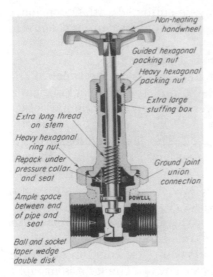

Figure 6.34   Gate valve. (*The Wm. Powell Company.*)

valve is used chiefly where the valve is to be operated either wide open or closed. When wide open, it offers very little resistance to flow, and consequently, pressure drop through the valve is minimized. The pressure acts on one side of the gate so that the gate is forced against the guides and requires considerable force to operate, at least for the larger valves. The gate valve is a difficult valve to repair once the seats have been damaged. If a gate valve is kept in an intermediate or partially open position, the bottom of the wedge and the seat will become badly eroded in a short time.

For both the globe and gate valves, the body contains the valve seat. The valve bonnet is attached to the valve body by a threaded nut or bolted flange (Fig. 6.35), depending on the size of the valve. The flanged-fitting steel valve shown is used on a steam line and is

Figure 6.35   Flanged bonnet attachment and bypass on steel valve. (*Crane Company, Valve Division.*)

equipped with a bypass valve. The bypass valve is opened prior to opening the main valve, permitting the line to heat up and equalize the pressure on both sides of the valve, thus making the valve easier to open.

The valve stem extends through the bonnet and is threaded and fitted with a handwheel. The bonnet is provided with a packing gland that prevents leakage around the valve stem. The movable part of the valve element is carried on the end of the valve stem. Turning the valve wheel moves the stem in or out, opening or closing the valve.

The Breech Lock cast-steel-wedge gate valve (Fig. 6.36) with seal-welded bonnet was developed primarily to meet the need for a body-bonnet joint that would be pressure-tight at high temperature and would not require constant maintenance. The Breech Lock design transmits bonnet thrust to the body through interlocking breech lugs with final pressure-tight closure of the body-bonnet joint made by a small seal weld. This results in pressure tightness and maintenance-free service of a full-strength bonnet, providing ready disassembly and reassembly. For boiler feed conditions the valve is available in

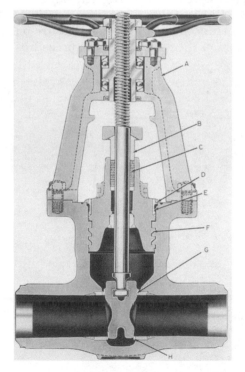

Figure 6.36 Breech Lock cast-steel-wedge gate valve. A = yoke. B = packing gland. C = stuffing box. D = seal weld. E = flexible steel ring. F = breech lugs. G = disk ring. H = seal rings. (*The Lunkenheimer Company.*)

pressures to 5800 psi at 525°F and for main steam conditions up to 2150 psi and 1050°F. This design is also available for globe and check valves. The unit that is shown in Fig. 6.36 is hand-operated but the unit is also available with gearing or motor unit for remote operation.

Some of the valve features are as follows: (1) The yoke is mounted directly on the body and independent of the bonnet, thus relieving the bonnet seal weld of the weight of yoke and operating mechanism and also relieving it of the effect of vibration and stem thrust. (2) The packing gland is secured to the bonnet by two swing bolts; packing is readily accessible. (3) The stuffing box is outside the high-temperature zone. (4) The seal weld provides pressure tightness and the weld can be disassembled and reassembled readily. (5) The flexible steel ring, permanently welded to the bonnet, prevents stress due to the seal welding. (6) Breech lugs lock the bonnet in the body, which carries the full internal pressure. (7) Both disk and seal rings have Stellite-faced seating surfaces. (8) The flexible disk prevents "sticking" or "freezing," resulting in less torque being required to open the valve.

Automatic-control valves, such as a valve operated by a feedwater regulator, must be designed so that little force will be required to operate them. This is accomplished by the use of a balanced valve (Fig. 6.37), which is similar to a two-seated globe valve. The balanced valve has two seats and two disks. One of these disks opens against, and one with, the pressure; i.e., a balanced valve has the pressure on the top of one disk and on the bottom of the other. Thus the pressures

Figure 6.37 Balanced valve used with a feedwater regulator.

are balanced, and it is possible to open or close the valve with a minimum of effort. Two-seated valves, such as the balanced valve, are not tight shutoff valves. They should be checked and inspected frequently. We can determine whether the valve leaks by using a listening rod.

The check valve (Fig. 6.38) is a modified globe valve without a stem. It is usually arranged so that it closes by gravity. The flow is directed under the valve to raise it from its seat; if the flow reverses, gravity plus the pressure above the valve closes it. The check valve is used where a flow in only one direction is desired. One of the main uses of this valve is in the feedwater line to the boiler.

Many forms of reducing valves are used to operate auxiliary equipment not requiring boiler pressures, such as for low-pressure heating systems or heat exchangers. The reducing valve consists of a balanced valve actuated by the pressure on the low-pressure side, the low pressure acting on a diaphragm to open or close the balanced valve.

The stem-pressure reducing and regulating valve (Fig. 6.39) is a single-seated, spring-loaded direct-acting diaphragm valve. This valve automatically reduces a high initial pressure to a lower delivery pressure, maintaining that lower pressure within reasonably close limits regardless of fluctuations in the high-pressure side of the line.

The inner valve assembly is easy to clean or replace by loosening the hex-head bottom plug. Major repairs are made without removing the valve from the line by having a bypass line around the valve. Pressure adjustments are readily made by simply turning the top adjusting screw. The valve is protected by a self-supporting built-in Monel strainer, or a strainer can be installed in the line ahead of the regulator. Preferred installation for this reducing-regulating valve is to provide a shutoff valve on each side of the regulator and then

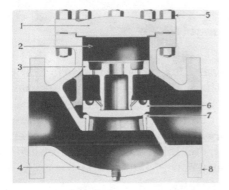

Figure 6.38   Check valve. 1 = valve cover. 2 = body bore. 3 = piston. 4 = valve body. 5 = studs. 6 = valve disk. 7 = valve seat. 8 = flange. (*Rockwell International, Flow Control Division.*)

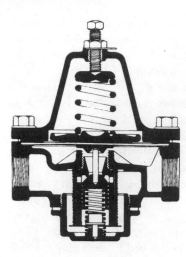

Figure 6.39 Steam-pressure-type regulator. (*Cash Valve, Inc.*)

a bypass line around it with pressure gauges at the entrance and behind the regulator so that one can observe the operation of the regulator at all times.

Stopcocks, or plug valves, are used frequently as blowoff valves, in gas and oil lines, for water softeners, etc. The valve consists of a circular, tapered plug that is ground fit in a hole in the valve body. There is a hole through this plug at right angles to its axis. When the plug is in one position, this hole lines up with the hole in the body of the valve, and the valve is then open. When the plug is turned, the holes are thrown out of line, and the valve is shut. The stopcock can be opened or closed very quickly. The chief advantage of this valve is that it is not easily affected by dirt in the substance handled. By various combinations of holes in the valve body and plug, several lines can be controlled with a single valve.

6.12   Instruments and Automatic Control Systems

In addition to the equipment necessary for the safe operation of a boiler and all power plant systems, there are many other accessories that add to the safety, reliability, efficiency, and convenience of operation. Instruments are available that indicate or give a complete graphic record of draft, pressure, temperature, flue gas analysis, stoker and fan speeds, steam and water flow, etc. In addition, automatic control systems are now common in most power plants, and these systems provide a safe and efficient operation.

A draft gauge is necessary on all boilers for indicating the furnace draft. Indication of the pressure under the stoker fuel bed (forced-draft-fan pressure) as well as draft at various points in the boiler is also desirable. This permits the operator to obtain proper furnace conditions, regulate the air supply to the fuel bed, and vary the capacity. When multiple-retort stokers are used, pressures under each section of the grate should be indicated, since proper air distribution can be determined accurately only in this manner.

Every boiler must have a pressure gauge to indicate the boiler steam pressure. Pressure gauges are also used to indicate or record the main steam line pressure or the pressure of water in the feed line. They also can be used in the feed or steam line to indicate flow, if reference is made to the pressure drop in the line. Pressure gauges are used for many other purposes and are perhaps the most common instrument available.

Temperature indicators and recorders are used for steam, water, flue gas, air, and fuel temperatures and for many other purposes. Also, there are carbon dioxide recorders, flowmeters for steam and water, indicators and recorders for boiler water level, and, required more recently, instrumentation that records stack opacity (clarity), sulfur dioxide (SO_2) emissions, and nitrogen oxide (NO_x) levels. The advantage of using a recorder is that we can make reference to it for changes that occur during operation and during emergencies. The record is then available for analysis and correction of the problem.

In many modern installations, digital displayed microprocessor control systems are being utilized for instant display of all controlled devices. These centrally located systems provide the operator with the ability to control, record, and locate potential problem areas. These systems also have the capability of being programmed to provide corrective procedures in cases of equipment breakdown or during emergencies.

Records obtained from instruments make it possible for the operator or management to determine whether the best operating practice is being maintained. Instruments and controls are usually expensive to install and maintain, and it therefore becomes a decision for each plant to determine just what is required or necessary to perform the task for which each is intended.

A distributed control system (DCS) automates the control of a process or of a plant. The system reads field device inputs such as thermocouples and pressure transmitters, calculates the optimal outputs, allows operator interface, and drives output devices such as valve positioners and pumps. An automated process is therefore less dependent on the experience and knowledge of operators but is dependent on previous operational experiences that are programmed into the system.

Processes were first controlled by operators using indicators, gauges, and manual actuators, and many plants still operate this way today, although much fewer in number. Several operators were needed to control each process, and the plant's operation could vary significantly based on the operator's experience and skill. Conventional analog control using pneumatics was implemented in the 1950s, and control strategies were limited and fixed. Today, microprocessor-based distributed control systems are used widely for process control, and many panelboards in the operator control room have been reduced to a cathode-ray tube (CRT) operator station.

Distributed control systems distribute the hardware physically throughout the plant. Control is done by microprocessors located in the environmentally protected enclosures near the field instruments. This reduces the amount of wiring to a central location. The DCS consists of two main pieces of hardware: the network processing unit (NPU) and the operator interface station (OIS). The NPU is located on the plant floor, and it controls the process and acquires process data. Wires from input-output (I/O) devices such as valves and pumps are brought into the NPU cabinet. The OIS is a CRT-based console that allows the operator to monitor all the values in the process. By means of interactive graphics, the operator can take manual control of the outputs. Process data are available at the station and give the operator a visual picture of the process values. All data can be stored for reports and future reference.

DCS systems are used for more than process control, since they are also used for management of the process. They are used to optimize processes and reduce costs.

Instruments alone do not improve economy; rather, it is the close control of operation and maintenance made possible by a knowledge of the conditions shown by the instruments that results in increased economy. Installation of proper instruments is only the first step toward the efficient operation of a power plant. Every instrument must be kept operating and properly calibrated if its cost is to be justified. Records should have a definite use in the control operation. Recorder charts filed away without being studied never result in increased economy but only money wasted. The correct instruments will pay for themselves in improved operation when they are operated properly and when the information obtained from them is put to proper use.

Instruments must be accurate, rugged, sensitive, and extremely dependable and must function as precision mechanisms.

Automatic control, developed from these simple instruments, goes a step further. Here, the instrument actually controls or operates the equipment. By means of automatic control, constant steam pressures and uniform furnace draft can be maintained and air or fuel can be changed to meet steam demand. At the same time the load can be varied from minimum to maximum capacity to obtain the most efficient combustion results. It is obvious that the control is no better than the instruments that do the metering and measuring. But by the installation of reliable instruments and with the proper operation of the controls, maintenance can be reduced, the operator can be relieved of repetitive duties, and fuel savings will result.

The problem with boiler control is one of coordinating the following factors: (1) steam pressure and temperature, (2) fuel quantity, (3) air for combustion, (4) removal of the products of combustion, and (5) feedwater supply. Feedwater control is indeed important but is not included in this discussion of automatic combustion control.

There are three types of control systems: (1) off-on, (2) positioning, and (3) metering. All three types of systems are designed to respond to steam-pressure demands, to control fuel and air for combustion so as to obtain the highest combustion efficiency.

Off-on control is applied to small boilers. A change in pressure actuates a pressurestat or mercury switch to start the stoker, the oil or gas burner, and the forced-draft fan. The control functions to feed fuel and air in a predetermined ratio to obtain good combustion. These results can be varied by manually changing the fuel or air setting. Off-on cycles do not produce the best combustion efficiency.

In a plant containing a number of boilers, each equipped with an off-on control operating from a pressurestat, the problem of one boiler appropriating most of the load is often encountered. This occurs because control pressures are difficult to adjust within close limits. Therefore, it is better to install a single master pressurestat to bring the units off and on at the same time. Or an off-on sequence control can be used that brings the units on and takes them off the line as required. A selector switch can be installed to vary the sequence in which boilers are placed in operation. Such systems also can be applied to the modulating control.

Positioning control is applied to all types of boilers. It consists of a master pressure controller that responds to changes in steam pressure and, by means of power units, actuates the forced-draft damper to control airflow and the lever on the stoker to adjust the fuel-feed rate. Such units usually have constant-speed forced-draft fans equipped with dampers or inlet vanes that are positioned to control the air for combustion. Furnace-draft controllers are used to maintain the furnace draft within desired limits. This control operates independently of the positioning-control system.

For the normal positioning-type control, the only time the airflow and fuel feed are in agreement is at a fixed point, usually where control calibration was made. This is so because the airflow is not proportional to damper movement. Variables that affect this relationship are the type of damper, variations in fuel-bed depth, variations in fuel quality, and lost motion in control linkage. The necessity of frequent manual adjustment to synchronize the previous control is readily apparent.

For the positioning-type control, the variables are in part corrected through the proper alignment of levers and connecting linkage between the power unit and the damper and fuel-feed levers that they operate, by installing cams and rods calibrated to alter the arc angularity of travel from the power-unit levers. These compensate for the movement characteristics of fuel-feed and air-damper control. In addition, this system can be provided with a convenient means for manual control, operation from a central point. This remote manual-control system can be used for changing the distribution of the load between boilers or for making adjustments in the fuel-feed rate to compensate for changes in fuel quality. The positioning-type control has an advantage over the off-on control in that the fuel and air can be provided in small increments to maintain continuous operation, therefore eliminating off-on cycling.

Metering control is used when the fuel rate and heat input (in Btu/h) vary widely because of variations in fuel supply and heat content and when combination fuels are burned. Here the fuel and air are metered, maintaining the correct air-fuel ratio for best combustion results, based on design and testing. The steam (or water) flow can be a measure of fuel feed. That is accomplished by measuring the pressure drop across an orifice, flow nozzle, or venturi. Air for combustion also can be metered by passing the air through an orifice, but most frequently airflow is measured by the draft loss across the boiler or air preheater (gas side) or across the air preheater (air side). The air side is frequently chosen as the point of measurement in order to prevent dust and dirt from clogging the lines and fouling the control system.

The metering-type control is more accurate than the positioning system, since compensation for variables is obtained through metering without regard for levers, linkage, lost motion, damper position, fuel variables, etc. Also, with the metering control, the fuel-air ratio can be readjusted from the air-steam flow relationship. The metering control usually incorporates a remote manual station wherein the control system can be modified and where hand or automatic operation is possible. There are many varieties of combustion-control systems. They may operate pneumatically, hydraulically, electrically, electronically, and sometimes in combination.

For the pneumatic system, all instruments in the control loop are air-activated measurements, to and from central points. Steam pressure is controlled by parallel control of air and fuel, and high-low signal selectors function to maintain an air-fuel mixture.

Airflow to the furnace is controlled by automatic positioning of the forced-draft flow inlet vanes. Furnace draft is maintained at the desired value by control or positioning of the induced-draft-fan damper. Feedwater flow to the boiler drum is controlled separately by feedwater-control valves.

Adjustment is provided at the panel board for steam-pressure set point, fuel-air ratio, furnace draft, and drum-level set points. The controls would include pneumatic switches for bumpless transfer from automatic to manual control, and manual control can be accomplished from the control panel.

Figure 6.40 is a schematic diagram for an all-electronic instrumentation system that is designed for a pulverized-coal-fired plant. Electronic transmission permits locating measurement transmitters and final control elements at long distances from the control panel.

As shown in the schematic, steam pressure is maintained at the correct value by controlled positioning of air dampers controlling airflow to the pulverizers. The fuel controller automatically corrects for variation in the number of pulverizers in service.

Airflow is controlled by positioning inlet vanes of the forced-draft fan. Steam flow, as an inferential measurement of fuel input, controls airflow. Fuel-air ratio is adjusted by the operator at a ratio station on the control panel.

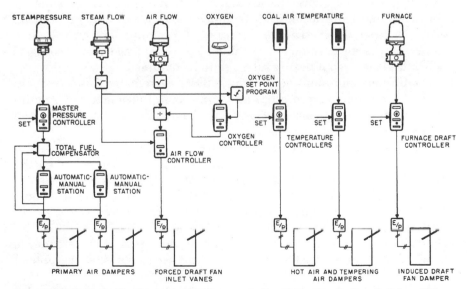

Figure 6.40 Schematic diagram of instrumentation for pulverized-coal firing. (*The Foxboro Company.*)

Coal-air mixture is controlled at a temperature set by the operator. Temperature controller output automatically positions hot-air and tempering-air damper operators to produce the desired temperature. A furnace draft controller positions the induced-draft-fan damper to maintain draft as set by the operator.

Long-distance all-electronic instruments enable fully centralized control, a significant improvement over decentralized arrangements with instruments scattered throughout the plant. The automatic combustion-control system functions to provide efficient operation, with safety.

Many combustion-control systems have been developed over the years, and each must fit the needs of the particular application. In each design, load demands, operating philosophy, plant layout, and type of firing have to be considered. However, in every case, a combustion-control system regulates the fuel and air input (the firing rate) in response to a load-demand signal. The demand for this firing rate is therefore a demand for input of energy into the system when, at some time, energy is being withdrawn from the system.

Automatic combustion control is justified by the benefits it provides: added safety, improved operation, reduction in manpower requirements, lower fuel costs, etc. Combustion-control systems are available to meet the needs of both the small and the large power plant. Selection should be made on the basis of justifying installation and maintenance cost by lower overall fuel costs. Each system should be as simple as possible to accomplish the purpose for which it was installed. Controls should be so located as to make them accessible for servicing and calibration and should be kept clean and in working order. Neither the instrumentation nor the combustion control alone improves the performance; they must be maintained and records of operation analyzed if best results are to be obtained.

Continuous-emission-monitoring (CEM) systems are used to help control the combustion process and emissions. In some areas, regulations require them to ensure that facilities are meeting their emission limits and preserving air quality.

Plant operating permits often require the continuous monitoring of SO_2, NO_x, CO, CO_2, and O_2 and opacity. In addition, depending on the type of fuel burned, some areas require the continuous monitoring of hydrogen chloride (HCl), ammonia (NH_3), and total hydrocarbons. Each constituent requires a specific instrument, reference test methods, and performance standards. In some cases, both inlet and outlet conditions must be measured in order to determine the percentage removal. A scrubber for the removal of acid gases is a good example of this.

A reliable CEM system is an important tool for the operation of a boiler and the air pollution control equipment. The CO and O_2 monitors provide the plant operator with essential information for boiler control, since these instruments are used to monitor the efficiency of the combustion process. They indicate to the trained operator when the required combustion air quantity is achieved for the incoming fuel.

The use of instrumentation at the economizer or air heater outlet and at the stack will supply information on acid gas removal (SO_2, HCl, etc.) that allows the effective control of these emissions. For example, the lime slurry injection rate for a dry scrubber can be adjusted as the inlet SO_2 or HCl quantities vary to the level required by the permit. This is a feature that is more adaptable to a relatively nonhomogeneous fuel such as municipal solid waste (MSW), where inlet conditions can vary significantly because of MSW composition. On a coal-fired unit, this may not be as important. A probe similar to that shown in Fig. 6.41 continually extracts flue gas from the stack or other required location and sends it to a CEM analyzer.

A CEM system is comprised primarily of three subsystems:

1. The sampling interface
2. The analytical instrumentation
3. The data-acquisition system (DAS)

The DAS has several data calculation and management reporting and recording capabilities. It calculates the emission averages and outputs them to a variety of units, such as recorders. The sampling system brings the flue gas into position for analysis. A typical data acquisition and reporting computer system is shown with a CEM system in Fig. 6.42. This system consists of a data logger and central computer system. The data logger handles all the input-output (I/O) points between the analyzer system and central computer system. Periodic

Figure 6.41 Probe for continually extracting flue gas from stack or other outlet and sending them to CEM system by a heated umbilical. (*Wheelabrator Clean Air Systems, Inc.*)

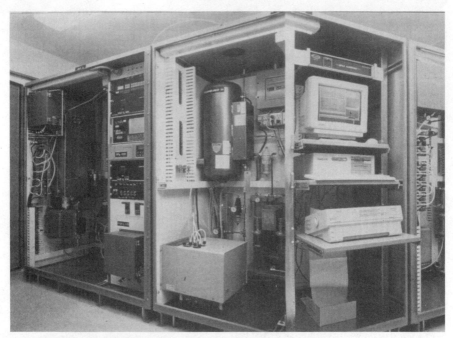

Figure 6.42 Continuous-emission-monitoring (CEM) system. (*Wheelabrator Clean Air Systems, Inc.*)

emission summary reports can be made on a monthly, quarterly, or annual basis (or other as required) together with calibration reports on a daily basis, if this is necessary.

Early power plant emission measurements focused on controlling the combustion process. The Orsat gas analyzer sampled the flue gas and determined the concentration of O_2, CO, and CO_2, and this device was used extensively. However, the Orsat analyzer is not a continuous monitor, since it requires manual operation.

As a result of increased regulation, many power plants must use continuous emission measurements, and this CEM equipment also provides input data to the environmental-control systems such as SO_2 scrubbers.

A typical CEM system is shown in Fig. 6.42, and such systems can be designed to measure particulate, opacity, SO_2, NO_x, CO_2, HCl, CO, and VOCs (volatile organic compounds). This particular system can measure 50 different gases, if desired, and up to 8 at once. The system provides accurate and reliable results, and it can effectively sample and measure highly reactive flue gases such as HCl and ammonia at elevated temperatures without any cooling or drying of the sample.

It is very important to evaluate the location and ambient conditions of the CEM equipment. The location can create reliability and maintenance problems due to dirty conditions, excessive heat or cold, vibration, and humidity. This equipment is most often found in closed, air-conditioned cabinets.

The instrumentation and control industry is changing rapidly from conventional analog controls to newer technologies such as digital, microprocessor-based, and distributed systems.

An overwhelming majority of electric capacity in the United States still comes from large, centralized steam power plants, although this reality is sometimes lost with the news of recently installed gas turbines and combined cycles. Most of these steam plants, both fossil and nuclear, will be 30 years old or older in the year 2000. New requirements and the electric power market are pushing these plants into extended service with very different operating requirements than those originally contemplated, including strict environmental requirements, more load variations, shorter maintenance outages, and decreasing budgets.

In response, power plants are being equipped with the newest technology in digital-based instrumentation and control (I&C). These systems offer

1. Faster plant startup and shutdown by programming control sequences
2. Higher availability by detecting and defining causes of probable malfunctions.
3. Higher thermal efficiency by moving variable set points closer to operating limits
4. Improved emissions by precisely controlling the combustion and downstream cleanup processes
5. Lower maintenance costs by eliminating outdated pneumatic, electromechanical, or electronic/analog devices
6. Lower operating costs by reducing staff requirements

Digital control systems have become more attractive because computer costs have decreased dramatically and computer systems have become more standard. It is now practical to monitor large amounts of data and analyze the data on relatively inexpensive personal computers or workstations.

Today's power plant operators must simultaneously maintain prescribed emissions, cycle the unit to maximize profitability, monitor production costs, including O&M expenses, and burn a variety of fuels. This is contrasted to the previous operator demands when a plant burned one type of fuel, met an emission limit that was tested annually, operated base loaded, and accumulated all the production costs just prior to the next review for a rate change.

One of the disadvantages to early digital systems was that the CRT screen, as the operator's interface, could show only part of the available plant information at one time. As compared with the control room's wall-to-wall instrument panels, CRTs offered only a restricted, narrow review of the process.

A design that has gained acceptance is the use of a large-screen display adjacent to a series of smaller CRTs. By displaying overview information, large screens provide a more complete view of the entire system with detailed information shown on CRTs. With these large-screen displays, all control room operators can view the situation simultaneously, which improves coordination and the making of decisions. This is extremely helpful during both normal and emergency operations.

The shifts from local control to central control, and from pneumatic to electronic technologies, took place over extended periods of time, both lasting nearly a decade. Currently, the acceptance of microprocessor-based technology is rapidly replacing the previous conventional centralized analog control with distributed digital control systems. Figures 6.43 and 6.44 show a modern control room using monitors and operator interface stations together with a computer that functions as an engineering workstation. The use of overhead displays is common in a modern control room.

The speed at which distributed control is gaining acceptance can be attributed to several factors:

Figure 6.43 Bailey INFI 90 open system with overhead display. (*Bailey Controls Company.*)

Figure 6.44   Bailey INFI 90 open system. (*Bailey Controls Company.*)

1. The need for more sophistication and precision in control systems to get maximum yield from a process
2. The need for clearer, simpler aids for the operator
3. The acceptance of microprocessor technology by industry

Microprocessor systems provide both modulating and sequential control, data acquisition, operator interface, and computer interface for industrial and utility applications.

Figure 6.45 shows another design arrangement of an operator station that includes display screens that can be used easily by operating personnel in the evaluation of system performance and during periods of equipment malfunctions. The diagnostic station displays all I&C fault alarms and includes information on the faulty component. The operator and engineering staff are rapidly guided to the display indicating the fault-generating component.

Microprocessor-based controls are rapidly replacing motor-driven switches. This method of monitoring system operation has increased safety and has provided better operating information. The latest development is the replacement of traditional limit switches with solid-state sensors. The sensor measures actual values rather than merely indicating a parameter is out of range.

Modern controllers can coordinate and share information on energy consumption and operating costs. Microprocessor-based controls can also link multiple units even from various sites, which permits data collection at one location. Boiler log reports can be generated automatically, and this allows operators to devote their time to more productive tasks. Changes in measured values can help identify problems prior to their becoming critical or causing a failure. The use of advanced controls have provided better ways of managing facilities at reduced costs and improved reliability.

Figure 6.45   Siemens Teleperm XP digital process-control system. (*Seimens Power Corp.*)

A microprocessor-based distributed control system is also useful in supporting the operator during power plant startups and shutdowns. Although equipment manufacturers provide operating procedures that describe the startup and shutdown methods for their equipment, these are generally based on optimal conditions. The operating personnel must blend these procedures with their actual operating experience to develop guidelines for safe and efficient operation for all the plant's systems.

Therefore, these developed plant operating procedures are incorporated into the power plant's automation system, and this results in repeatable startups and shutdowns in shorter periods without increasing process upsets. The system monitors process variables and equipment operation and displays the plant status. The status of sequence is organized and displayed with messages and symbols to inform the operator of the plant operation. When problems do occur, the plant automation system quickly identifies them to the operator and can shut a system down when necessary. The use of a control console interface provides a productive setting for the operator.

Questions and Problems

6.1 Why is the maintenance of a correct water level one of the most important criteria for safe and reliable boiler operation?

6.2 Describe a bicolor water gauge.

6.3 Describe the system that is widely used on modern boilers to transmit water-level signals to the control room.

6.4 A steam gauge is located 25 ft above the steam line. The gauge reads 245 psi. What is the pressure in the steam line?

6.5 A steam gauge is installed where it is 15 ft below the steam line. In checking this gauge, how would you set the pointer of the gauge so that it would indicate the correct steam line pressure?

6.6 What is a thermocouple, and how is it used for the measurement of temperature? Name a common use of thermocouples.

6.7 Explain the purpose and operation of a feedwater regulator.

6.8 Explain the difference between a two-and a three-element feedwater regulator. Which is the most commonly used system?

6.9 Why is the safety valve the most critical valve on a boiler? What makes it pop?

6.10 Explain the differences in use for a safety valve and a relief valve.

6.11 For a drum boiler with a superheater, how are the safety valves set? What sets the requirements for safety valves?

6.12 What are some of the essential requirements and rules governing safety valves and their installation and operation?

6.13 In what position should a safety valve be installed? Why?

6.14 For a superheater safety valve, approximately what percentage of the total boiler steam capacity must this valve relieve?

6.15 For a boiler with a superheater, what are the size requirements for the safety valves?

6.16 Which safety valve should be set to pop first, the one on the boiler or the one on the superheater? Why?

6.17 At what pressures must safety valves normally be tested?

6.18 When noise prevention is a concern at a plant, what do safety valves require to minimize noise when they are relieving steam?

6.19 To account for safety-valve discharge, what are the precautions that must be considered in the design installation?

6.20 A safety valve is $4\frac{1}{2}$ in in diameter and the boiler pressure is 250 psi. What is the total force on the valve?

6.21 What is the purpose of blowdown on a boiler? How is it accomplished?

6.22 How and why is heat recovered from a continuous blowdown system?

6.23 What is a nonreturn valve? Where is it located? What is its purpose?

6.24 In the design of steam piping, what is a critical evaluation factor in determining its proper size?

6.25 Prior to operation and after construction or repairs, what must be done to the steam piping?

6.26 What are sootblowers? What are their major components?

6.27 For sootblowers, what cleaning mediums are used, and which of these is preferred? Provide some advantages and disadvantages for the use of each.

6.28 Describe the various types of sootblowers, and provide examples of their locations and use.

6.29 When are sootblowers operated? For modern boiler designs, describe sootblower operation.

6.30 Explain the difference between a globe and a gate valve. Where would you use a globe valve in preference to a gate valve?

6.31 What is a check valve? Where is it normally used? How does it operate?

6.32 Name the most common instruments found in a steam power plant, and describe their purpose.

6.33 What is a distributed control system (DCS), and why is it important to operating a safe, reliable, and efficient plant?

6.34 In the control of a boiler, what are some of the major areas that must be coordinated?

6.35 Describe the three types of combustion-control systems.

6.36 Describe the benefits that result from an automatic combustion system.

6.37 What is a continuous-emission-monitoring (CEM) system, and what can it measure? Why is it an important tool to the effective operation of a power plant?

6.38 Compare the operation and advantages of a CEM system and an Orsat gas analyzer.

6.39 Digital-based instrumentation and control (I&C) systems are part of the modern power plant. What are the advantages gained with the use of these systems?

Operation and Maintenance of Boilers

The procedure to be followed in the operation and maintenance of a boiler plant depends to a large extent on its size, type of combustion equipment, operating pressures, steam requirements, and other factors pertinent to the specific plant. There are, however, standard practices that any operator should follow to ensure safe, continuous service and efficient operation. This chapter is intended to assist the operator in the correct use of the equipment, to recognize unsatisfactory conditions, and to take the necessary corrective measures before dangerous, costly emergencies develop.

Fluidized bed boilers have become popular in the burning of waste fuels and of those fuels which are difficult to burn by other combustion methods. As a result, the operating characteristics of fluidized bed boilers are discussed in this chapter. Although detail designs do vary between suppliers, the general features for operation apply to all designs.

All operators should be trained properly and should follow the operating procedures of the equipment manufacturers. These procedures are developed using the experience gained by the manufacturer from many operating plants, and they should be modified as required based on the operating experience and specific operating philosophy of the plant.

7.1 Boiler Startup

The necessary steps to be taken before placing a boiler in service depend on whether it is new, has been out of service a long time for repairs, or has been "down" for only a few days. New boilers and those on which extensive replacements of pressure parts have been made must be given a hydrostatic test before being placed in service. Such a test consists of filling the boiler completely with water (being careful to vent all the air) and developing a pressure $1\frac{1}{2}$ times the working pressure. Before the test is applied, a final inspection is made of all welding. Connections for flanged safety valves are usually blanked, and welded safety valves are closed with an internal plug. The water temperature must be at least as warm as the temperature of the air in the boiler room and in no case less than 70°F. A 150-psi pressure boiler would be hydrostatically tested at $150 \times 1.5 = 225$ psi. Treated demineralized water should be used for all hydrostatic testing. During hydrostatic testing, constant load supports for steam piping should be used in accordance with the manufacturer's instructions for the additional loads imposed by the water required for hydrostatic testing.

As part of the hydrostatic test procedure, maintenance and operations should review a checklist of major activities to ensure that the unit is operational. These activities should include the following for the particular boiler design involved:

1. Have all tubes been expanded in accordance with the tube-expanding procedure?
2. Have all welds been completed in accordance with the welding procedures? Have welds been stress relieved and nondestructive tested?
3. Following the test, have all gags, blanks, or hydrostatic plugs been removed from the safety valves and any other blanks that have been used to isolate the boiler.

New boilers, and even those on line, can accumulate a deposit of oil, grease, and paint and must be cleaned by boiling out with an alkaline detergent solution. This boilout can be accomplished with a combination of trisodium phosphate (Na_3PO_4) and disodium phosphate (Na_2HPO_4). The use of caustic NaOH and/or soda ash (Na_2CO_3) is not recommended. Chemical cleaning of a superheater and reheater is not necessary for initial operation. However, these systems should be given a steam blow, where high-velocity steam is used to blow any debris from the inside surfaces of the system.

After boilout and steam-blow flushing have been completed, corrosion products may still remain in the feedwater system and boiler in the form of iron oxide and mill scale. Chemical cleaning of the boiler system is generally delayed until the unit has operated at full load for a period of time and any loose scale or oxides have been carried from the feedwater system to the boiler. The boiler designer provides recommended procedures for these operations, and specialty companies can provide the cleaning and blowdown services, including the supply and installation of chemicals, tanks, and any temporary piping required.

Before putting a boiler in service for the first time, each of the safety valves must be checked for the correct setting by allowing the boiler outlet valve to remain closed and raising the boiler pressure until each valve opens. The pressure of the opening and closing of each valve must be recorded and the valve adjusted until specified results have been attained. Some plants make a practice of opening the safety valves by hand each time the boiler is put into service. In older plants, suitable cables attached to the valve handles and extending to the operating floor facilitate this operation. The safety valves should not be opened until sufficient steam pressure is available to prevent dirt from sticking under the seats.

When the boiler unit is turned over by the maintenance or construction department to the operating personnel, the procedure should be as follows: See that manhole and handhole covers have been replaced and that the pressure section of the boiler is ready for operation. Inspect the interior of the gas passages to see that all scaffolding, ladders, tools, etc., have been removed. Check the operation of the fans and dampers; if the fans and combustion equipment are supplied with a safety interlock system, check it at this time. After first making sure that no one is inside, close all doors and access openings to the boiler setting.

Most modern boilers are controlled automatically. However, before these boilers can be started up, the controls must be operated on manual until the automatic controls have been adjusted for specific conditions of the plant.

Prior to startup and operation, operators should be totally familiar with the equipment manufacturer's startup and operating instructions. For large and complex facilities, training programs are conducted to familiarize the operator with these critical activities.

After these precautionary measures have been taken, check over the various valves and arrange them for starting up as follows: Close the blowoff, the water-column gauge-glass drain, the gauge cock, and the feedwater valves. Open the drum vent and the cock to the steam gauge. When valves are used on the lines from the boiler drum to the water column, make sure they are open and locked or sealed. The drains should now be opened, except in cases in which superheaters are filled with water during the starting-up period.

Fill the boiler with water, using the auxiliary feed connections if they are installed. Do not fill the drum to the normal operating level, since the water will expand when heat is applied and cause the level to rise. Operate the gauge cock and blowdown valve on the water column and gauge glass to check for possible stoppage.

The boiler is now ready for a fire to be started in the furnace. For procedures in how to start a fire using the various types of combustion equipment, refer to Chap. 5. When the fire has been started, allow the steam to blow from the steam drum for a few minutes and then close the vent valve. Close all superheater drains except the outlet. This will allow a small quantity of steam to circulate through the superheater tubes and thus prevent excessive metal temperatures.

Before the firing of a boiler, there must be no combustible material lingering inside the unit. By purging the boiler, this ensures that the boiler is ready for firing. A general operating procedure is to purge the unit for about 5 minutes using at least 25 percent of the maximum airflow requirements.

After combustion is established, the required air-fuel rates must be maintained. With insufficient airflow, the formation of combustible gas pockets is possible, which creates the potential for explosions. Air should be furnished to match the combustion requirements of the fuel, and a small amount of excess air should be used to ensure proper mixing and to promote the correct fuel-air distribution.

In addition, it is important to verify boiler water levels and flows. Combustion should never be established until adequate cooling water is in the tubes and steam drum. Operating procedures should be followed carefully based on the designer's instructions.

The set point for each safety valve is normally checked and adjusted, if necessary, immediately after reaching the operating pressure for the first time with steam. Safety valve seats are susceptible to damage from wet steam or grit. For initial operation or after maintenance, this potential damage to the seats is a primary reason to clean the boiler and blow out the superheater and steam lines prior to testing the safety valves.

Safety valves on drum-type boilers are normally tested for both set-point pressure and the closing pressure. This requires that the boiler pressure be raised until the safety valve opens and it relieves enough pressure so that the safety valve closes properly.

The rate of combustion should be regulated to allow approximately 45 min for small and medium-sized 150-psi boilers to attain pressure. For large high-pressure boilers, allow from 45 min to $2\frac{1}{2}$ h depending on the size and superheater arrangement. Recommended operating procedures of the designers should be followed closely. When pulverized coal or oil is burned, it may not be possible to regulate the rate of combustion low enough to provide the necessary safe time for bringing the boiler up to pressure. When this condition is encountered, allow the burner to operate for a few minutes and then take it off to allow the heat to distribute through the unit, thus preventing excessive temperature differentials and the resulting unequal expansion.

In installations that have pendant nondraining superheaters, it is necessary that the warm-up rate be controlled very carefully, since all water must be evaporated from the superheater tubes before the boiler is placed on the line. Water remaining in the superheater would restrict the flow of steam and cause the tubes to overheat. With drainable superheaters the drains should be opened for a short blow before the boiler is "put on line."

It is essential that all water be drained from the boiler steam line and that it be filled with steam before the valve is opened and the flow established. This may be accomplished by opening the drain and backfeeding steam from the main steam line through the bypass around the main steam-line valve. When the pressure in the boiler approaches that in the main steam line, open the main steam valve and throttle, but do not close the drain. Before putting the boiler on line, open the superheater drains for a short blow to make sure all water has been removed. Check the water level, and if it is not sufficiently below standard (2 to 6 in) to compensate for expansion when the boiler starts to steam, open the blowdown and remove the excess water. When the boiler drum pressure is from 10 to 25 psi below the main steam-line pressure, unscrew the stem of the nonreturn valve so that it can open when the boiler pressure exceeds the steam-line pressure. Should waterhammer or vibration occur while the valve is being opened, close the valve at once, allow the pressure on the boiler to drop, and repeat the entire operation. When a boiler is not connected to a common steam line, it is advisable to open all drains and raise the steam pressure on the entire system at the same time.

With the boiler in service, the drain may be closed and the fuel feed regulated to maintain a low rate of steam flow. Consult the draft gauges and adjust the fans and dampers to establish the required flow of air to and gases from the boiler unit. Check the water level and feedwater-supply pressure; then put the feedwater regulator in operation.

7.2 Normal Operation

A boiler in service producing steam constitutes a continuous process. Fuel, air, and water are supplied while steam and waste products of ash, flue gases, and blowdown water are discharged. It is the operator's duty to keep these materials flowing in the correct proportion and as required to maintain the steam pressure. These operations must be done while maintaining the air pollution requirements of the operating permit with particulate and acid gas removal systems, as discussed in Chap. 12.

Most of these functions are performed automatically, but this does not relieve the operator of the responsibilities of an overseer of this equipment. Expert supervision is required because fuel costs are high and process steam demand or electric production depends on a continuous steam supply. The best automatic equipment may malfunction and require hand operation. Details of the various types of combustion equipment during standby, normal operation, and banking are discussed in Chap. 5. The following paragraphs explain the correct coordination of the various tasks that must be performed for satisfactory boiler-plant operation.

7.2.1 Combustion control

A drop in steam pressure as indicated by the steam gauge shows the operator that the fuel supply must be increased; an increase in pressure shows that too much fuel is being supplied. On boilers not equipped with automatic controls the operator must watch the gauge continually and adjust the fuel feed. If automatic control is used, the change in steam pressure adjusts the fuel feed, relieving the operator of this repetitive task. However, there is more to firing than simply admitting the correct amount of fuel; consideration must be given to fuel-bed thickness and contour with solid fuel and to flame shape and travel when pulverized coal or liquid or gaseous fuels are used.

The draft differential, flue gas analysis (carbon dioxide, oxygen, and carbon monoxide), appearance of the fire, emission of smoke, and furnace slag accumulation are all means by which the operator can determine the relative quantity of air being supplied to the furnace. Each time the rate of fuel supply is changed, the air supply to the furnace should be adjusted in proportion. The automatic control makes these repetitive adjustments to compensate for load changes and to maintain the correct fuel-air ratio under normal conditions. The operator must make adjustments for changes in coal size, moisture content, heating value, etc. In any case, the combustion control is a device to aid, rather than a substitute for, experienced operation. The operator should be capable of operating the boilers "on hand" at any time. In many plants, hand control is used when putting a boiler in service, and this is changed to automatic only after normal operating conditions have been established. The operator should learn to use all the instruments provided, since they are intended to be helpful in making adjustments. The draft gauge is a simple instrument that is not often used in the most effective manner. Note and memorize the draft-gauge reading (wind box, furnace, boiler outlet, etc.) when the combustion is satisfactory. These values should be used as a means of quickly establishing normal conditions, detecting trouble, and changing capacity. The draft gauge indicates the airflow to and the products of combustion from the furnace. (See Chap. 4 for an explanation of draft as applied to combustion.)

7.2.2 Ash handling

The removal of ashes from the furnace is the responsibility of the operator. Ashes are removed either as a molten liquid or in the solid state. The design of the equipment determines the method to be used. It is an operating responsibility to produce ashes in the solid or liquid state in accordance with the design of the equipment. Early methods involved hand operation for dragging the ashes from the furnace and the use of hand dumping grates. A later development, still in use on small and medium-sized units, consists of power-operated dumping grates. These use a steam or air cylinder that operates the grates, dumping the ashes into the pit. This system creates a periodic disturbance in the operation when ash dumping is in progress and until stable conditions have been restored. Continuous ash discharge systems are used to improve operations.

When operating a furnace equipped with dumping grates, the ash should be removed as often as practice shows it to be necessary to prevent the formation of clinkers, loss of capacity, and too heavy a fuel bed. Definite time schedules cannot be established due to changes in capacity and the percentage of ash in the coal. It requires judgment on the part of the operator to determine when the material on the grate has been burned down sufficiently to permit it to be dumped. A long burning-down period will result in loss of capacity on the boiler (failure to maintain steam pressure) and lowering of efficiency due to large amounts of excess air. Insufficient burning down allows large amounts of unburned carbon to be discharged with the ash. Dumping should be performed as quickly as possible to prevent prolonged interruption of stable operation.

The continuous dumping devices are adjusted to remove ashes at the same rate as they are being produced. Skill is required to ensure as complete burning as possible without introducing large amounts of excess air. Continuous ash discharge is applied to the various types of stokers as well as to slag-tapped pulverized-coal furnaces. The molten slag is allowed to run continuously from the furnace into a tank of water that chills the slag, causing it to shatter into small pieces. The tank of water forms a seal that prevents air from entering the furnace.

The effective use of fossil fuels depends to a great extent on the capability of the boiler to handle ash, which is the inert residual of combustion. In a pulverized-coal-fired boiler, most of the ash is carried out of the furnace by the flue gas, and it is called *fly ash*. Abrasive ash particles that are suspended in the flue gas can cause erosion problems on convection pass heating surfaces.

However, the most significant operation problem is the ash accumulating on the heating surface. During the combustion process, ash is released from the coal at approximately 3000°F, which is considerably above the melting temperature of most ash. Ash can be released in a molten fluid or in a sticky state. A portion of the ash that is not cooled quickly and changed to a dry solid can adhere to the furnace walls and other heating surfaces. Because there is such a large quantity of ash involved in the burning of coal, a small percentage of this total can seriously interfere with operation of the boiler.

The accumulation of ash deposits on the furnace wall causes a decrease in heat transfer, which results in an increased flue gas temperature leaving the furnace. This higher temperature at the furnace exit increases the steam temperature and the tube metal temperature of superheaters. This causes additional steam attemperation, and it could require different materials in the superheater tubes. The boiler design must take these operational problems into consideration.

The higher flue gas temperature also can extend the ash-deposition problem to pendant-type superheaters and other heating surfaces in the convection pass of the boiler. If the ash deposits remain uncontrolled, flue gas passages in the tube banks can become blocked, which prevents proper flow of the flue gas, and shutdown of the boiler is required for manual removal of the ash deposits, causing not only high maintenance costs but also loss of operational revenue because of the outage. These deposits also can cause corrosion problems on the tubes.

Sootblowers are the primary means in correcting the slagging of furnace walls and the fouling of convection passes. Sootblowers should be applied as a prevention measure, not as a corrective measure. These devices are most effective in controlling dry and loosely bonded deposits. It is for this reason that a sootblower system is programmed to operate in a specific time schedule or when boiler performance monitors indicate that this operation should be initiated.

7.2.3 Water supply

The water supply is a most important consideration in power plant operation. The operator should be familiar with every detail of this system. Is the water quality being maintained to the proper requirements? Where is the supply obtained, and what are the possibilities of failure? What types of pumps are used, and how are the pumps driven? What type of feedwater regulator is used, and where are the bypass valves located? Is there an auxiliary feedwater piping system?

The operator's job is to supply water to the boilers in the quantity required to replace what is being evaporated. Usually the boilers are equipped with regulators that control the flow to maintain the right level, thus relieving the operator of the tedious task of making repetitive adjustments. These automatic devices do not, however, relieve the operator of the responsibility of maintaining the water level. Frequent monitoring of the system must include water levels in the heaters and softeners and water temperature and pressure at the boiler feed pump discharge. In many cases these important quantities are recorded for the operator's convenience and to provide a permanent record. Improperly set water columns or foaming, leaking, or stopped-up connections will cause water columns to show false levels. The water column must be blown down periodically, and the high- and low-water alarm must be checked. The feedwater regulator should be blown down and checked according to the manufacturer's instructions. The recorders and other devices are convenient but must be checked frequently with the water level in the gauge glass. It is desirable to use hot feedwater, and the temperature should be maintained as high as feasible with the equipment available and within design limits.

The task of maintaining the water level is difficult because during operation the boiler is filled with a mixture of water and steam bubbles. When the water level in the drum drops, there is an obvious tendency to add water. This tendency occurs with both hand and automatic control. The addition of water, at a temperature lower than the water in the boiler, causes the steam in the bubbles to condense. This action decreases the volume of the steam and water mixture in the boiler and results in a further drop in level and a tendency to add more water. Then when the normal ratio of water to steam bubbles is restored, the level in the drum will be too high. The result is a cyclic condition in which the water level in the drum is alternately high and low. This condition is avoided by the use of two- or three-element feedwater regulators. Two-element feedwater regulators sense both the drum level and steam flow rates. This combined signal is used to actuate the feedwater flow control valve. The three-element feedwater regulator senses the drum level and the steam and feedwater flow rates and uses the combined signal to actuate the feedwater flow control valve. The object of these controls is to keep the feedwater flow equal to the steam flow. If because of blowing down or other irregularities the drum level fails to remain within the desired range, the signal from the drum level corrects the rate of flow of feedwater until the unbalanced situation is corrected. (See Sec. 6.5.)

Power producers are well aware of the economic penalties that occur when a component failure causes a plant shutdown. One of the heaviest financial burdens is attributable to steam-cycle corrosion, which accounts for about half the plant forced outages. Attractive financial returns are possible for improving water chemistry.

Proper water treatment and chemistry control result in the following benefits:

1. Deposits of impurities and corrosion products are minimized, thus reducing accelerated corrosion in the boiler.

2. Premature failure or damage to boiler components is decreased.

3. Excessive carry-over of impurities in the steam is prevented, which lessens any damage to superheaters, turbines, and process equipment that uses the steam.

Maintaining high water quality is of utmost importance to the proper operation of a steam plant because impurities can quickly destroy the boiler. Poor water quality can damage or plug water-level controls and cause unsafe operating conditions. During normal operation of a boiler plant, the water must be conditioned by heating, by the addition of chemicals, and by blowdown to prevent operating difficulties. Inadequate water conditioning results in scale formation, corrosion, carry-over, foaming, priming, and caustic embrittlement.

Scale consists of a deposit of solids on the inside of the heating surfaces such as tubing. The formation of scale is caused by a group of impurities initially dissolved in the boiler feedwater. As the water becomes concentrated and exposed to boiler pressures and temperatures, the impurities become insoluble and deposit on heating surfaces. Water that contains scale-forming impurities is termed *hard water* and will consume a quantity of soap before a lather is produced. These impurities are found in varying amounts in almost all water supplies. The amount of hardness in a given water can be determined by adding a standard soap solution to a measured sample of the water and noting the amount required to produce a lather. Hardness may be classified as temporary (or carbonate) or sulfate (or noncarbonate). Carbonates produce a soft, chalklike scale, and this temporary hardness may be removed by heating. Sulfate hardness produces a hard, dense scale and requires chemical treatment.

Scale deposits on the tubes of a water-tube boiler and the shell of a fire-tube boiler form an insulating layer that retards the flow of heat to the water and can cause the metal to become overheated. This results in a dangerous situation, since the metal is now subjected to design pressure at an elevated temperature. Heat is more injurious to a boiler than pressure. The metal in the shell of a fire-tube boiler, when weakened by overheating, yields to the pressure, causing a protrusion known as a *bag*. Bags provide pockets for the accumulation of sludge and scale. When boilers are operated for an extended period of time (without adequate boiler feedwater conditioning), these deposits of scale can weaken the boiler metal sufficiently to cause an explosion. A combination of imperfections in the boilerplate and overheating causes layers of metal to separate and form a "blister."

Corrosion is the result of a low-alkaline boiler water, the presence of free oxygen, or both. The boiler metal is converted into red or black powder (iron oxide), which is readily washed away by the water. This action is accelerated at points of greatest stress, and as the corrosion proceeds, the metal thickness is reduced and the stress is further increased.

Carry-over is the continual discharge of impurities with steam. These impurities may be in the form of moisture that contains dissolved solids or of solids from which the moisture has been removed. The amount of carry-over of impurities is determined by condensing a sample of steam and noting the conductivity of the condensate.

Carry-over of moisture with the steam is an objectionable condition. In many industrial processes the moisture may interfere with a drying process, or the solids carried by the moisture may be unable to be tolerated in the manufactured product. When the steam is used to operate high-pressure turbines, the silica carried by the moisture in the steam forms deposits on the turbine blades, reducing both the turbine's capacity and its efficiency. This silica is carried through the superheater and deposited on the turbine blades in a pressure area where the steam becomes wet. These deposits are very difficult to remove from the turbine blading.

The solution to the silica scale problem is to reduce the silica content of the boiler feedwater by the use of demineralizers and to provide the boiler drums with efficient moisture separators. Satisfactory turbine operation requires steam to have dissolved solids not in excess of 0.1 parts per million (ppm).

Foaming is the existence of a layer of foam on the surface of the water in the boiler drum. This condition is a result of impurities in the water that cause a film to form on the surface. Oil and other impurities such as excessive alkalinity that may enter the condensate in an industrial plant produce conditions in the boiler that cause foaming.

Priming is a condition in which slugs of water are suddenly discharged from the boiler with the steam. The condition is caused by boiler design, impurities in the water, the capacity at which the boiler is operated, or a combination of these factors. Priming is a serious condition both for the safety of the boiler and for the steam-utilizing equipment. Immediate steps must be taken to correct the condition. Foaming may result in priming, but it is possible to have priming without foam on the surface of the water in the boiler drum.

Caustic embrittlement is the weakening of boiler steel as the result of inner crystalline cracks. This condition is caused by long exposure to a combination of stress and highly alkaline water.

Water treatment includes such processes as deaerators for the removal of oxygen, water softeners, chemical additives, and boiler blowdown. Boiler water must be analyzed to determine its composition and the conditioning required.

The treatment of boiler plant water has advanced, from the hit-or-miss period of applying various chemical compounds, to a scientific basis. Water-control tests have been simplified to color comparisons, titrations, and other more sophisticated tests that can be performed quickly and easily, and the necessary adjustments in water treatment can be made by the plant operators. The results of these control tests are expressed in parts per million, grains per gallon, or some constant that tells the operator when the treatment is incorrect and how great a change must be made.

Water conditioning may be accomplished either by treating the water in a separate unit before it enters a boiler or by adding chemicals directly to the boiler. The combination of these two methods produces the most satisfactory results. The nature of the impurities, quantity of water to be treated, plant operation, pressure, etc., are all factors to be considered in selecting water-treating equipment, procedures, and control. This is a job for a specialist in this field. (See Chap. 11.)

External treatment is required when the level of feedwater impurities cannot be tolerated by the boiler system. Since pure water rarely exists, and since natural waters contain some amounts of dissolved and suspended matter, nearly all plants require a water treatment system. This can be accomplished by passing the water through a zeolite softener or demineralizer or by distilling the water in an evaporator. The zeolite softener consists of a bed of special granular material through which the water passes. The zeolite bed does not actually remove the scale-forming impurities (hardness) but converts them into non-scale-forming impurities, thus softening the water. When the zeolite bed has been exhausted, the softener unit is taken out of service and regenerated by introducing a salt (sodium chloride) solution. After regeneration, the bed is washed free of excess salts and returned to service.

A demineralizer is similar in external appearance to a zeolite softener, except that the water passes through two tanks in series, referred to respectively as *cation* and *anion* units. These systems can be designed to provide very pure water with a low silica content. This demineralized water is suitable for makeup in high-pressure plants, but normally water of this quality is not required for low- and medium-pressure applications. The cation unit is regenerated with acid and the anion unit with caustic.

The hot-process lime–soda-ash softener produces water satisfactory for use as boiler feedwater. Heating of the water before the addition of the chemical removes some of the temporary hardness, reduces the quantity of lime required, and speeds the chemical action. Softeners of this type provide a means of heating the water with exhaust steam, introducing chemicals in proportion to the flow, thoroughly mixing the water and chemicals, and removing the solid material formed in the softening process. After leaving the softener, the water passes through a filter and into the plant makeup system.

An evaporator consists of a steam coil in a tank, and this is a distillation process. The water is fed into the tank, and the heat supplied by steam in the coil causes the water to boil and leave the evaporator in the form of vapor. The vapor enters a heat exchanger, where the relatively cool condensate passing through the tubes causes the vapor to condense, producing distilled water. The impurities in the original water remain in the evaporator and must be removed by blowing down. Evaporators are connected to the plant condensate-and-steam system to produce distilled makeup water with a minimum of heat loss.

Internal treatment is direct boiler water treatment and is accomplished by adding chemicals to the boiler water either to precipitate the impurities so that they can be removed in the form of sludge or to convert them into salts that will stay in solution and do no harm. Some of the more common additives for drum boilers are hydroxide, phosphate, and chelant. Hydroxide (NaOH) is used on very low pressure industrial-type boilers. The use of phosphate is common for industrial boilers operating below 1000 psi. Phosphate salts are used extensively to react with the hardness in the water and thus prevent scale deposits. However, the phosphate reacts with the hardness in the water to form a sludge, which in some cases may result in objectionable deposits. Organic compounds are then used to keep the sludge in circulation until it can be removed through the blowdown. The amount of sludge formed depends on the amount of hardness introduced to the boiler in the feedwater. Therefore, it is desirable to have the hardness of the feedwater as low as possible. The phosphate treatment should be supplied directly to the boiler drum with a chemical pump. If introduced into the suction of the boiler feed pump, most phosphates will react with the impurities in the water and cause deposits in the pumps, piping, feedwater regulators, and valves. Therefore, the phosphate should be dissolved in the condensate to prevent deposit in the chemical feed pump and lines. The amount of treatment is controlled by analysis of the boiler water for excess phosphate. This analysis consists of a color comparison of a treated sample with standards. Generally less than 40 ppm is satisfactory to ensure removal of the hardness. Since this phosphate treatment removes the hardness, it is not necessary to run a soap hardness test on boiler water if the specified amount is maintained.

Care must be exercised in introducing phosphate into a boiler that contains scale. For example, on a fire-tube boiler, the old scale may be loosened from the tubes by the action of the phosphate and collected in a mass on the heating surface, causing bags, overheating, and ultimate failure. When phosphate is supplied to a boiler that already contains scale, the boiler should be inspected frequently and the amount of excess maintained at about 20 ppm. In time the phosphate will remove the old scale, but it is better to start the treatment with a clean boiler.

For industrial boilers, another approach to internal boiler water conditioning is the use of chemicals that prevent the precipitation of scale-forming materials. These chemicals have chelating power in that calcium, magnesium, and other common metals are tied up in the water and are eventually removed by continuous blow down. This action prevents the formation of scale and sludge in boilers, heat exchangers, and piping and is effective over the normal range of alkalinity encountered in boiler plant operation. Chelant chemicals are sold under a variety of trade names.

The treatment is introduced into the boiler feed line by means of a standard chemical feed pump. The pump and piping should be of corrosion-resistant stainless steel for high pressure and of either stainless steel or plastic for low pressure. A continuous feed is desirable, and a slight excess should be maintained in the boiler water at all times. However, if there is a deficiency for a short time, the deposits will be removed when the excess is restored.

The choice between the use of these chelating materials, hydroxide, and phosphate depends on the condition in the specific plant, the quantity of makeup water, the operating pressure, and the amount and types of impurities that it contains. The relative cost of the methods and anticipated results should be determined by a water-conditioning consultant.

Corrosion may be caused by either low alkalinity or the presence of oxygen in the boiler water. The alkalinity may be regulated by control of the lime-soda softener, by varying the type of phosphate used, by introduction of alkaline salts (sodium sulfite), and by treating the makeup water with acid. The alkalinity is indicated by testing (titrating) the boiler water with a standard acid, using phenolphthalein and methyl orange as indicators. Some alkalinity in boiler water is essential, but if allowed to get too high, alkalinity will result in priming and foaming. High concentrations of alkaline water also contribute to caustic embrittlement. The correct value must be specified for individual plants with due consideration for the many factors involved.

Free oxygen is most effectively and economically removed from boiler feedwater by heating both the return condensate and makeup water in a vented open-type heater called a *deaerator*. However, it is good practice to use chemicals to remove the last trace of oxygen from feedwater. Sodium sulfite or hydrazine is added to the deaerated water. It is not advisable to use sodium sulfite to react with large quantities of oxygen because of the cost and the fact that it is converted into sodium sulfate and therefore increases the concentration of solids in the boiler water. It is also only used in low-pressure boilers. Hydrazine is another chemical used to neutralize the corrosive effects of free oxygen. It combines with oxygen to form water and nitrogen and therefore does not increase the concentration of solids in the boiler water. Some of the ammonia formed from the hydrazine is carried from the boiler with the steam, where it neutralizes the carbon dioxide, reducing corrosion in the condensate return lines. Hydrazine, being alkaline and toxic, is best handled as a dilute solution.

Caustic embrittlement of boiler metal is the result of a series of conditions that include boiler water having embrittlement characteristics, leakage and the concentration of the solids at the point of leakage, and the depositing of these solids in contact with stressed metal. It follows that embrittlement can be avoided by preventing the presence of at least one of these factors.

Leakage and the resulting concentration of solids in joints have been eliminated by use of welds in place of riveted joints. All boilers should be inspected for leaks, but special consideration should be given to where tubes are expanded into drums and headers. Boiler water is rendered nonembrittling by lowering the alkalinity and adding sodium sulfate or sodium nitrate in controlled proportion to the alkalinity and boiler operating pressure.

Impurities enter the boiler with the feedwater, and when the steam is discharged, these impurities are left in the boiler water. This action makes it necessary to remove the solids by allowing some of the concentrated water to flow from the boiler. There are two general procedures for removing this water from the boiler. One is to open the main blowoff valve periodically, and the other is to allow a small quantity of water to be discharged continuously. The continuous blowoff is generally favored because it provides a uniform concentration of water in the boiler, may be closely controlled, and permits reclamation of a portion of the heat. Heat is reclaimed from the blowdown water by discharging it first into a flash tank and then through a heat exchanger. The flash steam is used in the feedwater heater or as a general supply to the low-pressure steam system. The heat exchanger is used to heat the incoming feedwater (see Chap. 11). The continuous-blowdown connection is so located as to remove the water having the highest concentration of solids. The location of this connection depends on the type of boiler; in most cases it is located in either the steam or lower drum.

The operator must exercise careful control over the blowdown, since too high a concentration may result in priming and foaming, in caustic embrittlement, and in scale and sludge deposits, while excessive blowdown wastes heat, chemicals, and water. Blowdown is effectively controlled by checking the chloride concentration of the feedwater and boiler water. Chloride salts are soluble and are not affected by the heat of the boiler. When the boiler water contains 120 and the feedwater 12 ppm of chloride, the concentration ratio is 10. That is to say, the boiler water is 10 times as concentrated as the feedwater. This is equivalent to taking 10 lb of feedwater and evaporating it until only 1 lb remains. All the salts that were in the original 10 lb are now concentrated in the 1 lb. For low- and moderate-pressure boilers under average conditions, 10-fold concentration is deemed satisfactory, but all factors in an individual case must be considered. Other methods of determining the concentration of boiler water and thus providing a means of controlling blowdowns are (1) a hydrometer, which measures the specific gravity in the same manner as that used in checking storage batteries, and (2) a device for measuring the electric resistance of the sample of water, since the resistance decreases when the amount of impurities increases. The hydrometer method is satisfactory when high concentrations are encountered. The electric-resistance measurement is advisable when the chlorides in the feedwater are low or subject to wide variation.

In plants where chemical control is not used, some general rules apply. Blow down at least one full opening and closing of the blowoff valve every 24 h. When continuous blowdown is used, the main blowoff should be opened once every 24 h or more often if necessary to remove sediment from the lower drum. It has been found that several short blows are more effective in removing sludge than one continuous blow. This is also a good precaution to take when blowing down a boiler that is operating at high capacity. In general, boilers should be blown down when the steam demand is lowest.

When the boiler is in service, the skilled operator will devote time to careful observations and inspections and thus detect possible irregularities before serious situations develop. This ability to discover minor difficulties before real troubles are encountered is a measure of the competence of an operator. It is this skill that prevents boilers from being damaged when the feedwater supply fails. The water level in the boiler should be observed frequently to check on the operation of the feedwater regulator. When in doubt and in all cases at least once per shift, blow down the water column and gauge glass, open the try cocks, and check the high- and low-water alarms. Check the feedwater pressure frequently, since this may enable you to detect trouble before the water level gets out of hand. Failure of the feed pumps or water supply will be indicated by a drop in pressure before the water gets low in the boilers. The temperature of the feedwater should be checked and adjustments made to supply the necessary steam to the feedwater heater to maintain standard conditions.

With hand operation, almost constant observation of the steam pressure is necessary. When automatic controls are used, the steam pressure should be observed frequently as a check on the control equipment. Steam-pressure recorders are helpful in checking the operation of an automatic combustion control. Difficulty in maintaining the steam pressure may indicate trouble developing in the furnace, such as clogged coal-feed mechanism, uneven fuel bed, or clinker formation. Many a disagreeable job of removing clinkers from the furnace could have been prevented by closer observation of the fuel bed and proper ash removal.

The delivery of the required amount of fuel to the furnace does not ensure satisfactory operation. The operator must consider these questions: Is the fuel bed of the correct thickness? Are there holes through which the air is bypassing the fuel? Are clinkers forming that blank off the airflow? Is the flame travel too short or too long? Is smoke, in an objectionable amount, being discharged from the stack?

The competent operator is always on the alert to detect leaks in the pressure system. An unusual "hiss" may be caused by a leaking tube. A sudden demand for feedwater, without an accompanying increase in load, may be the first indication of a leak. When these symptoms occur, be on the alert, advise plant management, and investigate further.

Leaks in boiler tubes and piping lead to maintenance costs and plant unavailability that result in lost revenues because of the loss of production or of electric power. Generally, when leaks first occur, they are undetected because they are not audible or they are concealed by insulation. Acoustic monitoring equipment has now been developed and is used extensively for leak detection. This technology permits earlier identification of a leak, and this can reduce the repair costs as well as the outage period for the repair.

Operators should be constantly on the alert for leaks in the pressure systems and settings of boilers. If the main or water-column blowdown valves and lines remain hot, it is an indication of leakage. Low concentrations of salts in the boiler water without a change in operation or procedure indicate leaking blowdown valves. High exit gas temperatures or a decrease in draft differential across the boiler may indicate leaking baffles. Report loose brick found in the ashpit when removing ashes or on the tubes when hand-lancing, since these may indicate the beginning of serious failure, and their origin should be determined.

Furnace waterwall tube failures are a leading cause of forced outages in both electric utility and high-pressure industrial boilers. These failures can result from such things as erosion, corrosion, overheating, and fatigue. Some of these are a direct result of the waterside scale and deposits that can build up inside the tubes. The periodic chemical cleaning of boilers is critical in the elimination of deposits that can cause tube failures and the resulting expensive plant outage.

There are two primary ways in which waterside deposits lead to tube failures. First, the deposit acts as an additional insulating layer and prevents the water in the tubes from adequately cooling the tube metal. This increases the tube metal temperature, as shown in Fig. 7.1 which in turn accelerates the rate of deposition inside the tubes. If the deposit buildup becomes very thick, it can obstruct the flow of cooling water through the tubes, causing an even more drastic increase in tube metal temperatures. Stress-rupture tube failures eventually can occur due to overheating, as in the typical failure shown in Fig. 7.2.

Waterwall deposits also play an important role in the onset of waterside corrosion such as hydrogen damage and caustic corrosion. Under normal conditions, boiler water chemistry usually can be maintained at the proper levels to prevent the occurrence of waterside corrosion. However, any porous deposits in waterwall tubes provide a location where corrosive boiler chemicals can concentrate.

Chemical cleaning of the boiler should be considered as part of the plant's preventive maintenance program to avoid costly forced out-

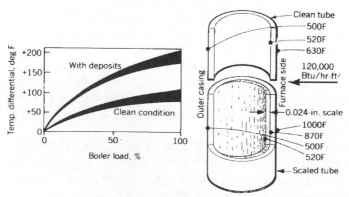

Figure 7.1 The harmful effect of waterwall deposits on tube metal temperatures. (*Babcock & Wilcox, a McDermott company.*)

Figure 7.2 Typical furnace waterwall tube overheat failure. (*Babcock & Wilcox, a McDermott company.*)

ages resulting from waterside tube failures. Tube samples are often removed at routine intervals to monitor deposit buildup. This, together with newly developed nondestructive techniques such as ultrasonic examination of tubes, provides the operator with critical information about potential problems. The ultrasonic testing (UT) of tubes and waterwall panels does identify the wall thickness. In order to have a baseline measurement, an initial UT scan should be made, with periodic scans during shutdowns to evaluate whether (and where) metal loss has occurred and if tube replacement is necessary.

7.2.4 Boiler efficiency

Instruments and controls are for the benefit of plant operators and should be checked, adjusted, calibrated, and kept in good operating condition. Draft and pressure gauges, steam flowmeters, flue gas temperature recorders, and water-level indicators are useful in making routine adjustments for changes in load as well as in detecting difficulties. Learn to use the instruments available. Keep the automatic control adjusted and operating.

Boiler efficiency is expressed as a percentage and is calculated by dividing the output of the boiler by the input to the boiler. Boiler performance may be expressed as the number of pounds of steam produced per pound of fuel. The total pounds of steam generated divided by the pounds of coal burned in a given time equals the pounds of steam per pound of coal. This formula may be valuable in comparing daily performance in a given plant, but it does not take into consideration variations in feedwater temperature, the changes in steam pressure and superheater temperature, or the heating value of the fuel. To calculate boiler efficiency, one must know the amount of steam generated and fuel burned, the steam pressure, the steam quality at the boiler outlet or the superheat temperature at the superheater outlet, and the feedwater temperature. (See Sec. 3.9 for the procedure for calculating the heat content of wet or superheated steam.)

$$\text{Boiler efficiency, } \% = \frac{W\,(h_g - h_f)}{Q \times H} \times 100$$

where W = steam flow (lb/h)
   Q = fuel flow (lb/h)
   H = heating value of fuel (Btu/lb)
   h_g = heat content of steam (Btu/lb)
   h_f = heat content of feedwater (Btu/lb)

Example   A boiler generates 50,000 lb/h of steam and burns 4880 lb of coal per hour having a heating value of 13,700 Btu/lb. The steam is 150 psig, dry, and saturated, and the feedwater is 220°F. What is the boiler efficiency?

Solution   Heat content of 150 psig, dry and saturated steam (h_g) = 1195.6 Btu/lb (see the steam tables in Appendix C). Heat content of 220°F feedwater (h_f) = 188.1 Btu/lb.

$$\text{Boiler efficiency} = \frac{50,000\,(1195.6 - 188.1)}{4880 \times 13,700} \times 100$$

$$= 75.35\%$$

In determining accurate boiler efficiencies, boiler designers develop a heat balance around the boiler system by calculating the heat losses. This is done at all expected operating capacities.

The efficiency of boilers changes as the capacity at which they are operated varies. At low capacities the combustion process tends to be inefficient because the amount of excess air must be increased to ensure complete combustion. This additional air carries heat to the stack and lowers the efficiency. As the capacity increases, combustion can be accomplished with less excess air, reducing stack loss and increasing efficiency. However, as the capacity increases, combustion space and heat-absorption capacity of the tubes begin to be exceeded. The combustion equipment does not develop all the heat in the fuel, which results in unburned carbon loss. The convection bank of tubes absorbs a smaller percentage of the heat developed, resulting in a rise in exit gas temperatures at the boiler outlet.

For a given boiler and set of operating conditions there is one output rate at which the efficiency is maximum. The use of heat-recovery equipment (economizers or air heaters) tends to decrease this variation in efficiency. In addition to operating efficiency, overall maintenance and boiler outages must be considered in scheduling plant operation.

When the plant consists of a series of boilers, a sufficient number should be operated to keep the output of individual units near their points of maximum efficiency. In these instances it is frequently advisable to regulate one boiler to meet the change in demand for steam while the others are operated at their most efficient capacity. As the variation in steam demand exceeds the amount that can be satisfactorily obtained with the regulating boiler, others are either banked or brought into service to meet the change. In utility plants these wide variations occur in daily cycles. Keeping the boilers operating near their most efficient load results in a fuel savings.

The addition of economizers and air preheaters and improvement in combustion equipment have resulted in boiler units that have a wide range of output at nearly maximum efficiency. This is referred to as having a *flat efficiency curve* and is essential to large plants when one boiler supplies all the steam for a turbine generator. For peak demands, boilers can be operated beyond their most efficient output as the design allows. Consideration of boiler loading improves plant economy and decreases outages and maintenance.

The operators should reduce the air leakage into the boiler setting to a minimum. The air that enters through leaks must be handled by the stack or induced-draft fan and frequently limits the maximum output of the boiler. The heat required to raise the air to the exit gas temperature is lost. A check for air leaks should be made while the boiler is operating by placing an open flame near points where leaks are suspected. The flame will be drawn into the setting by the draft. A torch made with a pipe fitting and wicking, and burning kerosene, is useful in testing for leaks.

Another method of testing a boiler setting for air leaks is to analyze both the gases leaving the furnace and those at the induced-draft fan inlet. The increase in excess air as indicated by these two analyses represents the air leakage in the setting. However, caution must be exercised in making these analyses because of the flue gas temperature. A series of samples must be taken at each location to ensure that the results represent average values.

Air leaks in boiler settings should be sealed by the application of a commercial compound manufactured for this purpose. The material selected should remain pliable to compensate for the movement resulting from expansion and contraction of the boiler.

The use of the correct amount of excess air is an important consideration for the best overall results. Too much air wastes heat to the stack, while too little causes overheating of the furnace, unburned carbon in the ash, smoke, and unburned gases. (See Chaps. 4 and 5 for a discussion of excess air.) Study the conditions in tho plant and regulate the excess air by observing furnace temperature, smoke, clinker formation, and combustible in the ash to give the best overall results. After the correct amount of air has been determined by experience, use draft gauges, carbon dioxide, carbon monoxide, or oxygen recorders, and steam-flow–airflow meters as guides in regulating the supply. In general, the excess air should be lowered until trouble begins to develop. This may be in the form of smoke, clinkers, high furnace temperature or stoker maintenance, slagging of the boiler tubes, or high combustibles in the ash. The combustible material dumped from the furnace with the ash represents heat in the original fuel that the furnace failed to develop. It represents a direct loss and must be considered by the operator. The designers of combustion equipment are continually striving to design units that will reduce this loss. In the cases of hand firing and with stokers, the amount of unburned carbon in the ash depends, to a large extent, on the skill of the operator. Dump the ash or clean fires as often as required for best overall results. Maintain an even distribution of fuel on the grates so that the combustible matter will be burned nearly completely before dumping. Learn to operate so that the formation of clinkers will be at a minimum. Avoid excessively high rates of combustion. It is difficult to determine the amount of unburned carbon in the ash by observation; the only sure way is to collect a representative sample and send it to a laboratory for analysis.

A deficiency of air or failure to obtain mixing of air and combustible gases will result in burning the carbon to carbon monoxide rather than to carbon dioxide. When this occurs, approximately 10,185 Btu of heat per pound of carbon passes out of the stack unburned (see Sec. 4.2). A flue gas analysis provides the only sure means of detecting and evaluating the amount of carbon monoxide. High percentages of carbon monoxide gas are usually accompanied by unburned hydrocarbons, which produce smoke. Because of particulate control equipment, smoke may not be readily observed. However, instrumentation such as continuous-emissions-monitoring (CEM) equipment will identify the situation for the operator.

After the heat has been developed in the furnace, the largest possible amount must be absorbed by the water and steam in the boiler. Scale formation on the water side of the heating surface and soot and ash on the gas side act as insulators and cause high flue gas temperature. It has been noted how water treatment can eliminate scale from the water side of the heating surface. Failure on the part of the water-treating system to function, due either to an inadequate system or to unskillful application, will result in the formation of scale. This scale must be removed either by mechanical methods or by acid cleaning. Soot and ash are removed by means of sootblowers.

To operate mechanical sootblowers, open the drain of the sootblower header to remove the accumulation of water. Then open the steam supply valve. Slowly turn each sootblower through its entire arc of travel by means of the chain or handwheel. It is advisable to start with the element nearest the furnace and progress toward the boiler outlet. Sometimes it is found desirable to increase the draft during sootblowing to prevent a pressure from developing in the furnace and to carry away the soot and ash removed from the tubes. Care should be exercised in draining the steam lines to the sootblowers, since water will cause warping and breakage of the elements.

On boilers equipped with fully automatic sootblowers, the operators first turn on the steam supply or start the air compressor (if air is used for blowing) and then initiate the sootblowing cycle by a switch that is usually mounted on the boiler control panel. The sootblower elements are inserted, rotated, and retrieved by power drives. The operation proceeds from one element to another in a programmed sequence, and indicating lights on the panel inform the operator of the progress being made and warn if trouble should develop in the system. Many plants find it advantageous to operate the sootblowers once every 8-h shift. Boilers are generally equipped with a gas outlet temperature measurement, and, therefore, the effect of blowing soot can be readily determined by the resulting drop in flue gas temperature. Sootblowers require an appreciable amount of steam or air and should not, therefore, be operated more often than necessary as indicated by the outlet gas temperature.

7.2.5 Insulation and lagging

The heat loss from a boiler setting is reduced by insulation. A typical membrane wall panel with insulation and lagging is shown in Fig. 2.32. The insulation is designed both to provide safety for personnel and to minimize heat loss and thus improve boiler efficiency and operating costs. Indoor units require ventilation for operator comfort as well as for normal room air changes. The materials that are used most frequently for insulation are mineral wool, calcium silicate block, high-temperature plastic, and ceramic fiber.

In order to maintain satisfactory working conditions for personnel around a boiler, a cold face temperature of 130 to 150°F is satisfactory. It is necessary to have proper air circulation around all parts of the boiler to prevent excessive temperatures in areas that are frequented by personnel. With the use of grating instead of solid floors and with ample aisle space between boilers, air circulation can be improved. Lagging is an outer covering over the wall (see Fig. 2.32) and is used for protecting the insulation from water or mechanical damage.

7.3 Operating Characteristics of Fluidized Bed Boilers

Fluidized bed boilers are becoming more popular throughout the world because of their ability to burn fuels with low heating values, such as waste fuels, and to operate with low-acid gas emissions without the use of expensive scrubbers. There are various designs, as discussed in Chap. 2, but each has similarities with regard to their operating characteristics. Because fluidized bed boilers have unique differences from boilers that have more conventional firing methods, these characteristics require separate mention for both circulating fluid bed and bubbling fluid bed designs.

7.3.1 Circulating fluid bed (CFB) boilers

Although in many areas the CFB boiler is constructed and operates similar to a conventional drum-type boiler, the CFB boiler has some significant differences:

1. The mass of solids in the bed, called the *bed inventory*

2. The size distribution of the bed material

3. The circulating rate of the bed

These characteristics significantly affect

1. Thermal performance, i.e., heat transfer

2. Combustion efficiency

3. Furnace absorption

4. SO_2 control

There are two distinct sections in the bed, the dense bed and the dilute bed. The term *bed density* is a measure of the weight of the particle material per volume of flue gas. In the primary zone of the furnace, there is a higher density and inventory of particles, and in the upper zone of the furnace, there is a lower density and inventory of particles. Between these two zones there is a transition area, called *freeboard,* where there is a distinct change in particle density.

In order to operate properly, the CFB design must include additional instrumentation that provides information to the operator for management of the bed. This information provides the following:

1. *Pressure drop of primary zone.*   The differential pressure is measured across the primary furnace zone, and is an indication of bed density and a control variable for operation of the CFB boiler. The differential pressure transmitter is often calibrated from 0 to 50 in of water gauge (in of wg), representing the bed density from 0 to 100 percent.

2. *Pressure drop of secondary zone.*   This measurement is an indication of the density in the upper zone of the furnace, and this is

used by the operator to evaluate the material size of the bed. This is not used as a control variable.

3. *Bed temperature.*   Multiple thermocouples are inserted into the bed. This allows the bed temperature to be input as a primary control variable, which ensures optimal combustion efficiency and control of emissions.

4. *Furnace exit gas temperature.*   At the entrance to the particle-separation device, thermocouples are installed to measure the furnace exit gas temperatures.

On a conventional boiler, all these variables generally are not available, although the overall operation of a CFB boiler and a conventional boiler is similar. The fuel bed is set by steam pressure requirements, and total airflow follows the fuel flow. Bed density is used to control the bed temperature, which is the most important variable. The bed temperature is usually controlled to approximately 1550°F, a temperature that results in optimal combustion efficiency as well as good SO_2 capture.

Air admission to the furnace is controlled differently in a CFB boiler than in a conventional boiler because the total airflow is divided into primary, secondary, and tertiary portions. Since the total airflow is based on the fuel flow, the splitting of the air between the three zones is another variable that the operator uses to ensure bed stability. The air split has an impact on the bed density because the primary air velocity determines how much bed material is released, called *elutriated,* and then recovered by the particle-separation device.

7.3.2 Bubbling fluid bed (BFB) boilers

As with the CFB boiler, management of the bed of a BFB boiler is an important skill of the operator. Operating procedures must reflect the integration of

1. Bed inventory
2. Bed size
3. Bed makeup
4. Bed drainage

The thermal performance of a BFB boiler is affected most by the bed height. The bed height can be minimized by inventory or by reducing the airflow to zero, called *slumping,* at selected areas in the bed.

The BFB boiler is also equipped with additional instrumentation to assist the operator in controlling the operation. These include

1. *Pressure drop measurement of primary zone.*   The differential pressure is measured across the primary furnace zone and is used as an indication of bed height.

2. *Bed temperature measurement.*   Multiple thermocouples are installed in the bed. The optimal bed temperature of approximately 1550°F also applies to BFB boilers, and this is a primary control variable.

The operation of a BFB boiler is similar to that of a conventional drum-type boiler because fuel and air quantities are set by the steam requirements. The bed height is used to control bed temperature.

The height of the bed is controlled by adding solids from the make-up system or by removing solids through the drain system. Bed inventory depends on the balance between feed solids and the bed drain purge flow, plus the solids that are released from the bed (elutriated) and lost from the system as they are captured in a particulate collector such as a fabric filter. In order to lower the bed temperature, the height of the bed is increased, and for an increased bed temperature, the bed height is decreased. The operator must control this function because it is *not* automatic. During startup testing, the optimal height for specific steam flow requirements is determined, and this is made part of the plant's operating procedures.

7.4   Abnormal Operation

When emergencies arise, analyze the situation, decide on the necessary action, and proceed quickly. It is essential that an operator know the plant thoroughly to handle emergencies. A knowledge of the plant not only will make it possible to act wisely but also will give the operator the necessary confidence. Learn what constitutes safe practices and how they are applied to your plant. Do not shut down unless it is absolutely necessary, but do not hesitate to shut down rather than resort to unsafe practices. Well-operated plants should include written operating procedures and training programs so that operators can respond properly during any operating situation.

Thought should be given to the procedure to be followed if difficulty is experienced in maintaining the water level in the boiler. Small oil- and gas-fired boilers equipped for automatic operation have low-water fuel-cutoff devices. In the event of low water, the fuel valve automatically closes and shuts down the unit until the difficulty has been corrected and the water level restored. However, industrial and utility plant shutdowns must be avoided if at all possible. Alarms are provided, but in the event of high or low water, the operator must decide on the action to be taken. When emergencies of this nature arise, quick and decisive action is important, and a thorough knowledge of the plant and a preconsidered line of action are essential. Study the various means available for supplying water to the boiler, including pumps, auxiliary piping systems, regulators, bypasses, and sources of water.

When the water level in the drum drops below the minimum required, as specified by the boiler designer, stop the fuel and air supplies. The procedure to be followed depends on the type of combustion equipment used: For hand-fired units, cover the fire with coal and ashes; for stokers, stop the fuel feed, shut off the air, and open the furnace doors; and with pulverized coal, oil, and gas, simply shut off the fuel and air supplies. Do not change the feedwater supply, open the safety valve, or make any adjustments that might result in changes in the stress to which the boiler is subjected. Take the boiler out of service for a thorough internal check by an authorized inspector to determine the extent of damage and repairs necessary as a result of the low water.

Priming and foaming are caused by boiler water high in solids, high capacities, high water level, or a combination of these. Foaming is serious, since it makes it difficult to determine the water level. Priming can result in a mixture of steam and water in the steam lines, causing waterhammer, which may result in rupture of the piping system. When this trouble develops, reduce the capacity and blow the boiler down through the surface blowoff, if one is provided; otherwise, use the main blowoff. If the priming becomes so serious that the water level can no longer be detected, close the steam valve long enough to determine the correct level. Blow down the gauge glass and water column to make sure the water level indicated is correct. If conditions do not return to normal, take the boiler out of service. Check for possible contamination of the feedwater by material leaking into the condensate-return system from industrial processes. Oil from pumps is a frequent cause of foaming. If the condition persists, it is best to have a consultant study the problem.

With the development of modern water-treatment systems, these conditions have been minimized on present-day designs. However, poor water treatment does exist at times, and such situations must be handled.

In the case of tube failure, stop the fuel feed and proceed to reduce combustion. Maintain the water level unless this affects the feedwater supply so that you cannot maintain the water level in the other boilers. This is a matter that requires instant decision on the part of the operator, and the correct action depends on the size of the leak and available reserve supply of water. The pressure should be reduced as quickly as possible to prevent the blow of water and steam from cutting other tubes. Except for the fact that tube failures cause boiler outages, they are not considered serious. An investigation should be made to determine the cause of the tube failure and necessary steps taken to prevent a recurrence.

Operating a boiler with a known tube leak is not recommended. The steam or water that escapes from the leak can cut other tubes by impingement and cause a chain reaction of tube failures. With the loss of water or steam, a tube failure can change boiler circulation and possibly cause overheating in other circuits, which can lead to further failures.

Small leaks can be detected at times by the loss of water from the system, by the loss of chemicals from a drum-type boiler, or by noise that is created by the leak. If a leak is detected, the boiler should be shut down as soon as operating procedures permit. After a hydrostatic test is applied and the location of the leak is detected, the leak is repaired.

Following the shutdown and cooling of the boiler, maintenance personnel should make a complete inspection of the boiler for evidence of overheating and possible cracks. If tube wear is apparent, or if a maintenance schedule dictates, an ultrasonic test (UT) may be required to evaluate the life of the tubes and determine the appropriate time for replacement or repair of the tubing.

An investigation of the tube failure is very important so that the cause of the failure can be eliminated and future failures prevented. This evaluation should not only include a visual inspection but also may require a laboratory analysis. All information preceding the failure also must be evaluated, and this could include such things as the location of the failure, length of operation, steam capacity, startup and shutdown conditions, feedwater treatment, etc.

Induced-draft fans are used to create the necessary draft for boiler operation, and when they fail suddenly, a pressure develops in the furnace and setting that causes smoke and sometimes fire to be discharged into the boiler room. This creates a dangerous condition, especially with pulverized coal, gas, and oil. The immediate remedy is to stop the supply of fuel and air to the furnace until the induced-draft fan is again operating. Modern units are equipped with a safety interlock system that shuts off the forced-draft fan and fuel feed in case of induced-draft fan failure. This interlock system also makes it necessary to start the induced-draft fan before starting the fuel feed and forced-draft fan. When starting up a boiler equipped with an interlock system, always trip the induced-draft fan out and make sure that the forced-draft fan and fuel feed stop. If the induced-draft fan trips out during operation and the automatic fails to operate, trip the forced-draft fan by hand.

Failure of the fuel supply, low capacity, too much primary air, and disturbance from blowing soot may result in loss of ignition when fuel is burned in suspension. If the equipment is supplied with a flame detector, as all modern boilers are, the fuel feed will be shut off automatically. When this safety device was not provided, the fuel supply had to be shut off by the operator. After the flame is out, regulate the airflow to about 10 percent of the maximum and allow 5 min for the pulverized coal, oil, or gas to be purged from the furnace. Then proceed to relight as in normal starting.

Power failures or other emergencies make it impossible to remove all the coal from the pulverizing mill before shutting down. When this occurs and a mill is to be returned to service within 3 h, the partly pulverized coal may be allowed to remain. However, if the outage is to be longer than 3 h, the coal should be cleaned out to avoid the possibility of fire. If a fire should develop while the pulverizer is out of service, close all the outlets and, one at a time, open the cleanout or access doors and drench the interior with water, chemical fire extinguisher, or steam. Do not stand in front of the doors or inhale these gases because they may be poisonous. Guard against possible explosion of the gases formed in the mill. After the fire is out, the material in the mill has been reduced to a temperature below the ignition point, and the pulverizer room has been ventilated, remove all the coal from the mill. Never use an air hose or a vacuum system in cleaning the pulverizer mill.

Fires in pulverizers during operation are very infrequent, but they sometimes occur as a result of fire originating in the raw-coal bunker, failure to clean up after welding, or using inlet air to the pulverizer that is of too high a temperature. The general procedure in this emergency is to reduce the supply of air so that the mixture in the mill will be too rich to support combustion. Fire in a pulverizer may be detected by the rapid rise in outlet temperature of the coal-and-air mixture. When this occurs, increase the fuel feed to maximum and supply the pulverizer with cold air. Do not increase the air supply or reduce the fuel feed. If the temperature does not begin to return to normal within 15 min, shut down the pulverizer and close off the air supply; then follow the procedure explained above for a fire in a pulverizer out of service.

When a safety valve will not close but continues to leak at a pressure lower than that for which it is set, try to free it by operating the lift levers. If this fails to stop the leak, the valve must be repaired or replaced. Never attempt to stop a leaking safety valve by blocking or tightening the spring.

When a gauge glass breaks, shut off the flow of steam and water. While the gauge glass is out of service, check the water level by using the try cock. Before inserting the new gauge glass, blow out the connections to make sure that no broken pieces remain in the fitting, and check the glass and the gasket for the correct length and size. Pull up the packing slowly and not too tightly to avoid breaking the glass.

When the new glass is in position, open the top or steam valve first and allow the drain valve to remain open. This permits steam to circulate through the glass, gradually heating the entire surface to a uniform temperature. Then partly open the lower valve. After the water has reached the normal level, open both valves all the way, and the glass is in service. When the lower valve is first opened, the glass is subjected to a variation in temperature that may result in breakage. Always keep a supply of gauge glasses and gaskets available as part of the spare-part inventory. Replace gauge glasses that become discolored and difficult to read.

7.5 Idle Boilers

Since there are no practical means available for storing steam for an extended period of time, it must be generated as it is required. When there are wide variations in steam demand, boilers must be put on and taken off the line to meet the load conditions. (See Chap. 5 for banking fires with the various types of combustion equipment.) Boilers also must be taken out of service for inspection and repairs.

Before taking a boiler out of service for an extended period of time, allow the coal in the bunkers to be used up. In some cases coal may overheat and start to burn in 2 weeks when allowed to remain in the bunkers. The characteristics and size of the coal and the temperature to which the bunkers are subjected affect the time necessary to cause overheating and burning.

To take a boiler out of service, reduce the fuel feed and slowly decrease the output. With solid fuel this consists of allowing the fuel on the grates to burn out. When oil or gas is used, the fuel supply must be shut off after the capacity has been reduced. Pulverized coal requires careful handling. Shut off the coal feed to the pulverizer to allow it to run empty. Observe the flame closely, and shut down the pulverizer as soon as the flame goes out. Continued operation of the pulverizer might result in a continuing supply of coal after the fire is out and in the formation of an explosive mixture in the furnace.

When the combustion rate is sufficiently reduced, the steam pressure will drop, the boiler will cease to deliver steam, and the nonreturn valve will close. Screw the stem down on the nonreturn valve or, if there is no nonreturn valve, close the main steam stop valve nearest to the boiler. Now close the valve to the steam line and open the drain connections to this line between the two valves. Finally, close the feedwater-supply valve.

Allow the boiler to cool slowly to prevent injury from rapid contraction. Forced cooling by blowing down and refilling with cold water or using the induced-draft fan to circulate cold air through the unit should be resorted to only after the pressure has been reduced by normal cooling. When there is no pressure in the boiler, open the drum vent to prevent the formation of a vacuum. Allow the boiler to cool as much as possible before draining. If the boiler is drained when too hot, sludge may be baked on the surfaces or the unequal contraction may cause tubes to leak. When there are multiple boilers, check carefully to see if the proper blowoff valves are opened and closed.

In the case of a water-tube boiler, proceed to wash the drums and tubes. Use the highest water pressure available to remove any sludge and soft scale. On a bent-tube boiler, the tubes may be washed by working from inside the drum. To wash straight-tube boilers effectively, the handhole covers must be removed. This is seldom advisable because of the labor required to remove and replace the handhole covers. If, however, these covers are taken off, the deposits should be removed by passing a mechanical cleaner through the tubes. Wash the tubes of the return tubular boiler from both the top and the bottom openings. Allow the wash water to drain through the blowoff lines to the wastewater-treating system. Exercise care to prevent the water from getting on the brickwork. If this is unavoidable, dry out the brickwork slowly before putting the boiler back into service.

There are two ways to proceed in laying up a boiler: One is to keep the interior dry, and the other is to fill the interior with water. The choice of methods depends on how long a boiler is to remain out of service. If for a long period of time, the dry method is recommended. If for a short or rather indefinite period, it is better to fill the drums with water.

To take a boiler out of service for an extended period of time, remove the ashes from the interior, especially where they are in contact with metal parts. Ashes contain sulfur and tend to collect moisture and form acid, which is corrosive. Make sure that all connections, including the steam outlet, feedwater, and blowoff valves, are closed and holding tightly. Repair or blank off leaking valves. After washing, allow the interior to dry. Then place unslaked (quick) lime, silica gel, or other moisture absorbent in the drum in suitable open boxes or containers to draw off the trapped moisture. Notify the insurance company that the boiler is not in service, and ask the company to adjust the insurance coverage accordingly. The insurance company must again be notified when the boiler is returned to service. The interior should be inspected every few months, and the moisture absorbent should be replaced when required after its effectiveness has been reduced by the absorption of moisture.

An alternative method of storing a boiler in readiness for almost immediate use is to condition the water as recommended by the boiler manufacturer, allowing the water to remain at normal level, and provide a "blanket" of nitrogen gas above the water. The nitrogen is supplied from pressure tanks through a regulating valve that maintains pressure in the boiler slightly above the atmospheric level. This prevents air from leaking into the boiler and contaminating the water with oxygen and causing corrosion. This method is suitable for use with a nondrainable type of superheater, which should not be flooded.

The boiler should be tagged properly, and appropriate warning signs should be attached as identification that the boiler is stored under nitrogen pressure. Before anyone enters the boiler, the nitrogen must be exhausted completely; otherwise, suffocation will result. This procedure also applies to the superheater and economizer.

If a boiler is out of service temporarily, or if it must be available for standby operation, a wet storage method is used. This procedure has the boiler stored completely wet with the recommended levels of chemicals to minimize any possible corrosion. Volatile chemicals are generally used to avoid the increase of dissolved solids in the boiler water.

This wet storage method has a very real possibility of corrosion, and the boiler should be inspected periodically for any corrosion damage. Boiler water analysis also should be conducted periodically, and when necessary, chemicals should be added. Recommended procedures are provided by the boiler manufacturer.

Boilers prepared for storage can be readily returned to service by restoring the water level, starting the fire, and venting in the usual manner. In outdoor units where a freeze problem exists, it is necessary to drain all drainable surfaces. It will be necessary to heat pendant superheaters and reheaters (e.g., external electric heating coil) to alleviate the freezing problem. The unit should be refilled with the recommended treated water when the time period for possible freezing has passed.

7.6 Maintenance

It has been stated that when a boiler is clean and tight, it is properly maintained. At first this may appear simple, but "clean and tight" applies to the entire field of boiler maintenance. *Clean* applies to both the interior and the exterior of the tubes, shell, and drums, as well as to the walls, baffles, and combustion chamber. *Tight* refers to the entire pressure section, setting, baffles, etc. When the heating surface of a boiler is free from scale on the water side and from soot and ash deposits on the gas side, it will readily absorb heat. When the boiler is free from steam and water leaks, air leaks, air leakage into the setting, and gases leaking through baffles, it is in excellent condition. However, in the overall operation, the combustion equipment, accessories, and auxiliaries also must be considered.

Boilers are constructed of different materials to withstand the conditions encountered in service. Any refractory in a furnace has little tensile strength, but it can withstand high temperatures and resist the penetrating action of the ash. On the other hand, steel used in pressure parts of a boiler has high tensile strength. However, care must be exercised to ensure that its temperature limits are not exceeded. The designer must select materials for the various parts of the boiler that are suited to the specific requirements. If selection of materials is correct and the unit is maintained and operated in accordance with recognized good practice, the service will be satisfactory. When faulty operation, excessive temperatures, or other abnormal conditions cause the safe limits of the material to be exceeded, failures will occur rapidly, and the equivalent of years of normal deterioration may take place in a short time.

By conducting a program of routine maintenance on a boiler and other power plant equipment, the production and life of the equipment are optimized. At one time, maintenance focused on the repair of damaged equipment; however, this operating philosophy has evolved into a sophisticated set of programs that include assessing the condition of the equipment, techniques for predicting the life of the equipment, developing corrective measures, preventing future problems, and evaluating the overall operation. It is the goal of a maintenance program to optimize production, availability, and safety, while at the same time to minimize operating costs and any effect on the environment.

Rapid changes in temperature cause unequal expansion, which may result in steam and water leaks in the pressure section of a boiler and air and gas leaks in the setting and baffle. It has been found that high frequency and rapidity of taking boilers out of service and putting them into service increase maintenance more than long service hours and large quantities of steam generated. In order to reduce outages and maintenance, always allow time for the temperature to change slowly and uniformly.

Unequal distribution of combustion in the furnace, sometimes referred to as *flame impingement*, can result in several difficulties. Although the average combustion rate may not exceed the manufacturer's specification, localized portions of the furnace can be overloaded. This will cause high temperatures, slagging, and rapid failure of the furnace. Severe cases of flame impingement, when accompanied by poor water circulation in the boiler or by scale deposits, will result in burned-out water tubes. The exterior surfaces of the tubes have been found to corrode until they become thin enough to burst. This type of failure has occurred when furnaces were operated at high temperature and with low or zero percent excess air. Flame impingement is prevented by correct operation and maintenance of the combustion equipment.

When scale deposits on the tubes of boilers or on the shell of a fire-tube boiler, the water cannot remove the heat, and the metal may reach a temperature high enough to reduce its tensile strength. This may cause tube failure, a bag, or possibly a boiler explosion depending on conditions. Overheating creates maintenance costs. The best method of avoiding these difficulties is to condition the boiler water to prevent the formation of scale (see Chap. 11).

When a lack of equipment or careless operation has permitted scale to form, it must be removed or loss of efficiency and boiler outages will result. Boilers should be scheduled out of service for cleaning before they are forced out by tube failure. There are two methods of removing scale from the heating surface of a boiler: mechanical, consisting of passing a power-driven cutter or a knocker through the tubes, a method that is very limited, and chemical, which uses materials that will partly or totally dissolve the scale, thus removing it from the surfaces.

Both of these methods are very costly in terms of the actual maintenance costs as well as the loss of revenues that results from the outage. This latter effect has a far greater financial impact than the maintenance costs. Therefore, in order to ensure safe and dependable operation, the boiler water must be maintained at a high quality, which requires periodic testing of the water and the addition of recommended chemicals when necessary.

In addition to checking the water chemistry, some modern boilers can determine internal water or steam cleanliness by the use of a chordal-type thermocouple that is installed in a tube and measures the temperature gradient through the tube wall. Thus, by measuring the differential temperature in the tube, the extent of internal deposits can be determined, and this can identify the need for any chemical cleaning of the boiler.

Mechanical cleaners consist of small motors driven by steam, air, or water. The motor is small enough to pass through the tubes that it is to clean. A hose attached to one end of the motor serves the dual purpose of supplying the steam, air, or water and of providing a means by which the operator "feeds" the unit through the tube and withdraws it when the cleaning is complete. Figure 7.3a shows a tube cleaner fitted with a *knocker head* for use in fire-tube boilers. The rotation of the cleaner motor causes the head to vibrate in the tube with sufficient force to crack hard scale from its exterior. Such a knocker head is sometimes effective in removing very hard scale from water-tube boilers. The *cutter head* shown in Fig. 7.3b has been used in cleaning the scale from water-tube boilers, but it is limited to tube lengths. Heads of this type have a number of cutters made of very hard tool steel. The head is rotated at high speed, and the resulting force presses

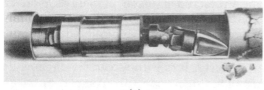

(a)

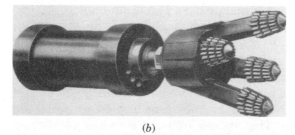

(b)

Figure 7.3 Tube cleaners. (a) Vibrating-head cleaner for fire-tube boiler. (b) Cutter-head-type cleaner for water-tube boiler. (*Liberty Tool and Engineering Corp.*)

the cutter against the inner tube surface, crushing and cutting the scale.

When mechanical tube cleaners are being used to remove scale from tubes, several precautions must be taken to obtain best results. The tube-cleaner motor must be of the correct size to fit the tube. Special arrangements of motors and universal joints are available for cleaning curved tubes. Although steam and water are sometimes used, compressed air is most satisfactory for operating these motors. A lubricator is attached to the inlet hose to supply oil for lubricating the motor. A stream of water should be introduced into the tube during the cleaning operation to cool the motor and cutter head and to wash away the scale as it is removed from the tubes. The operator feeds the cleaner into the tubes as the scale is removed. The rate of feed depends on the speed of scale removal and must be determined by trial. Experienced operators can tell by the sound of the motor when the cutter has removed the scale and is in contact with the tubes. Inspect each tube by shining a light through it and observing the interior. Never allow the cleaner to operate in one place for even a short time, or the tube will be damaged. In bent-tube boilers, the operator remains in the upper drum while cleaning the tubes. In a straight-tube boiler, the handhole covers must be removed to gain access to the tubes. The lower drum, lower header, and blowoff line must be cleaned thoroughly to remove the scale before the boiler is returned to service. As one can see, this mechanical procedure is very labor intensive and is not the preferred method in today's modern power plants.

Chemical cleaning procedures have been applied successfully to the interior of boilers and heat exchangers. Mechanical cleaning is tedious and time-consuming, but even large units can be chemically cleaned in a matter of hours. This decrease in outage time is an important factor.

Chemical cleaning is done satisfactorily and safely by specialists who furnish a complete service. Samples of scales are obtained from the tubes by use of a mechanical cleaner or other suitable means. These samples are analyzed, and after the type of scale has been identified, the correct cleaning solution is specified. These solutions consist of acids with added materials known as *accelerators* to ensure attack on scale and other materials known as *inhibitors* to lessen attack on the boiler metal. That is, the solution is made selective so that it will dissolve the particular scale without corroding the metal surface. The selection of the cleaning solution involves a knowledge not only of the scale but also of the material from which the boiler or heat exchanger is constructed. Classifications for chemical cleaning include the type and strength of solvent solution to be used, the amount and type of accelerators and inhibitors to be used, the temperature at which the solvent is to be applied, the time the solution is to remain in contact with the material, and the method of application. Safety precautions involved are prevention of the solvent and neutralizing solution from coming into contact with the personnel and removal of possible danger due to the formation of poisonous or explosive gases.

To complete the process of chemical cleaning of a boiler, the following general steps are required:

1. The internal heating surfaces are washed with a solvent to dissolve or disintegrate the deposits.

2. Clean water is used to flush loose deposits and any solvent that adheres to the surface.

3. The boiler is treated to neutralize the heating surfaces by producing a very thin protective film on freshly cleaned heating surfaces.

4. The unit is flushed with the clean water to remove any remaining loose deposits.

There are two methods for chemically cleaning that are used more frequently:

1. Continuous circulation of the solvent

2. Filling the boiler with solvent, allowing it to soak, and then flushing the boiler for cleanliness.

Special equipment in the form of tank trucks, pumps, and heat exchangers is required for effective application. The unit to be cleaned is filled with the solvent in the least possible time; then the solvent is circulated through the unit and heat exchanger to maintain the necessary temperature. The solution is analyzed during the operation to check on the progress in cleaning. When the analysis shows that the unit is clean, the solvent is removed and a neutralizing solution introduced. After the unit has been drained and flushed with water, it is ready for service.

Acid cleaning produces good results. Not only is the usual type of scale readily removed, but even difficult forms of silica and plated copper have been removed successfully. Deposit of silica compounds in high-pressure boilers has been very difficult to remove with mechanical cleaners, discounting the labor-intensiveness of this operation. Copper dissolved from the tubes of feedwater heaters has been deposited in boilers, causing scale and corrosion. Unlike mechanical cleaning, this acid cleaning removes the scale from the headers, tubes, and drums. Chemical cleaning has gone a long way to reduce the time and drudgery involved in scale removal and is the primary cleaning method today.

Boilers are designed and built for a maximum output, and when this is exceeded, maintenance is increased. High rates of steam generation may disrupt the circulation in the boiler. When this occurs, the heat will not be carried away, and the metal may become overheated. The high rate of steam flow may cause moisture to be carried with the steam that leaves the boiler drum. High gas temperatures may burn or distort baffles and flues. Slag may form in the furnace or on the tubes, limiting combustion and obstructing the flow of gases. Higher gas flows may increase erosion. Though a high capacity may be obtained for a short time, it may be necessary to take the boiler out of service because of the formation of slag or the failure of vital parts of the combustion equipment. Equipment should be operated within design limitations, and excessive capacities should be avoided.

The correct amount and distribution of combustion air are important considerations in reducing maintenance (see Chaps. 4 and 5). Too much combustion air results in serious loss of heat in the stack gases, while insufficient air causes excessive furnace temperatures. This may cause slagging and early failure of such areas as the furnace walls, arches, and grates.

Although sootblowers are valuable in removing soot, ashes, and slag from the gas side of the tubes, they also can be the cause of boiler outages and maintenance. When sootblowers are adjusted incorrectly, steam from the nozzles will erode the boiler tubes and cause failure. To prevent possible damage, the sootblower elements should be adjusted so that the nozzles do not blow directly against the tubes and baffles.

Slag on the boiler tubes restricts the flow of gases and reduces the amount of heat that can be absorbed. This tends to reduce the maximum output of the boiler and cause high furnace temperatures. The formation of slag may be caused by incorrect operation of the combustion equipment or by the fuel being unsuited for the furnace, such as higher-ash coal than the design coal. The conventional sootblowers are not effective in removing slag from the tubes, and specially designed elements are installed in large boilers to remove this slag. A pipe attached to a compressed-air hose and operated through the cleanout doors can be effective in removing the slag from the tubes. This hand-slagging operation should be conducted frequently to prevent the slag from bridging over between the tubes. Water is sometimes sprayed on the tubes to assist in removing the slag, but this may damage the baffles and brickwork in some older boiler designs and is not recommended. When the slag is deposited through the entire first pass and bridged between the tubes, it is advisable to take the boiler out of service for a thorough cleaning.

Boilers depend on auxiliary equipment to supply the necessary water, fuel, and air. These auxiliaries include pumps, stokers, pulverizers, fans, etc., which have bearings that must be replaced occasionally under normal conditions and frequently if they are not lubricated. Induced-draft fan rotors and casings are worn by the abrasive action of fly ash unless they are preceded by flue gas cleaning equipment such as electrostatic precipitators or baghouses. Boiler feed pumps wear and must be overhauled at regular intervals to ensure continuous operation. Piping, valves, controls, and instruments require adjustments and replacement of parts.

Incorrect selection and application of materials by the designer and manufacturer may result in premature failure of parts and in high maintenance. When there are repeated failures that cannot be attributed to poor operation, the design or material, and in some cases both, should be changed. If an arch of a furnace wall must be replaced frequently, it would be economical to use a better grade of material or to install waterwalls as part of a retrofit to a more modern design.

Boiler plant maintenance work should be performed on an established schedule to prevent equipment from being forced out of service by failures. Scheduled maintenance work makes it possible to have other boilers available to carry the load or to perform the work when the boilers are not required. Furthermore, the necessary spare parts, tools, and maintenance personnel can be made available to accomplish the work efficiently. In the case of forced outages, on the other hand, production is lost, the necessary parts are not available, and there are probably insufficient maintenance personnel on the job. Large boiler units are now used in both utility and industrial plants, and since their failure causes loss of a large percentage, and in some cases all, of the plant capacity, it is essential that forced outages be prevented.

The question arises as to how forced outages can be prevented or reduced to a minimum. Details that appear unimportant often cause or contribute to major difficulties. Stoppage of the chemical feed line to the boiler drum is not important in itself, but if the boiler is operated for a period without water treatment, the resulting scale deposits may cause tube failures. A defective low-water alarm can result in a failure to maintain water level and a boiler explosion. Failure of cooling water or oil on the induced-draft fan bearings may result in bearing failure and an outage of the complete boiler unit.

There is nothing that adds more to the impression that a plant is well maintained than cleanliness. Good housekeeping is a part of preventive maintenance. Maintaining a clean plant requires teamwork on the part of management, maintenance, operation, and cleanup personnel. The maintenance personnel must realize that the repair job is not complete until the old and surplus material has been removed. It is desirable to assign specific equipment to individual operators and make them responsible for keeping it clean. In this way the individual can take pride in doing the work and expect to be reprimanded if neglectful. These assignments must be given careful consideration to see that the work is equally distributed among those responsible. The work must be coordinated to prevent one cleaning job from depositing dirt on equipment already cleaned. In some places the dust is blown out of motors with an air hose; obviously, it would be a waste of time to wipe off machinery in the vicinity prior to cleaning motors by this method. With the use of floor grating in power plants, any air cleaning should be conducted from the top of the unit downward. Even this method does not solve the problem of high-pressure air removing dirt, making it airborne, and returning it to its original area.

Management should see that all help possible is given to keeping the plant clean. It is difficult for operators to keep up interest in cleaning when improperly designed or maintained equipment continually allows ash, coal, oil, or dust to be discharged. Often the design does not take into account accessibility for cleaning. Adequate equipment, tools, and supplies should be made available for this work. Vacuum cleaners are useful, and in a pulverized-coal plant they are often classed as a safety necessity. Cabinets for tools, supplies, and lubricants improve the appearance of a plant and facilitate the work.

Establish a lubrication schedule and preventive-maintenance program. Although some operators depend on memory to determine when an inspection should be made, bearings lubricated, etc., and although in some cases this is satisfactory, in large, rapidly expanding plants, memory is generally inadequate. Sooner or later something is neglected, and an outage results. Base the maintenance and lubrication schedule on manufacturers' recommendations and plant operating experience. Do not depend on memory. Maintenance should be incorporated into the plant's operating plan.

7.7 Boiler Internal Inspection

Boilers must have at least one internal inspection per year by a qualified representative of the company carrying the insurance. During the year, the insurance inspector will examine the boilers one or more times depending on conditions. At the time of inspection, someone representing the owner and operator should be present to confer with the official inspector. During these inspections, specific notes should be made of scale deposits, corrosions, erosions, cracks, leaks, and other irregularities. Care of a boiler cannot begin and end with a yearly inspection. The operating personnel must inspect the boiler continually and be on the alert for possible conditions that might lead to trouble. External examinations are made primarily to check on care and maintenance. An internal inspection should include examination of previous repairs, of the suitability of the unit for the operating pressure, and of the condition of the components of the boiler; a check for possible deterioration; a listing of necessary repairs; and a decision as to the advisability of continuing the boiler in service.

When it becomes necessary to determine the possible remaining safe and useful life of a boiler, nondestructive tests can be used to an advantage. X-ray and ultrasonic examinations are useful as well as other types of nondestructive tests that can be used to determine the condition of existing boilers.

The most common nondestructive tests used in boiler inspections include the following:

1. *Magnetic particle testing (MT)* is used to detect surface and near-surface flaws by observing any discontinuity that shows in the pattern of iron particles that are applied to the test piece. Because a magnetic field is used on the test piece, these tests are applicable only to ferrous materials. The shape of the test piece also determines whether it can be MT tested.

2. *Liquid penetrant testing (PT)* also detects surface cracking in a component, and it's not dependent on the magnetic property of the material or its shape. The PT test detects surface flaws by capillary action of the liquid dye penetrant, and it is only effective where the discontinuity is open to the component surface.

3. *Ultrasonic testing (UT)* is used extensively for measurement of the wall thickness of tubes. The loss of material due to corrosion and erosion is a primary cause of failure of pressure parts. In many areas of the boiler where corrosion and erosion are expected, a baseline UT test is conducted. In subsequent tests, these test results are compared with the baseline information, and the life of the component can be projected. Therefore, at planned outages, tubes having a high rate of wear can be replaced, which prevents an unscheduled outage due to tube failure.

4. *Radiography or x-ray testing (RT)* is the most important nondestructive test during the field erection of a boiler. It is also used in assessing the condition of pressure parts such as piping. The major disadvantage of radiography is excessive exposure to radioactive rays and the potential harmful effects to personnel.

When a boiler is taken out of service for inspection, it should be allowed to cool down slowly. The sootblowers should be operated to clean the heating surfaces before the temperature has dropped low enough to allow the formation of moisture on the tubes. When the boiler is cool enough, open the blowoff and drain. Remove the manhole covers and some of the handhole caps. Wash the inside of the drum with high-pressure water. Before entering a boiler, close the blowoff, the main stop valve, and feedwater valves, and either lock them in the closed position or tag them to ensure against the possibility of their being opened while maintenance and inspection personnel are in the boiler. Check extension cords to see that they are in good condition. The inspector should have available a list of the dimensions and other important data, as well as records of the history and previous inspections and repairs to the boiler.

Remove the accumulation of ash, slag, and soot from the furnace and passes of the boiler. If the boiler has refractory lining, care must be used in removing hard slag from furnace walls to prevent damage to the lining. The slag adheres tightly to the lining, and it is better to allow part of the slag to remain than to remove a portion of the refractory. Remove the soot and ashes from the horizontal baffle, hoppers, etc. Do not create dust by careless handling or by blowing ashes, fine coal, and soot with compressed air. The careful use of stack or fan draft while cleaning will reduce the amount of ashes discharged into the boiler room. Do not step into piles of ashes while cleaning, since they may be hot enough to cause serious burns.

An inspection of the inside of a fire-tube boiler includes an examination of the shell both above and below the tubes. Note the feedwater inlet and steam connections, including the dry pipe, for possible stoppage. Examine the shell and tube surfaces for scale deposits, bags, blisters, oil, and corrosion. Check for possible pitting or wasting away of the metal. When a fusible plug is used, note its condition and replace if corroded. In any case, the fusible plug must be replaced once a year. Arrange a light so that it is possible to look down between the tubes, and check for possible obstructions that might interfere with the water circulation. Interior inspections of vertical and other small fire-tube boilers must be made by observing conditions through the inspection holes.

For externally fired units, inspect the exterior of a fire-tube boiler shell from inside the furnace. Cracks in the shell plates are dangerous and should be investigated and repaired. While in the furnace, check the setting for cracks, and note the condition of the grates and bridge wall. If the boiler is of the flush-front type, inspect the front baffle for possible leaks. See that the blowoff pipe is adequately protected from hot gases. Open the doors at both ends of the boiler shell and inspect the tube ends for leaks and corrosion and the tube sheets for bulging tendency. Note the exterior appearance of the setting, and check for air leaks where the setting joins the shell.

To inspect a water-tube boiler, it is necessary to enter each of the drums. An inspection of the steam drum should include an examination of feedwater inlet, chemical feed, continuous blowdown, and water-column connections for possible stoppage; drum and tubes for scale and corrosion; tubes for thinning or weakening; the dry pipe or steam separator for corrosion or scale; and drum plates and tubes for signs of caustic embrittlement. In the lower drum, note the nature of the deposits and the condition of the tubes; check the blowoff connections for scale deposits and possible stoppage. In a straight-tube boiler, it is necessary to remove the handhole caps to inspect the interior of the tubes. Sometimes a mechanical cleaner is run through the tubes to determine the extent and nature of the scale.

Enter the furnace and various passes of the boiler to inspect the setting, baffle, and exterior of the tubes. Examine the tubes for possible erosion from flame impingement and the action of the sootblowers. Have the sootblowers rotated, and observe the angle of travel to see that they do not blow against a wall or baffle and that the nozzles do not blow directly on the tubes. Have the necessary adjustments made to the blowing angle and alignment with respect to tubes. Examine all baffles to see that the gases are not short-circuiting and thereby failing to come into contact with all the heating surface. Open the stack or fan inlet damper to see that it opens wide and closes tightly.

Observe the condition of the exterior of the boiler, piping, and accessories. Check the arrangement of valves, and inspect the following piping systems: main steam line, nonreturn valve and drains, feedwater piping, steam-gauge connections, sootblower supply lines, main and continuous blowoff, safety valves, discharge and drains, and chemical supply to the boiler drum. Inspect the induced-draft fan for any excessive erosion of fly ash. Operate the combustion-control mechanism that moves the damper and coal-feed device. Check the stoker for burned, plugged, or worn grates and wear or damage to the fuel-feed mechanism.

Check the sootblowers when the boiler is being inspected. Look for missing or defective bearings, bent or broken elements, and missing nozzles. Operate each element and compare the *blowing angle* with the specifications. Make sure that the blowing angle is adequate for cleaning but avoids direct "blow" on the tubes, which can result in erosion. Make the necessary repairs and adjustments.

After the necessary work has been completed, the boiler is ready to be closed and filled with water. Make sure that all maintenance equipment has been removed and that all personnel are out of the boiler. Always put a hydrostatic test on a boiler after it has been opened for inspection or repairs. Unless extensive repairs have been made to the drum or shell, it is not necessary to test at $1\frac{1}{2}$ times the working pressure. The pressure developed by the boiler feed pumps is sufficient to check for leaking tubes and handhole and manhole gaskets. If there are no leaks in the pressure section of the boiler, it is ready for service.

7.8   Evaluating the Condition of a Boiler

The operating staff must prepare proper records of all maintenance to ensure that any problem components of the boiler are known. Examination methods such as nondestructive testing are then used to evaluate the remaining life of the boiler component in question.

The condition of the steam drum is a critical component to evaluate. There are two types of steam drums:

1. The all-welded design used predominantly in utility boilers, where the operating pressure exceeds 1800 psi

2. Drums with rolled tubes, such as the two-drum industrial-type boiler

Since the steam drum operates at saturation temperature, the steam/water temperature is less than 700°F, and, therefore, the drum is made of carbon steel and is not subject to significant creep. *Creep* is the slow deformation of continuously stressed metal over time, which could lead eventually to a fracture.

Damage to a drum is primarily due to internal metal loss. This can be caused by corrosion and oxidation, which can occur during extended outages if proper precautions are not taken. Damage also can occur from mechanical and thermal stresses on the drum, which concentrate at nozzle and attachment welds. These stresses occur often in boilers that are cycled frequently in an on/off mode of operation. The feedwater penetration area has the greatest thermal differential because incoming feedwater can be several hundred degrees below the drum temperature.

Steam drums with rolled tubes (e.g., a two-drum boiler) have problems with tube seat leakage, where there is a slight seeping of water through the rolled joint. Caustic embrittlement can result if the leak is not stopped.

In lower drums of industrial-type boilers, large thermal differential or mechanical stresses are not present; however, rolled-tube seat leakage can occur, with similar problems resulting as with the steam drum.

Superheater and reheater tubes are affected by both erosion and corrosion. In addition, the high temperature results in increased stress on the tubes. These factors lead to tube cracks and eventual leaks.

Water-cooled tubes such as the furnace walls, boiler bank, and economizer operate at or below saturation temperature and therefore are not subject to significant creep. Proper water chemistry is important in maintaining tube life, and if necessary, water-side deposits can be cleaned by chemicals when required. Erosion and corrosion are the primary problems of the tubes, assuming that good water chemistry is practiced. If not, deposits form on the inside of the tubes, and heat transfer is affected, which can result in high metal temperatures and eventual failure of the tube.

These and other components and auxiliaries of the boiler must be examined and maintained properly during planned outages. This is proper operation, which will lead to high plant availability.

7.9   Making Repairs

After the condition of a boiler is evaluated by monitoring its performance and performing both visual examinations and nondestructive testing, a decision must be made by operations on the amount of repair that is required to have the boiler returned to dependable, safe, and efficient operation. Although the designer and supplier of the equipment can provide recommendations, it is the responsibility of the operating staff to decide on the repairs to be made and the schedule for performing them.

All repairs to pressure parts must be made in accordance with the procedures established by the *ASME Boiler and Pressure Vessel Code.* Welding repairs to a boiler must be conducted with the approval of an authorized inspector, and these are usually provided by insurance companies.

There are three general classifications on the types of welds:

1. Pressure-part welding
2. Pressure-part to non-pressure-part welding
3. Non-pressure-part welding

For any pressure-part welding including those to nonpressure parts, the requirements of applicable codes must be followed. These codes establish the minimum requirements that must be followed, including

- Material requirements
- Welder qualification
- Welding procedures
- Heat treatments required both before and after welding
- Nondestructive testing required

All welds for nonpressure parts should meet the requirements of the American Welding Society. These types of welds can include flues, ducts, hoppers, casing, and structural steel.

The replacement of tube sections requires a minimum of 12 in of tube to be replaced. The tube can be removed by a saw or cutoff tool. However, generally because of limited access, the tube section is removed with an acetylene torch. The tube ends must be prepared properly for welding of the new tube section. The use of backing rings is a common procedure in the welding process. Welding techniques and procedures are generally furnished by the boiler manufacturer, and the welding is performed by qualified welders who have passed the required code testing. The use of postweld heat treatment is required to relieve residual stresses that are created in the welding process.

Field welding generally cannot be applied to boiler repair unless the welded section is stress relieved according to code requirements and nondestructive tested (NDT). Welding without stress relieving and NDT testing is restricted to use for items such as a seal where it is not required to strengthen the pressure parts of the boiler.

Some types of boilers require baffles to effectively direct the gases for maximum heat transfer to the tubes (see Figs. 2.10, 2.11, 2.18, 2.27, and 2.36). Boiler baffles have been made of refractory, alloy, or standard steel, depending on the temperature of gases to which they are to be exposed. Refractory baffles are used near the furnace when high temperatures are encountered. They are constructed of special refractory shapes that are made to fit the exact location. Some baffles lie on a row of tubes, and others run at an angle across the tubes. To make repairs, it is necessary to obtain the refractory shapes for the specific application. It requires skill and patience to insert these tile shapes in baffles that are at an angle with the tubes (when the tubes pass through the baffle).

One-piece (monolithic) baffles are shown in a straight-tube boiler in Fig. 7.4. These baffles are made from moldable refractory material. Forms are constructed to retain the refractory and produce baffles that will direct the flow of gases.

When the tubes pass through the baffles, forms can be made by passing wooden strips down both sides of the space where the baffle is to be installed. These strips retain the refractory material as it is worked down between the tubes. The material air sets, and the wooden strips burn out when the boiler is placed in service. In some cases it is advisable to wrap heavy paper around the tubes before applying the baffle material. This burns out, leaving a space around each tube to allow for expansion and tube replacement. Some baffle material shrinks enough to provide the necessary clearance around the tubes.

Figure 7.4 Monolithic baffle installed in a longitudinal-drum water-tube boiler. (*Plibrico Co.*)

Baffles are also made of heat-resistant metals. One method is to drill a single metal plate to receive the tubes. These single-plate baffles are difficult to repair and cause interference when it is necessary to replace tubes. It is therefore preferable to use baffles composed of metal strips shaped to fit the tubes. These strips are inserted between the tubes and bolted together with tie bars to form a complete baffle. These baffles can be maintained by replacing individual strips. Sections can be taken out to facilitate tube replacement and then reassembled.

Many older boilers and furnaces that remain in operation today were designed with refractory walls (see Figs. 2.5, 2.6, 2.10, and 2.11). These refractory walls account for an appreciable portion of boiler maintenance, and it is therefore important that an effort be made to reduce this expenditure.

Solid refractory walls were lined with first-quality firebrick. In many instances this lining could be patched or replaced without rebuilding the entire wall if the repairs were made before the outer section had become too badly damaged. The lining had to be "keyed" to the outer wall, and expansion joints had to be provided.

Air-cooled refractory walls are constructed to provide an air space between the outer wall and the refractory lining. This lining may be a separate wall constructed of first-quality firebrick, moldable refractory material, or tile supported by cast-iron hangers. The circulation of air through the space between the two walls carries away some of the heat, resulting in a reduction in wall temperature and a possible decrease in maintenance. The lining must be adequately tied to the outer wall to prevent expansion and contraction from causing it to bulge and fall into the furnace. The cast iron hangers provide adequate support for the tile lining. However, the initial cost of these walls is high, many tile forms must be stocked for replacements, and once the lining has failed, the cast-iron hangers are quickly destroyed by the heat of the furnace.

Careful consideration should be given to the quality of the refractory material. Operating temperatures, ash characteristics, slagging, and flame impingement are all factors to be considered when selecting refractory lining for furnaces. If the fusion temperature of lining is not high enough for the operating temperatures, the inner surface will become soft and in extreme cases even melt and flow, resulting in rapid deterioration. The fusion of ash on the surface of the wall causes the inner face to spall, i.e., to crack and fall off. The removal of clinkers either with the unit operating or with it out of service for maintenance will invariably crack off some of the furnace lining. Flame impingement results in a localized high temperature and subsequent failure of the lining.

Furnace walls (either solid or air-cooled) and arches may be constructed or rebuilt by the use of moldable refractory material, which is furnished in a moist condition for ramming into place. Keys made of heat-resistant metal, refractory tile, or a combination of both materials are used to hold the lining in place. The refractory material is rammed into place by an air-operated hammer. Expansion joints must be provided at intervals to prevent excessive forces from developing.

Figure 7.5 shows the steel angles in place for supporting a moldable refractory arch. The hangers are attached to these angles and embedded in the refractory arch. Wooden forms are used to support the arch during construction. The refractory material is rammed into place around the inserts. Adequate expansion joints must be provided.

Where service is especially severe, causing outage and costly repairs, it is advantageous to use special high-quality, high-cost brick or tile. These materials have been used in the walls just above underfeed stokers where there is a combination of high temperatures and contact with ash and clinkers. They also have been used in spreader-stoker front-wall arches and in the piers between the feeders.

Refractory furnace walls caused major maintenance problems and costs, and, therefore, modern boilers are designed with water-cooled furnace walls, primarily of membrane wall design. Waterwalls are used to reduce maintenance and, unless misapplied or abused, seldom require repairs. However, corrosion and erosion reduce tube metal thickness, and repairs and replacements are required. The maintenance procedure depends on the type of waterwall construction. If the tubes themselves fail, the repair is similar to that for replacing any other boiler tube. The cast-iron blocks used on some waterwalls may have to be replaced after long, continuous service. In this case

Figure 7.5 Supporting steel for a plastic refractory arch in a boiler furnace. (*Plibrico Co.*)

replacements are made by first cleaning the tube surface, then applying a heat-conducting cement or bond, and finally bolting the new block securely in place. In many cases, a weld overlay on the furnace wall panels is made where the thickness of the tubes has been reduced from either corrosion or erosion or, in some cases, from a combination of both.

Make repairs and replacements to boiler equipment often enough to prevent small jobs from developing into major projects. Repair the furnace wall before the steelwork is damaged. Do not remove and replace serviceable parts. Some thinning of the furnace wall and burning of stoker castings are expected and permissible. Careful inspection and maintenance records should be kept for the planning of future repairs.

Repairs to steam-generating equipment must meet exacting requirements. All materials and construction used in repair work must meet code requirements. No repairs by welding are to be made to a boiler without the approval of an authorized inspector. Such welding repairs must be accompanied by the necessary nondestructive testing and hydrostatic tests to ensure that weld integrity has been met.

Questions and Problems

7.1 To what pressure is a hydrostatic test pressure applied? What test procedures should be followed?

7.2 How can a boiler with oil and grease in it be cleaned?

7.3 Identify the procedures that are necessary during startup prior to the firing of a boiler.

7.4 In the burning of a solid fuel, what significant effect on the boiler operation does ash have? How is this corrected or managed?

7.5 What are the benefits that result from proper water treatment and chemistry control?

7.6 What is scale, and how can it affect the operation of a boiler? What is corrosion?

7.7 What is meant by priming and foaming? What is the difference between them?

7.8 Describe external and internal water-treatment systems, and identify when they are used.

7.9   How is free oxygen removed from boiler feedwater? Why is its removal important?

7.10 Impurities are often contained in the boiler feedwater. How are these effectively removed?

7.11 Leaks in boiler tubes occur often. Name several ways in which these can be detected. Provide some of the major causes of these leaks and methods for prevention.

7.12 Describe how ultrasonic testing (UT) of tubes can be an effective method for evaluating the condition of tubes.

7.13 What is boiler efficiency, and why is it important? What is meant by a boiler being 80 percent efficient? For the same steam capacity, what occurs if the preceding efficiency declines to 75 percent?

7.14 Why does the addition of economizers and air heaters in the overall boiler design result in higher boiler efficiencies?

7.15 Name several situations that cause high flue gas temperature. What are the major changes in boiler operation that can change the boiler efficiency?

7.16 List the benefits that result from a boiler that is properly insulated.

7.17 As compared with a conventional drum-type boiler, what are some of the more important characteristics of a circulating fluidized bed (CFB) boiler?

7.18 In order to properly operate a CFB boiler, what additional instrumentation does the operator require?

7.19 How is the combustion air of a CFB boiler divided? Why is this split so critical?

7.20 In a bubbling fluid bed (BFB) boiler, what operating procedures are different from a conventional boiler? What additional instrumentation is required?

7.21 What should an operator do if the water drops below the minimum required?

7.22 What procedures should be followed in the case of tube failures?

7.23 If an induced-draft fan fails, describe the system that safely shuts the boiler down.

7.24 For a pulverizer, what precautions must be taken for its shutdown?

7.25 If a safety valve is leaking, can the spring be tightened? Why or why not? What are the proper procedures for repairing a leaking safety valve?

7.26 Describe a procedure for taking a boiler out of service.

7.27 What procedures should be followed for laying up a boiler?

7.28 Develop a tagging procedure for a boiler that is out of service. Identify whether this should be part of the plant's operating procedures.

7.29 For a boiler that is temporarily out of service, describe the procedures for wet storage.

7.30 The periodic condition assessment of equipment and a good maintenance program have proven to optimize operations. Why?

7.31 Why is chemical cleaning preferred over mechanical cleaning of the boiler internals?

7.32 What are the two most frequently used chemical cleaning methods for a boiler?

7.33 What are some of the major consequences of operating a boiler in excess of its design capacity?

7.34 What potential problem can result from improper sootblower operation?

7.35 Why are forced outages so important to be avoided if possible? What preventive measures can be taken to minimize this risk?

7.36 Why is plant cleanliness so important? Why should a preventive maintenance program be part of the plant's operating plan?

7.37 How should you get a boiler ready for inspection?

7.38 How should an inspection of a boiler be made? What are the most important things to look for?

7.39 Describe the most common nondestructive tests that are used in boiler inspections, boiler construction, and boiler repair.

7.40 In the inspection of a water-tube boiler, what are the major areas that require investigation?

7.41 How would you inspect a sootblower installation to make sure that it was in good operating condition?

7.42 Describe the two types of steam drums and the common problems that must be evaluated periodically.

7.43 What are the primary maintenance problems found in all tubes of a boiler?

7.44 Following most welding procedures on pressure parts, what must be done to relieve the stresses that occur during the welding process?

7.45 What are the reasons that refractory furnace wall designs have been replaced with designs incorporating water-cooled furnace walls?

8

Pumps

Pumps are used for many purposes and a variety of services: for general utility service, cooling water, boiler feed, and lubrication; with condensing water and sumps; as booster pumps, etc. There is a pump design best suited to each purpose and individual service.

Turbines and boilers have increased in size, requiring larger boiler feed pumps. Many of today's utility boilers operate above the supercritical pressure of 3206 psi. With increased pump reliability, many generating stations use fewer pumps, perhaps a single pump for each boiler-turbine unit. Steam-turbine-driven pumps are often used as compared with motor-driven pumps. The steam-turbine drive's advantages are (1) decreased electric power consumption, (2) ideal speed operation with the elimination of hydraulic couplings, and (3) exhaust steam that can be used to improve the station heat balance.

Although there are a variety of pumps found in a power plant, the basic steam power plant cycle includes a combination of a condensing and a feedwater heating cycle, and this requires a minimum of three pumps:

1. A *condensate pump* that transfers the condensate from the condenser hot well into the deaerator
2. A *boiler feed pump* that transfers feedwater from the feedwater heaters to the economizer or the boiler steam drum
3. A *circulating water pump* that provides cooling water through the condenser to condense the exhaust steam from the turbine

Pumps that are found in power plants come in a variety of sizes and designs that depend on the fluid and the service. However, pumps are divided into two major categories: dynamic and displacement pumps.

1. *Dynamic pumps* are those in which energy is continuously added to increase fluid velocities. These pumps include centrifugal pumps.

2. *Displacement pumps* are those in which energy is added periodically by the application of force. These pumps include reciprocating and rotary-type pumps.

Pumps also can be identified in four general classifications as follows:

1. *Reciprocating pumps.*   These include piston, plunger, and diaphragm pumps, and these can be of simplex or multiplex design. Power and vacuum pumps are also part of this classification.

2. *Rotary pumps.*   These include gear, screw, and vane pumps.

3. *Centrifugal pumps.*   These include radial-flow, mixed-flow, and axial-flow pumps, and the designs can be single or multiple stage.

4. *Special pumps.*   These include jet, gas-lift, and hydraulic-ram pumps.

Pumps that are used primarily in steam power plants will be described in this chapter.

8.1 Pumps

There are a great variety of pumps from which to make a selection, and each pump has its specific advantages that need to be analyzed for a specific application. The simplest pump is the injector or jet pump, which has been used on small boilers and portable units and has provided a low first cost and a simple design. Reciprocating pumps find ready acceptance, particularly in the smaller plant, where first cost is a factor. They are simple in construction, easy to repair, and reliable in operation. Rotary pumps find application in handling oil and lubricants. Centrifugal pumps are available for a variety of services and purposes having an almost unlimited range of industrial applications.

Pumps may be required to lift or raise water on the suction side. The extent to which this can be done is determined by the type of pump and the atmospheric pressure. Actually, the pressure of the atmosphere pushes the water up into the pump suction; the pump creates a vacuum by movement of the tight-fitting piston or action of the impeller into which the water (or other liquid) rushes. The height to which water can be lifted is then influenced by the atmospheric pressure and water temperature. Consideration must be given to these variables when pumping installations are being designed.

The total head developed by the pump, usually called the *total dynamic head,* is made up of the following:

1. Total static head of discharge above the level of the suction water

2. Pressure at the point of discharge

3. Pipe friction, including fittings

In addition, the total dynamic head contains the velocity head, which is usually unimportant except in large-capacity pumps. The total head that a pump can develop is a matter of pump design and is influenced by its mechanical condition and the factors enumerated.

8.2 Injectors

The injector (Fig. 8.1) is perhaps the simplest pump. It is a device designed to lift and force water into a boiler that is operating under pressure. It operates on the principle of steam expanding through a nozzle, imparting its velocity energy to a mass of water.

The injector is often called a *jet pump,* and is a pump that has no moving parts and utilizes fluids in motion under controlled conditions. The motive power is provided by a high-pressure stream of fluid that is directed through a nozzle designed to produce the highest possible velocity. The resulting jet of high-velocity fluid creates a low-pressure area in the mixing chamber, causing the suction fluid to flow into the chamber.

The boiler injector is a jet pump that uses steam as a motive fluid to entrain water, and it is also used as a boiler feedwater heater and pump. The now obsolete steam locomotive was the largest user of this type of injector. At present, it is sometimes used as a backup to a regular boiler feed pump on small installations.

The essential parts of the injector are the steam jet, the suction jet, the combining and delivery tube and ring, and the overflow and discharge tube. In operation, steam flows from the nozzle (8). The steam pressure drops as it passes through the jet but gains in velocity. As the steam passes between the steam jet (8) and the suction jet (7), a vacuum is created in the "suction" chamber. As a result, water is drawn into the suction chamber from either the main water supply, the overhead water-storage tank, or the underground water reservoir. The high-speed steam jet picks up the water as it crosses the space between (8) and (7), forcing the water along with the steam into the combining tube and finally into the delivery tube (6). Here the steam

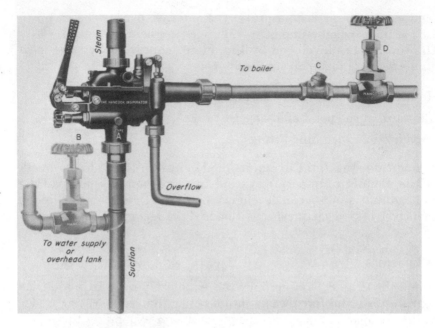

(a)

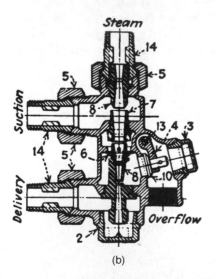

(b)

Figure 8.1 (*a*) Automatic injector installation. (*b*) Sectional view. (*Ingersoll Dresser Pump Co., no longer manufactured.*)

is condensed; the delivery tube receiving the water (and condensate) is designed to change a considerable amount of the velocity energy of the jet into pressure. The heat energy in the expanding steam not only provides sufficient energy to force the feedwater into the boiler but in addition heats the water, thus providing both a pump and a heater in one operation.

The injector is installed as shown in Fig. 8.1a. The main lever is used in regulating the inspirator; a check valve (C) is located in the delivery line ahead of the stop valve (D) to prevent the return of water from the boiler. The steam line should be of the same size as the injector connection and should be connected to the highest point on the boiler to ensure dry steam. A stop valve is provided so that the injector may be removed. The overflow should be as large as or larger than the injector connection and as straight as possible, and it must be open to the atmosphere and not piped below the surface of the water. The suction line must be tight. It should be as large as the takeoff of the injector unless the suction lift is more than 10 ft or unless the horizontal length is more than 30 ft, in which case the line should be two sizes larger than the injector takeoff. The piping is then reduced in size to match the injector size. If water enters the injector under pressure from an overhead storage tank, equip the line with a globe valve (B).

The delivery line should be at least as large as the injector connection. It should be equipped with a check valve (C) and a globe valve (D). The globe valve is always positioned wide open except when repairs to check valve or injector become necessary.

To operate, the main lever is pulled back until the resistance of the main steam valve is felt. This will lift the water. When water appears at the overflow, the lever is pulled back slowly and steadily as far as it will go. To stop, the lever is pushed all the way forward.

The maximum water temperature that an injector is considered capable of handling is 130 to 150°F. Increased lifts must be accompanied by a decrease in water temperature. When the water comes in contact with steam, the heat causes some of the impurities to drop out in the injector. This tends to scale up the nozzles so that the injector will fail to function properly.

Injectors are very inefficient pumping units. They are practical only on small boilers, and they are not entirely reliable. Since modern power plant practice favors high feedwater temperatures, and since an injector cannot handle these high water temperatures and because it is also unreliable, this method of feeding boilers has been almost completely discontinued. Injectors operate satisfactorily where load and pressure are somewhat uniform. With varying load conditions and fluctuating pressures, they become unreliable.

The injector, however, offers several advantages. It is very simple and has no moving parts to get out of order and require replacement. It is compact and occupies little room, and both the initial cost and the installation cost are low. It heats the feedwater without the aid of a heater, and thermally it is very efficient.

8.3   Duplex Pumps

The duplex pump (Fig. 8.2) has two pumps mounted side by side. The operating medium can be water, compressed air, or steam. The pump described here is steam actuated. The pump is direct acting in that the pressure of the steam acts directly on a piston to move the second fluid, in this case, water. Action is obtained through the motion of a piston or plunger reciprocating in a bored cylinder or a cylinder fitted with a liner. The action here is positive in that a definite amount of water is displaced per stroke. The quantity or volume delivered is reduced by leakage and slippage due to pump wear and the condition of the valves.

Pump dimensions are given in this manner: 3 by 2 by 3. The first figure refers to the diameter of the steam cylinder, the second to the diameter of the water cylinder, and the third to the length of the stroke. All dimensions are in inches.

Each of the two pumps of the duplex pump has a steam cylinder on one end and a water cylinder on the other. The rocker arm of one pump operates the steam valve of the opposite pump. A cross section of a duplex piston pump is shown in Fig. 8.3, and Fig. 8.4 shows a plunger pump. The steam cylinders are fitted with pistons. They are equipped with self-adjusting iron piston rings fitted to the cylinder bore. Above each steam cylinder are four ports; the two outside are

Figure 8.2 Exterior view of a duplex pump. (*American-Marsh Pumps, Inc.*)

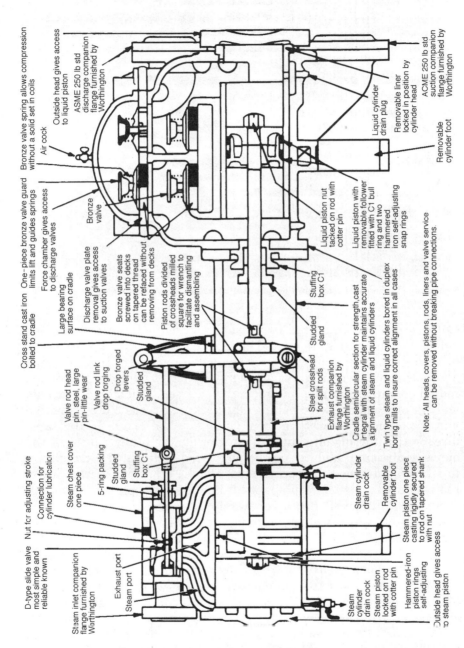

Figure 8.3 Horizontal duplex piston pump. (*Ingersoll Dresser Pump Co.*)

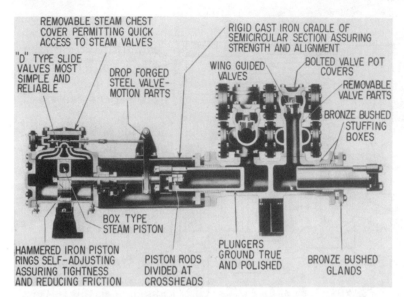

Figure 8.4 Horizontal duplex plunger pump. (*Ingersoll Dresser Pump Co.*)

called *steam ports,* and the two inside, *exhaust ports.* A D-type slide valve controls the admission and exhaust of steam.

An inside guide is provided for the valve rod where the valve rod enters the cylinder head. A packing gland is provided for the piston rod where the piston rod enters the valve chest. The gland prevents leakage of steam at these points. One rocker arm has a direct motion and the other an indirect motion. One valve rod moves in the same direction as the piston that is operating it; the other moves in the opposite direction. Midway on the piston rod is mounted a crosshead, to which is attached the rocker arm. On the other end of the piston rod is located a piston (or plunger in the case of the plunger pump) in the water cylinder. It has a renewable liner and is actuated by the steam piston, which transmits its power to the water side. The water piston is fitted with a removable follower using fibrous or metallic packing rings and moves back and forth in a bored cylinder. The ends of the cylinder are fitted with drain plugs similar to those of the steam cylinder. Above the cylinder are located two decks of valves, the lower group being the suction valves and the upper deck the discharge valves.

The seats of these valves are screwed into the decks on tapered thread and can be refaced without being removed from the decks. The lower valves are set over the suction inlet, and the discharge goes directly to the discharge pipe through the upper ones. An air cock is mounted on top of the chamber and is called a *vent.* This vent is used to remove entrapped air from the pump when it starts.

The slide valve is very simple and is operated by direct-lever connection. Motion is imparted to the valve by a rocker arm actuated by the piston rod on the opposite cylinder. After admitting steam to the other cylinder, the piston completes its stroke and waits for its own valve to be operated on by the other piston so that it may return on the next stroke. One of the valves is always open. Dead centers are not encountered.

The lost motion between the lugs on the back side of the valve and the nut does not permit the valve to move until the piston that actuates it (on the other side) has traveled some distance. The amount of lost motion permitted enables the valves to close slowly and quietly before the pump reverses the stroke. The flow of water, however, is not interrupted. While the one piston is being slowly brought to a stop, the other continues in action. This prevents fluctuations in the pressure and ensures a uniform flow. The lost motion usually allowed in the valve mechanism permits the piston to travel one-half its stroke before operating the valve.

Lost motion is provided in several ways. In one model, the valve rod has threads on which is screwed a small block. The block fits between the lugs on the back of the valve. Lost motion is the distance from the edge of the block to the lug. With this arrangement the amount of lost motion cannot be altered unless the block is replaced or the width changed. On another model the valve rod is threaded and the nuts are adjustable outside the lugs on the valve. Lost motion can then be adjusted by moving the locknuts. In a third type the lost-motion adjustment lies outside the steam chest, between the link connecting the valve rod to the rocker arm and the valve-rod headpin. The advantage of this arrangement is that the valve can be adjusted for lost motion with the pump running.

Clearance space has been defined as that volume which the steam occupies when the piston is at the end of the stroke, i.e., the space between the face of the piston and the underside of the valve. The steam occupying this space does no useful work, and hence a pump short-stroking increases clearance volume and as a result increases the steam consumption.

The duplex pumps shown in Fig. 8.5a and b are designed to reduce this clearance and increase the economy. The two large admission ports are eliminated, and a much smaller port is substituted. Steam enters this port, slowly moving the piston forward until the main body of steam strikes it. This reduction in clearance decreases the steam consumption and results in economical operation.

On some high-pressure duplex pumps a piston valve is used instead of the usual slide valve. This is a balanced valve, and its advantages are that it is light, perfectly balanced, and very simple. Valves on a duplex pump have neither lap nor lead. Since lap is necessary to cut

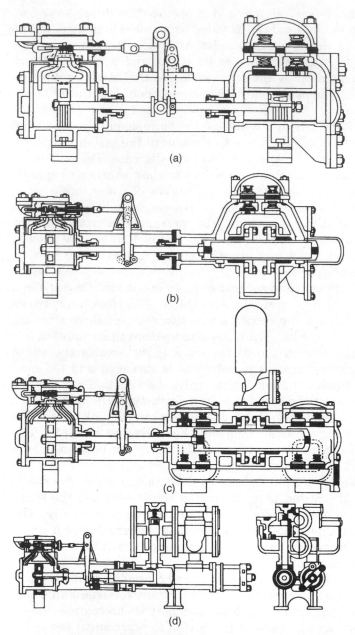

Figure 8.5 Duplex pumps. (*a*) Packed piston pattern, valve plate type. (*b*) Outside center-packed plunger pattern, valve plate type. (*c*) Outside center-packed plunger pattern, turret type. (*d*) Outside end-packed pot valve plunger pattern. (*American-Marsh Pumps, Inc.*)

off the steam before the piston reaches the end of the stroke in order to obtain expansion, it is evident that a duplex pump *must* take steam the *full length of the stroke.*

Some pumps are fitted with "cushioned valves." These are small valves fitted to duplex pumps, as in Fig. 8.6, and open a connection from the steam port to the exhaust port at each end of the cylinder. This pump is provided with a hand-operated regulating valve located between the steam and the exhaust port. This valve provides a means for releasing the excess cushioning at each end of the stroke after the piston has closed the exhaust port. The usual arrangement when the pump is being operated is to throttle or nearly close the valve at slow speed and completely open it at high speed.

To set the valves on the duplex pump shown in Fig. 8.3, proceed as follows: Remove the steam-chest cover. Push the piston forward (this can be done by placing a bar behind the crosshead on the piston rod) until the piston strikes the head end. Make a chalk mark on the piston rod at the point where the rod goes through the studded gland. Move the piston rod in the opposite direction until the piston strikes the opposite head, and again make a mark on the piston rod (be sure to make the mark at the same end of the studded gland). With a pair of dividers, locate a *mark* midway between these two points, and place *this mark* at the end of the gland. The piston should now be located exactly in midposition, and the rocker arm should be vertical. If the rocker arm is not in this position, it should be adjusted accordingly. The valve-rod headpin should now be removed from the valve rod of the opposite cylinder and the valve placed in midposition (so as just to cover the ports). The nut for adjusting the lost motion should be held exactly in the center of the space between the lugs on the back of the valve. The valve rod should now be screwed through the nut until the headpin again fits into the knuckle. The setting of one valve has now been completed, and it becomes necessary to repeat this operation for the other side. Before replacing the steam-chest cover, move one valve to uncover one of the ports; otherwise, the pump will not start. Once the pump has been started, it is impossible for it to stop on dead center.

When locknuts are used to adjust the lost motion, the same general procedure is followed. However, it is not necessary to disconnect the rod from the valve-rod head. Hold the valve in midposition, and adjust the locknuts (Fig. 8.6) so that they will be at equal distances from the valve lugs, allowing about one-half of the width of the steam port for lost motion on each side. If the valve has outside adjustment, follow the same procedure as the one just described. Because the jam nuts are on the outside of the steam chest, it is not necessary in this case to dismantle or stop the pump to adjust the valves.

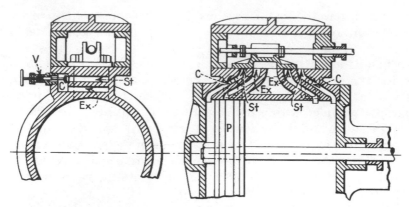

Figure 8.6 Duplex pump steam cylinder cushion valve. (*Ingersoll Dresser Pump Co.*)

Too much lost motion lengthens the stroke and may cause the piston to strike the head. *Too little lost motion* causes the pump to *short-stroke*. Short-stroking results in an increase in steam consumption and a decrease in pumping capacity.

Valves on the water end are called *suction* and *discharge*. The suction valves are for the purpose of admitting water to the cylinder and preventing a return of water to the pump suction. The discharge valves are for the purpose of discharging the water from the cylinder and preventing its return. Several different arrangements can be used to advantage:

1. The valve decks placed side by side over the cylinder (see Figs. 8.4 and 8.5*d*). This arrangement is used very frequently on high-pressure units.

2. The suction valves placed below the cylinder and the discharge above (Fig. 8.5*c*). This arrangement enables the flow to pass directly through the pump with a minimum amount of resistance because no reversing of flow is encountered.

3. The valve decks located one above the other. In this case both decks are above the cylinder (see Figs. 8.3 and 8.5*a* and *b*). The discharge-valve plate is removable, giving easy access to both suction and discharge valves. The position of the suction valves over the cylinder ensures a full cylinder of water at all times and consequently a more uniform flow than is normally produced. Pumps of this design are especially adapted to low and medium pressures. The disadvantage of this valve arrangement is that water flowing through the pump must change its direction twice.

Many different kinds of valves are used, the most common being the *flat-disk* and the *wing-disk* types. The flat-disk type is guided by a stem screwed into the valve seat (see Fig. 8.3). Here the valve seats are screwed into decks on tapered thread and made of bronze. These can be refaced without removing them from the decks. Flat-disk valves are also frequently made of rubber or composition. Wing-type valves (Fig. 8.4) are also circular in form. They are usually fitted to bronze seats, although sometimes they are made of fiber, leather, or hard rubber.

For cold (low-pressure) and warm water, the usual procedure is to use rubber valves (Fig. 8.7). For hot water and high pressure and temperature, bronze or steel should be used. (Practically all bronze and steel valves are made in the disk-poppet form.) Up to 100 psi, it is general practice to use soft rubber or fiber material; for 200 psi, hard rubber or composition; above this, steel or bronze, the bronze valves being used more and more even for low pressure. For liquids other than water, special types of valves may be necessary. A number of small valves are preferable to one or two large ones, since small valves are more positive in action, are easier to replace, are cheaper, do not warp, reduce leakage, and give less trouble from pounding. Moreover, they are usually accepted as being more reliable than large valves.

The difference between the theoretical and the actual displacement, expressed in percentage of the theoretical, is called *slip*. Slip may be due to leaky valves (both suction and discharge) or to packing and piston leaks. In most pumps the slip varies from 5 to 10 percent, as compared with a slip of 2 to 3 percent for pumps in good mechanical condition.

An *air chamber* is a device located immediately ahead of the suction valves or directly behind the discharge valves. In the latter case it is placed directly above the water end of the pump. Air chambers on the suction side are recommended for short-stroke pumps, pumps with high suction lift, or those running at high speed. Air chambers on the discharge side are provided especially when the line is long. An air chamber absorbs shocks and surges in the discharge line and relieves the pump from excessive strains. Air chambers ensure a steady supply of water and reduce pounding.

The manner in which this is done is as follows: The upper section of the chamber is filled with air. As the water leaves the pump, some enters the chamber and compresses the air therein. When the piston momentarily hesitates in its stroke, the flow of water would be expected to remain stationary, but the compressed air expands, and the flow continues. The air chamber thus forms a cushion to steady the flow. The volume of the air chamber is approximately $2\frac{1}{2}$ times the piston displacement.

All pumps require packing to prevent leakage of water past the piston or plunger. Different kinds of packing, based on pressures and temperatures of the liquid, are used on different pumps.

In piston pumps, the piston works back and forth in a bored cylinder (see Fig. 8.3). Here the packing rings are fastened securely to the

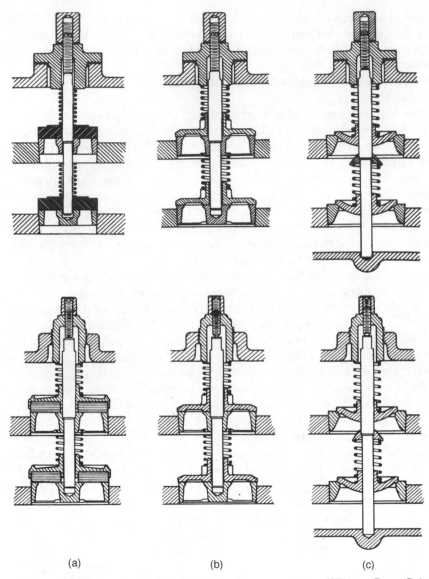

(a)   (b)   (c)

Figure 8.7 Valve construction of single direct-acting pump. (*Ingersoll Dresser Pump Co.*) (*a*) *Top*: rubber valve for fresh cold water, *Bottom*: enkens composition valve, of same construction, for hot water. (*b*) Metal valves for handling feedwater at high temperatures, or light liquids which attack rubber. (*c*) Bolve valves for heavy, viscous fluids; there are no supporting webs to impede the flow of liquid.

piston and move with it. The type of ring depends on the nature of the liquid to be pumped and on the pressure and temperature. Pistons used for general service use soft canvas packing. This packing consists of a cotton fiber, square in cross section. Some pistons are fitted with grooved metallic packing rings.

In plunger pumps the plunger moves back and forth in a packing gland (see Fig. 8.4) instead of in a bored cylinder. The plunger type of pump can be distinguished easily from the piston type by this characteristic. Moreover, plungers are longer than pistons and are long in comparison with the length of stroke. Plunger pumps are packed in different ways:

1. Outside-end-packed, Fig. 8.4 clearly showing the arrangement of stuffing boxes, glands, and packing.

2. Outside-center-packed, Fig. 8.5*b* and *c* showing how the plunger slides through the two packing glands. Access to the packing gland is made by removing the gland nuts, as it is outside-packed.

3. Inside-center-packed, in which the plunger slides back and forth in a packing gland. This gland consists either of a metal packing ring or of some form of hydraulic packing.

The inside-packed type has an advantage over the outside-packed in dusty and dirty locations, since the abrasive action of the dirt does not affect it as it affects outside-packed plungers. However, leakage in the inside-packed type is difficult to detect and repair, whereas with the outside-packed plungers leakage can be detected very easily and the packing replaced without dismantling the pump.

Several difficulties may be encountered when operating duplex pumps, particularly when handling hot water. Hot water, since it comes to the pump by suction, is difficult to lift, since it vaporizes when the pressure is reduced below atmospheric. As a result, the expansion of the vapor fills the suction chamber and pump cylinder. This vapor is compressed and reexpands, the pump thus failing to discharge a full cylinder of water. This results in considerable fluctuation in flow and pressure. Approximately 150°F is the maximum temperature at which water can be raised by suction. Hot water should *flow* to the pump, ensuring full pump cylinders. For pumping hot water, a pump should be selected to run at low speed; the pump is usually placed 10 to 30 ft below the source of supply, and the suction line is made large enough to ensure ample delivery and the minimum interference with vapors formed.

Pumps generally should be provided with a strainer and a foot valve on the suction end of the line. The strainer prevents any foreign matter from clogging the line or lodging under the valves, while the foot valve prevents the water from draining out of the suction line. If there is no foot valve and water comes to the pump under suction, it becomes necessary to prime the pump. In this case the pump is equipped with an aspirator that will exhaust the air in the suction line and draw water to the pump.

Thick liquids should always flow to the pump under pressure. Suction and discharge lines should be fitted with gate valves in preference to globe valves, since the latter offer too much resistance to the flow. The steam line to the pump should be fitted with a throttle valve and a drain at the pump inlet.

8.4 Power Pumps

A power pump is one in which the piston or plunger is operated by a unit not a part of the pump itself. The motive power in duplex pumps is the steam end of the pump. Power pumps, however, are geared, belted, or direct connected to gas engines, electric motors, chain drive, etc. Motors direct connected by gears, belts, or chain drives are used most commonly. These pumps may have their cranks coupled at 90°, an arrangement that provides a steady and uniform flow. The crank setting depends on the number of pistons or plungers used.

The pump shown in Fig. 8.8 can be classified as a power pump. Pumps of this variety are used for chemical feed and offer a precise volume control by means of a variable-stroke adjustment mechanism. General methods of adjustment are provided, the one shown being by means of a screw. The microadjustment is equipped with a vernier dial to permit quick and convenient resetting for any predetermined stroke length. This pump uses the step-valve design shown.

Power pumps also can be classified as metering pumps. This type of pump is a positive displacement chemical-dosing device with the ability to vary capacity manually or automatically as process conditions require. It is capable of pumping various types of chemicals such as acids, bases, corrosive or viscous liquids, and slurries. In power plants, the major applications of metering pumps include (1) cooling tower water treatment to prevent corrosion and fouling of heat-exchanger surfaces, (2) internal boiler water treatment to control scale deposits and corrosion, and (3) external boiler water treatment

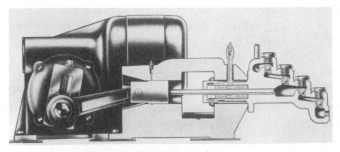

Figure 8.8 Volume control with step-valve design. (*Milton Roy Co.*)

to maintain quality of low-pressure feedwater being recirculated or discharged, including demineralizer regeneration.

Figure 8.9a and b shows the basic components of the metering pump that consists of the driver, drive mechanism, flow adjustment, and liquid end. The pump is usually driven by an ac constant-speed motor, although variable-speed, pneumatic, and hydraulic drivers are also utilized.

The drive mechanism translates the rotary motion of the driver into a reciprocating movement. Industrial-duty pumps will submerge this portion of the pump in an oil bath to ensure reliability during continuous operation. The pump flow rate is adjustable by varying stroke length or stroking speed. Most metering pumps are supplied with a micrometer screw adjustment, as shown in Fig. 8.9a and b. Instead of the micrometer, an electronic or pneumatic actuator can be used to adjust the pump flow rate in response to a process signal. The liquid end of this pump design is shown in Fig. 8.10. The design and materials of construction are selected for the specific service conditions. Required flow and pressure ratings are considered as well as the physical and chemical properties of the liquid. In this design the liquid is drawn into the pump by the rearward motion of the piston and expelled by its forward motion. In order to achieve this, the metering pump is supplied with check valves at the suction and discharge connection points.

During the suction portion of the stroke, the motion of the piston lifts the suction-ball check valve from its seat, which allows liquid into the pump. At the same time, the piston's motion and system's backpressure hold the discharge check valve closed. This is then reversed during the discharge stroke.

Metering pumps are characterized by three major performance features—accuracy, adjustability, and pressure. A metering pump is unique in terms of accuracy because it can deliver liquid at a rate that varies no more than plus or minus 2 percent throughout a wide operating range. Other classes of pumps generally maintain delivery rates in the plus or minus 5 percent range.

Adjustability is another major feature because the pump output can be varied by means of a micrometer indicator at 1 percent increments. To achieve output variation in most other types of pumps, a variable-speed motor must be used. A turndown ratio of 10 to 1 is standard in metering pumps. This means that performance is maintained from 10 to 100 percent of capacity. Depending on the type of metering pump used, pressures to 7500 psig can be generated.

Displacement power pumps are used for low flow rates and high heads. The efficiency of power pumps depends to a large extent on the character of the driving motor and the mechanism transmitting this

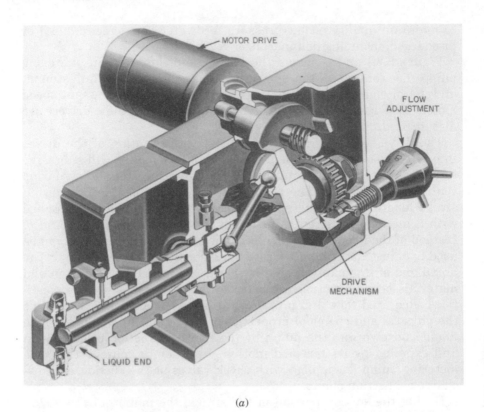

(a)

(b)

Figure 8.9 Metering pump. (a) Major components. (b) Polar crank drive. (*Milton Roy Co.*)

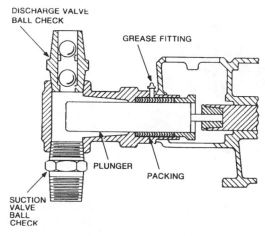

Figure 8.10 Liquid-end design of metering pump. (*Milton Roy Co.*)

power. Their design is simple as well as practical, and they can be located anywhere because they are independent of belts or pulleys.

8.5 Vacuum Pumps

Vacuum pumps are provided for a variety of services. The wet vacuum pump (Fig. 8.11) can be used for heating systems and process requirements because it operates at vacuums up to 26 in (relative to a 30-in barometer) and is available in a variety of sizes and capacities. This pump is steam driven and piston packed.

The manifold duplex water pump (Fig. 8.12) is used for boiler feed service and is capable of handling water at high temperatures without vapor binding or cavitation. The duplex pumps are operated by

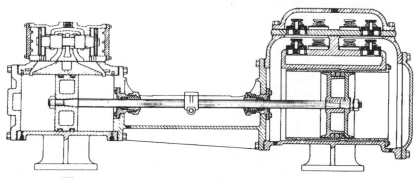

Figure 8.11 Wet vacuum pump. (*American-Marsh Pumps, Inc.*)

Figure 8.12 Manifold duplex vacuum heating pump. (*Nash Engineering Co.*)

two float switches set to start the water pumps at low and high water levels in the receiver. If one pump cannot handle an abnormally high flow of condensate, the second pump comes into operation. Of interest is the vacuum pump attached to this unit to remove entrained air from the system.

The vacuum pump (Fig. 8.13) is a centrifugal displacement pump consisting of a round multiblade rotor revolving freely in an elliptical casing partially filled with liquid. The curved rotor blades project radially from the hub to form with the side shrouds a series of pockets or buckets around the periphery. The rotor revolves at a speed high enough to throw the liquid out from the center by centrifugal force. This results in a solid ring of liquid revolving in the casing at the

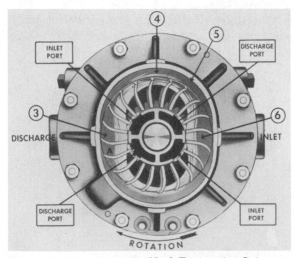

Figure 8.13 Vacuum pump. (*Nash Engineering Co.*)

same speed as the rotor but following the elliptical shape of the casing. This alternately forces the liquid to enter and recede from the rotor buckets, twice in each revolution.

The cycle of operation is as follows: The cycle starts at point A with the rotor bucket (3) full of liquid. The liquid, because of the centrifugal force, follows the casing. It withdraws from the rotor, pulling gas in through the inlet port, which is connected with the pump inlet. At (4) the liquid has been thrown entirely from the chamber in the rotor and replaced with gas. As rotation continues, the converging wall (5) of the casing forces the liquid back into the rotor chamber, compressing the gas trapped in the chamber and forcing it out through the discharge port, which is connected with the pump discharge. The rotor chamber (6) is now full of liquid and ready to repeat the cycle. The cycle takes place twice in each revolution.

Therefore, a vacuum pump is essentially a compressor. It takes its suction at low absolute pressure, compresses through a large number of compressions, and generally discharges to the atmosphere.

8.6   Rotary Pumps

A rotary pump is a positive-displacement pump. Instead of spinning out the liquid as a centrifugal pump does, a rotary pump traps the liquid, forces it around the casing, and expels it through the discharge. Unlike the reciprocating pump, the rotary pump discharges a smooth flow of liquid. This is accomplished by the use of a rotating shaft or shafts. In these pumps, the main pumping action is covered by the relative movement between the rotating and stationary elements of the pump. The rotary motion of these pumps is the design feature that is different from reciprocating positive-displacement pumps, where the main motion of moving elements is reciprocating.

The rotating shaft or shafts are rotors consisting of gears, vanes, or screws that operate in a close-fitting casing. Normal rotary-pump designs do not incorporate the use of valves, which permits the pump to operate efficiently on both low- and high-viscosity liquids with a low *net positive suction head* (NPSH) requirement (see Sec. 8.9 for an explanation of NPSH and Sec. 8.12 for an explanation of head).

Since the rotary pump is a positive-displacement machine, its theoretical displacement is a straight horizontal line when plotted against pressure with speed constant. A variation in displacement as pressure varies is not present as it would be with a centrifugal pump. When low-viscosity liquids are being handled, however, there is a loss in delivery because of slip at higher pressures. The actual pump capacity at any given speed and viscosity is the difference between the theoretical displacement and *slip,* slip being the leakage from the discharge back to the suction side of the pump through pump clearances.

Low-viscosity liquids can short-circuit more easily as the pressure increases.

Rotary pumps have the following advantages: They are self-priming, are capable of high suction lifts, have low NPSH requirements, can handle high-viscosity liquids at high efficiency, have a wide speed range, and are available for low-capacity, high-head or high-capacity, high-head applications. There are many different designs, of which a few follow:

1. *The gear pump* has two or more gears that mesh and provide the pumping action. There are two types of gear pumps, external and internal.

   a. *The external gear pump* is probably the most widely used rotary pump. It consists of two meshing gears in a close-fitting housing. Gear rotors are cut externally, and in this type of pump, fluid is carried between the gear teeth and displaced when they mesh. The gears can be spur, single-helical, or double-helical (herringbone type). Figure 8.14a and b shows a herringbone gear design. In operation, the liquid enters between the gear teeth and housing on each side of the pump casing, as shown in the schematic in Fig. 8.14c.

   b. *The internal gear pump* design has one rotor with internally cut gear teeth that mesh with an externally cut gear. On the outer sideplate is a stationary crescent. As the internal gear rotates, the idler (external) gear follows, and liquid is displaced between the internal gear in the crescent and between the idler and the crescent. This type of pump is generally used for lower-pressure applications at low speeds. Figure 8.15 shows a cross section of an internal gear pump.

2. *The vane-type pump* is a slow-speed pump design that has lower-viscosity limits than gear pumps. The vanes are self-compensating for wear, and this pump can be used on liquids that are slightly erosive, unlike a close-clearance gear pump design.

3. *The screw pump* is a special type of rotary positive-displacement pump where the flow through the pumping elements is axial. The liquid is carried between screw threads on one or more rotors and is displaced axially as the screws rotate and mesh. In all other rotary pumps, the liquid is forced to travel circumferentially. This gives the screw pump, which has its unique axial-flow pattern and low internal velocities, a number of advantages in applications where liquid agitation or churning is objectionable.

Since the rotary pump is a positive-displacement pump, a relief valve is usually required. On small pumps, integral relief valves are sometimes provided, while on larger pumps a relief valve should be

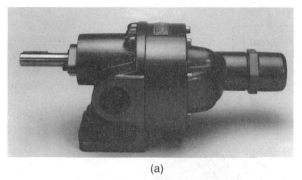

(a)

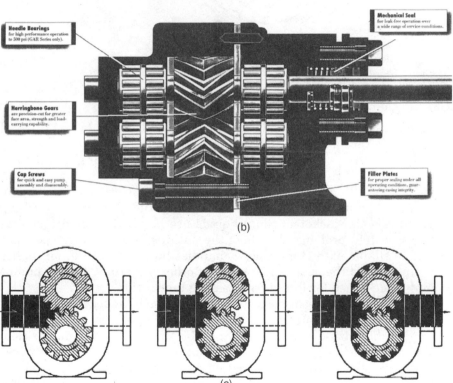

Needle Bearings
for high performance operation
to 500 psi (GAR Series only).

Mechanical Seal
for leak-free operation over
a wide range of service conditions.

Herringbone Gears
are precision-cut for greater
face area, strength and load-
carrying capability.

Cap Screws
for quick and easy pump
assembly and disassembly.

Filler Plates
for proper sealing under all
operating conditions, guar-
anteeing casing integrity.

(b)

(c)

Figure 8.14 (a) Gear pump, external view. (b) Gear pump, internal view of major components. (c) Gear pump, schematic of fluid flow through pump. (*Ingersoll Dresser Pump Co.*)

located in the discharge piping and set approximately 10 percent above the pump discharge pressure. Since rotary-pump performance depends on the maintenance of close internal clearances, and since many pumps of this type utilize internal bearings, it is important to prevent foreign matter from entering the pump. A strainer should be

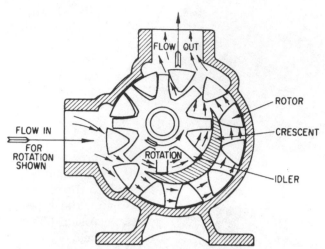

Figure 8.15 Cross section of internal gear rotary pump.
(*Deming Division, Crane Co.*)

installed on the suction side of the pump, and it should be of liberal
size to prevent a high pressure drop.

Rotary pumps are equipped with a packed stuffing box, or the
gland can be fitted with a mechanical seal, as shown in Fig. 8.14*b*. A
jacketed stuffing box for high pumping temperatures has two sets of
glands that are separated by a lantern gland. The inner packing
gland is water-jacketed. Since liquids are handled at high temperatures, cooling prolongs the life of the packing. The bellows-type shaft
seal available with Buna bellows handles liquids to 240°F; for higher
temperatures the seal can be made of other materials. Rotary pumps
may be designed for forced-feed bearing lubrication.

The mechanical seal consists of a stationary lapped face, integral
with the gland; a rotating assembly with a lapped carbon ring; synthetic rubber bellows on the shaft; and a spring to hold it in position.
The mechanical seal is effective. Its main limitation is that abrasives
may get between the lapped sealing faces and cause leakage. On the
basis of the type of liquid handled and its temperature and pressure,
the bellows-type shaft seal has materials capable of handling the service for which it is designed.

The internal-gear pump (Fig. 8.15) has two moving parts, the rotor
(internal gear) and the idler gear. Each revolution of the internal gear
means a definite rated output. The teeth separate at the suction port
and mesh together at the discharge port, at the top. In Fig. 8.15, at
approximately the 10 o'clock position, the rotor and idler form a barrier
just prior to flow coming into the pump. At about the 9 o'clock position, the idler withdraws from the rotor and creates a suction opening

that becomes filled with liquid between the rotor and idler. Then, at about the 2 o'clock position, the rotor and idler come together, which forces the liquid out of the pump through the discharge opening.

There are many applications for the rotary pump. In the power plant it is used chiefly for handling oil and lubricants. At times, pumps of this type handle many forms of chemicals in solution.

8.7 Centrifugal Pumps

A centrifugal pump is one that depends on centrifugal force and rotation of an impeller for its action. The type of pump used depends on the type of service for which it is intended and varies with the capacity required, the variations in suction and discharge head, the type of water handled (whether clean or dirty, hot or cold, corrosion effects, etc.), the nature of the load, and the type of drive to be used.

End-suction vertical split-case centrifugal pumps have an almost unlimited range of applications. They are available in capacities ranging to more than 4000 gpm (gal per min) and heads ranging to more than 550 ft. The single-stage double-suction centrifugal pump is used most often for general-service purposes. Sump pumps are most frequently vertically mounted. Boiler feed pumps are usually of the split-case variety, both single and multistage. Condenser pumps are of either horizontal or vertical design, depending somewhat on the suction lift and the headroom available for installation and removal. These pumps are of the low-head variety.

Centrifugal pumps may be classified as vertical or horizontal, depending on the direction of the pump shaft; volute or turbine, depending on their construction; single- or double-suction, depending on the manner in which the water is made to enter the pump; single-stage, double-stage, or multistage, depending on the number of stages of impellers in the pump; and open- or closed-impeller.

A *volute* pump (Fig. 8.16) is one in which the impeller rotates in a casing of spiral design. The casing is designed to enclose the outer extremity of the impeller. The volute chamber changes velocity head to pressure head.

A *turbine* pump (Fig. 8.17) is one in which the impeller is surrounded by diffusion rings (Fig. 8.18). The diffusion ring takes the place of the spiral casing in the volute pump. It takes the water from the impeller and changes velocity head to pressure head. The quantity of water pumped depends on the size of the impeller and its speed.

A *single-suction* pump is one in which the water enters on only one side of the impeller (Figs. 8.19 and 8.20). In Fig. 8.20 is shown a two-stage, hydraulically balanced opposed-impeller pump.

A *double-suction* pump (see Fig. 8.16) is one in which the water enters on both sides of the impeller.

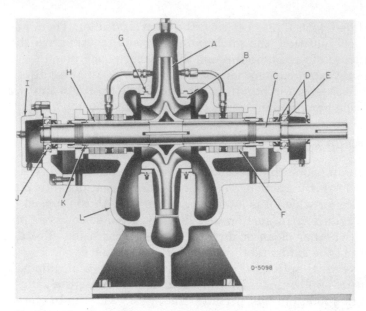

Figure 8.16 Cross section of double-suction volute pump. A = impeller. B = impeller ring. C = shaft. D = seal rings. E = ball bearing (radial). F = packing stuffing box. G = casing ring. H = shaft sleeve. I = bearing end cover. J = ball bearing (thrust). K = O-ring. L = casing complete. (*Ingersoll Dresser Pump Co.*)

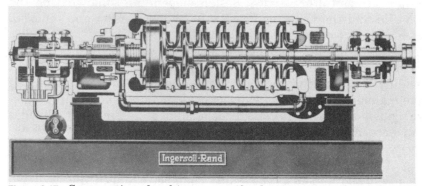

Figure 8.17 Cross section of multistage centrifugal pump equipped with forced-feed lubrication. (*Ingersoll Dresser Pump Co.*)

A *single-stage* pump is one in which only one impeller is mounted on a shaft (see Figs. 8.16 and 8.19).

A *double-stage* (two-stage) pump is one in which two impellers are mounted on a shaft (see Fig. 8.20). This pump has opposed impellers.

A *multistage* pump (see Fig. 8.17) is one in which two or more impellers are mounted on a shaft, the water passing from the discharge of one impeller to the suction of the next.

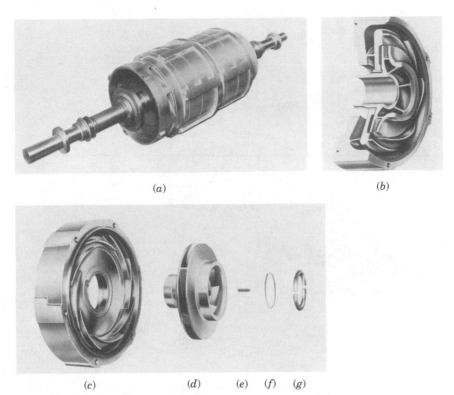

(a)   (b)

(c)   (d)   (e)   (f)   (g)

Figure 8.18   Centrifugal pump parts. (a) Inner assembly. (b) Channel ring and impeller. (c) Channel ring. (d) Impeller. (e) Key. (f) Retaining ring. (g) Split ring. (*Ingersoll Dresser Pump Co.*)

The single-stage pump shown in Fig. 8.19 is cradle-mounted, flexible-coupled, and driven by a standard motor. Pump and motor are mounted on a steel plate providing maximum rigidity to prevent distortion and misalignment. The back-pullout design permits fast, easy access to stuffing box, impeller, and casing rings without disturbing suction or discharge piping. When a spacer coupling is included, the pump can be completely inspected and serviced without moving the driver. These pumps are available for 1000-ft heads and capacities to 1300 gpm.

To protect the casing, replaceable casing rings are provided at both the front and back impeller clearances. When wear occurs, they can be replaced easily. The closed impeller is dynamically balanced for vibration-free operation and hydraulically balanced to reduce stuffing-box pressure and to minimize thrust loads on the bearings. The impeller is keyed to the shaft, preventing the impeller from backing off if rotation is reversed. The dual-volute casing reduces radial hydraulic thrust on the impeller and holds shaft deflection to a minimum.

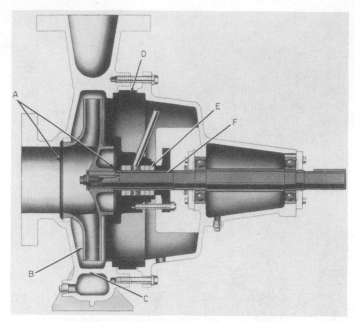

Figure 8.19 Sectional view of single-stage pump. A = casing ring. B = balanced, closed impeller. C = dual volute casing. D = self-centering stuffing box cover. E = stuffing box. F = shaft sleeve. (*Ingersoll Dresser Pump Co.*)

The unit has a self-centering stuffing-box cover. The extradeep stuffing box can accommodate either packing or mechanical seals. The shaft sleeve is replaceable, and bearings are permanently lubricated. These units are designed for pressures of 450 psi.

A cross section of a double-suction volute pump is shown in Fig. 8.16. This pump is of the horizontal split-case type, divided at the centerline, with suction and discharge nozzles cast integrally with the lower half. The upper half of the casing can then be removed without disrupting the piping or the pump setting. The pump is designed to handle hot or cold liquid, capacities to 20,000 gpm, pressures to 300 psi, and temperatures to 300°F.

The advantages of this pump are as follows: (1) It has a wide range of applications, sustains high efficiency, and permits high suction pressure. (2) It requires minimum stuffing-box maintenance and allows for longer wear of rings and bearings. (3) It has prelubricated, grease-packed, or oil-lubricated bearings and requires only periodic maintenance.

A cross-sectional view of a two-stage, hydraulically balanced centrifugal pump with opposed impellers is shown in Fig. 8.20. This

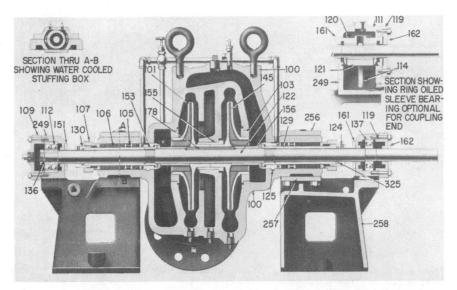

Figure 8.20 Sectional view of two-stage, hydraulically balanced opposed-impeller pump. 100 = casing. 101 = impeller (suction). 103 = casing wearing ring (suction). 105 = seal ring. 106 = stuffed-box packing. 107 = stuffed-box gland. 109 = end cover (thrust end). 111 = bearing housing cap. 112 = ball bearing (thrust end). 114 = oil ring. 119 = end cover (coupling end). 120 = upper bearing shell. 121 = lower bearing shell. 122 = shaft. 124 = shaft sleeve packing nut, right-hand. 125 = stuffing-box throat bushing. 129 = shaft sleeve (discharge). 130 = shaft sleeve packing nut, left-hand. 136 = locknut (for ball bearing, thrust end). 137 = ball bearing (coupling end). 145 = impeller (discharge). 151 = liquid deflector (thrust end). 153 = shaft sleeve (suction). 155 = casing center bushing. 156 = spacer sleeve. 161 = liquid deflector (coupling end), inner. 162 = liquid deflector (coupling end), outer. 178 = impeller key. 249 = bearing housing. 256 = stuffing-box plate (upper). 257 = stuffing-box plate (lower). 258 = pump pedestal. 325 = shaft-sleeve-nut packing. (*Goulds Pumps, Inc.*)

arrangement reduces thrust to a minimum. The pump shaft has ball bearings. Details of the oiled sleeve bearings are shown.

A cross section of a multistage centrifugal pump with forced-feed lubrication is shown in Fig. 8.17. It is designed for applications in which liquids must be handled at high pressures. This is a single-suction multistage diffuser-type unit featuring cylindrical *double-case* construction and *unit-type* rotor assembly. Such pumps are available for pressures to 7000 psi, capacities to 23,000 gpm, and heads to 14,000 ft.

The unit shown consists of two vertically split, concentric cylindrical casings, a high-strength outer casing built for full discharge pressure, and a segmented inner casing formed by interlocking channel rings. The rotor (see Fig. 8.18) is contained within the inner casing. The symmetrical design of both casings permits equalized expansion in all directions, thus eliminating any stress or distortion due to temperature changes. Since the space between the inner and outer cas-

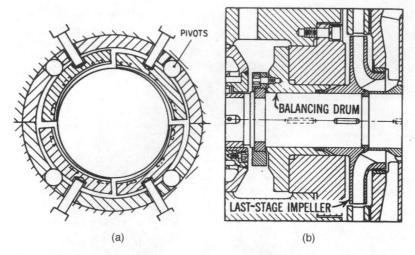

(a)   (b)

Figure 8.21   Centrifugal pump. (*a*) Tilting-shoe bearing arrangement. (*b*) Balancing drum. (*Ingersoll Dresser Pump Co.*)

ings is under discharge pressure, and since this pressure acts on the discharge end of the inner casing assembly, interstage gasket sealing is ensured by keeping the assembly under a compression force. The entire inner casing (and rotor) can be pulled out without disturbing piping connections or driver.

The outer casing is made of forged steel proportioned to withstand maximum operating pressures. Suction and discharge nozzles are welded in position. Individual channel rings are interlocked with aligning rings. Adjacent parts have ground joints that make metal-to-metal contact when the pump is assembled. The outboard end of the rotor is equipped with a balancing drum (Fig. 8.21*b*) to take the axial thrust of the rotor. A thrust bearing is also provided (Fig. 8.21*a*).

Mounted at the outboard end of the pump is a gear oil pump located in the oil reservoir and driven through gears from the pump shaft. Oil from the pump is made to pass through a filter and cooler before reaching the bearings. An auxiliary motor-driven centrifugal pump is mounted on top of the oil reservoir for startup purposes or for use if oil pressure falls below a safe level.

Impellers of the enclosed type are usually made of chrome steel. Alloy steels are also used in impeller design. The materials used are based on the water or fluid handled and conditions imposed by the liquids that are pumped.

The impellers have curved radial passages or vanes that connect with the hub. Each impeller mounted on the shaft is provided with an impeller ring and casing ring, which reduce to a minimum the leakage from the discharge to the suction side. The casing ring is held stationary in the casing, while the impeller ring is attached to the impeller and rotates with it.

Around the impeller (Fig. 8.18b) is placed a diffusion ring that is held stationary in the casing. The ring contains slots or openings that receive the water from the impellers at a high velocity. By means of the gradually increasing area of the opening, the velocity head is converted to pressure. This procedure then makes it possible for the water to advance from one impeller to the next, with little loss of energy.

In operation, the water enters at the right and proceeds to the first impeller, which is revolving at high speed. Immediately on entering the impeller, it comes under the influence of the centrifugal force resulting from the rotation and is moved to the outer edge of the impeller at a gradually increasing speed, finally leaving at a high velocity. It then immediately enters the diffusion ring.

It enters the ring at a high velocity, which is to be changed into useful pressure and also reduced so that the turn of 180° at the top of the ring can be accomplished without serious friction loss and without the water striking the surface at a velocity that might be destructive even if pure water were handled. The diffusion vanes accomplish both these results.

The water is now flowing at a reduced velocity (and at the pressure gained in the first stage) through the ring into the second stage. This process is repeated until the last stage has been reached, each impeller adding its pressure. In the final stage the diffusion vane is omitted and a volute chamber substituted. The vane is omitted here because it is not necessary for the water to make a turn, since it passes directly to the discharge line.

Centrifugal pumps have vanes or impellers, some of which are radial or inclined forward in the direction of rotation. With this type of vane, the head will increase as the capacity increases. If, however, the vanes are moved backwards, the head will remain relatively constant or decrease (see Fig. 8.27) with an increase in capacity. Naturally, best results are obtained by basing the design on plant requirements.

The most common ways to control stuffing box leakage are by means of packing and mechanical seals. The mechanical seal is the most adaptable to the wide variety of operating conditions encountered with pump operation.

Conventional packing is held in the stuffing box by means of the gland. Its purpose is to throttle the fluid that leaks between the rotating shaft and packing. Some leakage is necessary for lubrication, cooling, and sealing and thus to prevent scoring of the shaft and burning of the packing.

To ensure lubrication between the shaft and the packing and to prevent overheating, a liquid such as water is introduced into a seal cage located in the stuffing box. The packing gland is adjusted to allow the proper amount of leakage for lubrication and sealing.

The basic mechanical seal consists of two parts, stationary and rotating, each with a polished sealing surface at right angles to the shaft. The stationary member is fastened to the casing, and the rotating member is fastened to the shaft. Highly polished wear surfaces on both members do the sealing. The sealing faces are held in contact by means of a spring, and, usually, only an adjustment is required at installation and startup.

Mechanical seals have several advantages over packing:

1. *Controlled leakage.*   A mechanical seal requires some lubrication of the sealing faces to operate properly.

2. *High suction pressure.*   A mechanical seal can be designed to operate successfully at pressures that are higher than packing can withstand.

3. *Resistance to corrosives.*   A mechanical seal is available with most corrosion-resistant materials and is not limited to a few basic materials, which is a limitation with packing.

Packing glands or stuffing boxes should be deep to facilitate packing and to ensure a minimum of leakage. The amount and kind of packing depend on the design of the pump and on the operating conditions.

Soft metallic packing is usually preferred. This is so because the centrifugal pump rotates at high speed, and hard packing would score the shaft. Moreover, with hard packing it is impossible to make as tight a joint, with a minimum of pressure exerted, as with soft packing. Graphite packing is also used. Packing should be pulled up just enough to prevent air leakage, and no attempt should be made to prevent some leakage of water past it. For the high-pressure pump (see Fig. 8.17), the type of stuffing-box arrangement used will depend on the service for which the pump is designed. Stuffing boxes are usually of two types: (1) packed-water-cooled and (2) floating-ring, arranged for cold-condensate injection. The latter is throttled between shaft sleeve and a self-adjusting floating ring. Mechanical seals frequently replace the conventional packing.

For water-cooled packed boxes (Fig. 8.22), the inbound and outbound boxes are water-jacketed to provide efficient cooling. They are designed to accommodate the packing arrangement that best meets the suction conditions under which the pump operates. For the normal range of boiler-feed application, the box is packed solid with eight

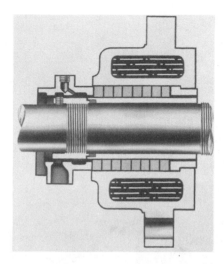

Figure 8.22 Water-cooled packed boxes. (*Ingersoll Dresser Pump Co.*)

rings of packing. At high suction pressures, a bleed-off bushing may be inserted to reduce the pressure on the packing.

The discharge stuffing box is contained in a removable extension, bolted to the casing with a gasket fit. This stuffing-box extension serves to hold the balancing-drum sleeve in place and permits its easy removal without opening the casing.

Mechanical seals are also used. As noted previously, they provide a practical solution to the most difficult stuffing-box problems, since effective mechanical sealing is maintained without packing and without wearing parts that require frequent attention and maintenance. Pump and seal design is integrated so that the seal can be serviced in the field without dismantling the pump. A spacer coupling with a taper fit on the shaft is recommended for all mechanical seal applications.

There are many varieties of mechanical seals. Their design is based on pump pressures, capacity, temperatures, and rpm. The mechanical seal (Fig. 8.23) is used for heavy-duty centrifugal process pumps with operating pressures to 720 psi, capacities to 7500 gpm, temperatures to 800°F, and speeds to 3600 rpm. This type of pump has built-in seal alignment and built-in recirculation; no gland is required. Properly adjusted, the mechanical shaft seal requires no further attention.

Thrust is eliminated or reduced in a number of ways. The pump shown in Fig. 8.16 has a double-suction inlet. While the water flow may not always be exactly equal on both sides, the inlet does go a long way toward neutralizing the imbalance that occurs with a single inlet. In the same pump is built a balanced port so that variations in

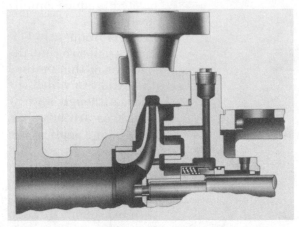

Figure 8.23   Built-in seal for high-duty centrifugal process pump. (*Ingersoll Dresser Pump Co.*)

pressure on either side of the impeller will be equalized, thus neutralizing the thrust which might be developed.

For the pump in Fig. 8.17, radial balance is obtained through multivolute design, while axial balance is achieved by a hydraulic balancing drum (see Fig. 8.21*b*). The axial thrust of the rotor toward the suction end, which is developed by the sum of the unbalanced pressure differentials across the individual impellers, is counteracted by the balancing drum located next to the last impeller at the outboard end of the rotor. Here the chamber to the left of the drum is connected to suction pressure. Discharge pressure from the last stage bleeds along the drum to act on the face of the drum, and the difference in pressure on the two sides positions the rotor. Radial balance is obtained by having the discharge from each impeller directed to the succeeding stage through a series of volutes (see Fig. 8.18*c*) equally spaced around the entire circumference of the impeller. Such an arrangement provides for radial thrust.

In addition to the balancing drum, stabilized bearings are usually furnished to prevent oil whip and possible vibration in high-speed pump operation (5000 to 10,000 rpm). A Kingsbury thrust bearing (see Fig. 8.21*a*) is incorporated with the outboard bearing to maintain longitudinal alignment of the rotor and to take up any thrust that may be set up by abnormal operating conditions. The thrust bearing contains a revolving collar attached to the shaft. The collar transmits any thrust that is developed to the stationary thrust shoes, which are made so that they lift slightly at one end (facing the direction of rotation), admitting a wedge-shaped film of oil under pressure between the surfaces. Four individual shoes centrally pivoted produce equally positioned oil wedges and stabilize any shaft movement.

To eliminate balancing devices, hydraulic thrust is frequently balanced by opposing single-suction impellers in equal groups so that the thrust of one neutralizes that of another. The success of this method depends on the impeller arrangement and on the extent to which the various factors previously mentioned are satisfied. Although a great number of combinations and arrangements of stages are possible, only a few are worthy of consideration. Figure 8.24 shows some of the common methods used.

With opposed impellers, the most important factors are stuffing-box pressure, interstage leakage, and casing design. To satisfy these conditions, the first and second stages are placed next to the stuffing boxes in order to have the lowest pressure on them. To keep interstage leakage to a minimum and to make possible the simplest casing arrangements, the stages must be in series. Hydraulic balance requires that one-half the stages face opposite to the other half. Since no stage arrangement satisfies all these conditions simultaneously, each arrangement must be a compromise that favors some particular point.

Staging has been resorted to because pumps with one impeller are limited as to the head against which they are capable of pumping. Volute pumps of a single stage are usually limited to about 300 psi.

A good system of lubrication is essential for the centrifugal pump. The reason is readily apparent when it is considered that these pumps operate at high speeds and very frequently handle high-temperature water. Not only should the pump have an excellent lubricating system, but the lubricant should be of high quality. Bearings are both oil and grease lubricated; generally the grease-lubricated bearing is less expensive. If the grease-lubricated bearing has a disadvantage, it is that it can be overlubricated, resulting in overheating. Since most pump manufacturers designate their preference on the basis of experience, indicating whether oil or grease is to be used and the quality of the lubricant, their recommendations should be rigidly followed.

Another major disadvantage that centrifugal pumps have is that of operating pumps in parallel. A number of pumps to be operated in parallel are selected with design characteristics that are similar. However, after they are in operation for a while, wear of the moving parts changes the design pattern so that the pumps no longer perform as they were designed to perform. If the head at a given capacity will not match that of the next pump in line, the first pump will back the second pump off the line until it virtually shuts the second pump down. At

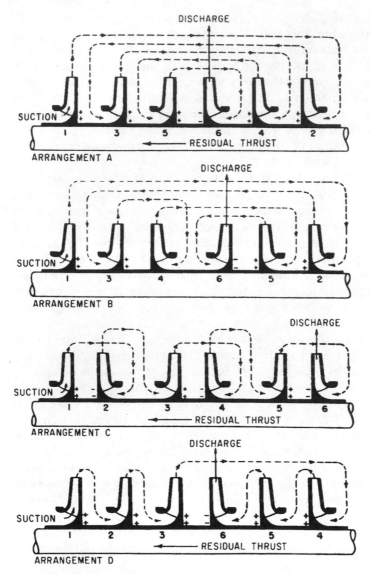

Figure 8.24 Four opposed single-suction impeller arrangements to obtain balance. This design results in arrangements A and B providing the lowest pressure on the stuffing box. (*Ingersoll Dresser Pump Co.*)

any rate, the pump with the highest head will tend to hog the load. Pumps therefore should have the same or very nearly the same performance characteristics to avoid loss of load and difficulties in operation.

In most cases, boiler feed pumps are of the centrifugal pump design. The capacity requirements of a boiler feed pump generally are determined by the addition of a margin to the maximum boiler feed-water flow. This margin is important to successful operation because it accounts for both boiler swings (or upsets) and the anticipated reduction in capacity resulting from pump wear that is expected over time. This margin can vary from as much as 20 percent in small power plants to as little as 5 percent in large utility power plants. This variation results from the requirements of the plant design, such as multiple pumps and spare pumps.

The total feedwater capacity must be handled by a single pump or by several smaller duplicate pumps that operate in parallel. In industrial plants, a spare boiler feed pump is generally included as part of the plant design, while in large utility power plants, two half-capacity pumps are often the selection, with a spare pump of half capacity available for redundancy. The actual plant design reflects design preference, operating and redundancy philosophy, and perhaps more important, experience.

Condensate pumps are also of centrifugal pump design. These pumps take their suction from the condenser hot well and discharge to the low-pressure feedwater heaters. They must be designed to operate at very low suction pressures, often 1 to 3 in Hg (inches of mercury), since the turbine operates with a vacuum at its exhaust end.

A comparison of the major features of centrifugal and other types of pumps is shown in Table 8.1.

TABLE 8.1   Comparison of Centrifugal Pumps with Other Pumps

Advantages	Disadvantages
Absence of parts such as valves and packing	Lower efficiency than that of piston pumps
Simplicity	Frequent difficulty with the end thrust
Application of motor or turbine drive	Not suitable for high head requirements at low flow rates
Ability to maintain a uniform flow	
Small floor-space requirement	Care necessary to align high-speed pumps properly
Low initial cost	
Ease of regulation	Possibility of overloading motor due to certain load characteristics
Economy of operation	Difficulty of regulating with wide fluctuations in load
Ease of maintenance	
	Difficulty of operating at very low speed

8.8 Facts about Fluids and Pumping

Pumps and their efficient and reliable operation are critical to the successful operation of a steam power plant. Not only is it important to know the various types of pumps that are used in a power plant and how to select them for a specific purpose, but it is also necessary to understand the more important definitions that reflect this technology, which is generally classified as hydraulics and pumping. The following are some of the more important definitions covering this subject that are used in this book and which are found in a detailed study of fluid technology.

1. *Hydraulics.*   This is a study of fluids at rest or in motion, with the fluids including both liquids and gases.
2. *Density.*   This is sometimes referred to as *specific weight,* and it is the weight per unit volume of a substance. The density of water is 62.4 lb/ft^3 at sea level and 60°F.
3. *Specific gravity.*   This is the ratio of the density (or specific weight) of a substance to that of a standard substance. For liquids, the standard used is water at 60°F. Since most pump performance characteristics are determined using water and are based on a specific gravity of 1.0, it is very important to know the specific gravity of the liquid to be pumped so that the proper corrections can be made. Specific gravity is a dimensionless number. Specific gravities of less than 1.0 indicate liquids that are lighter than water and therefore will float on water.
4. *Pressure.*   This is the force exerted per unit area of a fluid. The most common unit for designating pressure is pounds per square inch (psi).
5. *Vacuum.*   This term is used to refer to pressures that are below atmospheric pressure. With the use of a gauge having a column of mercury to measure vacuum, vacuum is usually expressed in inches of mercury (14.7 psi atmospheric pressure is equal to 30 in Hg at sea level).
6. *Head.*   This is the amount of work necessary to move a liquid from its original position to its final position and includes the extra work necessary to overcome the resistance to flow in the line, called *pressure drop.* A liquid has three kinds of energy:
   a. *Potential head:* The energy of position, which is measured by the work that is possible in dropping a vertical distance.
   b. *Static pressure head:* The energy per pound due to pressure, i.e., the height to which liquid can be raised by a given pressure.
   c. *Velocity head:* The kinetic energy per pound or the vertical distance a liquid would have to fall to acquire the stated velocity.
7. *Friction.*   Since piping and tubing are not smooth, some of the fluid's energy is lost due to friction. Thus, in flow from one point to another, a loss of head results due to friction.

8. *Vapor pressure.*   The vapor pressure of a liquid at a specified temperature is the pressure at which the liquid is in equilibrium with the atmosphere or with its vapor in a closed container. Perhaps the best way to understand vapor pressure is to consider a container that is completely closed and half filled with a liquid. If the other half of the container is completely evacuated of air, a portion of the liquid will vaporize and fill the upper half of the container with vapor. If the pressure of the vapor in the upper half of the container is measured, that pressure is equal to the vapor pressure of the liquid at that temperature. For water at 60°F, the vapor pressure is 0.256 psia. At the boiling point of water, 212°F, the vapor pressure is equal to the atmospheric pressure of 14.7 psia (refer to the steam tables). Other liquids have vapor pressure characteristics that are different from water. Vapor pressure can result in cavitation with an impact on pump operation.

9. *Cavitation.*   When the pressure in the suction line to a pump falls below the vapor pressure, vapor is formed and moves along with the fluid stream. These vapor bubbles or cavities collapse when they reach regions of higher pressure on their way through the pump. The most obvious effects of cavitation are noise and vibration. These are caused by the collapse of the vapor bubbles as they reach the high-pressure side of the pump. The larger the pump, the greater are the noise and vibration. If operated under cavitating conditions for a long enough period of time, especially on water service, pitting will take place on the impeller vane. Noise and vibration also can cause bearing failure, shaft breakage, and other fatigue failures in the pump. The other major effect of cavitation is a drop in pump efficiency, which results in a reduction in flow capacity. Cavitation can be avoided by ensuring that the system has more suction pressure (NPSH) than is required by the pump.

8.9   Factors Affecting Pump Operation

A factor affecting the operation of a centrifugal or a rotary pump is the suction conditions. Suction performance (the relation between capacity and suction conditions) involves a quantity called *net positive suction head* (NPSH). Net positive suction head is the amount of energy in the liquid at the pump datum. It is referred to as either *available* or *required* NPSH.

Required NPSH is the energy needed to fill a pump on the suction side and overcome friction and flow losses from the suction connection to that point in the pump at which more energy is added. The NPSH is a characteristic of the pump and varies with pump design, pump size, and operating conditions; it is supplied by the manufacturer.

Available NPSH is a characteristic of the system and is defined as the energy that is in a liquid at the suction connection of the pump, over and above the energy in the liquid due to its vapor pressure.

For a centrifugal pump, the required NPSH is the amount of energy (in feet of liquid) required:

1. To overcome friction losses from the suction opening to the impeller vanes

2. To create the desired velocity of flow into the vanes

For a rotary pump, the required NPSH is the amount of energy (in psi) required:

1. To overcome friction losses from the suction opening into the gears or vanes

2. To create the desired velocity of flow into the gears or vanes

Since the NPSH varies with pump design, the special operating conditions of the pump must be provided to the pump manufacturer. Abnormally high suction lifts cause a serious reduction in the capacity and efficiency of a pump, thus leading to problems during operation.

If the static pressure at the impeller vanes falls below the vapor pressure corresponding to its temperature, a portion of the water will flash into steam, and cavitation will occur. It is therefore important to ensure that the proper NPSH is available at the pump.

Piping to and from the pump is also important. The suction line to a centrifugal pump should be as short and straight as possible. If at all possible, long suction lines should be avoided. If a long suction line is unavoidable, it should be made several diameters larger than the inlet to the pump to keep friction loss to a minimum. Directly after the pump, a check valve and a gate valve must be located. A gate valve should be placed ahead of the pump. The check valve relieves the pressure from the pump when it is starting up and prevents the waterhammer from damaging the pump. When the pump stops, the gate valve should be closed to relieve the strain on the check valve.

Hot water requires an increase in suction head. Since head varies with the temperature of water, sufficient head is required to compress the vapors and to prevent the pump from becoming steam-bound. For a given condition, increasing the water temperature decreases the head and capacity.

The output of centrifugal pumps can be controlled by throttling the discharge valve or by changing the speed of the pump (speed regulation). If the load is reduced to such an extent that the flow is almost completely stopped (throttled), the pump may become overheated.

Where it becomes necessary to throttle the flow to such an extent, it is advisable to install a bypass orifice.

The bypass line is installed on the discharge side of the pump and ahead of the check valve. The bypass line contains an orifice, with a gate valve ahead and behind the orifice, to permit removal for cleaning, if necessary. There should be a minimum of 3 ft of straight run of pipe on either side of the orifice to reduce turbulence and noise. The recirculating line should enter the feedwater heater at a point below the minimum water level and as far away from the pump suction as possible.

If performance curves are provided (see Fig. 8.27), they will show the relation of head, power consumption, and the efficiency of the pump to the capacity of the pump. As capacity increases, the total head that the pump is capable of developing is reduced. The highest head generally occurs at a point where there is little or no flow through the pump, i.e., when the discharge valve is closed and the pump is running. Depending on design, these characteristics will vary for each pump.

Difficulties are frequently encountered in the operation of pumps. If no water is being delivered, the following conditions may exist: speed too low, discharge head too high, pump in need of priming, suction lift too high, or pump motor running in the wrong direction. If not enough water is being delivered, the following conditions may be the cause: air leaks in the suction line or stuffing box, speed too low, motor running the pump in the wrong direction, obstruction in the suction line or foot valve too small, pump in need of repair in that wearing rings are worn and impeller damaged, water too hot and insufficient head, and suction lift too high. If the pump is not developing enough pressure, the speed may be too low, the pump may be in need of repair, or air leaks may be present.

8.10 Considerations for Pump Selection

In determining selection of the proper pump for a particular application, the following are major areas that must be considered:

1. *Type of liquid.* Information must be known about the type of liquid, which includes
   a. Fresh or salt water, acid or alkali, oil, or slurry
   b. Temperature
   c. Specific gravity
   d. Viscosity
   e. Suspended material in liquid
   f. Chemical analysis, pH value, and corrosive properties

2. Capacity
3. Suction and discharge conditions
4. Total head requirements

5. Pump arrangement—horizontal or vertical
6. Drive—motor or turbine
7. Space limitations
8. Installation location—elevation, indoor/outdoor, ambient temperature

Although all the preceding characteristics are important to making the proper pump selection, two of them require special attention: entrained solids and corrosive properties.

Entrained solids   The amount and type of solids that are entrained in the liquid influence the characteristics of the pumped fluid and therefore influence the pumping system and the pump application. In heavy solid concentrations, mixtures of solids and liquids are called *slurries*. Slurries not only have higher specific gravities, but the viscosity of the fluid is increased considerably because of the solids content.

Corrosion properties   Corrosion characteristics generally have little effect on fluid flow or the hydraulics associated with pumping. However, they do have a severe effect on the life of the system. When corrosive fluids are to be handled, extreme care must be given to the selection of proper materials of construction.

8.11 Pump Installation and Operation

Considerable attention must be paid to proper pump installation (Figs. 8.25 and 8.26) to ensure trouble-free operation. The following factors must be taken into consideration: The pumps should be accessible for inspection and maintenance; they should be removed from areas that are subjected to flooding, water leakage, and corrosion; they should be located as close as possible to the source of supply; and headroom should be provided for lifting the rotor and casing. If the pumps must be located in areas of high humidity, the motors must be designed for this purpose.

The pump is provided with a steel baseplate to which the pump and motor are attached. This baseplate must be bolted securely to the foundation. In most cases, larger bases are grouted into place, but this may not be required on smaller units if the foundation is rigid and the baseplate can be bolted down securely.

The piping should run to a pump as directly as possible, avoiding sharp bends. Entrance to the pump should be provided with long-radius ells or bends to reduce inlet friction to a minimum, and the suction piping should run straight into the pump. Piping must be supported to take the strain from the pump and provide for expansion and contraction to avoid pump misalignment. Expansion joints or loops should be used when hot liquids are being handled.

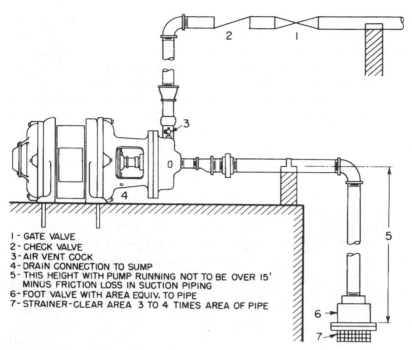

1 - GATE VALVE
2 - CHECK VALVE
3 - AIR VENT COCK
4 - DRAIN CONNECTION TO SUMP
5 - THIS HEIGHT WITH PUMP RUNNING NOT TO BE OVER 15'
      MINUS FRICTION LOSS IN SUCTION PIPING
6 - FOOT VALVE WITH AREA EQUIV. TO PIPE
7 - STRAINER - CLEAR AREA 3 TO 4 TIMES AREA OF PIPE

Figure 8.25   Installation of a small pump.

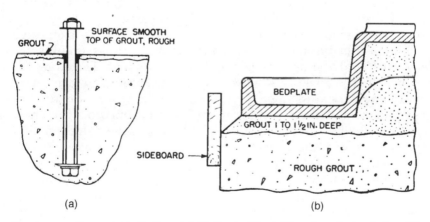

Figure 8.26   Pump installation. (*a*) Method of installing foundation bolts. (*b*) Method of grouting in bedplates.

The suction piping should be large enough to minimize the friction losses. Short suction lines are recommended but are not always possible. In order to avoid problems experienced with long intake pipes and high suction lifts, an individual suction line should be provided for each pump. If this is impractical, the suction-header size should be approximately 50 percent larger than the pump connection and should be reduced at the pump flange. The use of tees or right-angle fittings should be avoided, and long-radius ells or bends should be used instead. The suction header should be sloped upward from the reservoir to the pump, and high spots in the header, where air might collect, should be avoided. The suction end should be flared, and the suction inlet should be well below the low-water level to eliminate the possibility of air entrainment. The suction should not be located close to the point of makeup or returning water, since eddy currents may interfere with pump operation. When supplying two or more pumps from a long intake header, a tapering header with Y branches to the pump suction should be used. For pumps operating under high suction lift, stuffing boxes should be sealed, using water taken from the discharge header or from a separate source of supply.

The suction lift should be limited to 10 to 15 ft, since only clean cold water can be raised this distance without experiencing operating difficulties. For hot water, it is necessary to bring the water to the pump under a positive head. The velocity should be low, and a suction head of 10 to 20 ft is desirable.

When starting up new boiler-feedwater or high-temperature-water circulating pumps, a strainer is frequently installed to prevent foreign material from getting into the pumps (and lines) or clogging the suction intake. The strainers can be removed when the system is cleaned. All strainers should be inspected periodically.

For a pump installation such as shown in Fig. 8.25, a foot valve should be installed in the suction connection below the normal water level. A check valve should be installed near the pump discharge, followed by a gate valve. This valve arrangement prevents the pump from running backward when shut down and facilitates startup. For large pumps and with dirty water, the installation of twin strainers is recommended. This arrangement will minimize the shutdowns and delays that may be experienced when a single strainer is used.

The small pump installation shown in Fig. 8.25 shows an arrangement of piping and fittings. The pump strainer should not rest on the floor. Turning the pipe down prevents objects from falling into the suction line. In many piping arrangements, a valve should be provided on both suction and discharge lines to isolate the pump when it is necessary to remove it or for repairs.

A pump that has its suction line below the pump centerline requires priming; i.e., the air must be evacuated from the pump suction before water will enter the pump and fill the piping and casing. Priming may be accomplished in a number of ways: (1) if the foot valve is tight, by filling the suction piping with water from some other source; (2) by using an ejector; and (3) by priming with a vacuum pump. The ejector may be air or steam operated or hydraulically operated; it is installed on top of the pump casing. When water is ejected in a steady stream, it may be assumed that the pump is full of water, and the pump may be started, after which the discharge valve may be opened slowly. If a vacuum pump is used, it should preferably be a wet vacuum pump, since water may damage a dry vacuum pump.

To ensure successful operation of a pump, it should be set on a foundation that is substantial and rigid enough to permit the pump to absorb vibration. Good foundations are usually made of concrete. Before the concrete is poured, the foundation bolts are secured as in Fig. 8.26 and surrounded by a pipe sleeve several sizes larger than the bolt. This permits the bolt to be shifted or moved to meet bedplate drillings while the bolt is held securely in place by the washer, which is anchored down by concrete. The pipe surrounding the bolt must not be permitted to extend above the level of the concrete. The top surface of the foundation should be approximately 1 in below the level at which the bedplate is to be set in order to allow for grouting.

To prevent the bedplate from springing out of line, the pump must be aligned carefully. Pumps that come already mounted on a bedplate should be leveled before they are placed in operation. Pumps that are to be set on a rough foundation should be set approximately in location and leveled with shims before the bedplate is grouted in position. The grout should be $\frac{3}{4}$ to $1\frac{1}{4}$ in thick, and the shims should preferably be tapered. Next, check suction and discharge flanges to determine if they are level.

Build a wood form around the outside of the bedplate to hold the grout. Use 1 part cement and 2 parts sand and sufficient water to permit material to flow in and around the bedplate. This mixture should be permitted to set approximately 2 to 3 days, after which the holding-down bolts can be tightened. Be sure to recheck the coupling for alignment. Commercial grouting is readily available with recommended curing periods.

After suction and discharge flanges have been bolted up, the alignment should be checked again. If connecting of piping has caused misalignment, repeat the procedure followed previously. Check pump and motor rotation and insert bolts in flanges and connect up. A short time after the pump has been in operation, the alignment should be checked again. After it is certain that all is satisfactory, the unit can be doweled. The doweling is done with tapered pins usually provided with the pump. Dowel pins are located in the feet of the pump and driver as well as in the bedplate. Permit dowel pins to extend above the feet.

Before operation, after all piping has been connected and after the baseplate has been bolted in place, the pump and motor should be aligned with a dial indicator to ensure that operation is free from vibration. Any misalignment will cause vibration that could lead to bearing and mechanical seal failure.

When the alignment is being checked, and with the coupling disconnected, the direction of rotation should be checked. This check also can be made with the coupling connected; however, the pump must be filled with liquid to prevent running the packing or seal dry. A mechanical seal can fail in a short period of time if it is run dry.

Before placing the pump in operation, make certain that bearings are lubricated. Rotate the motor by hand to make certain everything is free. Prime the pump by one of the methods explained previously. Fill the pump full of liquid, and with suction valve open and discharge valve closed, start the pump, noting suction and discharge pressure. (All pumps should be fitted with pressure gauges on both the suction and discharge sides.) After running the pump this way for a minute or two with the air vent open, close vent valves and open the discharge valve slowly. Observe gauges closely, and if anything unusual occurs or pressures seem unusual, stop the pump and check thoroughly.

Start and stop the pump several times and observe performance. If all looks satisfactory, continue to operate for $\frac{1}{2}$ to 1 h, meanwhile observing bearing temperatures and watching gauges, lubrication, and general actions of pump and motor for overheating. Then shut the pump down, recheck alignment, tighten all bolts, and give the entire installation a careful check.

Place the pump again in operation; observe the packing gland, gland seals, lubrication, and overheating. Never run a pump for an extended period of time with the discharge valve closed unless the vents are open or the pump is equipped with a bypass to permit some water to circulate through the pump at all times.

The pump manufacturer provides recommended installation, start-up, and operating procedures. These recommendations have been developed over many years based on the experience gained with many different pumps in many different services. The plant operating staff should use these procedures as well as those procedures which have been used successfully in their own operating plant.

8.12   Pump Testing and Calculations

The pressure against which a reciprocating pump will operate depends on the dimensions of the pump and the steam pressure. The pressure against which a centrifugal pump will operate depends on design, speed, and the number of stages used. The maximum head against which the pump will operate can be determined from the pressure.

Centrifugal pumps generate velocity in the liquid to move it from place to place or from one level to another or to raise the pressure from the suction to the discharge of the pump. This difference in level (static head) and the difference in pressure (pressure head) must be taken into consideration in calculating losses. Then there are losses due to friction and velocity, which also must be converted to *head* in feet.

A column of water 1 ft high exerts a pressure at the base of this column of 0.433 psi. Assume that a pump operates against a head of water 275 ft high. The pressure in pounds per square inch at the base can then be determined in the following manner: since a column of water 1 ft high exerts a pressure of 0.433 psi, a column 275 ft high will exert $275 \times 0.433 = 119$ psi. If the pressure is known, it is possible to determine the head in feet or the height to which the water can be pumped. Suppose that the gauge at the pump reads 119 psi; the head to which this water can be pumped can be determined as

$$119 \div 0.433 = 275 \text{ ft}$$

The head to which a pump can raise water includes both suction and discharge. Water usually comes to the pump under suction. The distance that the water can be raised (vertical lift) is limited. The theoretical lift is approximately 34 ft, as calculated below. This height, however, is not possible in practice, since leaks in the suction line, packing leaks, etc., reduce the actual lift to about 24 ft. In fact, 10 to 15 ft is more usual because of friction in the line.

At atmospheric pressure of 14.7 psi (approximately equivalent to a 30-in barometer), the theoretical lift is approximately

$$14.7 \div 0.433 = 33.9 \text{ ft}$$

As the temperature of the water is increased, the possible suction lift is decreased. At atmospheric pressure, the boiling point of water is 212°F. If the suction line is under a vacuum, the boiling point is reduced. As a result, vapor from the water would fill the space and partially destroy the vacuum. Water above 130°F should come to the pump under the head. This ensures a full cylinder of water and prevents the pump from becoming steam-bound.

Water at its greatest density weighs approximately 62.5 lb/ft^3. Since there are 1728 in^3 in one cubic foot, 1 in^3 weighs $62.5 \div 1728 = 0.0361$ lb/in^3. In addition, 231 in^3 is equivalent to 1 gal of water. Since 1 in^3 weighs 0.0361 lb, 231 in^3 will weigh $231 \times 0.0361 = 8.33$ lb, the number of pounds in one gallon. If 1 ft^3 of water weighs 62.5 lb and one gallon weighs 8.33 lb, it is apparent that there are $62.5 \div 8.33 = 7.5$ gal in 1 ft^3 (approximately).

Therefore, the following are major conversions used in determining pump performance:

Density of water	62.5 lb/ft^3
Volume of 1 gal of water	231 in^3/gal
Weight of 1 gal of water	8.33 lb/gal

A pump delivering water requires a certain amount of power. The amount depends on the rate at which the water is pumped and the height to which it is lifted. The number of pounds of water pumped per minute times the number of feet that the water is lifted equals the number of foot-pounds of work expended per minute.

Pumps with a given capacity may vary in characteristics to a considerable extent. A pump is said to have *steep* characteristics when the *shutoff* head is considerably above the *operating* head; a pump is said to have *flat* characteristics when the shutoff head is only slightly above the operating head. Figure 8.27 shows pump performance curves for a double-suction single-stage pump operating at constant speed. The shutoff pressure is that shown by checking the head, or pressure, at zero flow or capacity. In Fig. 8.27a, the head would be approximately 115.

It is obvious from the performance curves (see Fig. 8.27) that as the pump capacity increases, the head decreases (there is a drop in pressure); and that an increase in pump capacity requires more horsepower input. The efficiency of the pump increases until it reaches some maximum point, after which it drops off.

A centrifugal pump will develop a given total head regardless of the specific gravity of the liquid, but because of the specific gravity of the liquid, the discharge pressure will vary. For example, when hot water is being pumped, the head in feet will be the same as when cold water is being pumped, but the pressure will be less.

A centrifugal pump operates most efficiently under a head and a speed that approximate design conditions. For a given pump,

1. The power varies as the cube of the speed, i.e.,

$$\frac{\text{Power}_1}{\text{Power}_2} = \left(\frac{\text{speed}_1}{\text{speed}_2}\right)^3$$

2. The head varies as the square of the speed, i.e.,

$$\frac{\text{Head}_1}{\text{Head}_2} = \left(\frac{\text{speed}_1}{\text{speed}_2}\right)^2$$

3. The quantity pumped varies directly with the speed, i.e.,

$$\frac{\text{Capacity}_1}{\text{Capacity}_2} = \frac{\text{speed}_1}{\text{speed}_2}$$

where the power is in horsepower, the speed is in revolutions per minute, the capacity is in gallons per minute, and the head is the pressure head, in feet.

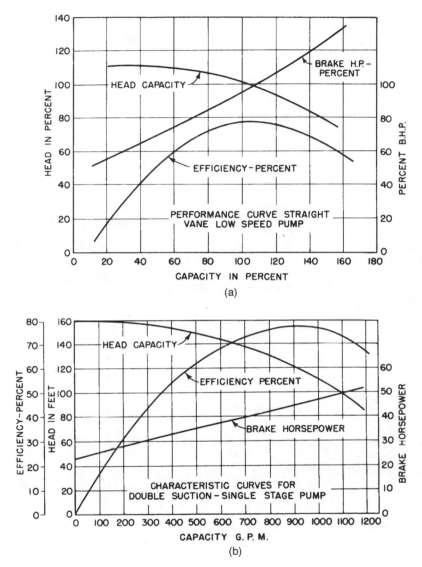

Figure 8.27 Performance curves for centrifugal pumps. (*a*) Head and brake horsepower versus capacity. (*b*) Efficiency, head, and brake horsepower versus capacity.

The power required to pump water is determined as follows:

$$\text{Theoretical hp} = \frac{Q \times 8.33 \times H \times \text{sp gr}}{33,000}$$

$$\text{Required hp} = \frac{Q \times 8.33 \times H \times \text{sp gr}}{33,000 \times E_p}$$

$$\text{kW input} = \frac{Q \times 8.33 \times H \times \text{sp gr}}{33,000 \times E_p \times E_m \times 1.341}$$

where Q = capacity (gpm)
  H = head (ft)
  sp gr = specific gravity of fluid (water = 1.0)
  E_p = pump efficiency expressed as a decimal
  E_m = motor efficiency expressed as a decimal

and 1 gal of water at 62°F = 8.33 lb; 1 hp = 33,000 ft-lb/min; 1 kW = 1.341 hp.

Example If we assume a pump handling 1000 lb of water per minute against a head of 200 ft, what is the theoretical horsepower required?

Solution 1 gal of water weighs approximately 8.33 lb; the pump handles 1000 lb/min.

$$\text{gpm} = \frac{1000}{8.33} = 120 \text{ gpm}$$

$$\text{Theoretical hp} = \frac{120 \times 8.33 \times 200 \times 1.0}{33,000} = 6.06 \text{ hp}$$

Note: To this must be added the pump losses to obtain the required pump horsepower; for example, if the pump efficiency is 70 percent, then

$$\text{Required hp} = \frac{6.06}{0.70} = 8.6 \text{ hp}$$

If the electrical input to the motor drive is required, the motor efficiency must be included. For example, if the motor efficiency is 90 percent, then the required motor input is

$$\text{kW input} = \frac{8.6}{0.90 \times 1.341} = 7.12 \text{ kW}$$

Example A pump is discharging 50 gpm against a head of 300 ft. What horsepower is required if friction and losses are neglected?

Solution

$$\text{Theoretical hp} = \frac{50 \times 8.33 \times 300 \times 1.0}{33,000} = 3.79 \text{ hp}$$

Example Consider a duplex pump 8 by 6 by 10, with a 2-in piston rod, making 50 strokes (1 stroke of a duplex pump is both pistons moving once over their path) per minute against a head of 150 ft. The pump is 75 percent efficient and has a slip of 6 percent.

Determine the capacity of this pump in gallons, pounds, and cubic feet per minute; also determine the horsepower developed.

Solution

$$\text{Area of head end} = \frac{\pi(6)^2}{4} = 28.27 \text{ in}^2$$

$$\text{Area of crosshead end} = 28.27 - \frac{\pi(2)^2}{4} = 25.13 \text{ in}^2$$

$$\text{Displacement of head end} = 28.27 \times 10 \times 25 \text{ strokes/min}$$
$$= 7067.5 \text{ in}^3/\text{min}$$

$$\text{Displacement of crosshead end} = 25.13 \times 10 \times 25$$
$$= 6282.5 \text{ in}^3/\text{min}$$

$$\text{Displacement of one side of pump} = 7067.5 + 6282.5 = 13,350 \text{ in}^3/\text{min}$$

For the duplex pump,

$$\text{Displacement of pump} = 13,350 \times 2 = 26,700 \text{ in}^3/\text{min}$$

$$\text{Volume of 6 percent slip} = 26,700 \times 0.06 = 1602 \text{ in}^3/\text{min}$$

$$\text{Actual displacement of pump} = 26,700 - 1602 = 25,098 \text{ in}^3/\text{min}$$

$$\text{gpm} = 25,098 \text{ in}^3/\text{min} \times \frac{1 \text{ gal}}{231 \text{ in}^3} = 108.65 \text{ gpm}$$

$$\text{lb/min} = 108.65 \text{ gpm} \times 8.33 \text{ lb/gal} = 905 \text{ lb/min}$$

$$\text{ft}^3/\text{min} = 905 \text{ lb/min} \times \frac{1 \text{ ft}^3}{62.5 \text{ lb}} = 14.48 \text{ ft}^3/\text{min}$$

$$\text{hp} = \frac{905 \times 150 \times 1.0}{33,000 \times 0.75} = 5.48 \text{ hp}$$

The total head developed by the pump (see Sec. 8.1) is composed of (1) static head, (2) pressure at the point of discharge, and (3) pipe friction.

Example   Find the total head required for a pump to operate under the following conditions: capacity, 40 gpm; discharge pressure, 15 psi; point of discharge, 30 ft above pump; suction lift, 5 ft; pipe and fittings, 2-in size; pipe, 150 ft long; 6 elbows in line.

Note: Elbows offer resistance equivalent to running length of pipe; friction loss is given from tables not included in text.

Solution

$$\text{Static head} = 30 \text{ ft (discharge head)} + 5 \text{ ft (suction lift)} = 35 \text{ ft}$$

$$\text{Discharge pressure} = 15 \text{ psi} \times \frac{1 \text{ ft}}{0.433 \text{ psi}} = 34.6 \text{ ft}$$

Given six 2-in elbows = 48 ft equivalent pipe length,

$$\text{Total pipe length} = 150 + 48 = 198 \text{ ft}$$

Friction loss for 2-in pipe = 6.6 ft per 100 ft

$$\text{Pipe friction loss} = \frac{198 \times 6.6}{100} = 13.1\,\text{ft}$$

Total head required = 35 + 34.6 + 13.1 = 82.7 ft

Note: Velocity head here is considered minor and is not included. If fluid comes to the pump from above or is delivered to the pump under pressure, this suction head or pressure is deducted, whereas in the example it is added.

8.13 Pump Maintenance

When a *reciprocating* pump is packed with canvas packing, the rings should be cut about $\frac{1}{8}$ to $\frac{1}{4}$ in short of coming together. The packing should be cut diagonally and each joint lapped. The number of rings varies from one to four, and the width depends on the design of the pump. Before packing a pump with rings of canvas, the packing may be expanded to its working size by soaking it in water for several hours prior to packing. If this is not done, short stroking frequently results because of swelling that occurs shortly after the pistons have been packed. Moreover, the packing should not be jammed or crowded into position. Allow enough room for expansion when the pump packing becomes wet.

Scored piston rods are a sign of improperly installed packing, packing that is too tight, or incorrect packing. Leaks around the packing glands are always difficult to detect in the early stages, as in valve leakage. Leakage past the piston is usually detected in the operation of the pump. A careful operator can detect any unusual conditions that occur from time to time. If it is required to run the pump considerably faster than usual to maintain the same service conditions, the packing or valves are evidently leaking. The first procedure, then, is to inspect the valves. After these have been examined, the actual capacity of the pump can be determined and compared with the displacement volume. In these calculations about 5 to 10 percent should be allowed for slip.

Leakage in the water-suction valves can be determined by the appearance of the valves on examination, by the erratic action of the pump, and by the failure of the pump to deliver the required amount of water or the desired water pressure (by leakage of water back into the suction chamber). At times this type of leakage also may be detected by the fact that the pump seems to speed up at points in the stroke rather than maintain a uniform speed. Leakage of the water-discharge valves also can be detected by their appearance on examination, or the valves may be tested by blocking the piston in midstroke and closing the main suction valve on the pump. This will trap water between the head and the underside of the discharge valve. The drain cock on the bottom of the pump on the side being tested should be opened. If water continues to pass through the cock after the water present in the cylinder has been drained, it is reasonably sure that the discharge valve or valves on that side of the pump are leaking, assuming that the piston is packed satisfactorily.

The leakage of the steam valves can be detected by the erratic pump operation. Leakage can frequently be determined from the general appearance of the valve.

Steam valves wear in the seats, and after a while leakage occurs on both ends of the piston simultaneously. It is far less expensive to replace a valve that is thought to be giving trouble than to spend a great deal of time trying to determine how badly it leaks. Usually visual examination should suffice; when there is any doubt, the valve should be replaced. Valves stick because of broken springs and foreign matter getting under them. When leakage occurs, it can sometimes be detected by the jerky action of the piston. If the piston can be moved back and forth with the pump stopped, the gate valve closed, and the discharge valves in good condition, leaking of the suction valves naturally is indicated.

Difficulty is frequently experienced with slippage of the crosshead. The crosshead can be reset without removing the steam-chest cover by disconnecting the rocker arm from the piston rod and pushing the piston to the head of the cylinder. Make a mark on the rod where it passes through the packing gland. Now push the piston until it strikes the opposite head of the cylinder. Again mark the piston rod at the packing gland. Find the center of these two positions marked on the rod. Mark and place this position at the packing gland. The piston is now centered up in the cylinder. Drop the rocker arm vertically, and fasten to the piston rod. The pump is now ready to run.

Another operating difficulty is the deposit of solids on water valves, causing leakage. This can be removed by scraping but more readily by cleaning with dilute acid solution. Also, piston rings wear and must be replaced. Though difficulties are more frequently encountered on the water end of the pump, generally, any part of the pump can give trouble.

Many pump problems, such as vibration, bearing failure, shaft failure, and short coupling life, may be traced to misalignment between pump and driver. The flexible coupling does not provide the adjustment for misalignment. It is important that the manufacturer's instructions be read carefully before starting up the pump. The alignment record should become part of the permanent maintenance record for the pump.

Assuming alignment to be satisfactory on startup, this in no way ensures that the pump will remain in this condition. It is therefore desirable to check the alignment periodically or whenever trouble is experienced. Alignment should be checked when the pump and shaft are at room temperature and again when they are operating at elevated temperatures, since expansion (or a shifting foundation) frequently can change the alignment.

Bearing life depends primarily on *lubrication* and keeping the bearings free from dirt and foreign matter. Bearing life is extended in proportion to the care exercised in providing an excellent lubricant free from moisture and impurities. A *preventive-maintenance schedule* should ensure that the lubricant is drained and the bearing flushed at regular intervals. Use oil recommended by the pump manufacturer. In all cases, stock spare bearings, although bearings and thrust shoes can be repaired. If bearing linings are renewed, provide both lower and upper halves, since if one face is worn, the new shoe will have to carry the entire thrust. Ball bearings must be lubricated separately from the babbitt bearings. They should be lubricated with oil, not grease. There are so many different types of bearings that no effort will be made to cover them here. Rather, it is important to read the instructions provided with each pump, in which are presented the procedures to be followed for maintaining the bearings of that specific pump.

An indication that a pump is operating with too much thrust in one direction is overheating or wearing out of thrust shoes and damage to the impeller. Overheating of the bearing can be detected by placing the back of the hand on the bearing housing. If it is impossible to keep the hand in contact with the bearing, the bearing is running too hot, and the pump is out of alignment or has excessive clearance.

In single-stage pumps, the impeller should be in the center of the casing rings. This can be determined with the top half of the casing removed.

The life of a *centrifugal* pump as well as its reliable operation depends on the sealing or wearing rings and the thrust imposed. As the sealing or wearing rings begin to wear between high and low pressure, the pump begins to lose efficiency; it is then time to renew the rings. The amount of wear is influenced by the amount of clearance originally permitted, by the accuracy of the pump alignment, and by the nature of the liquid pumped.

Packing used for a specific pump should follow the recommendations of the manufacturer. Exercise care in installing this packing, first making certain that all old packing has been removed. In adjusting the gland nuts, pull up evenly on the studs, preferably while the pump is running. Permit a slight leakage to lubricate the packing. Pulling up a gland too tightly will cause excessive shaft-sleeve wear and ruin the packing. If the packing gland has been pulled up too tightly, this will be indicated by an increase in temperature and power consumption. If the packing is installed properly and the gland adjusted correctly, the shaft should turn freely by hand. Should the packing gland leak excessively, do not pull up the gland nuts tightly to stop the leak; rather, take the pump out of service and repack. Nor should time be wasted in trying to pack a pump that has a scored shaft sleeve. Scored piston rods on reciprocating pumps and shafts on centrifugal pumps can be repaired by metallizing.

In packing against high suction pressures, the packing must fit the box exactly, and the shaft sleeves must run true and not be scored. Likewise, in using metallic packing, care must be exercised in order that the packing fit the box exactly. Furthermore, the metallic packing must be lubricated thoroughly, or difficulty with leaks and scored shafts will be experienced. Metallic packing should never be used where soft packing will do.

In some stuffing boxes, a sealing liquid is introduced through the lantern ring to the packing ring. This liquid may be used for cooling or to prevent the escape of the liquid being pumped. In suction pumps, water sealing is frequently used to keep air from leaking in while keeping packing lubricated.

In order to ensure continuous and uninterrupted service, it is necessary to stock a sufficient number of spare parts, and in order to make certain that one obtains the correct part, a record identifying both the pump and its component parts should be kept. For plants that contain many pumps and varieties, it is well to have a suitable storage place. The parts should be catalogued and classified (serially and with reference to catalog number and manufacturer) so that no time is lost in procurement and outage of equipment is prevented. In many modern facilities, the spare parts inventory and records of maintenance are computerized, and these systems significantly assist operations.

For reciprocating pumps, it is usually necessary to have only a number of valves and packing. Inspection generally will reveal any additional maintenance requirements before the parts are actually needed so that corrections can be made after the necessary spare parts have been obtained.

Periodic maintenance on a centrifugal pump usually consists of attention to the stuffing box or seal, proper lubrication, and occasional replacement of bearings. The life of packing or mechanical seals depends largely on the severity of the service, and as operating experience is gained, a preventive maintenance schedule can be developed.

For centrifugal pumps it is necessary to carry such items as packing, a set of bearings, a shaft sleeve or a spare shaft, and a set of impeller wearing rings. If a number of similar pumps are used, it might be well to carry two sets of the above, depending on past experience and the importance of uninterrupted operation.

There is no standard rule that can be applied as to the number and variety of spare parts to be carried in stock. In normal times, the manufacturer may be able to supply the necessary spare parts and deliver them almost as quickly as they can be identified and located in the average stockroom. The essential thing to remember is to keep the spare pumps in first-class condition at all times.

Just as high maintenance costs reflect poor operating or poor mechanical skill, so also delay in making necessary repairs is a sign of carelessness and indifference and poor management. Each pump should be repaired not immediately after it has failed or given trouble in operation but rather before failure or trouble occurs. This can be done by establishing a *preventive-maintenance schedule,* wherein each pump is given a thorough inspection to determine, in advance of failure, whether maintenance is required. Preventive maintenance does not imply that the pump must be dismantled at a scheduled time, since there are many other ways of determining whether a pump is in good operating condition. Once it has been determined whether the pump is or is not in need of repair, it can be either dismantled or restored to service. If in doubt, dismantle the pump.

If a pump is set up on a preventive-maintenance schedule, the schedule should be determined less by time than by the number of operating hours. These figures are best obtained by having the operators keep a log of the running time, from which the maintenance schedule can be fixed and lubrication needs satisfied. Too many oil changes are made when they are absolutely unnecessary.

Preventive-maintenance schedules are best established on the basis of experience. Periods of maintaining the various pumps should be varied so that they do not occur all at the same time. If the operators work in shifts, it is well to spread the work by specifying certain types of pumps as the responsibility of a certain individual (by designation) and so to avoid the shifting of responsibility for carelessness in maintenance. The schedule should be so arranged that the proper work necessary to administer the schedule is held to a minimum.

In setting up such a schedule, it is well to write out all the various inspections that need be made together with the estimated time required to complete each. After this has been done, the total number of manhours can be divided by the mandays over the extended period in which the work is to be completed. Otherwise the schedule may be so filled that should other difficulties be experienced at the same time, there would be little time to devote to this maintenance job. Again, the use of computers can assist operations in this record keeping and in organizing an effective maintenance program.

When examination reveals that the pump requires maintenance, it is usually better to do a complete and thorough job rather than a temporary repair even if the schedule does not demand it, since labor cost is a major item. On the other hand, the practice of replacing parts and making major overhauls just because the schedule indicates them is both foolish and costly. After the pump has been overhauled, the capacity should be checked against the head or, if this is impossible, an approximate test of the pump's condition can be made by determining the static head at shutoff pressure and making a comparison with the design data.

There are many practical methods whereby it is possible to determine whether a pump is performing properly. If the pump is unusually noisy, this may be a sign of vibration, of water recirculation, of undue friction due to wear, or of packing trouble. If the pump casing gets unusually warm (for the liquid being pumped), this is another sign of trouble inside the pump. If the overflow in bearing-cooling water indicates low flow or hot cooling water, shut the pump down and inspect it.

A preventive-maintenance schedule might be initiated, such as this one for boiler feed pumps:

Weekly: Operate under normal load for ½ h; inspect lubrication and cooling-water system; check for noise, unbalance, etc.

Monthly: Repeat above; check bearings and shaft; clean and inspect the motor; feel the bearing to determine heat due to friction, etc.

Yearly: Dismantle pump—overhaul if necessary; check wearing rings, impellers, packing, and alignment. (At the same time, inspect the motor or the turbine drive.) The pump should be checked for clearance of bearings, wear, pitting, corrosion, and alignment.

Since lubrication usually determines the life of the bearing and bearing failure frequently causes many other operating difficulties, lubrication is important. Not only should the lubricant be maintained at the proper level, but the correct lubricant for the particular service must be provided.

Since there are so many varieties of lubricants and special uses to which they can be applied, the manufacturer of the equipment involved should be consulted. Follow instructions implicitly, and do not experiment with lubricants that have not been proved in the field. Lubricants must be free from foreign material, dirt, and water, when contamination exists, change the oil. Usual recommendations, depending on the grade of lubricant and the type of bearing, would be to change the lubricant after every 700 to 1000 h of operation—this is approximately once a month. Manufacturers of pumps provide their recommendations based on the operating experience of many pumps in many different types of operating plants, with each pump having different service experience. Ball bearings are frequently grease-packed with a short-fiber soda-base grease, or a high-grade petroleum is used. Which to use depends on the type of bearing and the speed at which it operates. Follow the manufacturers' instructions. Do not fill the bearing completely, since an excessive amount of lubricant causes overheating. Never run cooling water at such a rate as to keep the oil too cold. Lubricant should be maintained at approximately 120°F or at such a temperature that one can place the back of one's hand on the bearing at all times. When the pump has been stopped, shut off the cooling water to avoid condensation in the bearings. From time to time flush out the cooling system to make certain that it is in good operating condition.

Questions and Problems

8.1   For what types of services are pumps used in a power plant?

8.2   What pumps are required in a basic steam power plant cycle? Describe their purpose.

8.3   What are the two major categories of pumps? Provide examples of each.

8.4   Describe how an injector operates. What are its advantages and disadvantages?

8.5   Describe a duplex pump. What do the numbers 6 by 4 by 5 refer to in reference to a duplex pump?

8.6   What is meant by the percentage of slip of a duplex pump?

8.7   What is the difference between a piston and a plunger?

8.8   Describe the primary difference between a duplex pump and a power pump.

8.9   How are power pumps used in a power plant, and what is their advantage?

8.10   Why are power pumps also classified as metering pumps?

8.11   Why is a metering pump considered unique in terms of accuracy compared with other pumps?

8.12   What is a vacuum pump?

8.13   How does a rotary pump operate? What are its advantages?

8.14 Name the various types of rotary pumps. What application do they serve in a power plant?

8.15 What is a centrifugal pump? How does it operate?

8.16 Describe the classification of centrifugal pumps.

8.17 What is the difference between a volute and a turbine pump?

8.18 What is the purpose of the diffusion ring on a centrifugal pump?

8.19 How is leaking controlled along the shaft of a pump?

8.20 What advantages do mechanical seals have over packing?

8.21 Provide examples of pumps that are found in a power plant and are of the centrifugal pump design.

8.22 If a liquid has a specific gravity of less than 1.0, what does this mean?

8.23 As related to a pump design, describe *head.*

8.24 What is vapor pressure, and what impact does it have on pump operation?

8.25 Describe the effect of cavitation on the operation of a pump. How can it be avoided?

8.26 Describe the net positive suction head (NPSH) for both a centrifugal and a rotary pump.

8.27 How can the output of a centrifugal pump be controlled without causing the pump to overheat?

8.28 Before selecting a type of pump for a particular application, what are the major characteristics that must be considered?

8.29 What are slurries, and how do they affect pump design?

8.30 What should be considered when determining the proper location of a pump?

8.31 Why must a pump be primed?

8.32 Would you prefer that a water supply tank be located above or below a pump? Why?

8.33 Provide a general startup procedure for a pump.

8.34 A gauge is connected to the bottom of a column of water that is 100 ft high. What will the gauge read in pounds per square inch?

8.35 A pressure gauge reads 125 psi at the pump discharge. How high can the water be lifted?

8.36 A tank is $10\frac{1}{2}$ ft long, 4 ft wide, and 20 ft high and is filled with water. How many cubic feet of water does it contain? Cubic inches? Pounds? Gallons?

8.37 A tank has a diameter of 60 in and is 40 ft high. How many gallons of water can it hold?

8.38 A pump 6 by 4 by 8 makes 111 strokes per minute. How many gallons of water will it deliver per minute? Per hour?

8.39 How long will it take the pump in Problem 8.38 to empty the tank in Problem 8.36?

8.40 With reference to the pump in Problem 8.38, how long will it take to empty a water-filled tank that is 15 ft in diameter and 20 ft high?

8.41 A pump 8 by 5 by 10 uses steam at 100 psi. What pressure will be developed on the water side? How high can this water be pumped?

8.42 A pump 3 by 2 by 3 runs 75 strokes per minute with a slip of 15 percent. Calculate the capacity in gallons per 24 h.

8.43 If the atmospheric pressure were 14.3 psi, what would be the theoretical lift?

8.44 A pump 4 by 3 by 6 delivers water at 200 psi. What steam pressure would be required to operate this pump?

8.45 A pump with a 5-in steam piston operating with 150 psi delivers water at 300 psi. What is the size of the water piston?

8.46 How fast will a duplex pump 5 by 3 by 8 run to supply 25,000 lb of water per hour?

8.47 A pump delivers 5000 lb of water to a height of 100 ft in 1 min. What is the theoretical horsepower required?

8.48 A centrifugal pump delivers 300,000 gal of water per hour to a height of 250 ft. The pump is 65 percent efficient. What horsepower is required to deliver this water?

8.49 If the motor drive for the pump in Problem 8.48 is 92 percent efficient, what commercial size motor would be selected for this application? What is the electrical input to this motor?

8.50 What severe maintenance problems can occur when there is a misalignment between a pump and its drive?

8.51 How is the life of a bearing extended?

8.52 For centrifugal pumps, what would a recommended set of spare parts include? For this type of pump, where should periodic maintenance be concentrated?

8.53 Why is a preventive-maintenance program important?

9

Steam Turbines, Condensers, and Cooling Towers

9.1 Steam Turbines

A *heat engine* is one that converts heat energy into mechanical energy. The steam turbine is classified as a heat engine, as are the steam engine, the internal combustion engine, and the gas turbine. Steam turbines are used in industry for several critical purposes: to generate electricity by driving an electric generator and to drive equipment such as compressors, blowers, and pumps. The particular process dictates the steam conditions at which the turbine operates.

The turbine makes use of the fact that steam, when passing through a small opening, attains a high velocity. The velocity attained during expansion depends on the initial and final heat content of the steam. This difference in heat content represents the heat energy converted into kinetic energy (energy due to velocity) during the process.

The fact that any moving substance possesses energy, or the ability to do work, is shown by many everyday examples. A stream of water discharged from a fire hose may break a window glass if directed against it. When the speed of an automobile is reduced by the use of brakes, an appreciable amount of heat is generated. In like manner, the steam turbine permits the steam to expand and attain high velocity. It then converts this velocity energy into mechanical energy. There are two general principles by which this can be accomplished. In the case of the fire hose, as the stream of water issued from the nozzle, its velocity was increased, and because of this impulse, it struck the window glass with considerable force. A turbine that makes use of the impulsive force of high-velocity steam is known as an *impulse turbine*. While the water issuing from the nozzle of the fire hose is increased in velocity, a reactionary force is exerted on the nozzle. This reactionary force is opposite in direction to the flow of the water. A turbine that makes use of the reaction force produced by the flow of steam through a nozzle is a *reaction turbine*. In practically all commercial turbines, a combination of impulse and reactive forces is utilized. Both impulse and reaction blading on the same shaft utilize the steam more efficiently than does one alone.

An example of an impulse force is illustrated in Fig. 9.1a. Here water is seen striking a flat plate (*P*) and being scattered so that any energy remaining is splashed, lost, or dissipated. Only impulse force is used. Figure 9.1b shows a combination of impulse and reaction forces on the bowl (*B*). The increased energy effect can be seen for the identical initial conditions.

A simple illustration of an impulse turbine is shown in Fig. 9.2a. A unit of this type, using steam, was made in the seventeenth century by an Italian named Branca and is the first impulse turbine of which there is any record. Here pressure causes the steam to flow with high velocity from a small jet or nozzle. This steam is directed against the paddle wheel, and rotation results. All pressure drop takes place in the nozzle or stationary elements, and the moving paddles absorb the velocity energy in the steam issuing from the nozzles.

For best economy, the moving element should travel at *one-half* the velocity of the steam from which it is receiving its energy. To meet this requirement, the rotating element of large turbines, using high-pressure steam, would have to operate at excessive speed. The required velocity of the moving parts is reduced by applying the principle of *pressure staging*. Pressure staging consists of allowing only a

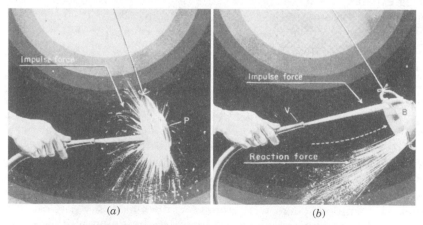

(a)   (b)

Figure 9.1 (*a*) Simple illustration of impulse force only. (*b*) Simple illustration of impulse and reaction forces. (*Tidewater Technological, Inc.*)

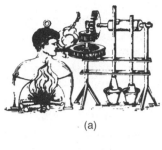

(a)

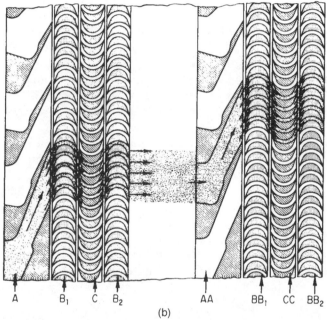

A   B₁   C   B₂   AA   BB₁   CC   BB₂

(b)

Figure 9.2 The impulse turbine. (*a*) Early impulse turbine
(Branca, seventeenth century). (*b*) Diagrammatic impulse turbine. (*Mobil Oil Corp.*)

limited pressure drop in one set of nozzles. After the steam from a set
of nozzles has passed the rotating element (blades), it is expanded in
another set of nozzles. Figure 9.2*b* shows how the steam expands
through nozzles (*A*) and after passing the blading is again expanded
further through additional nozzles (*AA*). A turbine may have many
sets of nozzles (pressure stages), each increasing in size to accommodate the increased volume of the steam.

The principle of *velocity compounding* is also used in the operation
of impulse turbines. This means applying the velocity energy in
steam coming from the nozzles to two or more sets of moving blades.

The operation makes use of a set of stationary blades, which reverse the flow of steam, between each two sets of rotating blades. In Fig. 9.2b the steam in the first pressure stage is expanded through nozzles (A). Then it strikes, in the following order, the moving blades (B_1), the stationary blades (C), and another set of moving blades (B_2). From B_2 the steam enters the second stage and expands through the nozzles at (AA). It then goes through moving blades (BB_1), stationary blades (CC), and another set of moving blades (BB_2) and into the exhaust or next pressure stage. Large high-pressure turbines usually have many pressure stages and in addition use velocity compounding in the high-pressure stages.

Figure 9.3 shows an application of velocity compounding in a small turbine. The steam is expanded in the nozzles and strikes the single row of buckets on the rotor. It then enters the steam-reversing passages in the casing and is again directed against the rotor buckets. This procedure is repeated several times until the velocity of the steam has been reduced. Such an arrangement is known as a *reentry-type turbine.*

If one set of nozzles is used and all the pressure drop occurs in this group of nozzles, and if all the energy is directed against a single wheel, we have a single-stage simple velocity turbine.

The *reaction* turbine is one in which the pressure drop takes place in the rotating element. Figure 9.4a shows a reaction turbine of the simplest type. A turbine similar to this type was made by Hero of Alexandria about 150 B.C. but was never put to any practical use. Steam generated in the boiler passes through trunnions to the hollow sphere. From here steam is discharged through nozzles attached to the sphere. The steam leaving the nozzles at high velocity causes

Figure 9.3 Action of steam in a small turbine.

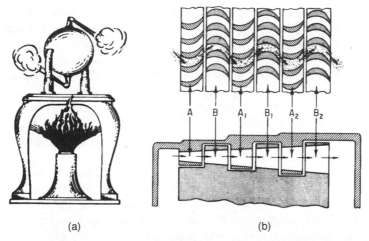

Figure 9.4 The reaction turbine. (*a*) Earliest reaction steam turbine. (*b*) Diagrammatic illustration of section of blading of reaction turbine. (*Mobil Oil Corp.*)

them to move in a direction opposite to the direction of the flow of steam; this causes the sphere to rotate on the trunnions.

The illustration in Fig. 9.1*b* shows a combination of impulse and reaction forces in action. Note that, with the same stream of water, the bowl (*B*) moves farther than the plate (*P*). The water from the jet produces an impulse force in the direction in which the jet is moving, and on leaving the bowl, the water produces a reaction force against the bowl. All turbine blades are shaped to give somewhat the same effect as the water in the bowl.

The modern reaction steam turbine consists of stationary and moving blades. The blades are similar, each being arranged so that the area through which the steam leaves is less than that through which it enters. Pressure drop occurs in both the stationary and the moving blades. The restricted area at the outlet of the blades causes the steam to increase in velocity as it leaves the blading. This same reaction principle is used today in launching rockets into space; the modern airplane uses the same jet reaction to propel its jet engines.

Figure 9.4*b* is a diagrammatic arrangement of the blading in a typical reaction turbine. *A*, *A*$_1$, *A*$_2$, etc., are the stationary blades mounted in the casing; *B*, *B*$_1$, *B*$_2$, etc., are the moving blades mounted on the rotor. The steam in passing through the turbine expands alternately through the stationary and moving blades. The stationary blades are designed to direct the steam against the next row of blades. Note that the area of the steam passages at the outlet of the blades is less than that at the inlet. As a result of this change in area of the steam passage, the velocity increases and the pressure decreases as the steam leaves each set of blades.

The reaction turbine is in reality an expansion nozzle in which the pressure of the steam is used to increase its velocity. This velocity is converted into mechanical energy by the rotating element. During this process of expansion, the steam increases in volume. The area of the steam passage through the blading must increase, from the high-pressure to the low-pressure end of the turbine, in order to accommodate this larger volume of steam.

Radiation and condensation losses in turbines are small. Leakage losses occur through clearances over the ends of reaction buckets and at labyrinths and glands. The efficiency of the turbine is affected by friction of high-velocity steam jets through nozzle passages and across buckets together with friction losses of high-speed revolving disks and idle buckets in steam-filled chambers.

The steam turbine is a prominent part of the overall system, whether it drives a generator for the production of electricity or is used as a drive for equipment. This chapter will concentrate on turbines used for the generation of electric power and on the major balance of plant equipment, such as condensers and cooling towers, that are so critical to the efficient performance of a turbine.

As described in Chap. 1, the production of power from steam involves the integration of many major components and systems. As shown in Fig. 1.2, this schematic shows a coal-fired power plant used for the production of electricity. The steam-system portion generally can be described to include the fuel handling and preparation equipment, the boiler and its associated systems, and the environmental control systems that handle the plant's emissions. For the generation of electricity, a steam turbine with a connected electrical generator is necessary for the conversion of the thermal energy of the steam into the mechanical energy of the turbine and finally into electrical energy by the connected generator.

A number of heat-exchanger systems are used to improve the overall cycle efficiency and to reduce fuel costs, which reduces the costs for the production of electricity. Some of these systems, as shown in Fig. 1.2, are air heaters for the heating of combustion air and feedwater heaters for the heating of the condensed water from the turbine prior to its reentry back into the boiler system. The water-cooled condenser and cooling tower reject the unused heat energy to the environment in an acceptable manner. The condenser also maintains a low backpressure on the steam turbine so that the performance of the turbine is optimized.

Power plants for the production of electricity typically have been in size ranges from 100 to 1300 MW for single-turbine generator installations. However, with the presence of independent power producers (IPPs), who often use waste fuels such as municipal solid waste (MSW), wood wastes, and waste coal, plant sizes often are in the ranges from 10 to 80 MW, and they operate with the same reliability criteria as a large electric utility.

Turbines operate at rotating speeds of 3600 rpm for 60-Hz applications and 3000 rpm for 50-Hz applications. The maximum exhaust moisture in the turbine is generally limited to between 10 and 15 percent steam by weight. Moisture in the steam is highly erosive, especially at the high rotating speeds of the turbine. Special blade materials and limitations on blade length are used to minimize any erosion and the resulting maintenance costs and loss of revenues due to plant outages that would be required for unplanned repairs.

9.1.1 Turbine types and applications

As described previously, a steam turbine takes the thermal energy of the steam, which is provided by a boiler, and converts it into useful mechanical work by means of the steam expanding as it flows through the turbine. Steam is introduced into the turbine through small stationary nozzles, where the steam expands and reaches a high velocity. This process converts the thermal energy in the steam to kinetic energy as it passes through the nozzle openings and moves the turbine blades, which are attached to the rotor.

Steam turbines are designed for a variety of applications and in a variety of sizes that range from 1-hp (0.75-kW) units, which are used as drives for process equipment such as pumps and compressors, to large 1500-MW capacity units, which are found in large electric power plants. A typical utility turbine arrangement is shown in Fig. 9.5, where 600-MW units are shown in a turbine room. Figure 9.6 shows a large turbine rotor for an electric utility with its casing removed.

Heat rate is a term that is frequently used in the power industry, and it defines power plant efficiency. The net plant heat rate is the total fuel heat input in Btu/h divided by the net power output leaving the power plant in kilowatts. The net output is obtained by subtracting all auxiliary electrical power needs of the plant from the gross output of the generator. The following equation expresses this relationship:

$$\text{Net plant heat rate (NPHR)} = \frac{\text{total fuel heat input (Btu/h)}}{\text{net electrical generation (kW)}}$$

The total energy efficiency of a plant can be expressed as follows:

$$\text{Efficiency, \%} = \frac{3413 \text{ Btu/kWh}}{\text{NPHR}} \times 100\%$$

Figure 9.5 Turbine room of utility power plant showing three 570-MW steam turbine generators. (*Babcock & Wilcox, a McDermott company.*)

The NPHR usually varies with the plant load and is expressed in Btu/kwh.

Turbines are classified in two ways: (1) steam supply and exhaust conditions and (2) casing or shaft arrangement. These are further described below.

Steam supply and exhaust conditions. When classifying steam turbines by their steam supply and exhaust conditions, they are categorized as condensing, noncondensing or backpressure, reheat-condensing, and extraction and induction.

1. *Condensing turbine.* This type of steam turbine is used primarily as a drive for an electric generator in a power plant. These units exhaust steam at less than atmospheric pressure to a condenser. Figure 9.7 shows a typical arrangement of a condensing turbine that is designed for an output of 100 MW and has one extraction point. Figure 9.8 shows an industrial-sized condensing turbine designed for

Figure 9.6 Turbine rotor for utility power plant. (*Siemens Power Corp.*)

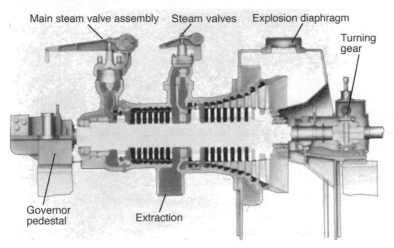

Figure 9.7 Single-casing 100-MW extraction condensing turbine. (*Westinghouse Electric Corp.*)

outputs ranging between 2 and 65 MW when supplied with steam at approximately 1450 psi and 930°F. The rotor for a condensing turbine of this type is shown in Fig. 9.9.

2. *Noncondensing or backpressure turbine.* This type of turbine is used primarily in process plants, where the exhaust steam pressure is controlled by a regulating station that maintains the process steam at

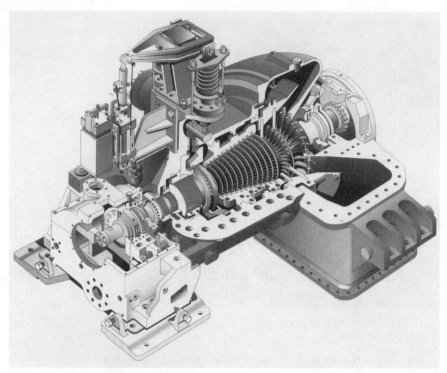

Figure 9.8 Condensing steam turbine for approximately 65-MW output. (*Siemens Power Corp.*)

the required pressure. Figure 9.10 shows a typical arrangement of a backpressure turbine. An industrial-sized backpressure turbine is shown in Fig. 9.11, where the total steam flow is exhausted and generally used for some process. The unit shown can be designed for initial steam conditions of up to 1450 psi and 930°F, and it can produce outputs between 2 and 28 MW.

3. *Reheat-condensing turbine.* This type of turbine is used primarily in electricity-producing power plants. In these units, the main steam exhausts from the high-pressure section of the turbine and is returned to the boiler, where it is reheated with the associated increase in steam temperature. The steam is now at a lower pressure but often at the same superheat temperature as the initial steam conditions, and it is returned to the intermediate- and/or low-pressure sections of the turbine for further expansion.

4. *Extraction and induction turbine.* This type of turbine is also found primarily in process plants. On extraction turbines, steam is taken from the turbine at various extraction points and is used as process steam. In induction turbines, low-pressure steam is introduced into the unit at an intermediate stage to produce additional

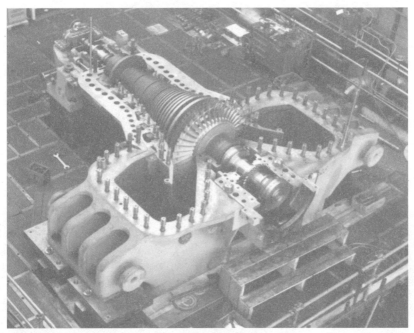

Figure 9.9   Turbine rotor for condensing steam turbine. (*Siemens Power Corp.*)

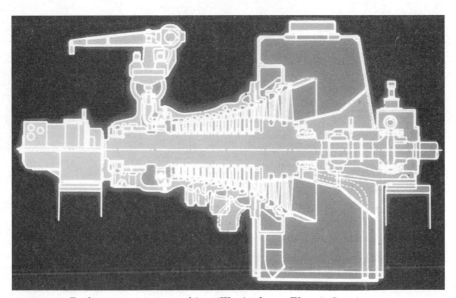

Figure 9.10   Backpressure steam turbine. (*Westinghouse Electric Corp.*)

Figure 9.11 Backpressure steam turbine for approximately 28-MW output. (*Siemens Power Corp.*)

power. Figure 9.12 shows an industrial-sized extraction condensing turbine where some of the steam is extracted as it passes through the turbine. This extraction steam can be used for feedwater heating or for some process. This particular unit can be designed to handle steam with initial steam conditions as high as 2000 psi and 1000°F and provide electrical outputs from 2 to 75 MW.

For industrial-sized turbines as described above, the design must match the particular requirements of the plant. Some turbines are required to meet electrical requirements and therefore are considered to be similar to large turbines for utility use in that they are condensing turbines. These units can be designed for electrical outputs of up to approximately 150 MW and are housed in a single casing. Other industrial turbine designs furnish a specific quantity of exhaust steam or extraction steam for some industrial process or for heating. These turbines are called *backpressure turbines* and *extraction-condensing turbines*. It is this type of turbine that is used in combined cycle systems.

Figure 9.12 Extraction condensing steam turbine. (*Siemens Power Corp.*)

Casing or shaft arrangement. Steam turbines are also classified by their casing or shaft arrangement as being single, tandem-compound, or cross-compound and are described as follows:

1. *Single casing.* This is the basic arrangement for smaller units, where a single casing and shaft are used.

2. *Tandem-compound casing.* This arrangement has two or more casings on one shaft that drives a generator.

3. *Cross-compound casing.* This arrangement has two or more shafts that are not in line, with each shaft driving a generator. These units are found in large electric utility power plants.

9.1.2 Turbine stage design

The efficiency of a turbine is optimized as the steam expands and does work in a number of steps or stages as it flows through the turbine. These stages are identified from the manner by which the energy is removed from the steam.

There are two types of turbine stages, impulse and reaction, and most turbines combine features of both, as shown in Figs. 9.2 and 9.4. Figure 9.13 shows a diagram of simple impulse and reaction stages of a turbine.

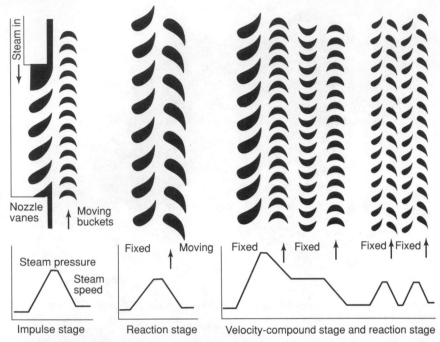

Figure 9.13 Diagram of simple impulse and reaction turbine stages. (*Power Magazine, a McGraw-Hill publication.*)

Impulse turbine. This stage design is often compared with a water wheel because nozzles direct the steam that flows through high-velocity jets. These steam jets, which contain kinetic energy, flow against the moving turbine blades or buckets. This energy is converted into mechanical energy by rotating the shaft. In a pure impulse turbine, when the steam passes through the stationary blades, it incurs a pressure drop. There is no pressure drop in the steam as it passes through the rotating blades. Therefore, in an impulse turbine, all the change of pressure energy into kinetic energy occurs in the stationary blades, while the change of kinetic energy into mechanical energy takes place in the moving blades of the turbine.

Reaction turbine. This design uses the reaction force resulting from the steam accelerating through the nozzles. The nozzles are actually created by the blades, as shown in Fig. 9.13. Each stage of the turbine consists of a stationary set of blades and a row of rotating blades on a shaft. Since there is a continuous drop of pressure throughout each stage, steam is admitted around the entire circumference of the blades and, therefore, the stationary blades extend around the entire circumference. Steam passes through a set of stationary blades that direct the steam against the rotating blades. As the steam passes through these rotating blades, there is a pressure drop from the entrance side to the exit side that increases the velocity of the steam and produces rotation by the reaction of the steam on the blades.

9.1.3 Major components of a turbine

The major portions of a turbine are shown in Fig. 9.14. The turbine consists of a shaft, which has one or more disks to which are attached moving blades, and a casing in which the stationary blades and nozzles are mounted. The shaft is supported within the casing by means of bearings that carry the vertical and circumference loads and by axial thrust bearings that resist the axial movement caused by the flow of steam through the turbine. Seals are provided in the casing to prevent the steam from bypassing the stages of the turbine.

Blades. On the outer portion, or circumference, of each disk located on the shaft are blades where steam is directed and converted into work by rotation of the shaft. There are many blades in each turbine stage, and larger turbines have more stages. Blades generally are made from low-carbon stainless steel; however, for high-temperature applications and where high moisture is expected, alloy steels are used to provide the strength and erosion resistance needed. Special coatings on the blades are often used where high erosion is anticipated.

As the steam flows through the turbine, it expands and its volume increases. This increased volume is handled by having longer blades and thus a larger casing for each stage of the turbine. Figure 9.15 is a schematic showing how the blade size varies as the steam flows through the turbine.

The turbine efficiency, as well as its reliable performance, depends on the design and construction of the blades. Blades not only must handle the steam velocity and temperature but also must be able to handle the centrifugal force caused by the high speed of the turbine. Any vibration in a turbine is significant because there is little clearance between the moving blades and the stationary portions on the casing. A vibration of the moving blades could cause contact with the stationary components, which would result in severe damage to the turbine. Vibration has to be monitored continuously and corrected immediately when required.

Rotor shaft and bearings. The rotor shaft is supported at each end by bearings. These are normally ball bearings on small turbines; however, on the larger turbines, a pressure-lubricated journal bearing is used. Because of the axial thrust along the shaft that results from the

High Pressure Turbine

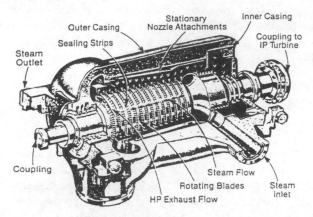

Intermediate Pressure Turbine

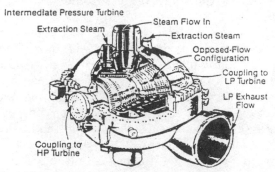

Low Pressure Turbine

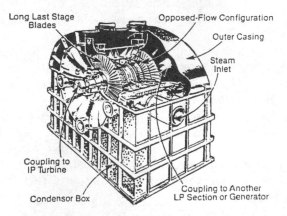

Figure 9.14 Schematics of typical high-, intermediate-, and low-pressure steam turbine sections. (*Power Magazine, a McGraw-Hill publication.*)

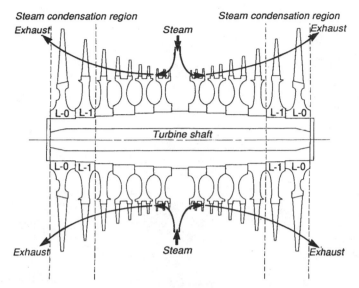

Figure 9.15 Double-flow low-pressure turbine showing variation in blade size. (*Power Magazine, a McGraw-Hill publication.*)

difference in steam pressure across the stages of the turbine, thrust bearings are used to maintain the clearances between the moving blades and the stationary portions in each stage of the turbine.

Casings and seals. Casings are steel castings whose purpose is to support the rotor bearings and to have internal surfaces that will efficiently assist in the flow of steam through the turbine. The casing also supports the stationary blades and nozzles for all stages.

At the turbine inlet, steam enters through a stop valve and steam chest. In high-temperature turbines these components are separate from the main turbine structure. In smaller units, the steam chest is usually mounted directly on the casing. At the outlet of the turbine, an exhaust hood guides the steam from the last stage to the condenser inlet.

Figure 9.16 shows a cross section of a steam admission section. This design has the stop valves located in the steam chest, and when necessary, this can be removed without having to dismantle piping. The control valves are suspended on a cross-bar that is moved by two stems. Therefore, only two stuffing boxes are required for passing the stems through the casing, although four or five control valves are provided for nozzle group control.

The steam chest and valve assembly shown in Fig. 9.17 show another design, and the illustration identifies the major components,

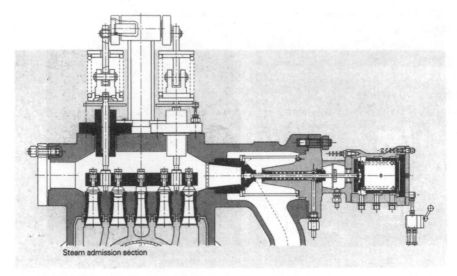

Figure 9.16   Turbine steam admission section. (*Siemens Power Corp.*)

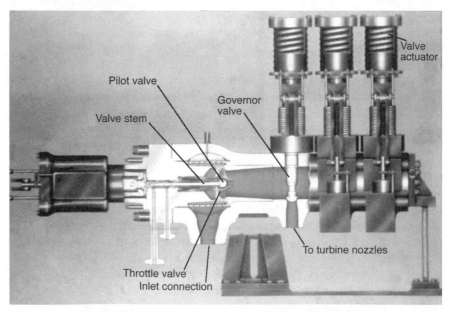

Figure 9.17   Turbine steam chest and valve assembly. (*Westinghouse Electric Corp.*)

including the steam inlet, the throttle valve, governor valve, and valve actuators.

In addition to being designed to support the weight of the stationary nozzles and blades, the casing also must resist the mechanical stresses that are caused by the reaction forces on these nozzles and blades as well as the thermal stresses that are caused by the steam temperature differentials that occur during operation in the various stages of the turbine.

Since the shaft penetrates through the casing as shown in Fig. 9.14, seals are necessary to minimize the leakage of steam. In small, low-temperature turbines, carbon packing ring seals are used. These seals are located directly on the shaft and are held in place by a spring assembly. In larger turbines, labyrinth seals are used to control steam leakage. In many turbine designs, a combination of the two types of seals are used at the ends of the shaft.

Pressure sections of turbines. Small turbines are housed in a single casing that admits high-pressure steam at one end, and low-pressure steam leaves at the back end of the turbine to the condenser or as steam for a process or heating. On large, high-pressure turbines, two or three separate casings are used, with the turbine having three sections: high pressure (HP), intermediate pressure (IP), and low pressure (LP). Refer to Fig. 10.2.

The steam from the boiler passes through the HP section of the turbine. The exhaust steam from this section is returned to the boiler's reheater, where it is reheated to generally the same superheated steam temperature as the HP steam but at a lower pressure. The steam is then returned to the IP section of the turbine. In some turbine designs, the HP and IP sections of the turbine are combined into one casing, and the steam flows in opposite directions in order to equalize the axial forces in the turbine.

The IP turbine is used with boilers that have a reheat cycle, and this is found in large utility power plants. The IP turbine can use single or double flow depending on the pressure and steam flow.

The LP turbine receives steam from either the HP turbine in nonreheat units or an IP turbine. Since the steam has expanded and its volume has increased significantly, the blades of the LP turbine are much longer than those of the HP and IP stages in order to handle the steam flow.

Steam flow control. The part load performance and responsiveness of a steam turbine are to a large degree dependent on the method used to control the steam flow to the first stage of the turbine. There are two admission methods used: partial arc and full arc.

Partial-arc admission requires the adjusting of the active nozzle area, where the nozzles in the first stage are divided into groups and controlled separately with throttling valves. When the load is increased, the throttling valves for each group of nozzles are opened in sequence to the full-throttle position until the desired load is attained.

Full-arc admission requires the steam pressure entering all the first-stage nozzles to be adjusted by either of two ways:

1. By operating the boiler at constant pressure and throttling the steam flow, with all valves being opened together until the required load is obtained

2. By varying the boiler operating pressure with the throttle valves basically full open, except at the lowest loads

The selected control system is based on the planned operation of the turbine, since both systems have disadvantages. A base-loaded turbine would have a different control system than would a turbine planned for variations in load.

Some of these disadvantages for each system are as follows:

1. Full-arc admission that requires constant throttle pressure is a relatively simple system, but when the turbine operates at part loads, pump power is wasted.

2. When the steam flow is throttled for full-arc admission, this results in a turbine inlet steam temperature drop.

3. When a partial-arc admission system is used that has a constant throttle pressure, this reduces the energy loss caused by steam throttling, but it is generally less efficient at full load.

4. By varying the boiler pressure, a less responsive system results for partial arc admission.

Therefore, each system affects the part load efficiency differently. Differences also exist on the turbine inlet steam temperature and on the system responsiveness. As a result, systems are designed with combined features of each control strategy in order to optimize the controls for the expected operation of the turbine.

9.2   Turbine Design and Construction

A simple turbine consists of a shaft on which is mounted one or more disks. On the circumference of the disks are located blades or buckets to receive the steam and convert it into useful work. The rims of the disks have dovetail channels for receiving the blades. The ends of the blades are made to fit these dovetail channels. A turbine requires bearings for support, a suitable housing or casing to enclose the rotor, a system of lubrication, and a device known as a *governor* to maintain control over the speed.

The efficiency and reliable performance of the turbine depend to a great extent on the design and construction of the blading. The blades therefore must be made to withstand the action of the steam and the centrifugal force caused by the high speed at which the turbine must operate. In designing turbine blading, a compromise must be made between strength and economy. The stationary and moving elements have very little clearance, and any vibration of the moving element will cause them to rub. For this reason, extreme care must be exercised to design the rotors so that they will not vibrate. The length and size of the blades must be increased as the steam pressure drops and the steam increases in volume as it flows through the turbine. Large condensing turbines have large rotors and long buckets in their last stages. This reduces the velocity of the steam leaving the turbine and as a result improves the efficiency of the turbine.

Turbine blades are forgings made of steel or alloys, depending on the conditions under which they are to operate. The use of superheated steam requires blades that are specially designed to prevent warping and deterioration. Steam in the last stages of a turbine becomes very wet (approximately 10 to 15 percent moisture), and this moisture erodes the turbine blades. Special materials are used to lengthen their life.

The blades are assembled on the disk, and a *shroud ring* is placed around their outer ends (Figs. 9.18, 9.19*f*, and 9.21). The tips of the blades pass through holes in the shroud ring. The ends are then welded so that they are held securely by the ring. When the blades are very long, extra lacing is sometimes used to tie them together to provide the necessary support.

Rotors for small turbines consist of a machined-steel disk shrunk and keyed onto a heavy steel shaft. The shaft is rust protected at the gland zones by a sprayed coating of stainless steel. The rotor is statically and dynamically balanced to ensure smooth operation throughout its operating range.

The small steam turbine (see Fig. 9.18) is used as a mechanical drive. Figure 9.19 shows the major component parts of this turbine. The steam chest is bolted to the base and is made of iron or steel. It contains a governor valve, a strainer, and an operating hand valve that is used for manual adjustment to obtain maximum efficiency. Regardless of whether or not a hand valve is provided, the steam is made to pass through the governor-controlled admission valve contained in the steam chest. These are only typical illustrations of a

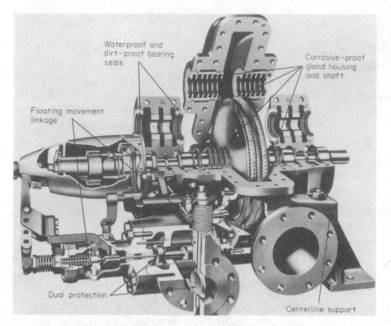

Waterproof and
dirt-proof bearing
seals

Corrosive-proof
gland housing
and shaft

Floating movement
linkage

Dual protection

Centerline support

Figure 9.18 Typical general-purpose turbine for mechanical drive.

small turbine design, since there are numerous designs with different features that vary between manufacturers.

The governor valve located in the right end of the steam chest is a double-seated balanced valve operating in a renewable cage (Fig. 9.19c). The valve and cage are made of noncorrosive material, and the valve stem is made of stainless steel. Packing is provided where the valve stem passes through the chest cover.

Speed regulation is maintained by means of the governor (Fig. 9.19b). This governor is of the centrifugal-weight type and is connected to the admission valve by the arm and linkage shown in Fig. 9.18. Movement of the governor weights is opposed by the compression spring to transmit its motion from the rotating spindle to a stationary sleeve through a self-aligning ball-thrust bearing. If the speed of the turbine exceeds a safe limit (10 percent above normal speed), an overspeed trip device (Fig. 9.19d) is provided. This trip is eccentrically mounted and restrained by a spring to trip simultaneously the governor valve (Fig. 9.19c) and the butterfly valve (Fig. 9.18). Hand reset is provided when the speed has been reduced to normal. In normal operation, the butterfly valve is held open by the trip linkage against the force of the coil spring shown. In some cases the butterfly trip valve is omitted, and overspeeding closes the governor valve.

Located in the casing are the steam-admission nozzles (Fig. 9.19c), which are cut into a solid block of bronze or alloy steel, depending on

Figure 9.19 Typical component parts of a general-purpose turbine. (*a*) Steam chest. (*b*) Speed-regulating governor. (*c*) Governing valve. (*d*) Overspeed trip. (*e*) Nozzles. (*f*) Rotors and blades.

steam conditions. Nozzles are so proportioned as to be contributory to efficient operation and are made of corrosion- and erosion-resistant materials. This nozzle block is bolted to the steam chest, which in turn is bolted to the base of the turbine casing. The entire assembly of nozzles for one stage is called a *diaphragm*. The casing assembly with the stationary blading or nozzles is referred to as the *turbine cylinder*. The cylinder of an impulse turbine is frequently referred to as the *wheel casing.* The turbine blades (Fig. 9.19*f*) are made from rolled and drawn sections of stainless steel. The blading, shrouding, and rotor rim are contoured to approximate the steam-expansion characteristics. The rotating blades are secured in dovetail grooves cut into the rotor disk and are regularly spaced by packing pieces.

Around the outer rim of the blades or nozzles is provided a shroud ring (Fig. 9.18*f*) to stiffen the blades against vibration and confine the steam to the blade path. Each group of blades is tied together. Round tenons at the blade ends are machine spun to secure shroud bands to the blades.

Bearings support the rotor. They are horizontally split and are often lined with high-grade tin babbitt. Babbitt is a soft, silvery antifriction alloy composed of tin with small amounts of copper and antimony. Access to the bearings is possible without raising the cylinder cover or rotor. The governor-end bearing is babbitt-faced at both ends and acts as a combined thrust and journal bearing. Rings suspended in oil roll on the shaft to provide lubrication to the bearing. Water jackets are provided for cooling the oil.

To prevent leakage of steam where the rotor shaft extends through the turbine casing, carbon-ring-type glands are provided. These consist of segmental rings held around the shaft by springs. Between the two outer rings of each gland is a leak-off connection that prevents steam that may pass the inner rings from leaking. The casing proper (Fig. 9.20) is bolted together at the horizontal joint. Flanges are frequently finish-ground, making possible a steamtight joint without the use of a gasket.

On units larger than that shown in Fig. 9.18, oil for governing and flood lubrication is supplied by a shaft-driven gear-type pump. If lubrication requirements exceed the capacity of a single pump, a separate gear-type pump is furnished to supply lubricating oil. The direct-acting oil governor with hand-speed changer for 3:1 maximum speed adjustment is used for variable-speed applications, such as driving fans, blowers, etc. For the mechanically driven turbine, the speed is frequently automatically controlled by the draft, air pressure, feedwater pressure, and other means. *Note:* Design standards for materials and governor performance are covered in publications of the National Electrical Manufacturers Association (NEMA).

On large turbines there are normally high- and low-pressure sections, with the steam chest integral with the high-pressure section. Cylinders are split along the horizontal plane. Blades may be assembled in separate blade rings or directly in the cylinder, depending on the turbine size and pressure and on temperature conditions. In order to minimize misalignment and distortion, turbines are so designed as to permit expansion and contraction in response to temperature changes.

(a)

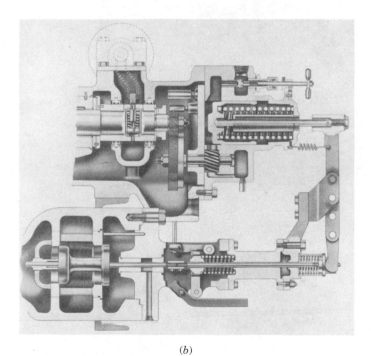

(b)

Figure 9.20 (a) Typical turbine casing. (b) Hydraulic orifice governor.

Larger turbines have their rotors (Fig. 9.21) formed from a single-piece forging, including both the journals and the coupling flange. Thrust-bearing collar and oil impeller may be carried on a stub shaft bolted to the end of the rotor. Forgings of this type are carefully heat treated and must conform to specifications. Rotors are machined, and after the blades are in place, they are dynamically balanced and tested.

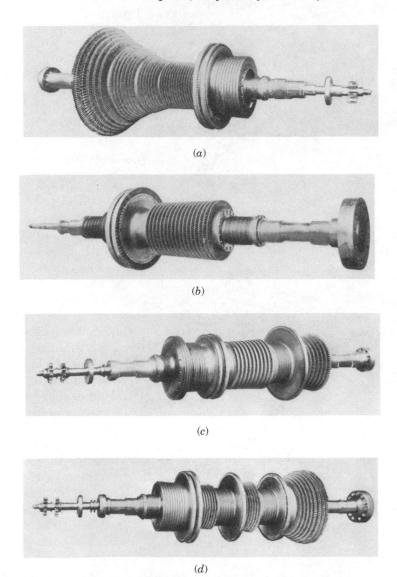

(a)

(b)

(c)

(d)

Figure 9.21 Rotors for various types of turbines. (a) Rotor for condensing turbine. (b) Rotor for noncondensing turbine. (c) Rotor for noncondensing single-extraction turbine. (d) Rotor for condensing double-extraction turbine. (*Westinghouse Electric Corp.*)

As with the small turbine, the efficiency and life of larger turbines are influenced chiefly by the form or shape of the blades, the manner in which they are fastened in place, and the materials from which they are made. *Rotating* blades are secured by a T-root fastening with lugs (Fig. 9.22) machined on the shank, straddling the blade groove. Blades are held against a shoulder in the groove by half-round sections caulked in place at the bottom. *Stationary* blades are anchored in straight-sided grooves by a series of short keys that fit into auxiliary grooves cut in the blade shank and in the side of the main groove. If high-pressure and high-temperature steam is used, steam leakage across impulse stages is controlled by thin sealing strips (Fig. 9.22*a*) that can be set with close running clearance. On reaction turbines of the larger sizes where high-pressure and high-temperature steam is used, *shrouded* blades (Fig. 9.22*b*) are used with radial seal strips to control leakage between stages. Seal strips are made very thin, permitting close running clearances.

Two types of nozzles are used (Fig. 9.22*a*): round nozzles with holes drilled and reamed in a solid block of steel and curved-vane nozzles. The interstage diaphragms are located in grooves that are accurately spaced and machined into the casing. The upper halves are attached to, and lift with, the casing cover. Labyrinth seals (see Fig. 9.27) minimize steam leaks along the shaft where it passes through the diaphragm. The seal rings are spring backed, and they are made of a material that permits close running clearances with complete safety.

Turbine blades are at times designed in different configurations in an attempt by the designer to optimize performance and minimize maintenance. Figure 9.23 shows an example of an advanced blade configuration on a rotor shaft that is designed to optimize the flow distribution over the length of the flow path. This blade design has reduced losses and has optimized the flow conditions for the blades. The advanced design shown has a forward-curved and twisted hollow blade. This design provides optimal flow distribution. The hollow blade provides suction slots for moisture removal. A diaphragm with the welded-in forward-curved and twisted blades is shown in Fig. 9.24. By research and operating experience, each designer of turbines attempts to improve the overall performance and efficiency of the turbine. As a result, different designs have evolved.

Main bearings (Fig. 9.25) consist of a cast-steel shell, split horizontally and lined with high-grade tin-base babbitt. They are supported in the bearing housing by steel blocks. Between the supporting blocks and the bearing shell are steel liners, where the bearings can be moved both vertically and horizontally to align the rotor accurately within the cylinder.

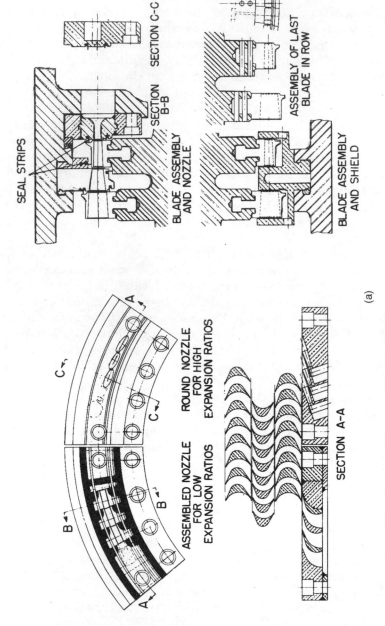

SECTION C-C

SECTION B-B

SEAL STRIPS

BLADE ASSEMBLY
AND NOZZLE

ASSEMBLY OF LAST
BLADE IN ROW

BLADE ASSEMBLY
AND SHIELD

A

C

C

B

B

A

ROUND NOZZLE
FOR HIGH
EXPANSION RATIOS

ASSEMBLED
NOZZLE
FOR LOW
EXPANSION RATIOS

SECTION A-A

(a)

Figure 9.22 Turbine blades. (*a*) Nozzles and impulse blading.

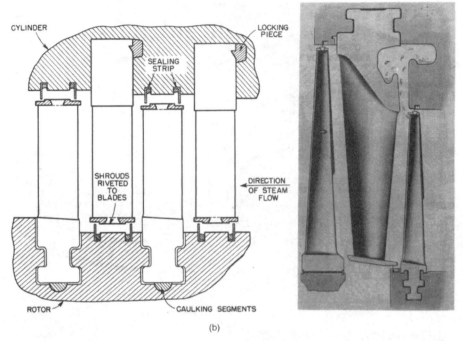

(b)

Figure 9.22 Turbine blades. (*b*) Reaction blading showing moisture catcher and seal strips. (*Westinghouse Electric Corp.*)

Figure 9.23 Advanced turbine blade design to optimize flow distribution through turbine. (*Siemens Power Corp.*)

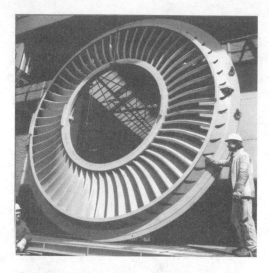

Figure 9.24 Forward-curved last-stage stationary blades. (*Siemens Power Corp.*)

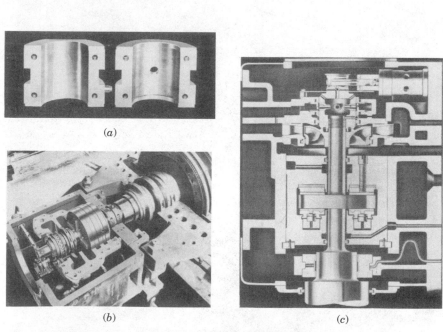

(a)

(b)

(c)

Figure 9.25 Main and thrust bearings. (a) Main bearing. (b) Section of thrust bearing and housing. (c) Thrust bearing cage in place. (*Westinghouse Electric Corp.*)

The conventional turbine bearing consists of a cylindrical shell divided into halves so that it can be assembled on the shaft. The outside of the shell has a spherical section at the middle. This fits into a similarly shaped seat in the bearing support pedestal and is an aid in properly aligning the bearing. In most cases the top half is grooved on the trailing side of the bearing in such a manner that there is a tendency to draw or aspirate oil into the bearing. The bottom half of a split bearing should be relieved on each side of the joint by scraping it clear of the shaft. The clearance of the bearing can be measured by placing a soft lead wire on the shaft and bolting the top half down tight. The flattened wire can be removed and the thickness measured by a micrometer. The clearance in the bearing should be 0.002 in per inch of shaft diameter. Each turbine manufacturer provides operating instructions and recommendations to ensure that clearances are established properly. These should be followed.

Oil is supplied through a hole drilled in the bearing housing that matches a similar hole in the lower supporting block. For large bearings, orifice outlets are provided for positive control.

The pressure drop through the rotating blades of the reaction turbine produces a force that tends to push the turbine rotor toward the low-pressure end. In some turbines this force is balanced by what is known as the *double-flow principle*. In this case, steam enters the middle of the turbine and flows in both directions. This produces two forces that balance each other. Another method is to use balancing or dummy pistons, as shown in Fig. 9.26.

9.2.1 Thrust bearing

The thrust bearing (Figs. 9.25 and 9.26) consists of a collar rigidly attached to the turbine shaft rotating between two babbitt-lined shoes. The clearance between the collar and the shoes is small. The piston is attached to the spindle, and steam pressure is exerted on one side and atmospheric pressure is exerted on the other side. The difference in pressure produces a force that balances the thrust exerted on the rotating blades. If the shaft starts to move in either direction, the collar comes into contact with the shoes, and the shaft is held in proper position. Larger thrust bearings have several collars on the shaft and a corresponding number of stationary shoes.

The *Kingsbury thrust bearing* (Fig. 9.25) is used when a large thrust load must be carried to maintain the proper axial position in the turbine cylinder. (The one shown in Fig. 9.25 is a combination of the Kingsbury and collar types.) The thrust collar is the same as that used in the common type of thrust bearing. The thrust shoes are made up of segments that are individually pivoted. With this arrange-

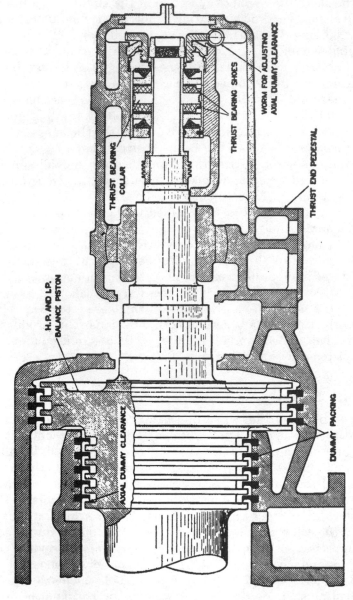

Figure 9.26 Turbine thrust end showing balance piston and thrust bearing.

ment, the pressure is distributed equally not only between the different segments but also on the individual segments. The openings between the segments permit the oil to enter the bearing surfaces. Almost 10 times as much pressure per square inch can be carried on the Kingsbury-type bearing as on the ordinary thrust bearing. Axial position of the bearing and turbine rotor may be adjusted by liners, located at the retainer rings, on each end of the bearing. The bearing is lubricated by circulating oil to all its moving parts.

The impulse turbine does not require as large a thrust bearing as the reaction turbine because there is little or no pressure drop through the rotating blades. However, the thrust bearing must be used to ensure proper clearance between the stationary and rotating elements. Reaction turbines that do not have some method of balancing the force caused by the drop in pressure in the rotating blades must be equipped with large thrust bearings.

Turbine bearings are subjected to very severe service and require careful attention on the part of the operator. Most turbines operate at high speed (3600 rpm) and are subjected to the heat generated in the bearing itself as well as that received from the high-temperature steam. These conditions make necessary some method of cooling. In some cases the bearings are cooled by water jacketing; in others the oil is circulated through a cooler.

9.2.2 Packing

The shaft at the high-pressure end of the turbine must be packed to prevent leakage of steam from the turbine. The one at the low-pressure end of a condensing turbine must be packed to prevent the leakage of air into the condenser.

Labyrinth packing is used widely in steam turbine practice. It gets its name from the fact that it is so constructed that steam in leaking must follow a winding path and change its direction many times. This device consists of a drum that turns with the shaft and is grooved on the outside. The drum turns inside a stationary cylinder that is grooved on the inside (Figs. 9.26 and 9.27). There are many different types of labyrinth packing, but the general principle involved is the same for all. Steam in leaking past the packing is subjected to a throttling action. This action produces a reduction in pressure with each groove that the steam passes. The amount of leakage past the packing depends on the clearance between the stationary and the rotating elements. The amount of clearance necessary depends on the type of equipment, steam temperatures, and general service conditions. The steam that leaks past the labyrinth packing is piped to some low-pressure system or to a low stage on the turbine.

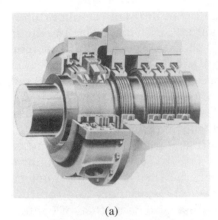
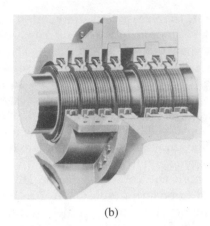

(a)   (b)

Figure 9.27 (a) Water-sealed glands and labyrinth seals as used on the high-pressure end of condensing turbines. (b) Labyrinth-type gland as used on noncondensing turbines. (*Westinghouse Electric Corp.*)

A water-packed gland consists of a centrifugal-pump runner attached to the turbine shaft. The runner rotates in a chamber in the gland casing. In some designs, water is supplied to the chamber at a pressure of 3 to 8 psi and is thrown out against the sides by the runner, forming a seal. Water seals are used in connection with labyrinth packing to prevent the steam that passes the packing from leaking into the turbine room. Such a seal is also used on the low-pressure end of condensing turbines. In this case the leakage to the condenser is water instead of air.

Figure 9.27 shows water-sealed glands and labyrinth seals as used on the high-pressure end of condensing turbines. They are used singly or in combination, depending on the service required. Each labyrinth consists of a multiplicity of seals to minimize steam leakage. The seal rings are spring backed and made of material that permits close running clearances with safety. The glands are usually supplied with condensate water for sealing to prevent contamination of the condensate water. Seal designs are continuously being improved to minimize steam leakage and thus improve turbine performance. The illustrated designs are typical of those found on operating turbines.

Carbon packing is composed of rings of carbon held against the shaft by means of springs. Each ring fits into a separate groove in the gland casing. When adjustments are made while the turbine is cold, carbon packing should have from 0.001 to 0.002 in of clearance per inch of shaft diameter. The width of the groove in the packing casing should exceed the axial thickness of the packing ring by about 0.005 in. Carbon packing is sometimes used to pack the diaphragms of impulse turbines. Steam seals are used in connection with carbon packing. This is essential when carbon packing is used on the low-pressure end of condensing turbines, because if there is a slight packing leak, steam instead of air will leak into the condenser. In operating a turbine equipped with carbon packing, a slight leak is desirable because a small amount of steam keeps the packing lubricated.

Flexible metallic packing is used to pack small single-stage turbines operating at low backpressure. In most cases the pressure in the casing of these turbines is only slightly above atmospheric pressure. The application is the same as when this packing is used for other purposes, except that care must be exercised in adjusting. Due to the high speed at which the shaft operates, even a small amount of friction will cause overheating.

9.2.3 Governors

A close control of turbine speed is essential from the standpoint of safety and satisfactory service. The same theory of centrifugal force that applied to steam engine flywheels also applies to the rotating elements of turbines. If the turbine runs at a speed far above that for which it was designed, the blading will be thrown out of the rotor. When a turbine is thrown apart in this manner, the resulting damage may be as great as, or even greater than, that caused by a boiler explosion. Turbines operating electric generators producing ac current must operate at constant speed. Some electrical appliances are seriously affected by a slight change of frequency in the power supply. The speed of the generator determines the frequency of the electric-current generator. Even with generators producing dc current, a small change in speed will affect the voltage.

There are two ways of changing the turbine supply of steam to meet the load demand. One consists of throttling the steam, by means of a valve, in such a manner that the pressure on the first-stage elements is changed with the load demand. Small turbines, like those shown in Figs. 9.18 and 9.28, have a main governor of the throttling type. The operating mechanism is driven directly by the main turbine shaft. Overload is taken care of by means of hand-operated valves that admit additional steam to the turbine. The other method uses several valves, governor-operated, that are opened separately to supply steam to secondary nozzles as the load increases. This arrangement is shown in Fig. 9.29; it is used only on large turbines.

Economical partial-load operation is obtained by minimizing throttling losses. This is accomplished by dividing the first-stage nozzles into several groups and providing a separate valve to control the flow of steam to each group. Valves are then opened and closed in sequence, and the number of nozzle groups in service is proportional to the load on the unit. Valve seats are of the diffuser type to minimize pressure drop, and these valve seats are renewable.

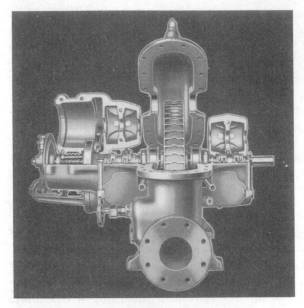

Figure 9.28 Cross section of steam turbine.

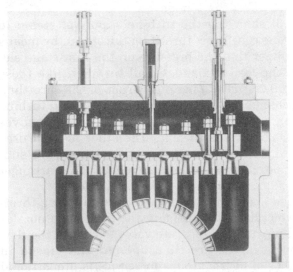

Figure 9.29 Simplified steam chest with multiple valves.
(*Westinghouse Electric Corp.*)

The multivalve steam chest (Fig. 9.29) is cast integrally with the cylinder cover with a cored passage from each valve to a nozzle group. Single-seated valves are used, arranged in parallel within the steam chest and surrounded by steam at throttle pressure. The governor mechanism raises and lowers the valve-lift bar in a horizontal plane, opening the valves in sequence, with an unbalanced force tending to close the valves.

In the *oil-relay* system, the governor operates a small valve that admits oil to and allows it to drain from a cylinder. This oil cylinder contains a piston that is connected, by means of a rod, to the steam-valve mechanism. The governor admits oil either above or below the piston, depending on which way the load is changing. There is no connection between the governor and the governor-valve mechanism except by means of the oil cylinder. The movement of the governor is transmitted to the governor valves by the oil pressure. Oil is supplied to the governor system by the same pump that supplies oil to the turbine bearings. If this oil supply should fail, the governor valve would close and stop the turbine, thus preventing damage to the bearings.

The hydraulically operated throttle valve (Fig. 9.30) is used to control the flow of steam when starting a turbine and in addition functions as an automatic stop valve in case of overspeed. It cannot be opened nor can the turbine be started until after normal operating pressures for the turbine oiling system have been established. If oil pressure falls to an unsafe point, the valve automatically closes, and the flow of steam to the turbine is interrupted. A strainer is located within the valve to protect both the valve and the turbine.

The emergency overspeed governor (Fig. 9.30) is separate and independent of the speed governor. It functions to protect the unit from excessive speed by disengaging a trip at a predetermined speed, permitting the throttle valve to close. This trip may be reset and the throttle valve reopened before the turbine speed returns to normal.

Large turbines are arranged frequently for extraction of steam at various points in the turbine. This extracted steam is used for feedwater heating or other heating or process purposes, and thus a more economical cycle is obtained. One turbine design uses a grid-type extraction valve (Fig. 9.31) at normal pressure and temperature. This valve consists of a stationary port ring and a rotating grid. The rotating grid turns against the stationary ring and opens the ports in sequence. Thus simple hydraulic interconnections are obtained between the several components of the control system, such as the accurate and positive control of speed or the load carried by the unit, and of the extraction steam pressures.

Where electric power generation is the prime consideration, condensing turbines are used. Openings are provided in the turbines for the extraction of steam for heating the feedwater to the boilers. A condensing turbine with thirteen stages and one extraction stage is

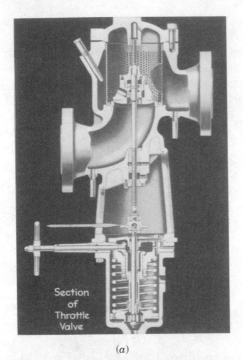

(a)

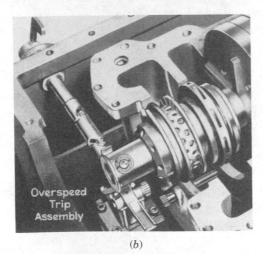

Overspeed
Trip
Assembly

(b)

Figure 9.30 (a) Section of throttle valve. (b) Overspeed trip assembly. (*Westinghouse Electric Corp.*)

shown in Fig. 9.32. This illustration is typical of an operating unit; however, turbine designs are continuously upgraded and vary between manufacturers. Figure 9.33 shows a condensing turbine with a single casing and with the casing removed. The unit shown will produce approximately 35 MW.

The change in speed of a turbine from no load to full load divided by the full-load speed is known as the *speed regulation*. A turbine with a

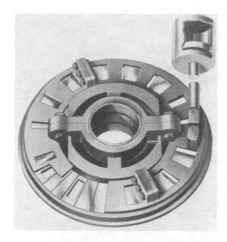

Figure 9.31 Grid-type extraction valve for lower extraction pressure. (*Westinghouse Electric Corp.*)

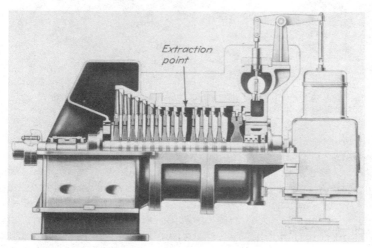

Figure 9.32 Condensing turbine. (*Westinghouse Electric Corp.*)

full-load speed of 1764 rpm and a no-load speed of 1800 rpm would have a change of 36 rpm, or regulation of 2 percent. These types of turbines are used to drive mechanical equipment such as pumps, compressors, etc. For electricity production, the speed must be constant at either 3600 or 3000 rpm depending on 60- or 50-Hz applications.

The governors of turbines operating electric generators are supplied with synchronizing springs. These are arranged to aid in moving the governor weights and lowering the turbine speed or to work against the governor weights and increase the turbine speed. The tension on the synchronizing spring is varied by means of a small motor (synchronizing motor). By adjusting the tension on this spring, the operator can change the load on the generator.

Figure 9.33 Single-casing condensing turbine for approximately 35-MW output. (*Westinghouse Electric Corp.*)

Dashpots are frequently used in connection with turbine governors. They are used to keep the governor from overtraveling because of the weight of the governor parts. When a governor adjusts the speed of a turbine and makes a change that an adjustment in the opposite direction is immediately necessary, the governor is said to "hunt" or overtravel. In other cases dashpots are used to prevent the governor from operating too quickly. Friction or lost motion in the valve mechanism or, in some cases, too heavy or improperly installed dashpots will cause a turbine to hunt. Rapid changes in speed and the corresponding variation in load are referred to as *surges*.

A governor mechanism must be adjusted as follows: The valve should be closed when the governor is in the closed position and opened the correct amount when it is in the open position. The speed adjustment must be such that the governor will be in the open position when the turbine is at rated speed, and the speed regulation (i.e., the change in speed from no load to full load) must be adjusted.

Turbine governors require very little attention on the part of the operator. They must, however, be kept oiled and the joints kept working freely. With the oil-relay system, oil leaks must be stopped as soon as possible. The overspeed trip must be checked at regular intervals. It is good practice to operate the overspeed every time that the turbine is placed in service.

9.2.4 Lubrication

Proper lubrication is of utmost importance in the operation of a steam turbine. High journal speed, heat conducted from the steam to the bearing, and possible water leakage into the oil are some of the conditions that make lubrication difficult.

There are two methods of lubricating turbine bearings. One utilizes oil rings in supplying oil to the bearings; the other consists of a pressure system that circulates the oil to the bearings.

The oil-ring system is used on small turbines. There is an oil well under each bearing. The oil level in the bearing is kept below the bottom of the shaft. The oil rings have a larger diameter than that of the shaft. They hang over the shaft, and a section extends into the oil. When the shaft rotates, the rings also turn and the oil that adheres to them is carried to the bearing. The bearings are grooved in such a manner that the oil is properly distributed to the bearing surfaces.

A typical self-contained oil-pumping system for a large turbine generator is shown in Fig. 9.34. This system is designed to supply bearing oil at 25 psig and hydraulic oil at 200 psig, approximately. It operates as follows: During the startup of the turbine, the steam-driven oil pump provides hydraulic oil to the control system, a portion of the hydraulic oil being fed to the booster-pump controls. Part of this oil pressure is reduced as the oil passes through the oil cooler to the bearing header for journal lubrication. The remainder of the oil to the booster pump drives an oil turbine coupled to the booster pump itself. The function of the booster pump is to provide suction pressure to the shaft-driven oil pump.

When the main turbine reaches near-rated speed, the shaft-driven pump supplies the hydraulic oil to the controls as well as hydraulic oil to the booster pump. At this point, the steam-driven oil pump can be put in standby service. The bearing-oil pump is used to provide bearing oil only when the turbine is shut down and when it is operating on turning gear. During this period, oil is required for the bearing journals.

With the governor in operation, the governor controls the pilot valve to position the control-valve piston that operates to open or close the steam-admission valves, thus regulating the speed of the turbine. Oil drained from journals and operating mechanism is returned to the oil reservoir, repeating the cycle.

A variation of the foregoing comprises replacing the steam-driven oil pump with an ac motor-driven oil pump and, for emergency or standby service, the use of a dc emergency oil pump operating in parallel with the ac pump. The dc emergency unit is used when the turbine is operating below rated speed or when ac power is lost.

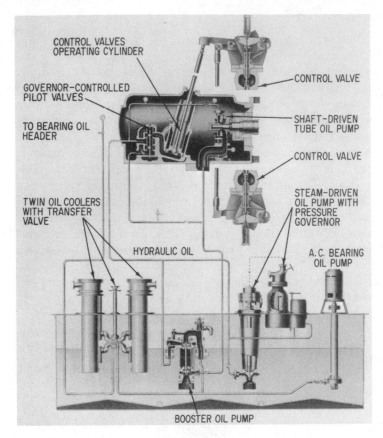

Figure 9.34 Schematic of an oil-pumping system for a steam turbine. (*General Electric Co.*)

9.2.5 Turbine control

Turbine controls have evolved over the years to accommodate the increasingly complex operating requirements as well as the necessity for monitoring all major operating characteristics. Figure 9.35*a* shows a modern turbine control and monitoring system that performs two main functions: control of turbine speed and control of turbine load. A feature of this system is a high degree of system reliability through redundancy and self-diagnostic capabilities. Failure in one area of the system will not cause failure of the entire system.

The modularity of the design allows the addition of functions in increments, as shown in Fig. 9.35*a*. The base-control system shown as level 1 provides speed and load control and includes the manual backup and calibration unit. Redundant *distributed-processing units* (DPU) are provided for overspeed protection and operator autocon-

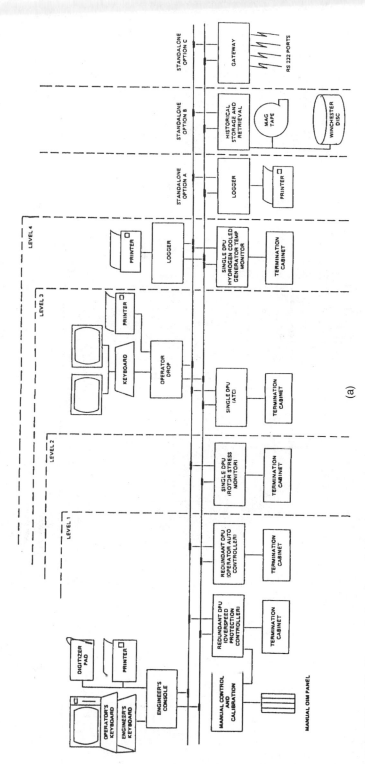

Figure 9.35 (a) Turbine control and monitoring system.

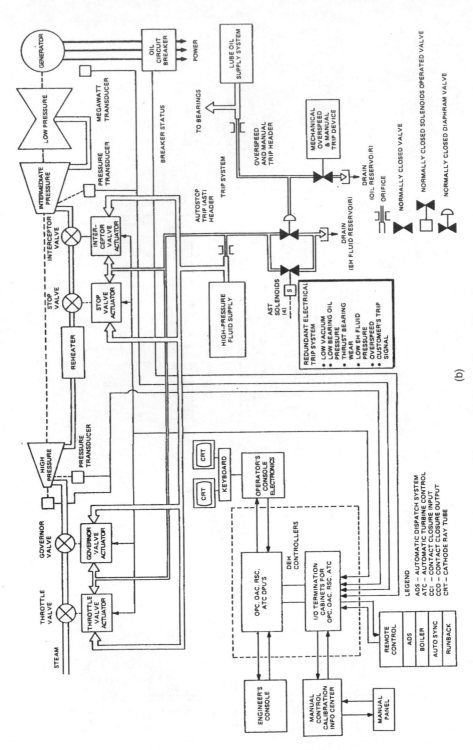

Figure 9.35 (b) Digital electrohydraulic (DEH) control system. (*Westinghouse Electric Corp.*)

troller functions. A schematic of a modern *digital electrohydraulic* (DEH) control system is shown in Fig. 9.35*b*.

Steam from the boiler enters the high-pressure turbine via throttle and governor valves. From the high-pressure turbine exhaust end, the steam flows to the reheater section of the boiler and reenters the intermediate-pressure portion of the turbine through reheat stop and interceptor valves.

The spring-loaded valves are positioned by hydraulic actuators that receive their fluid from a high-pressure fluid supply system. In the event of an overspeed condition, loss of pressure in the overspeed trip header is transmitted in the autostop trip header by means of a diaphragm valve.

The digital controller positions the throttle and governing valves by means of electrohydraulic servo loops. In the event of a partial loop drop, the interceptor valves are closed by energizing a solenoid valve on the appropriate valve actuators. The digital controller receives three feedbacks from the turbine: speed, generator megawatt output, and first-stage steam pressure, which is proportional to the load. The operator controls the turbine and receives information from the manual panel or the operator's *cathode-ray tube* (CRT) as well as an alarm and message printer.

9.2.6 Erection of the steam turbine

Steam turbines of less than 1 MW are usually shipped completely assembled. These units are assembled on their baseplates or bedplates and may be placed in position by means of rollers or a crane. In making hitches with a crane, care must be exercised to prevent appreciable deflection. The foundation must be roughed and swept clean to receive the grout. The turbine is set in position on iron blocks and tapered wedges. These must be arranged to hold the baseplate from $3/4$ to $1\frac{1}{2}$ in from the foundation to allow space for the grouting. The correct level is obtained by placing a sensitive level across machined bosses on the base. The level should be checked at all points.

Small turbine units are assembled complete on one bedplate, but in the case of medium-sized units the turbine is separate from the machine that it is to drive. When the machines are separate, they must be aligned. The shafts must be central and parallel. The central relation of the two shafts may be checked by placing a straightedge parallel to the shafts and across the rims of the coupling flanges. This check should be applied at four places about the flanges. The parallel relation of the two shafts can be checked by measuring the distance between the coupling-flange faces with a feeler gauge. If the two shafts are parallel, these measurements will be equal at all points on the coupling. Patent couplings that will compensate for a certain amount of misalignment are used. Even if these couplings are used, however, it is advisable to have the alignment as close as possible. *Manufacturer's alignment instructions should be followed explicitly.*

The turbine changes in temperature more than the generator does, and if the shafts are lined up cold, they must be set to allow for expansion. Even if they are properly aligned when cold, this does not necessarily mean that they will be in proper alignment when in operation.

The foundation bolts can now be tightened up and the level rechecked. If the level is found to be correct, the machine is ready to be grouted. A dam of boards must be built around the foundation to keep the grouting in place. A suitable place must be provided for admitting the grout. Some foundations have holes for this purpose. The grouting mixtures vary from a pure cement to 2 parts sand and 1 part cement. The foundation should be wet before the grouting is poured. The mixture should be thin enough to flow readily so that it will find its way under the base. It is customary to leave the wedges in place, but in some cases they are removed after $1\frac{1}{2}$ or 2 days. Commercial grouting is readily available with recommended curing periods.

Many medium-sized and large turbines are shipped completely disassembled because of equipment size limitations. Their erection should be supervised by a thoroughly trained manufacturer's representative. Turbines require careful adjustment, and each different type presents different problems. The bearings are aligned by means of a fine steel line that is stretched through what is to be the center of the shaft. Laser beams are also used on the large turbine to obtain proper alignment. The axial clearance between stationary and moving parts must be checked with a taper gauge. The packing clearance must be checked and adjusted carefully.

The piping to a steam turbine must be large enough to handle the steam flow at full load without excessive pressure drop. Steam velocities of 6000 ft/min were at one time considered the upper limit, but large high-pressure piping is now designed for velocities of 8000 or even 15,000 ft/min.

The piping must be arranged so that it will not produce strains on the turbine. This precaution is especially important for medium-sized and large turbines. Expansion is taken care of by means of loops or bends. The piping near the throttle valve is supported by springs that give with the expansion produced by the change from hot to cold condition yet hold the pipe firm enough to prevent vibration. A valve is placed in the steam line near the main header. This makes it possible to close off the line to the turbine, allowing work on the throttle valve while there is pressure on the main header. A strainer is often placed in the line to the turbine just ahead of the throttle valve. This prevents solid particles from entering the turbine.

9.2.7 Relief valves and rupture disks for turbines

A noncondensing turbine with the exhaust line connected to an exhaust header must have a shutoff valve located near the header. The exhaust line also must have an expansion joint to prevent strains on the turbine or exhaust header. There must be a relief valve between the turbine and the shutoff valve. This reduces the possibility of a turbine explosion if, by mistake, the turbine should be operated with the shutoff valve closed.

When a turbine is operated condensing, an atmospheric relief valve must be placed in the exhaust line between the turbine and the condenser. If the vacuum should fail, this valve would open and the turbine would exhaust to the atmosphere. It is necessary to keep a water seal on the atmospheric relief valve to prevent the leakage of air into the condenser. Atmospheric relief valves are costly, and *rupture disks* or *rupture diaphragms* are used on some designs.

A rupture disk (Fig. 9.36a) is a prebulged membrane made of various metals based on the service for which it is intended. It is located in the exhaust hood of the turbine to prevent excessive pressure buildup if the condenser loses its vacuum. The disk may be used in lieu of an atmospheric relief valve or installed ahead of the atmospheric relief valve provided (1) the valve has ample capacity, (2) the maximum pressure range of the disk designed to rupture does not exceed the maximum allowable pressure of the vessel, (3) the area is at least equal to the area of the relief valve, or (4) the disk unit has a specified bursting pressure at a specific temperature and is guaranteed to burst within 5 percent (plus or minus) of its specific bursting pressure. If the pressure is greater than 15 psi, the ASME code states that either a rupture disk or a safety valve or a combination of the two may be used (Fig. 9.36b), or the disk can be used in parallel with the safety valve so that the safety valve maintains all normal overpressure relief protection. Then the rupture disk is set at a higher pressure (approximately 20 percent above the safety relief valve). Therefore, if excessive pressure occurs, a safety system is ensured in case the safety relief valve fails. In all cases, the turbine designer must meet the ASME code requirements.

The rupture disk is fail-safe and ensures a seal-tight system with no moving parts and therefore nothing to malfunction, stick, or corrode shut. Therefore, it is a true safety fuse. A rupture disk cannot reclose itself, as can a safety relief valve.

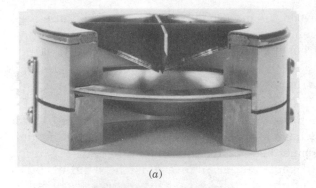

(a)

(b)

Figure 9.36 (a) Rupture disk–reverse buckling unit.
(b) Rupture disk installed in series with relief valve.

The rupture-disk assembly installed beneath the safety valve includes a telltale assembly consisting of a pressure gauge and an excess flow valve; therefore, leakage is immediately apparent. Jackscrews lift the discharge piping or outlet flange and provide for quick and easy rupture-disk replacement.

Automatic vacuum breakers are also used to prevent the turbine from overspeeding. When the overspeed trip closes the throttle, there is the possibility that enough steam will leak past to cause the unit to overspeed. The emergency governor operates the vacuum breaker, which eliminates any possibility that enough steam will be drawn through the turbine to cause overspeeding.

9.2.8 Turbine piping

The piping and turbine must be supplied with drains at all places where water can accumulate. The steam piping must have a drain located between the main steam line and the throttle. This drain is used to remove moisture from the line before placing the turbine in service. When saturated steam is used, a steam separator is placed in this line to remove the condensation. The drain is then connected to the separator. These drains are supplied with traps that are left open when the turbine is in operation and bypassed when it is being started.

The turbine casing must be drained, since an accumulation of water while the turbine is out of service would cause the blades to rust, and this would result in the rotor becoming unbalanced. A drain must be connected to the exhaust line of a noncondensing turbine between the turbine and the shutoff valve. Drains must be supplied to carry the water away from the gland seals. These lines must be of sufficient size to handle an excess of water, because it is advisable to have this water discharge into funnels so that the flow can be observed by the operator. The lines must be piped to a place where there is no backpressure.

Mechanical-drive turbines operated below 250 psig are frequently used to start up (automatically) emergency electric generators in case of a power failure. They are also used to automatically start up pumps, air compressors, etc. For automatic operation, the throttle valve opens to place the turbine in service. The exhaust line remains open, and casing and other low points are provided with drain traps.

Bearing cooling water is supplied either to an oil cooler, in the case of large turbines, or to water-jacketed bearings, in the case of small turbines. The cooling water for bearings should be as free from scale-forming material as possible. The water entering the drain lines should be visible to the operators. The discharge lines should be large enough to handle the discharge.

The grouting of a new turbine must be allowed to set until sufficiently hard before the turbine is tried out. The piping and drain should be inspected carefully to see that the valves are properly arranged. It is good practice to blow out the steam lines with compressed air or auxiliary steam to remove mill scale and any foreign deposits that may be in the line.

9.3 Electric Generators

An electric generator works by electromagnetic induction in that it uses magnetism to make electricity. The power source is the steam turbine or, in the case of a hydroelectric plant, a water-driven turbine. It spins a coil, which is contained in the rotor of the generator, between the poles of a magnet or an electromagnet produced by the stator of the generator. As the coil of the rotor passes through the lines of force, an electric current flows through the coil to the main transformer and eventually to the transmission lines.

The electric generator is a major piece of equipment in a power plant because it converts mechanical energy from the turbine to which it is connected into electrical energy. Figure 9.37 shows an assembled generator as it is being installed in a power plant and eventually connected to the turbine. This hydrogen-cooled generator produces approximately 200 MW. Each generator incorporates the following major components:

- Frame
- Stator core and winding
- Rotor and winding
- Bearings
- Cooling system

A cross section of a modern hydrogen-cooled generator is shown in Fig. 9.38. The major parts of the generator are identified. Of these major parts, the stator and rotor are particularly important components.

The stator has a slotted and laminated silicon-steel-iron core. The winding of the stator is placed in the slots and consists of a copper strand configuration. Most frequently the stator is hydrogen cooled;

Figure 9.37 Installation of 200-MW hydrogen-cooled electric generator. (*Westinghouse Electric Corp.*)

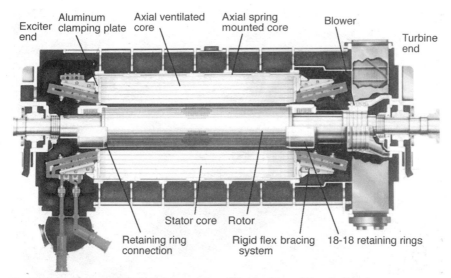

Exciter end   Aluminum clamping plate   Axial ventilated core   Axial spring mounted core   Blower   Turbine end

Stator core   Rotor

Retaining ring connection   Rigid flex bracing system   18-18 retaining rings

Figure 9.38 Inner-cooled electric generator. (*Westinghouse Electric Corp.*)

however, small units may be air cooled and very large units can even be water cooled. Figure 9.39 shows an air-cooled generator stator during final assembly, and this unit can produce approximately 120 MW.

The rotor is solid steel and has slots milled along the axis, as shown in Fig. 9.40. A copper rotor winding is placed in the slots and is also cooled by hydrogen for this particular design. Cooling of the rotor is

Figure 9.39 Stator for 120-MW air-cooled electric generator. (*Westinghouse Electric Corp.*)

Figure 9.40   Rotor for electric generator. (*Westinghouse Electric Corp.*)

improved by subslots and axial cooling passages. The rotor winding is restrained by wedges that are inserted into the slots.

The rotor winding is supplied by dc current, either directly by a brushless excitation system or through collector rings. Bearings are located on each end of the rotor to provide the necessary support. The hydrogen is cooled by a water-cooled heat exchanger that is mounted on the generator or installed in a closed-loop cooling system.

The dc current of the rotor generates a rotating magnetic field that induces an ac voltage in the stator winding. This voltage drives current through the load and supplies the electrical energy.

9.4   Condensers

The condenser is a major component of the power plant system. It receives exhaust steam from the last stage of a turbine and condenses it to water for reuse as feedwater in the boiler system. Refer to Fig. 1.1 for a schematic of the power plant cycle. For large utility turbines that have a design incorporating high-, intermediate,- and low-pressure sections, the exhaust steam comes from the low-pressure section of the turbine.

There are two general types of condensers:

1. A *direct contact* type of condenser, where the cooling water is sprayed directly into the exhaust steam from the turbine, and the mixture of the water with the steam condenses the steam

2. A *surface condenser,* in which the cooling water and exhaust steam remain separate. The vast majority of power plants use the water-cooled surface condenser, and this book will describe its operation and not the direct-contact condenser because it has limited application.

There are situations where air-cooled steam condensers are used instead of surface condensers, and these generally are necessary when cooling water is not readily available, such as in an arid region, or if a cooling tower is required, a plume from a tower is not acceptable. A description of the air-cooled condenser is given below in Sec. 9.4.2.

In the condensing of steam, a vacuum is created. The vacuum reduces the backpressure on the turbine, and this reduction in backpressure increases the efficiency of the turbine. The cooling water absorbs the heat contained in the steam, and the volume of steam is greatly reduced when it is condensed into water. When a space filled with steam is cooled until the steam condenses, the resulting water occupies only a small portion of the volume, and a vacuum is created. By continually condensing the exhaust steam, the pressure is reduced below that of atmospheric pressure.

Example What is the volume of steam at 14.7 psia as compared with the volume of water at 1 psia?

Solution Referring to Table C.1 in Appendix C:

Specific volume of steam at 14.7 psia = 26.8 ft^3/lb

Specific volume of water at 1 psia = 0.01614 ft^3/lb

The steam occupies over 1660 times more volume than does the water.

9.4.1 Surface condensers

The surface condenser is a closed vessel filled with many tubes of small diameter. Cooling water from a lake, river, or other natural water source or from a cooling tower flows through the tubes, with steam on the outside of the tube. The water flowing through the condenser may be once-through, or single-pass, or it may be made to reverse one or more times before being discharged.

The surface condenser offers the following features:

1. It provides a low backpressure at the turbine exhaust, which maximizes the plant thermal efficiency and reduces the heat rate, and therefore the plant's operating costs are reduced.

2. It allows the reuse of high-purity water in the boiler and turbine system, which also minimizes water treatment costs.

3. It deaerates the condensate, which minimizes the potential for corrosion.

4. It serves as a collection point for all condensate drains.

In order to achieve the preceding features, condensers should be as leaktight as possible. The design should not allow leakage of air or cooling water into the condensing steam space. In order to absorb heat released by the condensing steam, large volumes of cooling water are required, and this quantity can be 50 to 80 times the steam flow depending on cooling water temperatures. Condensers operate at a vacuum, i.e., pressures below the atmospheric pressure of 14.7 psia or 29.9 in Hg. These backpressures generally range from 1.0 to 4.5 in Hg, with higher pressures (actually higher vacuums) possible when there are high cooling water temperatures or when an air-cooled steam condenser is used.

Surface condensers are basically shell and tube heat exchangers. The turbine exhaust steam condenses on the shell side, and the cooling water flows through the tubes in one or more passes depending on the condenser design. A series of baffles and plates supports the tubes and distributes the flow of the exhaust steam, directs the noncondensable gas such as O_2 and CO_2 to the outlet, and minimizes flow-induced vibrations, which could result from high steam velocities. Water boxes are located on each end of the tube bundle and distribute the cooling water through the tubes. Most condensers receive steam from the top as the steam leaves the exhaust (or low-pressure) end of the turbine.

The condensate collects in the bottom of the condenser in what is called the *hot well*. The condensate pump removes it and pumps it through the low-pressure feedwater heaters. At this point the main boiler feed pump (BFP) increases the now-called *feedwater* and pumps it through the high-pressure feedwater heaters and finally to the boiler, usually through an economizer. This is shown schematically in Fig. 1.1.

The condenser tube bundle is arranged so that the steam flow moves the O_2, CO_2, and other noncondensable gases to a removal area, where they are removed by an ejector system that is at the end of the condenser. If these gases are not removed effectively, the gases will collect in an area and block the condenser tube surfaces, which results in lower heat transfer and thus lower efficiency.

Expansion occurs between the turbine and condenser as a result of the temperature difference between the two components. This expansion is accommodated by an expansion joint located between them.

Condenser performance is very important to having an efficient and reliable power plant. Leakage of air and cooling water can result in accelerated boiler corrosion and deposits. In addition, poor condenser performance results in high backpressure, which in turn results in lower electric output, lower efficiency, and therefore high operating costs.

The condenser shown in Fig. 9.41 has exhaust steam from the turbine entering at the top of the condenser and passing down and around and between the tubes in the tube bank. Cold cooling water flows through the tubes in sufficient quantity to condense the steam. Depending on the design, the cooling water makes one or more passes through the tubes before being discharged to a natural water source, such as a lake or river, or to a cooling tower. In condensing a large volume of steam into a small volume of water, a vacuum is created, reducing the backpressure for the turbine. The condensate (condensed steam) passes to a hot well, where the condensate is recovered and pumped back to the boiler as feedwater. Noncondensable gases and air finding their way into the condenser are removed by ejectors or other suitable air-removal devices. The removal of air and the oxygen not only reduces the backpressure but prevents the air and oxygen from reentering the system, thus reducing the possibility of corrosion in the piping and boiler.

Condensers such as that shown in Fig. 9.41 are adaptable to a wide range of turbine-condenser arrangements for industrial and utility plants. Some features of this condenser are as follows. (1) The shell, including the tube support plates, hot well, and connecting piece, is fabricated from heavy steel plate welded into one integral unit.

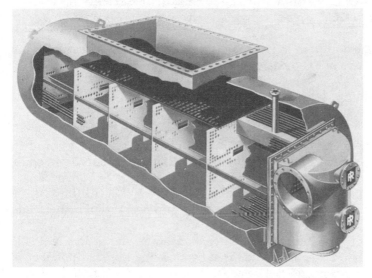

Figure 9.41 Surface condenser for industrial or utility power plants. (*Ingersoll-Rand Company.*)

Liberal flow area above and around the tube bundle allows complete distribution of steam to all tubes. (2) Water boxes are deep, with large nozzles to keep water velocities low. They can be furnished with a suitable lining for corrosive water applications. (3) The hot well has ample capacity for storage requirements. (4) Tubes are rolled and expanded into the tube sheets at both ends, forming a leakproof and mechanically strong tube to tube-sheet joint, and many designs use welded-tube to tube-sheet joints. (5) A diaphragm-type shell-expansion element provides for thermal expansion and contraction of the tube bundle under all operating conditions. (6) Arrangement provides removal of noncondensable gases. (7) An impingement plate, a large perforated baffle plate, is provided directly below the steam inlet to protect the tube bundle against moisture impingement.

9.4.2 Air-cooled condensers

Air-cooled condensers are used primarily when water availability is a problem. They are also used to meet the requirements for the discharge of water where so-called zero-discharge facilities are necessary. They also are used in areas where vapor plumes associated with cooling towers are to be eliminated. These vapor plumes might be objectionable if they ice up nearby roads or airport runways or cause foglike conditions or where communities find them aesthetically unacceptable.

An air-cooled condenser also can simplify plant design and construction; however, its disadvantage is that its use results in a lower plant efficiency than a plant designed with a condenser and a cooling tower.

The dry-cooled system is the most prevalent type of air-cooled condenser, and there are two basic types:

1. Direct acting
2. Indirect acting

The direct-acting air-cooled condenser is shown in Fig. 9.42, and it operates like a radiator on an automobile. Exhaust steam flows from the turbine outlet through external finned tubes. Air is blown across these tubes, and the steam is condensed. The condensate flows to a condensate storage tank, from which it is pumped through the feedwater heater chain and returned to the boiler.

With the indirect-acting air-cooled condenser, as shown in Fig. 9.43, the turbine exhaust steam is condensed by a cooling water loop in a conventional surface condenser. The cooling water then rejects heat to the atmosphere in a dry cooling tower by air flowing across external finned tubes. The direct-acting type of system is the design used most often.

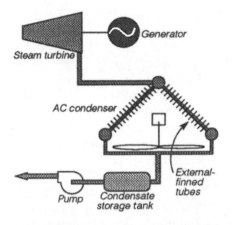

Figure 9.42 Air-cooled condenser, direct acting. (*Power Magazine, a McGraw-Hill publication.*)

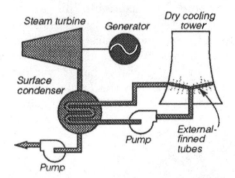

Figure 9.43 Air-cooled condenser, indirect acting. (*Power Magazine, a McGraw-Hill publication.*)

9.4.3 Condenser materials and construction

To accommodate large turbine generators, very large condensers are required, and at times, twin units are used. Shell flanges are made of quality steel plate, and all seams are welded. Shells are flanged or butt-welded for air takeoffs, condensate pump suction, return vents, drains, etc. Tube heads are made of Muntz metal, aluminum bronze, silicon bronze, copper nickel, or steel. Sheets are machined and drilled to receive the tubes. If the tubes are welded to the tube sheets, the tube sheets are made of silicon bronze to facilitate welding.

Tube supports are provided across the width and length of the condenser to hold the tubes in place and avoid the tube damage that may result from vibration. The support plates are made of steel, and the interior bolting and cap nuts exposed to the circulating water are made of corrosion-resisting composition or steel. The plates in turn are welded to the shell. The necessary drains and vents must be provided for removing condensate, for venting pumps back to the condenser, etc. Braces, drain plates, baffles, and supports in the steam space are welded in place. Water boxes are fabricated of steel for fresh water, although for unusual conditions, cast iron is used. For salt water, the boxes are made of cast iron or suitably lined fabricated steel in order to minimize the corrosion effect of the salt water. Figure 9.44 shows an end view of a partially shop-assembled condenser. This condenser contains 300,000 ft^2 of surface and is of the modified double-flow design with divided single-pass water boxes.

Condenser tubes are made of Admiralty Metal, aluminum alloys, and various stainless-steel compositions. Aluminum alloys are less expensive than the other materials, but experience has indicated some difference in fouling characteristics, with performance falling off more rapidly than for copper-base alloys. Although this fault is corrected by using a larger condensing surface and more circulating water, this practice has the disadvantage of increasing condenser and power costs. Where circulating water is corrosive and no reasonable life can be obtained with copper-base alloys, stainless steel is used. Stainless steel appears to have more favorable fouling characteristics than either Admiralty Metal or other copper-base alloys when used with acidic waters. Copper alloys present problems in high-pressure boiler systems because of the potential for copper transport into the boiler, where deposition can occur, and this results in a degradation of heat transfer.

Figure 9.44 Shell section of condenser under construction showing tube sheets and support plates. (*Foster Wheeler Corp.*)

Tubes are small, usually ½ or 1 in in diameter, and fit into tube-sheet holes that are drilled and machined to receive them. Condensers may have stationary tube sheets, or they may be constructed with a floating tube sheet on one end. The floating-tube-sheet design accommodates tube expansion.

In large condensers tubes are installed in the form of an S-shaped bow, the deflection curve of a uniformly loaded beam, to provide for unequal expansion. One end is higher than the other to eliminate any pockets where corrosive gases could collect and to afford positive drainage of circulating water during shutdown periods.

For the large condenser, the shell is manufactured in a minimum number of sections to minimize field erection. The maximum size of each prefabricated section is dictated by shipping clearances. The shell and exhaust connections have the various edges of the various sections chamfered, as though they were to be ready to be finally welded. Sections are then assembled in the shop, temporary lugs and bolts being used to fit the sections together. Tube support plates and struts are welded in place to act as braces. When the unit has been completely assembled, it is marked for future erection, dismantled, and shipped. When the unit is in its final location, all seams are welded to form a rigid structure free from air leaks.

Condensers are frequently designed to serve as a complete foundation for the turbine generator. Such a unit is shown in Fig. 9.45. This condenser is a two-pass unit, containing 7150 ft^2 of surface and using tubes ⅞ in in diameter. All auxiliaries are mounted on the side of the condenser, which provides a compact installation.

Condensers require large quantities of circulating water to dissipate the heat released when the steam is condensed. Therefore, power plants must be located where a large quantity of water is available as a natural source, or they must use cooling towers or air-cooled condensers as a means for condensing the exhaust steam.

The cooling water temperature significantly affects the turbine backpressure. The higher the temperature, the higher is the backpressure and therefore the lower is the thermal efficiency. For example, an inlet cooling water temperature of 55°F will provide a condenser backpressure of approximately 1.5 in Hg. A 95°F inlet temperature results in a backpressure of approximately 3.5 in Hg. The lower the backpressure, the higher is the turbine output and, when coupled to an electric generator, the higher is the electric output.

In passing through the condenser, and depending on station design and capacity, the temperature of the circulating water increases by 15 to 20°F. A nuclear power plant requires (for an equivalent power output) more circulating water than that for a fossil-fuel-fired plant because of the larger quantity of steam required. (Since nuclear

Figure 9.45 Condenser designed to serve as a complete foundation for a turbine generator. (*Ingersoll-Rand Co.*)

plants operate at lower steam pressures and temperatures, a higher steam flow is required for a comparable electric output.)

In instances where the natural water supply is of limited volume, the temperature of a portion or all of the water will be increased, giving rise to *heat or thermal pollution*. This condition may adversely affect aquatic and marine life, and environmental legislation has established limits on the use of natural waters to receive the rejected heat from power plants. It has been shown that in some locations, the warmer water has enhanced marine life. It therefore becomes necessary to increase plant investment by installing cooling towers, which reject this heat to the atmosphere instead of to natural water sources.

9.5 Cooling Towers

The amount of heat rejected from modern thermal power plants is significant in that it represents over 60 percent of the total heat input. Of this amount, 10 to 15 percent is rejected out the stack with the flue gas. Most of the balance (approximately 45 percent) results from the condensing of the exhaust steam from the turbine. Eventually, all this rejected heat is absorbed by the atmosphere, although the heat may first be rejected to a body of water such as a lake, river, or ocean.

Heat-rejection systems are generally classified as once-through or closed:

1. *Once-through systems.* Water is withdrawn from a lake, river, or ocean and then pumped through the condenser, where its temperature is increased by 15 to 20°F. The warmer water is then discharged back to its source. Evaporation from the natural water source to the atmosphere eventually cools this water.

2. *Closed-loop systems.* Heat is rejected to the atmosphere through the use of either a cooling tower or some form of outdoor body of water such as a spray pond or cooling lake.

In many areas, the once-through cooling system is unacceptable for a new power plant. Either the site is already developed and the natural source of water cannot support another plant, or environmental restrictions prevent the use of this system. Therefore, nearly all new power plants use the closed-loop system for heat rejection. Cooling ponds and lakes are found primarily at existing sites.

The cooling pond, which is often called a *spray pond,* is the simplest type of closed-loop system. The circulating water is pumped into a pond or basin, where it provides water storage in addition to the cooling.

The cooling pond is converted to a spray pond by locating a series of sprays above the water surface. The cooling water flows through piping and then vertically from the spray to form the shape of an inverted cone. This method of spraying provides uniform water distribution, which increases the area of exposure and improves the efficiency of the cooling as the spray droplets come into contact with the air and thereby are cooled.

The problem with the spray pond is that the water is sprayed into the air; water particles in varying amounts are carried away by the wind, which results in the loss of water and can create a nuisance in a congested area. This problem has been reduced by the installation of a louvered fence around the pond. However, although spray ponds are still operational, they are limited due to location and to environmental issues because of the airborne spray. Modern facilities predominantly use either natural or mechanical draft cooling towers.

In wet cooling tower systems, cooling water is circulated through the condenser and absorbs heat from the exhaust steam from the turbine. The heated cooling water is then circulated through a cooling tower, where the absorbed heat is rejected to the atmosphere by the evaporation of some of the circulating water. The cooled water is then returned to the condenser by a circulating water pump. Refer to Fig. 1.1. Makeup water must be provided to replace the water lost during evaporation and during blowdown, where contaminants are controlled.

The cooling tower therefore performs the following major functions:

1. Removes the heat that the cooling water absorbed in the condenser
2. Minimizes the use of cooling water
3. Provides cooling water to the condenser to obtain high plant thermal efficiency

9.5.1   Types of cooling towers

Cooling towers are special direct-contact heat exchangers where the warm cooling water from the condenser is brought into direct contact with the relatively dry air. The heat-transfer rate depends on maximizing the contact area between the water and air and the length of time for this contact.

All cooling tower designs have the following common features:

1. An air circulation system
2. A water distribution or spray system
3. Packing or fill to maximize the contact between the water and air
4. A cooling water collection and discharge basin
5. Mist eliminators that minimize droplet carry-over and water loss

Cooling towers are generally configured with the air and water in a counterflow or cross-flow arrangement. In counterflow units, water falls down through the fill while the air moves upward. In cross-flow units, water cascades downward while the air moves horizontally, which is perpendicular to the water flow. These cooling towers are generally classified by the method used to move the air. These are commonly known as (1) mechanical-draft units and (2) natural-draft or hyperbolic units.

Mechanical-draft cooling towers.   Mechanical-draft cooling towers use either single or multiple fans to provide a known volume of air through the cooling tower. Therefore, their thermal performance is generally more stable and is affected by fewer air variables than with natural-draft cooling towers.

Mechanical-draft cooling towers are either forced draft or induced draft:

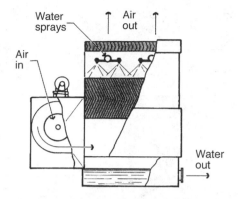

Figure 9.46 Schematic of a forced-draft mechanical-draft cooling tower. (*Marley Cooling Tower, a United Dominion company.*)

1. *Forced draft.* The fan is located in the ambient air stream entering the cooling tower, and the air is blown through the unit as shown in Fig. 9.46.

2. *Induced draft.* The fan is located at the exit of the airflow and draws air through the tower as shown in Fig. 9.47.

Mechanical-draft towers are characterized by high air entrance velocities and low exit velocities. Therefore, they are extremely susceptible to recirculation and are considered to have less performance stability than the induced-draft tower. In addition, when plant sites are located in cold weather climates, since the fans are located in the cold ambient air stream (see Fig. 9.46), the fans can become subject to severe icing with resulting imbalance. As a result, these fans are often located in a specially designed enclosure to prevent this.

Fans on mechanical induced-draft towers are not subject to recirculation and therefore are more stable. Their location within the warm air stream also provides protection against the formation of ice. These advantages lead to the widespread use of mechanical induced-draft towers.

Mechanical-draft cooling towers also can be classified by their shape, either rectilinear or round towers. Figure 9.48 shows towers that are constructed in cellular fashion, and these have been increased linearly to the length and number of cells necessary to meet the required thermal performance.

Two configurations of round towers are shown in Fig. 9.49 and 9.50 with fans clustered as close as practicable around the centerpoint of the tower. These tower arrangements can handle large heat loads but with considerably less site area requirements than multiple rectilinear towers.

A double-flow-principle mechanical-induced-draft tower shown in Fig. 9.51 with a transverse cross section shown in Fig. 9.52 serves a

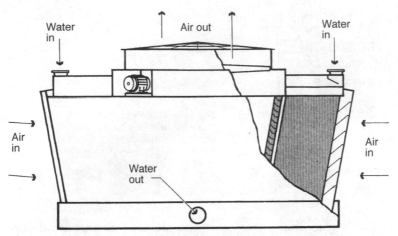

Figure 9.47 Schematic of an induced-draft mechanical-draft cooling tower. (*Marley Cooling Tower, a United Dominion company.*)

Figure 9.48 Multicell cross-flow mechanical-draft cooling tower in a rectilinear shape. (*Marley Cooling Tower, a United Dominion company.*)

100-MW plant, cooling 50,000 gpm. This unit features 28-ft cells, 24-ft fans, 6-ft fiberglass fan cylinder bases, and a 12-ft laminated-wood velocity-recovery stack extension. Special mechanical equipment-handling devices include removal davits and dollies and a monorail system for transporting heavy equipment to tower end walls.

The double-flow tower consists of two identical sections divided at the tower center by a partition that extends from the water level in the basin to a point close to the fan inlet housing. The housing consists of vertical columns made from sturdy timbers that are spaced on close centers for added strength. The vertical members are supported

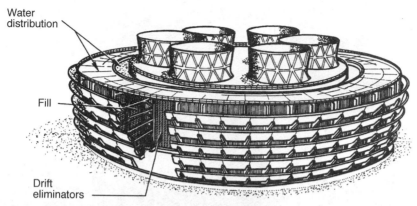

Figure 9.49 Multicell mechanical-draft cooling tower in a round shape. (*Marley Cooling Tower, a United Dominion company.*)

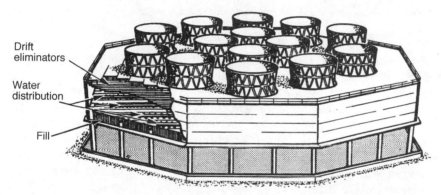

Figure 9.50 Multicell mechanical-draft cooling tower in an octagon shape. (*Marley Cooling Tower, a United Dominion company.*)

mechanically by bracing to provide rigidity where required. Transverse bracing as well as longitudinal bracing is provided to give maximum strength.

To avoid the corrosive influences to which cooling towers are subjected by water conditions and atmospheric contamination, moldings are made of glass-reinforced polyester. Structural-ceramic rings are applied in conventional connector-ring joints where values exceeding those available in bolted joints are required. The ceramic rings are made of a complex porcelain. This construction and use of materials eliminate the possibility of bearing failure in the wood under the inroads of rot in a rust-deposit area. Cement board covers much of the exterior of the tower, the covering including end-wall casings and board louvers that form the tower sides. This adds to fire safety and structural sturdiness.

Figure 9.51 Double-flow induced-draft mechanical-draft cooling tower. (*Marley Cooling Tower, a United Dominion company.*)

The material under each "section" is filled with splash bars set in fiberglass supports, which are impervious to all corrosive conditions. These high-strength grids are on close centers so that there can be no sagging fill or channeling of water flow. Splash bars are securely retained without nails or other corrosive fasteners and can elongate or shrink without distortion or cracking. Between the filling and the center longitudinal partition are herringbone drift eliminators for removing water entrained in the air. The drift eliminators also function as effective diffusers, equalizing pressure through the cooling chamber.

For large fans, blades of glass-fiber-reinforced polyester are used to eliminate corrosion. Aluminum blades are also available. The gear reducer contains an enclosed lubrication system with a renewable cartridge filter.

The use of glass-fiber-reinforced grid supports and molding together with the method of construction eliminates corrosion or degradation effects resulting from contaminants in the circulating water; therefore, there is no galvanic corrosion or rusting, as occurred with ferrous parts in the past. Structural-ceramic rings are applied in conventional connector-ring joints; the ceramic material is porcelain. The combination of the foregoing reduces deterioration by corrosion regardless of the presence of acid, caustic, salt, or other contaminants in the circulating water. Water treatment for the circulating water is also commonly used to minimize corrosion.

Directly over the center of the tower are located the induced-draft fans. The motors are removed from the airstream and set outside the

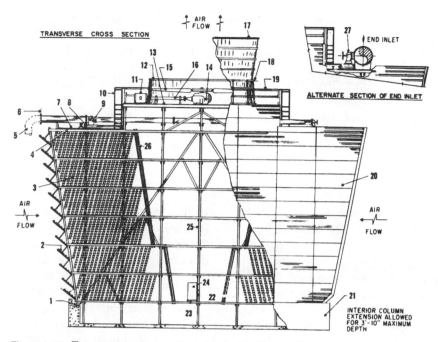

Figure 9.52 Transverse cross section of a double-flow cooling tower. 1 = perimeter anchorage. 2 = corrugated louvers. 3 = splash bars in fiberglass grid supports. 4 = diffusion decks (above fill). 5 = side-inlet pipe system located as shown. 6 = pipework stops at face of inlet flange. 7 = open distribution basin (removable nozzles). 8 = covered distribution box. 9 = flow-control valve. 10 = ladder. 11 = motor mounted on transverse centerline of each cell. 12 = vertical laminated fan cylinder. 13 = unitized steel mechanical equipment support. 14 = gear reducer. 15 = multiblade fan. 16 = driveshaft. 17 = velocity-recovery cylinder. 18 = access door in fan cylinder opposite motor for gear reducer removal. 19 = handrail around fan deck. 20 = horizontal corrugated end-wall casing. 21 = concrete basin. 22 = access opening through longitudinal partition at each cell (no doors). 23 = walkway (one side only). 24 = partition and end-wall access door. 25 = longitudinal partition. 26 = herringbone drift eliminators. 27 = flow-control valve. (*Marley Cooling Tower, a United Dominion company.*)

fan housing. Figure 9.51 shows seven individual towers making up one cooling tower unit. The louvered openings are on the side through which the air is drawn. Across the top of the tower are the open-distribution hot-water basins. The bottom of each basin contains a series of porcelain nozzles that are placed to provide uniform water distribution to the "filling" below. Flow-control valves are provided for each tower and are used to vary the flow if desired. In the center of the fan housing is the fan assembly with gear-reducer drive and flexible-drive shaft. Two-speed motors are available for low-load operation.

The level is held constant by automatic-control float makeup valves to replenish the water supply by automatic control of the blowdown from the basin and by the overflow pipe to the drain. If well designed, the water-distribution system should be readily accessible for regulation of flow and for cleaning and inspection. With the double-flow tower, one-half of any cell may be shut down while the other half remains in service.

Large cooling towers are usually provided with concrete basins. These basins should be watertight and deep enough to store an adequate amount of water. Sufficient space and clearance should be permitted for ease in cleaning and painting. Access doors should be conveniently located and large enough to permit equipment that is necessary for repairs and maintenance to be moved into the tower's interior.

In designing and locating a cooling tower, consideration should be given to the prevailing wind direction and to any obstructions surrounding the tower, since any interference will reduce the efficiency of operation and influence the performance. Cooling towers must be designed and built to support their own weight, together with the weight of the water and the force produced by the wind. Standard wind-pressure design is 30-psi wind load, though in certain areas where windstorms prevail, additional safeguards must be taken to withstand loading. Wood towers are to be recommended over steel towers to avoid corrosion. If bolts are used, they should be bronze or galvanized. Nails are never used to carry a load. Consideration should be given to fire prevention, to proper access to facilities, and to walkways for adjustments, servicing, and maintenance.

Natural-draft hyperbolic cooling towers. The natural-draft hyperbolic cooling tower shown in Fig. 9.53 utilizes airflow that is produced by the density differential that exists between the heated air inside the tower, which is less dense, and the relatively cool ambient air outside the tower, which is more dense. This density differential is such that no fans are required because natural draft results. These types of cooling towers tend to be quite large, since they often handle large quantities of cooling water, 250,000 gpm and greater. Because of the relatively small temperature and density differences of the inside and outside air, these cooling towers are generally very tall, in the range of 300 to 500 ft high.

The shape of the tower shell is hyperbolic, thus the name of this type of cooling tower. This shape has little effect on the natural-draft capabilities, but it offers superior strength and resists wind loads and therefore requires less material than other designs, thus being more cost-effective.

Natural-draft cooling towers operate most effectively in areas that have a higher humidity as compared with plants located in an arid region or one located at a high altitude. Such plant sites would most

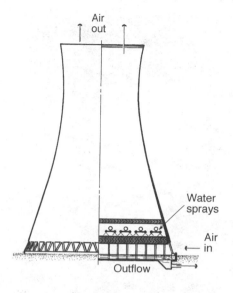

Figure 9.53 Schematic of natural-draft hyperbolic cooling tower. (*Marley Cooling Tower, a United Dominion company.*)

probably use mechanical-draft cooling towers or possibly an air-cooled condenser system.

Hyperbolic cooling towers are more expensive than mechanical-draft cooling towers, and they are used extensively in large utility power plants. However, because of their lower operating cost as a result of no fan power requirements, the overall costs over the plant life are lower, which justifies the higher initial capital cost. For any plant design, a careful evaluation of the initial capital costs and the operating and maintenance costs must be made of the various designs when the plant size is such that natural-draft towers are known to be a possibility.

An illustration of a hyperbolic natural-draft cooling tower is shown in Fig. 9.54. This unit is part of a 265-MW power plant and handles 120,000 gpm of cooling water from 110 to 87°F. It has a diameter of approximately 130 ft at the top and 245 ft at the bottom and is 320 ft high.

9.5.2 Design considerations

There are several design considerations that should be incorporated into a cooling tower design, whether it is a mechanical- or natural-draft tower.

1. *Tower fill or packing.* Cooling tower fill, which is often called *packing,* is one of the most important components of the cooling

Figure 9.54 Hyberbolic natural-draft cooling tower. (*Marley Cooling Tower, a United Dominion company.*)

tower. Its purpose is to increase the contact area between the air and water, as well as the water residence time. Fill is generally classified as film type or splash type.

a. *Film-type fill.* This design allows a thin water layer to be directed along a plate or sheet, and air is forced past the water. Several plates are placed together at fixed angles in order to maximize the air-water contact area and the water residence time.

b. *Splash-type fill.* By inducing a splashing action, the air-water contact area and the water residence time are increased. In these units, water enters the tower and falls on the splash bars, which divide the large waterdroplets into smaller ones and thus increase the surface area. The bars also slow the fall of the droplets, which increases the residence time. The materials used for splash bars are fiberglass, PVC (plastic), and redwood.

2. *Fire protection.* There is the potential for fire on cooling towers, especially when wood or other combustible materials are used. Wood towers are susceptible to fire after they have been out of operation for a period of time, which would allow them to dry out. In order to provide protection, designs incorporating wood have fire-protection systems as part of their design. In fact, during construction, the plant's fire-protection system must be operational to

protect against the possibility of fire as a result of welding operations, etc. With the use of more noncombustible material such as PVC and concrete, approval of designs without a fire-protection system has been granted.

9.5.3 Materials of construction

Wood is the predominant structural material used in cooling towers because of its availability, durability, and relative low cost. Douglas fir is the wood most often used today, having replaced the previously preferred redwood. No matter what the wood selection, all woods must be treated with a reliable preservative in order to prevent decay.

Steel is used for many components of the cooling tower where high strength is required and plastics are now used in increasing amounts to take advantage of their resistance to microbiologic attack, corrosion, and erosion; of their high strength; and of their relatively low cost.

The use of concrete is predominant in hyperbolic cooling towers, and its use is being expanded to large mechanical-draft towers. The higher first cost of concrete is often justified by a lower fire risk and, for large structures, higher load-carrying capacity.

9.6 Condenser Auxiliaries

Various auxiliaries are required for the proper operation of a condenser. Circulating pumps must provide cooling water to condense the steam and produce a satisfactory backpressure or vacuum at the turbine outlet. Condensate pumps must remove the steam that has been condensed. A vacuum pump or ejector must remove the air and noncondensable gases, and to avoid pressure in the condenser, an atmospheric pressure-relief valve is required. A typical piping installation of a large surface condenser and auxiliaries is shown in Fig. 9.55. The figure shows the arrangement of auxiliaries on a two-pass surface condenser.

Circulating pumps used for surface condensers are of three types: (1) The impeller type is used where high suction lift or pumping head is specified. This type of pump is usually double suction and horizontally mounted. (2) The propeller type is suitable where a moderately large quantity of water is required at a relatively low head. This pump is mounted vertically and operates with the pump submerged; no priming is required. (3) The axial type is used where moderate pumping head and suction lift are required. This pump too is set submerged and does not require priming.

Figure 9.55 Typical piping arrangement of a two-pass surface condenser with auxiliaries. (*Ingersoll-Rand Co.*)

Since the impeller pump is not much different from that described in Chap. 8, reference here is made only to the mixed- and axial-flow pumps, as illustrated in Fig. 9.56. These are vertical pumps built to meet specific requirements. Both operate with submerged suction and hence are self-priming. They are made adaptable to float-controlled start-and-stop operation. The axial-flow pump is a single-stage, low-head, large-capacity unit and is made in sizes up to 50,000 gpm at 25 ft. Axial-flow pumps with multistages are available for heads up to 460 ft and capacities of 60,000 gpm. If variable head and capacity conditions are encountered, an adjustable-vane impeller can be furnished. The mixed-flow pump combines axial- and centrifugal-flow characteristics. It is made in sizes up to 700,000 gpm at 180 ft. Pumps of this type can be designed for higher heads if desired.

The shaft of the vertical pump is held in alignment by column bearings spaced at approximately 9 ft or less. If water lubrication is provided, rubber bearings are used. If oil lubrication is used, bearing linings are either bronze or iron. Shaft sections are coupled with heavy threaded sleeves, self-locking even when flow through the pump is reversed.

The impeller is located in its most efficient position with respect to the casing by means of an adjusting nut in the top of the motor housing. The impeller is keyed and locked on the shaft to prevent movement relative to the shaft.

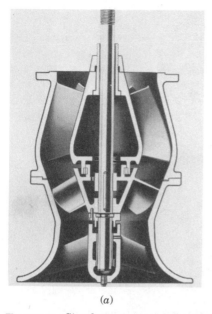
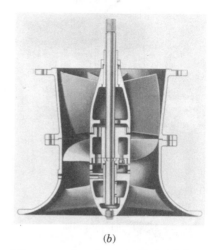

(a)   (b)

Figure 9.56 Circulating water pump. (*a*) Mixed-flow pumping element. (*b*) Axial-flow impeller pump. (*Ingersoll Dresser Pump Co.*)

The vertical pump, as normally installed, can be pulled out for inspection after the pipe has been disconnected. This must be done when the water level is such as to permit removal. Complete removal facilitates inspection and repairs.

Condensate pumps remove the water from the hot well and maintain a continuous water supply through the feedwater heaters to the boiler feed pumps. The head against which the pump is required to operate depends on the type of installation. As a result, some condensate pumps are single stage while other condensate pumps are two, three, or four stage. Pumps of this general description have been explained in Chap. 8.

High vacuum requires the removal of the noncondensable gases. Sometimes the pump that removes the condensate also removes the air. More often the condensers are provided with separate *vacuum pumps*. Vacuum pumps may be of the reciprocating type or of the hydraulic-vacuum type. In the former we have a positive-displacement unit exhausting the air from the condenser at a point close to the water line in the hot well. In the hydraulic-vacuum unit, water is recirculated from a water tank through a revolving jet wheel, causing air entrainment. Both water and air are discharged through the jet to the tank, where the air is liberated.

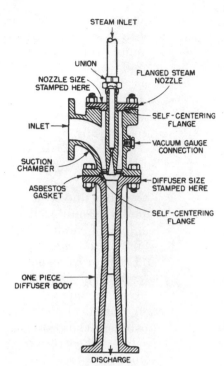

Figure 9.57   Steam-jet ejector. (*Ingersoll Dresser Pump Co.*)

The *steam-air ejector* (Fig. 9.57) consists of a steam nozzle, a suction chamber, and a diffuser. Steam enters the nozzle and discharges a jet of high-velocity steam across a suction chamber into a venturi-shaped diffuser. Air or gases to be evacuated enter the ejector suction, become entrained by the moving jet of steam, and are then discharged through the diffuser, where the velocity energy is converted into pressure, thereby compressing the mixture to a lower vacuum. The steam pressure is variable depending on the size of the nozzle used and vacuum to be produced.

Steam-jet ejectors are used for air removal on surface condensers. They are simple in operation and rugged in construction, can use a wide range of steam pressures, and will handle wet or dry mixtures of air, gases, or vapors at a near-perfect vacuum. However, several ejectors may be required to obtain the desired vacuum.

Where lower vacuums are desired, or where it is not economical to do the entire job of evacuation in one stage, multistage ejectors are used. Here the discharge of the primary stages enters the suction chamber of the succeeding stages either directly or through an intercooler placed between the stages. Precoolers are sometimes placed ahead of the first-stage ejector to remove condensable vapors before they enter the ejector, thereby reducing its size and steam consumption. Single-stage ejectors are suitable where moderate vacuum is desired, after which two-stage ejectors are used. Ejectors of this type require little maintenance and are relatively trouble-free in operation, and, consequently, they have been universally accepted for this operation.

The steam-air ejector has virtually replaced all other forms of air-removal devices. In most condenser installations, the vacuum to be maintained is greater than can be obtained economically by a single-stage ejector. Two-stage ejectors usually have some type of condenser between them to condense the steam and condensable vapors from the first stage.

With large quantities of cooling water being handled and when natural water sources are used, difficulty is frequently encountered with foreign matter clogging up the tubes and openings of condensers as well as interfering with the proper operation of circulating and condensing pumps. In some localities, leaves and fish prove especially troublesome during certain seasons of the year. To avoid such operating difficulties, large *revolving* or *traveling screens* are placed in the entrance to the intake water tunnel. The screen consists of a string of baskets which are joined together to form a continuous chain. Baskets or screens are made of brass or galvanized wire to resist corrosion.

In operation, one end of the chain revolves about a sprocket placed at the bottom of the intake tunnel; the other end, the driving motor, and the sprocket are above the water level. As water flows through this revolving or stationary screen, it deposits the foreign matter on the wire basket. As the screen revolves on its way to the bottom of the pit, it encounters a spray of water directed against it from the inside that then washes off the foreign matter into a trough, from which point the foreign matter is carried away for disposal.

Relief valves are placed in a branch line leading from the main exhaust pipe to the atmosphere to avoid the possibility of having a pressure on the condenser. To prevent air leakage into the condensers, the valve is provided with a water seal. A relief valve is similar to a check valve. It is held closed by the atmospheric pressure on top of the valve. As long as a vacuum is maintained in the condenser, it remains closed. If the vacuum is lost, the valve promptly opens to discharge the steam to the atmosphere. The valve continues to remain open until the vacuum has been restored, at which time the valve closes automatically. To prevent pounding during intermittent exhaust, such valves are frequently fitted with dashpots. Leakage of air is prevented at the seat by placing a water seal at the raised lip. These valves usually have an arrangement attached to hold the valve wide open if it is found necessary.

Questions and Problems

9.1 What is a steam turbine, and how is it used?

9.2 Explain how a steam turbine operates.

9.3 What is the difference between an impulse and a reaction turbine?

9.4 How does steam impart energy to a turbine to develop work?

9.5 For an electrical output with 60-Hz applications, what must the rotating speed of the turbine be? In order to prevent severe turbine blade erosion, what is the moisture limit in a turbine?

9.6 What is the net plant heat rate? How is it expressed? What is it important?

9.7 How are turbines classified? Briefly describe them.

9.8 Describe the turbine stage design, and identify the major components.

9.9 What are the major components of a turbine?

9.10 Of what material are turbine blades made?

9.11 In the design of turbine blades, what unique characteristics must be considered?

9.12 What types of bearings are used on a turbine? Why?

9.13 What prevents steam from leaking at the point where the rotating shaft penetrates through the casing?

9.14 How is speed regulation on a turbine obtained?

9.15 What is the potential problem of having blades that are very long?

9.16 Describe the change in the steam as it passes through the turbine. What impact does this have on turbine design?

9.17 What is the purpose of a shroud ring on turbine blades?

9.18 What is the *double-flow principle* of a turbine design?

9.19 What is the purpose of a thrust bearing? How is a bearing cooled?

9.20 What is the purpose of having extraction steam from a turbine?

9.21 Name two methods for lubricating turbine bearings.

9.22 For piping connections to a turbine, list several design concepts to account for piping expansion.

9.23 What is the purpose of a rupture disk on the turbine?

9.24 Describe how a generator produces electricity.

9.25 What are the major components of an electric generator?

9.26 What is a condenser? Why is it a very important component in a power plant?

9.27 Describe the two types of condensers.

9.28 Which is the type of condenser that is the most common in a power plant? What are its features?

9.29 Why does the condensation of steam create a vacuum in the condenser?

9.30 What is the hot well of a condenser? Describe the operation of a condensate pump.

9.31 How are noncondensable gases removed from a condenser? Why are they removed?

9.32 When is an air-cooled condenser used? Describe its operation. What are its advantages and disadvantages?

9.33 Of what types of materials are condenser tubes made?

9.34 If a natural body of water is unavailable for condenser cooling water, what must be used to provide cooling water?

9.35 Why is a lower cooling water temperature entering the condenser important to power plant operation?

9.36 In a modern steam power plant, approximately how much of the total heat input is rejected to the atmosphere in some manner? How much of this results from the condensing of the exhaust steam from the turbine?

9.37 What are the purposes of a cooling tower?

9.38 How does a cooling tower work, and what are its major features?

9.39 Describe the two classifications of cooling towers and how they operate.

9.40 Of the two types of mechanical-draft cooling towers, which is preferred? Why?

9.41 Describe the operation of a natural-draft hyperbolic cooling tower.

9.42 Why is the natural-draft tower in a hyperbolic shape?

9.43 What is the advantage of a hyperbolic cooling tower over a mechanical-draft cooling tower that must be evaluated carefully prior to selection of a type of cooling tower?

9.44 What is cooling tower fill, and why is it so important?

9.45 Name some of the more important materials in cooling tower design.

9.46 What are circulating water pumps?

9.47 What is a steam-air ejector, and what is its purpose?

Operating and Maintaining Steam Turbines, Condensers, Cooling Towers, and Auxiliaries

10.1 Turbines

As described in Chap. 9, there are two general ways of classifying turbines: (1) by their steam supply and exhaust conditions and (2) by their casing or shaft arrangement. They are also identified by the equipment they drive, either mechanical equipment or an electric generator. The type of drive, either direct or geared, is also used in describing a turbine.

For large electric utility fossil-fuel-fired power plants in the United States, where plant sizes range from 100 to 1300 MW, most of the large plant designs are based on one of two system cycles:

1. Subcritical pressure systems of 2400 psig with 1000°F superheat and 1000°F reheat temperatures

2. Supercritical pressure systems of 3500 psig with 1000°F superheat and 1000°F reheat temperatures

However, with the presence of independent power producers (IPPs), where plant sizes are less than 100 MW and various waste fuels are burned, many different plant cycle designs are used, with many plant steam pressures less than 1000 psig and steam temperatures of 750°F. Yet the operating goals of these facilities are identical to those of the large electric utility—*produce electricity at the lowest possible cost with the highest degree of reliability*. The lower steam pressures and temperatures are often required because of inherent problems with burning a certain fuel. For example, when burning municipal solid waste (MSW), high steam pressures and temperatures are related to accelerated corrosion in the boiler, which leads to low availability and high maintenance costs.

Turbines are also used to drive mechanical equipment, and they often use low-pressure steam, less than 150 psig, which often comes from an extraction point within the main steam turbine. Thus turbine steam pressures and temperatures vary significantly depending on the application. However, for each design, the pressure and temperature of the steam supplied are important factors in determining the ultimate efficiency of the turbine. The materials used in the construction of the turbine also play an important part in the overall performance.

High-pressure, high-temperature steam turbines are used primarily in large industrial and utility power plants. Several types and applications are shown in Fig. 10.1. Pressures for these types of turbines generally range from 400 to 3500 psig, with steam temperatures to 1000°F. Most of the large units for electric utilities operate with reheat. Here, the steam, after passing through the high-pressure stages of the turbine, is taken back to a reheater in the boiler, where the steam is reheated to its initial temperature and then returned to the turbine. High-pressure turbines are sometimes used as topping units. This arrangement consists of installing a high-pressure turbine where the exhaust enters a low-pressure turbine (installed earlier and operating at some lower pressure). In essence, the high-pressure turbine acts as a reducing valve and, while so doing, generates power. The exhaust steam to the low-pressure unit produces the same amount of power as it did previously, provided steam conditions on entering and leaving remain the same.

The turbine in Fig. 10.2 is a tandem cross-compound unit. The upper portion shows a high- and intermediate-pressure (tandem) turbine on a single shaft. The lower portion is the low-pressure unit. Mounted on the right of each (not shown) are the electric generators.

In operation, initial steam enters the high-pressure turbine at 3500 psig and 1000°F through two inlets (top and bottom). It passes through the turbine to exit at the left (and below) at approximately 600 psig and 550°F and then passes to a reheater in the boiler, where the steam is again heated to 1000°F. On leaving the reheater, steam at a pressure lower than 600 psig, because of pressure drop losses, and 1000°F enters the intermediate unit (at bottom center) and passes through the double-flow turbine, exhausting at the top through two outlets. This steam is at approximately 170 psig and 710°F when it passes to both sections of the low-pressure unit, finally exhausting to the condenser.

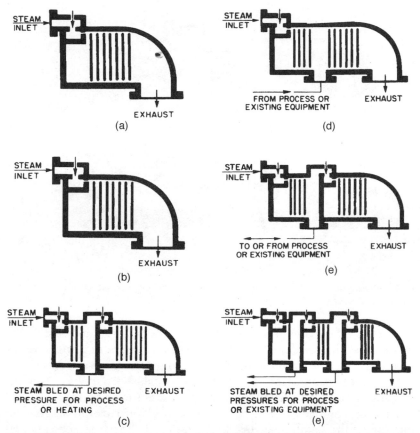

Figure 10.1 Steam turbine types and applications. (*a*) Condensing; used when exhaust steam from the turbine cannot be utilized and power must be generated on a minimum amount of steam. (*b*) Noncondensing; used when all or practically all the exhaust steam from the turbine can be used for process or heating. (*c*) Single extraction; used when process steam requirements are variable or intermittent. (A noncondensing extraction turbine can be used when process steam is required at two different pressures.) (*d*) Mixed pressure; used when excess steam is available at lower than inlet pressure and when this supply is intermittent. (*e*) Mixed-pressure extraction; used to supply process steam when necessary and to utilize a surplus of process steam when available. (*f*) Double-extraction; used when process steam is required at two different pressures (or at three different pressures if noncondensing).

This is an extraction turbine, steam being extracted from one stage of the high-pressure turbine and four stages of each low-pressure turbine. It is used for feedwater heating. Note that the intermediate-pressure unit and both low-pressure units have double-flow arrangements; i.e., steam enters the turbine at the center and flows in two directions. This unit has a capacity of approximately 900 MW.

Steam from a turbine may be exhausted into a condenser to obtain the maximum amount of energy in the steam, or it may be removed at any intermediate pressure by the use of a noncondensing or backpres-

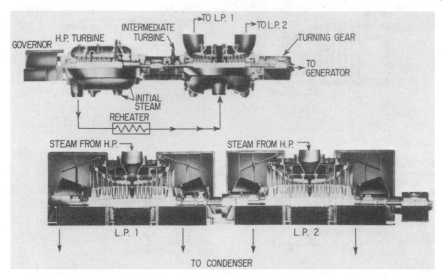

Figure 10.2 Tandem cross-compound 3600 rpm four-flow reheat steam turbine. (*General Electric Co.*)

sure turbine. While the latter reduces the amount of energy available to the turbine, it provides steam for process or space heating. A compromise between the condensing and noncondensing arrangements is the extraction turbine. Here, steam, having passed partway through the turbine, is removed from the casing at a point or points with the desired steam conditions. The pressure at a given turbine stage (i.e., connection to the casing) varies with the load. For this reason, two methods of extraction are used: (1) The uncontrolled extraction method consists of connection to the casing with varying pressures, increasing as the load increases and decreasing as the load decreases. This arrangement is used extensively in power plants for extraction feedwater heating. (2) The controlled extraction method maintains constant pressure by regulating the flow of steam through the turbine on the downstream side of the extraction point. This arrangement is used to obtain steam at constant pressure for process or heating. The amount of superheat (or steam quality) delivered at the point of extraction depends on the initial steam conditions and load on the turbine. The turbine is adaptable to the use of steam at various pressures and temperatures. It can supply maximum power in the condensing operation or a lesser amount of power in addition to steam at a reduced pressure. Several types of turbines together with their applications are shown in Fig. 10.1, as well as those described in Chap. 9.

Noncondensing turbines are used where process steam at only one pressure is required and where no condensing water is available or the cost of providing it is prohibitive. An example of such a turbine would be the topping unit referred to previously. Noncondensing turbines may be designed for one or more stages of extraction, and such turbines lend themselves to the mixed-pressure extraction-type unit. Extraction steam can be used for process and to drive plant auxiliaries such as pumps. Mixed-pressure extraction turbines are used in paper mills and other industries where large quantities of steam at varying pressures are required at certain times, while at other times a large amount of steam at a constant pressure is required.

In a single turbine, the steam expands from the initial condition to exhaust in one unit. The tandem compound turbine consists of two separate units arranged in line with their shafts connected end to end and with steam passing from the high- to the low-pressure turbine. The cross-compound unit also consists of two separate turbines. These are high- and low-pressure turbines mounted side by side with their shafts parallel. Refer to Chap. 9 for descriptions.

Steam turbine drives are used for pumps, air compressors, fans, and other mechanical equipment. To be efficient, a turbine must operate at high speed. When the driven unit also can be operated at high speed, the turbine shaft is connected by a coupling to the driven shaft. When, however, a driven machine is required to operate at a lower speed than that of the turbine, reducing gears are used to transmit the power.

When steam expands in a turbine, it imparts energy to the rotating shaft as it comes into contact with the blades. This energy is taken from the steam, and as a result of this heat loss, part of the steam is condensed, and moisture begins to form in the steam. The water thus formed can do no work in the turbine; in fact, it increases friction and the potential for blade erosion and actually hinders the flow of steam, resulting in a lowering of turbine efficiency. The formation of moisture is delayed by superheating the steam in the boiler. Superheating improves the economy of the turbine in two ways: The additional heat increases the energy available for conversion into work, and it also reduces friction. With superheat temperatures up to 100°F, the saving in steam may be estimated as 1 percent for each 10°F. At higher superheat temperatures, the savings are slightly less.

The greater the pressure range through which the steam is expanded, the more is the superheat required to prevent the formation of excessive moisture in the last stages of the turbine. The steam temperature is limited to about 1000°F by the metals available, but in some cases 1050°F steam temperatures are used.

Some turbines use a reheat cycle. This consists of passing the high-temperature steam through a turbine, then returning it to a special superheater (reheater), and resuperheating it before it is expanded further in the low-pressure turbine. This system makes it possible to have dry steam in all but the last few stages of the turbine. It would be possible, but is seldom economically advisable, to use more than one stage of reheating.

Turbines are well adapted to the utilization of high-pressure steam. At 100 to 150 psi pressure, there is a decrease in water rate of $1\frac{1}{4}$ percent for each 10 psi increase in pressure. From 150 to 250 psi pressure, this decrease becomes $\frac{3}{4}$ percent for each 10 psi increase in pressure. At higher pressures, the reduction in water rate is lower.

Condensing turbines are operated effectively and efficiently with low backpressure (vacuum). Surface condensers are used, and the use of these condensers decreases the water rate by approximately 5 percent for each inch of mercury improvement in vacuum within the range of 25 to 29 in Hg. Exhaust pressures of 3.5 to 1.0 in Hg are common in modern power plants. Thus, by operating at a vacuum as created by the condenser, the turbine can produce more power, just as it would with an increase in steam pressure, but much more economically.

The impact and importance of a condenser that creates a vacuum can be illustrated by the use of a simple equation for the thermal efficiency of a perfect heat engine or turbine that ignores any losses due to radiation or friction.

$$E = \frac{T_1 - T_2}{T_1} \times 100$$

where E = efficiency (%)
T_1 = absolute temperature of entering steam
T_2 = absolute temperature of leaving steam

Note: Absolute temperature = °F + 460.

In this example, assume that the entering steam has a temperature of 400°F and that the steam exhausts at atmospheric pressure, or 14.7 psia, which would have a steam temperature of 212°F. Compare the efficiency of this heat engine or turbine with that of a heat engine or turbine that has a condenser and a backpressure (vacuum) of 5 psia and a steam temperature of 162°F.

Without condenser:

$$E = \frac{860 - 672}{860} \times 100 = 21.8 \text{ percent}$$

With condenser:

$$E = \frac{860 - 622}{860} \times 100 = 27.7 \text{ percent}$$

Therefore, by the addition of a condenser, the efficiency is increased by nearly 6 percent. Although this is a simple example that ignores losses such as thermal radiation, it illustrates the goal for higher steam temperatures and for lower exhaust steam temperatures to obtain higher efficiencies and thus lower fuel costs and higher outputs.

Nothing is free, however, and with the use of a condenser, it is necessary to pump large quantities of cooling water through it and the condensate also must be pumped out of the condenser. Both these actions require power, which reduces the effect of improved efficiency, and additional capital costs are also necessary. These added costs must be evaluated against the improved efficiency. Since the majority of electric power plants have condensers, this improvement in efficiency and power output has far exceeded the additional costs in any economic evaluation.

10.1.1 Turbine operation

A well-organized and properly operated plant should have an operating plan that includes instructions for all plant equipment and for interacting systems, of which turbine operation is a part. Before starting the turbine, the operator should become familiar with the general piping layout, the operating characteristics of the unit, and the manufacturer's operating instructions. These instructions will utilize the following general procedures:

Starting a small noncondensing turbine (like that shown in Fig. 9.18):

1. Fill all grease cups and oil the governor and other miscellaneous parts.

2. Open drains on the header, separator, casing, and exhaust lines. If these are equipped with traps, open the bypass line.

3. Slowly open the exhaust-line shutoff valve.

4. If the turbine has a pressure lubricating system with an auxiliary oil pump, start the pump.

5. Inspect the bearings for ample flow of oil; check the oil pressure and see if the pump is operating properly; 3 to 5 psi pressure is sufficient.

6. Turn on the cooling water for the bearings or oil cooler.

7. Open the throttle quickly to start the turbine. During this time observe the turbine carefully for signs of rubbing (a small pipe or rod placed between the ear and the points to be checked will aid in detecting unusual noise and rubbing, or check the vibration monitoring equipment, if provided).

8. Slowly bring the turbine up to speed, approximately 300 rpm, and operate it at this speed for a period of 15 to 30 min.

9. Trip the emergency valve by hand to see that it closes properly.

10. Open the throttle wide and allow the governor to regulate the speed; keep a close check to prevent overspeeding. *Note:* If the governor will not control the speed at no load, the hand throttle may be used until the load is on.

11. Inspect the bearings to make certain that they are getting oil. Oil should be up to the level recommended by the manufacturer. Usually the level drops when oil fills all the various cavities in the system.

12. The drains, which were opened before the turbine was started, should now be closed or arranged to discharge through the trap.

13. Gradually increase the load on the turbine while keeping a close check on the oil, cooling-water, and bearing temperatures.

Starting a medium-sized or large condensing turbine:

1. Inspect the governor mechanism, fill all grease cups, and oil where necessary.

2. If the boiler stop valve is not open, open it to permit as much heating of the steam line as possible and avoid condensation in the line.

3. Open the following drains: header, separator, throttle, and turbine casing.

4. Open the stop valve in the steam supply to the auxiliary oil pump. *Note:* The operation of this pump is controlled by a governor to shut off the supply of steam after the main oil pump has delivered oil at normal pressure; it opens when the pressure fails.

5. Adjust needle valves to obtain 10 to 15 psi oil pressure on the main bearings and 15 to 20 psi oil pressure on the thrust bearing; make sure gauges are in operating condition and have been calibrated.

6. Start the condensing equipment, circulating pumps, and dry vacuum pump; operate the condensate pumps as found necessary to remove water during the warming-up period.

7. The turbine steam or water seal should be turned on and the vacuum maintained at 24 to 26 in Hg during the warmup period; maintain approximately 1 psi pressure in the packing chamber.

8. Turn on the water to the generator air cooler, and see that water flows properly to this and other points requiring water.

9. If the drain ahead of the throttle valve has been closed for any reason, open it again and keep it open until all water from condensation has been removed.

10. Now open the throttle valve quickly to set the rotor in motion.

11. As soon as the turbine is rolling, trip the overspeed by using the hand lever. This is to determine if the tripping mechanism operates properly and to prevent the turbine from accelerating too rapidly.

12. Reset the emergency overspeed valve, and before the turbine comes to rest, adjust the throttle so that the turbine will operate between 200 and 300 rpm.

13. While the rotor revolves slowly, use a metal rod or listening device to determine rubbing or mechanical difficulty, or observe the vibration monitoring equipment, if provided.

14. When the oil leaving the bearing reaches a temperature of approximately 110 to 120°F, start the circulating water through the oil cooler to maintain these temperatures. At this time, the bearing oil pressure should again be checked.

15. Gradually increase the speed. The rate depends on the size of the turbine. Follow the manufacturer's instructions.

16. Adjust the water seal on the turbine and the atmospheric relief valve.

17. When normal operating speed has been reached and the turbine is under the control of the governor, test the emergency governor by opening the valve in the oil line to it. See that all valves controlled by this tripping mechanism close promptly. Reset, open the throttle valve, and restore speed to normal.

18. If high-pressure packing is of the water-seal type, adjust the water to 15 psi pressure and shut off the steam.

19. Close the drains mentioned in step 3.

20. Open the leak-off from the high-pressure packing so that any excess steam may flow to the feedwater heater or to one of the lower stages of the turbine.

21. Synchronize the generator and tie it in the line.

22. With the throttle valve wide open, the speed is controlled by the governor. The turbine is now ready for load and is regulated from the turbine control panel.

Note: Coordination of effort on all portions of the power plant is essential in placing the turbine in operation.

It must be remembered that a large turbine has close clearances and that expansion or improper operation is likely to cause more damage than in the case of a small unit. Large turbines are provided with instruments, including oil-pressure gauges, thermometers, and vibration monitors. These instruments should be observed at frequent intervals and the readings recorded on the log sheet or plant computer. In addition to the turbine, the condenser and other auxiliaries require attention, and this phase of the operation must not be forgotten.

Note: Always read and follow the manufacturer's instructions. These should be part of the overall operating plan for the plant.

The turbine in motion:

1. Apply the load gradually.
2. Observe the oil level; check to see if an ample supply of oil is going to the bearings and the hydraulic cylinder. This can best be observed by watching the pressure gauge and sight indicator on the oil discharge.
3. Watch the oil-bearing temperature. This is always a good indication of overheating and mechanical trouble. Temperatures of approximately 140 to 150°F are desirable; above 175°F, serious operating difficulties may be experienced.
4. Observe the turbine for any unusual noise, vibration, etc.

Shutting the turbine down:

1. Gradually reduce the load to zero.
2. Start the auxiliary oil pump, and make certain that the proper pressure is maintained while the turbine is coming to a stop.
3. Trip the emergency valve. In most cases this valve also operates the vacuum breaker.
4. Close the leak-off from the high-pressure packing; admit steam to the chamber at approximately 1 psi, and shut off the water.
5. Shut off the supply of cooling or condensing water.
6. Shut down the condensing equipment, and open the drains on the turbine piping and casing.
7. Continue the auxiliary oil pump in operation until the turbine rotor has stopped.
8. If the turbine is to be left idle for a period long enough for it to cool to room temperature, operate the condenser air pump to dry it out. In this way corrosion can be avoided.

While a turbine is operating, it is good practice to keep a log sheet or a computerized record of the hourly readings of the instruments. These readings should be taken by the operator while making a regular inspection. On modern units such information is identified to the operator as part of the electronic control system, where the information is also permanently recorded. Such a procedure prevents the operator from neglecting some important inspection, and the data make a valuable record for future reference. The number of readings to be taken depends on the size of the turbine and the number of instruments installed. Some of the readings that might prove valuable are load on the generator in megawatts, throttle steam pressure and temperature, exhaust pressure, extraction pressure, temperature of cooling water entering and leaving the cooler, bearing oil pressure and temperature, and the throttle steam-flow rate.

The proper application of oil to the bearings and a continuous flow of cooling water are the main requirements of a turbine while it is operating. The operator detects trouble on the inside of the turbine by noise or vibration or both. An appreciable drop in oil pressure can be corrected by operating the auxiliary oil pump until there is an opportunity to investigate the cause of the trouble.

Some small turbines have a number of hand-controlled individual nozzles for admitting the steam. The operator can improve the economy by having a minimum number of these nozzles open for the load at which the turbine is operating.

10.1.2 Turbine maintenance

Proper maintenance is essential to the continuous and efficient operation of the turbine. Items requiring maintenance should not be permitted to accumulate but rather should be taken care of as soon as the trouble arises. The exception, of course, is the general overhaul, which is scheduled in advance and during which careful inspection and examination are made of the complete machine. Breakdown or emergency maintenance is best avoided by using a preventive-maintenance schedule. Maintenance is costly from an equipment and labor standpoint. However, because of the outage of the equipment, this cost is greatly exceeded by the loss of revenues that occurs while the unit is out of service.

Since each turbine is basically different, maintenance requirements must be discussed here in a rather general way. Specific recommendations are to be found in the instructions supplied by the turbine manufacturer, where details in the particular design are explained and described together with the proper maintenance of the unit. It should be remembered that timely replacement of parts and adjustment for wear may prevent a shutdown and save costly repairs.

Many operators find it economical and advisable to dismantle their turbines completely and give them a thorough internal inspection once a year. Although this may be desirable, experience has demonstrated that a turbine properly operated and maintained can run for years without dismantling. In fact, the turbine is considered to be one of the most reliable pieces of equipment in the entire plant. The merit of such inspections, however, together with their frequency, is a matter of operating experience and depends to a large extent on the age of the equipment and the service-hour record if it is assumed that proper operation, etc., has been provided.

An annual inspection of a turbine would consist of examination of blading and nozzles for wear, erosion and corrosion, cracks, and scale deposits; examination of main bearings and the thrust bearing for wear; inspection of clearance between stationary and moving parts and packing clearance; and thorough inspection of the oil pumps and the various governor mechanisms.

Because turbines remain in service for longer periods of time, factors such as thermal stresses, chemical attack from steam impurities, erosion, and corrosion all have become real concerns to an operator. Operational situations such as process and load swings, plant upsets, and extended turbine shutdowns with accompanying corrosion damage and occasional boiler carry-over can affect turbine performance and its maintenance costs.

Maintenance and operating problems on large turbines that are found in electric utilities are often attributable to

1. Operating the turbine beyond its design life
2. Operating at continued outputs higher than design
3. Increased cycling operation on the turbine

Although operating at higher outputs and for longer periods of time than originally planned can create fatigue stresses, cyclic operation is more of a problem. Solids carry-over to the turbine is a primary problem with cycling operation, and this can cause erosion in the high- and intermediate-pressure portions of the turbine and cause corrosion in the low-pressure section. It is therefore very important to maintain high-quality boiler feedwater so that carry-over is minimized. The major impurities that cause turbine corrosion are sodium hydroxide, sodium chloride, and various organic and inorganic acids.

The performance of steam turbines will deteriorate over time because of erosion, corrosion, deposition, foreign-object damage, and seal degradation. These can degrade performance by 1 to 15 percent.

The most significant cause of efficiency degradation is deposits along the steam path as a result of boiler carry-over. Deposits are not only detrimental to efficiency but also can have a drastic effect on turbine reliability because of increased thrust loading and the possibility of stress-corrosion cracking of components from high levels of caustic carry-over. Boiler water chemistry, therefore, is of critical importance. Water washing, detergent soaking, grit blasting, and hand polishing are typical methods of removing deposits.

Since turbine rotors rotate at high speed, one of the difficulties encountered is vibration. Vibration in many cases is caused by unbalanced weight in the rotating parts and misalignment of the shafts. Vibration due to an unbalanced condition changes with the speed at which the machine is operating. It may be caused by broken, eroded, or corroded blades, by a bent shaft, by distortion due to unequal heating, or by scale deposits on the blading.

Improper alignment of turbine and rotor shafts causes a vibration that does not change with the turbine speed but which increases as the load on the turbine increases. It might be caused by improper alignment of a new turbine. Foundation settling or strains due to pipe expansion could cause vibration in a turbine that has been in operation for some time.

Turbine vibration also occurs from water coming over from the boiler with the steam, which will cause the turbine to become noisy and, in severe cases, will cause vibration; electric-generator troubles, such as an unequal air gap or a loose coil in the stator; internal rubbing of parts caused by warping of bladings or diaphragms, improper adjustment, or worn thrust bearings; too much clearance in the main bearings; and overheating due to faulty lubrication or lack of cooling. Also, vibration may result if the shaft becomes distorted as a result of heat generated by having the packing rub.

The remedy for most of these difficulties immediately suggests itself: The different parts of the equipment must be supported rigidly and aligned properly, the bearing clearance must be within proper limits, and all bolts must be well tightened.

The rotor of the turbine is held in correct axial clearance by the thrust bearing (see Figs. 9.25 and 9.26) and is adjusted with shims. The space provided for the shims is greater than the internal clearance between the wheel and the nozzles on one side and the stationary buckets on the other. With shims removed, the striking limits of the rotor can be determined and measured at the bearing. If we assume a rotor with two rows of buckets on the first wheel and with a total clearance of 0.160 in or less, the thrust must be shimmed so that the clearance is divided equally and the rotor is in midposition between striking limits when washers are set. If wear occurs, a complete thrust bearing should be ordered as an assembly; a spare should always be carried in stock.

Carbon packing rings are used at the high- and low-pressure ends of turbines that operate condensing, on those turbines exhausting against high backpressure, and where high initial steam temperature and pressure are used. Metallic packing rings are placed at the shaft bore of diaphragms to minimize steam leakage along the shaft due to the difference in pressure between the stages.

The recommended packing (carbon) clearance of the bore of the ring in excess of the diameter of the shaft for 2- to 3-in-diameter shafts is 0.002 to 0.005 in. When diaphragm packing is used, the bore of the teeth of the packing ring must be 0.010 to 0.013 in larger than the shaft. Again, the manufacturer's recommendations should be followed.

Proper maintenance of the emergency governor is likewise important. To ensure proper operation, the governor should be tested at frequent intervals. All parts should be kept clean and pins and bearings oiled. Alteration in the tripping speed on most governors is accomplished by altering the number or thickness of shims. The emergency governor should be tested weekly or when being placed in operation after an extended outage.

Lubrication is of extreme importance to the successful operation of the turbine. Proper lubrication of a steam turbine depends on the proper application of oil to the bearings, the selection of a suitable oil, and maintenance of the oil in good condition after it is in the turbine. The proper method of application of the oil to the bearing is part of the turbine design. With use of the manufacturer's recommendations, the operator is responsible for selection of the proper oil and for keeping it in good condition after it is in service in the turbine.

Oil having the proper viscosity must be selected. A thin oil is recommended, but it must not thin out and lose its lubricating properties when heated to operating temperature. It should have the ability to separate readily from water. Some oils mix with water and form sludge, which is very objectionable in the lubricating system. Oil must be free from acid so that it will not corrode the highly polished surface of the bearings. It must not give off volatile gases when heated to temperatures of 325 to 350°F. Oils that have their gases distilled off at low temperatures lose their lubricating properties after they have been in use for some time. Some oils leave a sludge deposit in lines, tanks, etc.

Oils in service pick up particles of dirt mixed with water and tend to become acid. Medium-sized and large turbine installations have filters and cleaners for keeping the oil in good condition. Several different methods are used: *bypass filtration,* in which part of the oil is continually circulated through a filtration system; *continuous filtration,* in which the entire amount of oil returning from the turbine is filtered; *makeup system,* in which quantities of oil are drawn off at intervals and replaced with new or reconditioned oil; and *batch treatment,* in which all the oil in the turbine is removed and replaced with new or reconditioned oil.

The oil pump, as well as the oil cooler, strainer, and other appliances through which the oil passes, should be kept in first-class condition. The worm and worm-gear teeth of the pump should be inspected. The relief valve should be adjusted for 50 to 55 psi. To ensure the best lubrication, the oil should be withdrawn from the tank and the tank cleaned thoroughly. All openings, pipes, etc., should be blown out with air or steam.

Oil coolers require attention and should be removed from service for the necessary maintenance. This can best be accomplished by having available a spare tube bundle. After removing the heads, clean the tubes by passing a brush through them. The cleaning of the outside of the tube is accomplished by boiling it in a solution of hot water or cleansing agent, the kind of solution depending on the type of deposit on the tube. The entire tube bundle is then immersed in a tank with sufficient capacity to house the bundle conveniently. The task can be accomplished to better advantage if a pump is used to circulate the cleaning solution.

Preventive maintenance, which includes inspections at frequent intervals, will do much to keep the turbine in peak operating condition. The purpose of the above-mentioned inspections is to find possible sources of trouble before they have had an opportunity to cause serious damage. Testing the emergency governor and checking the operating governor are two points worth remembering.

A general inspection and maintenance procedure follows; however, specific procedures should be developed based on the manufacturer's recommendations and the plant's actual operating experience.

Weekly: Inspect the turbine exterior. Keep control-line valves free from dirt and grease at the guides. Clean and oil the spindle of the throttle valve and connections between levers of the governing mechanism, using a light oil to prevent gumming; repair all oil leaks.

Monthly: Inspect the oil reservoir to determine if any sludge is accumulating or any water is getting into the oil. Check the oil strainer. Check the operation of the governor and automatic tripping devices and vacuum breakers. Observe the governor operation to determine how it acts and whether it will carry the maximum turbine load. If the turbine is operating at condensing, inspect the condensing equipment and check the vacuum to see if it is the best obtainable with the equipment and circulating-water temperatures.

Yearly: Dismantle turbine and auxiliaries; check for corrosion, dirt, scale, and fouled passages and for encrusted, corroded, and eroded blading. Inspect oil and steam atomizers. Check all valves and seats. Check governor knife edges and bearing blocks and the emergency tripping device. Check vacuum pumps and ejectors.

Check for air leaks. (On modern turbine generators, outages for major inspections and maintenance generally occur every five years.)

At this time a test should be made to check the spring calibration and to determine the speed above normal at which the oil-tripped emergency governor will operate. In operation, the emergency governor should be tested weekly. Testing the emergency governor should be done after the turbine has been placed back in service or approximately once a month.

1. Attach a vibration tachometer to the turbine (or observe the permanent tachometer) so that it can be read by the person operating the throttle valve; the person at the other end of the turbine should be provided with a hand tachometer.
2. Operate the throttle valve slowly to bring the turbine up to the required speed, where the governor takes over.
3. Disconnect the governor beam from the vertical connecting linkage. Hold it firmly to prevent it from closing the controlling valve. Lift it slowly, permitting the control valve to open and the speed to increase slowly until the tripping point is reached on the emergency governor. Do not permit the speed to exceed the safe limit established.
4. Take tachometer readings to check the speed at which the emergency governor operates.
5. If and when the emergency governor acts, check to see that the throttle valve trips and speed decreases. If the emergency governor fails to operate, repeat all the previous check points to determine what the problem might be.
6. Trip the emergency governor two or three times to make certain speed check is correct and all parts are in working order.
7. Test the oil-tripped governor when shutting down the turbine by:
   a. Running the turbine without load, throttle valve wide open and turbine under control of the operating governor.
   b. Opening the oil cock in the pipe supplying oil to the nozzle and checking that the governor operates instantly when the oil jet is applied and the throttle valve trips, at which time the speed should drop, indicating the throttle valve has closed.
   c. When the governor trips, shutting the oil cock and determining whether the governor returns to normal; then resetting the tripping mechanism. Never keep a turbine in service unless you are certain that the emergency devices are reliable. A complete record should be kept of the changes and adjustments made when tests are conducted.

10.2 Condensers

The condenser reduces the backpressure, thereby increasing the output and efficiency of the turbine. The condensed steam becomes excellent feedwater and is returned to the boiler. The vacuum created in the condenser depends on the cooling water temperature entering the condenser, the effectiveness of the air-removal devices, and many other factors.

Some of the major areas of condenser operation that affect overall plant performance and turbine backpressure include cooling water temperature, temperature differential between exhaust steam and inlet cooling water, degree of tube fouling, and the velocity of the cooling water. Condenser fouling is common, especially when the cooling water is from a natural source, such as a lake, river, or ocean. The fouling results from sediment, scaling, corrosion, and biologic growth. When cooling water is part of a closed system with a cooling tower, fouling of the condenser is minimized because the water chemistry can be controlled and plugging from debris has been eliminated.

The operation of surface condensers can be enhanced by the following design and operational features:

1. Avoid flow-induced vibration, which results in tube failures.

2. Incorporate proper steam flow distribution to ensure proper use of heating surfaces.

3. Prevent air and circulating water leakage.

4. Maintain clean condenser tubes to ensure high performance.

When tubes become plugged with debris, proper backflush valve and piping arrangements can remove this debris. In addition, various mechanical and chemical water treatment procedures are available to minimize fouling.

10.2.1 Surface condenser operation

In the operation of surface condensers (see Fig. 9.41), steam passes over the bank of tubes through which the cooling water circulates. The source of this cooling water is a river, lake, ocean, cooling pond, etc., or a cooling tower. The steam is condensed and drops as condensate to the hot well, being removed by the condensate or hot-well pump. Air and noncondensable gases are removed by ejectors.

To place a surface condenser in operation:

1. Open the discharge and inlet valves to and from the condenser and start the circulating pump.

2. Open the steam-jet air-ejector valves and slowly bring up the vacuum.

3. Operate the turbine throttle to admit steam and permit steam to pass to the condenser.

4. Start the condensate pump; operate it to remove the water as it condenses. This can be accomplished by operating the pump on an intermittent basis.

5. As the turbine comes up to its operating speed, increase the vacuum by operating the steam valves to the air ejector.

6. If the load increases, speed up the circulating water pump if a variable-speed motor is provided. Operate the condensate pump on a continuous cycle to remove the water from the hot well.

The quantity of cooling water required to condense the steam can be computed. The heat absorbed by the circulating cooling water in passing through the condenser must equal that given up by the exhaust steam, if leakage and radiation are neglected. This quantity of cooling water can be approximated with the following formula:

$$Q = \frac{H-(t_0-32)}{t_2-t_1}$$

where Q = quantity of water to condense 1 lb of steam (lb)
H = heat content of exhaust steam (Btu)
t_0 = temperature of condensate (°F)
t_1 = temperature of cooling water entering (°F)
t_2 = temperature of cooling water leaving (°F)

Example Assume the heat in the exhaust steam to be 1100 Btu; the temperature $t_0 = 100$, $t_1 = 80$, and $t_2 = 90$. Calculate the amount of cooling water required to condense 1 lb of steam.

Solution

$$Q = \frac{1100-(100-32)}{90-80} = \frac{1032}{10}$$

$$= 103.2 \text{ lb} \qquad \text{(amount of water to condense 1 lb of steam)}$$

This is a calculation to determine the approximate quantity of cooling water required to condense the exhaust steam. For actual design calculations, the requirements are determined with the use of the enthalpies for the steam and water conditions.

In the preceding example, if the quantity of exhaust steam is 100,000 lb/h, then the quantity of cooling water required would be

$$103.2 \times 100,000 = 10,320,000 \text{ lb/h}$$

or

$$\text{gpm} = 10,320,000 \text{ lb/h} \times \frac{1 \text{ gal}}{8.33 \text{ lb}} \times \frac{1 \text{ h}}{60 \text{ min}}$$

$$= 20,648 \text{ gpm}$$

10.2.2 Surface condenser maintenance

The surface condenser can require significant maintenance to meet operating requirements. The most common problem is loss of vacuum caused by air leaking into the condenser through joints or packing glands. Another problem is cooling water leaking into the steam space through the ends of the tubes at the tube sheets or through tiny holes in the tubes that may be caused by corrosion or erosion. This water leakage also contaminates the condensate and results in the need for additional water treatment prior to it entering the boiler to ensure good water quality. With the use of a natural cooling water source such as a river, lake, or ocean, tubes can be plugged with mud, leaves, shells, debris, slime, or algae, reducing the supply of cooling water. Tube leaks require rerolling of the tube in the tube sheet, replacement, or plugging of the leaking tube.

Tube leakage can be determined in a number of ways. The condensate continues to exceed the normal or average requirements, causing the hot-well level to rise. This may be due to any one of a number of things: leaks around the tube ends, split tubes, or excessive leakage into the system from the sealing water. The unit must be inspected to determine the cause and location of the leak.

Inspection for leaks can be done by pulling a vacuum on the condenser and exposing the flame of a candle to each tube. This is a slow and tedious job but is usually effective. Or the inspection can be made with the condenser out of service and with circulating water passing through the tubes. By entering the space below the tubes, it is possible to detect drips or leaks. At times it is difficult to detect actual leakage and the location of the leaks because the tubes tend to "sweat," giving one the impression that a leak exists where there is none. This sweating is due to condensation of moisture in the air that is in contact with the cold surface through which the circulating water is passing. Still another method frequently used for locating leaks is as follows: With the condenser out of service, fill the steam space with water. The tubes can be inspected for leakage by entering the water boxes. Tubes found to be leaking can be replaced or plugged and repairs made where necessary.

The quantity of water leaking per unit of time can be determined with the condenser out of service by pulling a vacuum on the condenser while noting the rise of water level in the hot well. Or a condition may arise in which the leakage is not so easily noted, as in the former case. Small amounts of water leaking into the hot well are best determined by running a chemical analysis on the condensate from time to time or by testing the condensate to determine the resistance or conductivity of a given sample of water.

If a tube is split, temporary repairs may be made by driving a plug into the end of the tube or by replacing the ferrules with a special brass cap. Permanent repairs are made by replacing the tube. In the case of tubes expanded in the tube sheet, care must be exercised to obtain a tight joint without bending the tubes by overexpanding them. When packing is used to prevent leakage between the tubes and the tube sheet, the specified amount of packing is placed in the box, and the ferrule is screwed down tightly but with care to prevent crushing (necking) the tube.

Air leaks that develop are difficult to detect and locate because of the large surface area of the condenser and the many valves, fittings, and auxiliaries attached to it. These leaks are detected by (1) examining all the joints for leaks using a candle, (2) filling the entire condenser with water and maintaining a slight pressure on the system, and (3) using low-pressure air and checking for leaks using a soap solution on the joints. The leaks must be checked around all auxiliaries connecting the turbine seals, bleeder connections, etc.; this is very often a slow and tedious job. Plastic compounds are often used to seal the leaking joint when it is located.

A convenient method of determining the extent of leakage is to pull a vacuum on the system and close down, permitting the unit to stand for a while. If the vacuum drops more than 2 to 4 in Hg per hour, it will be necessary to hunt for leaks. Another method frequently used to determine whether the condenser is leaking air is to check the pressure and see if the pressure in the condenser corresponds to the condensate temperature. Should the pressure be higher, air leaks or noncondensable gases are present in the condenser. As an example, suppose the pressure in the condenser is 1.932 in Hg absolute (0.9487 psia) and the temperature in the hot well is 96°F. Referring to the steam tables, we find that at 96°F the pressure should be 1.711 in Hg absolute, or 0.8403 psia. Hence the added or excess pressure is due to air or noncondensable gases that have leaked into the condenser.

Condensers are sometimes equipped with metering devices to measure the air removed from the condenser. These meters are usually calibrated in terms of cubic feet of free air, at a given temperature, leaking into the condenser per unit of time.

Condensers are frequently troubled with scale and algae growth. These reduce the heat transfer and at times become so serious as to interfere with the circulation of water. Backwashing is sometimes effective in removing foreign material. Condensers frequently have their piping so arranged that this can be accomplished while the condensers are in service. Two-pass condensers often have a divided water box with horizontal decks; one-half of the surface can be cut out of service for cleaning, while the remainder carries the load, at reduced vacuum. This avoids interruption to operation and, if done at low load, causes little inconvenience.

Scale can be removed by washing the tubes. Sediment inside the tubes can be removed by blowing rubber plugs and compressed air through the inside of the tubes. Algae growth is inhibited or destroyed by using various chemical solutions, depending on the nature of the algae growth; at times, chlorine can be used very effectively.

Since tube leaks are common and the plugging of tubes is the quickest repair, most surface condensers are designed with excess surface so that plugged tubes will not reduce design performance.

When replacing tubes or making repetitive adjustments, it is well to keep a continuous record for future reference. Continued maintenance and repeated adjustments indicate trouble that must be taken care of before the situation becomes serious.

10.2.3 Air-cooled steam condenser operation

As explained in Chap. 9, an air-cooled steam condenser is a system that performs the same functions as a water-cooled condenser and the wet cooling tower. It is often called a *dry system*. The system provides low backpressure for the turbine, conserves high-purity water, and deaerates the condensate.

Instead of a condenser and a cooling tower that dissipates the heat absorbed by the cooling water in the condensing process, an air-cooled steam condenser has all the heat rejected from the turbine exhaust steam being absorbed in the form of heat gain in the air. This system may be mechanical or natural draft; however, most modern systems are mechanical draft.

These systems are attractive because of the following:

1. Makeup water is not required. This permits siting a power plant at a location that does not have a large supply of cooling water.

2. Maintenance costs are generally less.

3. Large amounts of water-treatment chemicals are not required, and the disposal of blowdown materials is minimized.

4. Fog, mist, and local icing have been eliminated as compared with a wet cooling tower.

There are significant disadvantages, however, since the condensate temperature and the associated steam condenser pressure are not as low as with a wet cooling tower system. Therefore, a higher turbine backpressure results that reduces plant efficiency and the electric power output. Not only is there lower output for the same initial steam conditions, but also operating costs are higher due to higher fuel costs. The backpressures of turbines are in the range of 3.7 to 9 in Hg with air-cooled steam condensers as compared to 1 to 4.5 in Hg for wet cooling tower systems under similar conditions.

10.3   Cooling Towers

Power plants are not always located adjacent to a plentiful supply of cooling water, and even when they are near a plentiful water supply, environmental restrictions sometimes prohibit the use of that water. At other times the source of supply is found to be contaminated by the presence of foreign matter that would prove injurious to the pumps, condensers, and circulating systems through which the water must pass. The cost of treating this water into a satisfactory condition for use would prove prohibitive. Yet for condensing turbines, cold cooling water is required.

Therefore, *cooling towers* (see Figs. 9.51 and 9.54) are provided where sufficient water is not available, their purpose being to discharge the heat absorbed by the cooling water in the condenser. In order to do this in the shortest possible time, the water must be broken up into a very fine spray so that as much water surface as possible is present to the air.

Cooling ponds and spray ponds are usually less expensive to install and maintain than are cooling towers; however, few sites today have these ponds, and, therefore, cooling towers are predominant on modern power plants. In any case, better performance is often obtained with a properly designed cooling tower, since it can provide water at a lower temperature, which in turn can produce a better or lower vacuum in the condenser and thus a higher output from the turbine. The cooling tower is very compact and can be located almost anywhere that space permits, even on the top of buildings for smaller installations. Where ponds are used, a spray pond requires considerably less area than a cooling pond.

The principle of cooling applied in both the cooling pond and tower is that of transferring sensible heat in the cooling water to the air, thereby lowering the water temperature. Most of the cooling is due to exchange of latent heat resulting from evaporation of a small portion of the water. In cooling (evaporation), some water is lost; it amounts to approximately 1 percent for each 12 to 14°F of actual cooling.

The capacity of the air to absorb additional water vapor depends on the wet-bulb temperature, the wet-bulb temperature being an indication of the total heat content of the air. The dry-bulb temperature is an indication of only the sensible heat contained in the water or substance and not the total heat. When the wet- and dry-bulb temperatures are the same, the relative humidity is 100 percent, and the air's capacity to absorb additional water vapor is zero. (Relative humidity is the ratio of the amount of water vapor contained in the air to the amount that the air could contain if it was saturated.) Hence the ability of a spray pond or cooling tower to give off heat contained in the water depends on the wet-bulb temperature or relative humidity of the air with which it comes into contact. Influencing factors that must be taken into consideration are wind direction and velocity. The extent of cooling also depends on the area of the water exposed to the air and the temperature of the water when it enters the basin. Cooling is brought about by evaporation and conduction.

The degree to which the temperature of the water approaches the air wet-bulb temperature is a measure of performance for the pond and cooling tower.

The wet-bulb temperature of the surrounding air is the theoretical point to which water can be cooled. Usually it is not economical to approach closer than 5°F of the wet-bulb temperature. The temperature drop (cooling range) will vary from 10 to 12°F for a spray pond versus 12 to 17°F for a mechanical-draft cooling tower. Since the wet-bulb temperature of air varies geographically, it follows that the performance of a cooling tower also will vary with its location. So the wet-bulb temperature record for a given location must receive consideration when a tower is being designed.

The cooling pond is used where space adjacent to the power plant is plentiful and where land is relatively inexpensive. Cooling ponds are also used in certain parts of the country where atmospheric conditions are favorable. As an example, a cooling pond located in an area where the relative humidity is usually low would prove more effective than one installed in an area where the humidity approaches 100 percent. However, as noted previously, sites that can incorporate a cooling pond are few.

The advantages of the *forced-draft* tower are that it is suitable for corrosive waters; the fan can be mounted, firmly, close to the ground;

and the fan is accessible. Its disadvantages are recirculation of air
with reduction in tower efficiency (air leaving the tower exits at low
velocity), higher fan power requirements and hence increased operating cost, and increased maintenance. Also, since the size of the fan is
limited, more fans and foundations may be required.

The *induced-draft* tower (see Fig. 9.51) discharges the air at a high
velocity, thereby avoiding recirculation; induced-draft fans are usually quiet in operation and permit more uniform air distribution.
However, they have several disadvantages: They require a heavy
superstructure, they must be perfectly balanced, and they are somewhat difficult to service when major repairs are required.

10.3.1 Operation of cooling towers

Where cooling or spray ponds are used, their operation is relatively
simple. Cooling water from the condenser is pumped (1) to the cooling
pond, where it circulates and cools, *or* (2) to the spray pond, where
the water is sprayed through nozzles located above the pond's surface.
In either case, the cooler water is then returned to the condenser.

In the operation of the induced-draft cooling tower, water from the
condenser is passed to the top of the tower (see Fig. 9.51) to a flow-
control valve at each section or cooling chamber. The flow-control
valve permits the water to fill the overhead water distribution basin,
in which are located nozzles for uniform distribution of water. Interior
decking spreads the water uniformly, permitting intimate contact
with air passing up and through the descending spray of water. The
water finally arrives at the bottom of the tower, where storage is provided. From this point it is returned to the condenser as required.
This cycle is repeated again and again.

Air enters the side louvers, passing up and over the interior decking and cooling the water. The air then is made to pass through the
drift eliminators, where moisture carry-over is reduced to a minimum. Figures 9.51 and 9.52 show a double-flow tower operated by
means of an induced-draft fan. Air and water pass in counterflow so
that the air is in contact with the hottest part of the water immediately before it leaves the tower. Power costs are kept low by holding
the flow of air to a minimum. Merely shutting off the valve to the
water distribution system reduces pumping costs, and air regulation
is obtained by operating the fan on a low-speed motor.

To place an induced-draft cooling tower in service, open the makeup
valves on the tower to provide a sufficient amount of water storage.
Open the valve to the tower outlet and inlet as well as valves on the
condenser outlet. Now slowly open the inlet valve to the condenser to
permit water to fill the condenser and discharge piping. Next, close

the valve on the pump discharge and start the circulating pump, after which the valve on the pump discharge can be opened slowly. This is to avoid the shock and vibration that might result if the pump were placed on the line suddenly. Adjust the gates on the top of the tower to obtain equal distribution of water to the various sections. Now start the fans. Make a routine inspection of fans and motors to make certain that they are operating properly.

If the water temperature leaving the tower is lower than required, or if the turbine load has been reduced so that not all the tower capacity is needed, reduce the fan speed if two-speed motors are provided rather than take a cell out of service. The reason for this is that it is less expensive to operate cells at half capacity or rating than to shut them down and permit other cells to operate at top fan speed because reducing the load to 50 percent capacity reduces the horsepower or energy required to operate the fan. This is so because the horsepower requirements vary with the cube of the speed of the fan, and thus reducing the fan speed by 50 percent drops the fan horsepower to approximately one-eighth full-load power requirements. This is an appreciable power saving and is the reason why two-speed motors are used to take care of seasonal changes in load.

Should the load be further reduced and more than one circulating pump be provided, take one pump out of service. Further load reductions will necessitate taking towers out of service or shutting fans down in multiple units and permitting water to cascade from top to bottom.

Cold-weather operation introduces a few hazards that must be watched carefully. To prevent icing, maintain the circulating water temperature as high as is practical. For induced-draft towers, several recommendations are provided.

1. Shut off the fans to the tower or run the motor at low speed.

2. Shut down cells and run the water over fewer compartments.

3. Bypass some of the water to the tower and reduce the number of cells operating.

4. If ice forms on the deck filling or louvers, shut down the fans temporarily while continuing to circulate the water.

As a further precaution against the accumulation of ice, the fans are sometimes made reversible. Also, the sump beneath the cooling tower can be provided with heating coils.

In the operation of cooling towers, an important item to consider is water treatment. High water concentrations will deposit scale in the condenser heating surfaces, in the piping transmission system, and in the tower. Low water concentrations result in higher operating costs because they waste water. Since most cooling towers are installed in locations where water is a scarce or costly commodity, it is very easy to see why excess blowoff is uneconomical.

Water concentrations build up because part of the original water is evaporated in the heat transfer, while additional water is lost as drift or as entrained moisture in the air passing through the tower. Evaporation accounts for a loss of approximately 1 percent of the water for each 10°F range of water circulated. In a good cooling tower, drift is usually less than 0.02 percent. The amount of blowdown water wasted depends on the hardness of the circulating water, the type of treatment used, and the drift loss. To control the scale-forming solids at a satisfactory level, blowing down is resorted to. This is done by testing the hardness of the water and by blowing down to keep the water below the point where it would produce scale or where it would become corrosive. Other factors that must be considered are the amount of dirt in the water or other foreign material (depending on the location).

The pH value is important in controlling the chemical properties of cooling tower water. Neutral water has a pH value of 7. Below this value the water is acid and tends to be corrosive, while above 7 the water is alkaline and scale-forming. For cooling towers, the pH of the circulating water is preferred to be between 6 and 8. If the pH value is maintained too high, serious damage may result to the wooden cooling tower because of delignification. This results in the dissolving of the lignin that binds the wood fibers together, a condition that reduces the structural strength of the wood. This condition first appears at points where the timbers are alternately wet and dry because here the water concentration is increased due to the rapid evaporation. Sulfuric acid, zeolite, and sodium hexametaphosphate are the chemicals most frequently used for treatment. If the concentration of solids increases, the effective wet-bulb temperature will be raised.

Algae formation is also frequently experienced. This growth collects in the tower and circulating system, reduces the heat transfer, and is a source of considerable trouble and annoyance. It can be controlled by a number of chemicals, such as chlorine, copper sulfate, potassium permanganate, and others. The type of algae growth determines the treatment to be used. In general, proper water treatment requires the services of a qualified water quality consultant to ensure that the proper treatment is used for the specific condition.

Inspection of equipment should be made at least once a shift. Usually mechanical failures give first warning by excessive noise or vibration. Overheated motors can be detected by feeling the motor.

Vibration in fans is detected by peculiar noises and by observing the vibration after placing a listening rod between the point under observation and the ear.

The overall size and performance of a cooling tower are affected by the heat-removal requirements, the cooling-water temperature change, and the humidity at the site, all of which vary seasonally. A hyperbolic cooling tower is also affected by the humidity and other atmospheric conditions.

An operator must be prepared to handle potential operating problems, which include

1. Fill performance degradation

2. Buildup of silt on the fill

3. Icing during winter weather

4. Wind effects

Operators also must ensure that the water loading is balanced and that fans and motors are running properly.

As noted previously, during the operation of cooling towers in cold climates, the formation of ice is an important concern. Special care should be given to the air intakes and fill, where buildups of ice will jeopardize performance. One method of minimizing severe buildup of ice on a mechanical cooling tower is to reverse the direction of the fans, which shifts relatively warm water over the ice formation.

Also, some mechanical-draft cooling towers incorporate variable-speed fans. This feature can reduce power and thus operating costs when climatic conditions do not require the design output of the fans. An economic evaluation of increased capital costs versus lower operating costs and higher electric outputs must be performed in order to make the correct decision.

10.3.2 Maintenance of cooling towers

Mechanical maintenance is reduced first of all by keeping the equipment clean and adjusted after frequent inspections. Grease the motor bearings, and add oil to the gearbox and speed reducers as recommended by the equipment manufacturers. Keep all the clamps and bolts tightened. Inspect fans, motors, and housing to avoid undue vibration; they should be painted yearly, although the necessity for painting is most frequently determined by experience. Redwood does not require painting for protection but for appearance only. Other wood that has been pretreated may require periodic treatment applications.

Remove all debris, scale, etc., from the decking and distribution system. Drain and wash down the storage basin and overhead deck from time to time. Keep all parts of the tower in alignment. Inspect nozzles for clogging (weekly). Constant vigilance is important; observe anything unusual in the form of high water temperatures that may indicate faulty operation, scale in the system, etc. Check gearboxes weekly; locate any unusual noise or vibration immediately and correct it. Check the oil level weekly and change the oil as recommended by the manufacturer or as the operating experience of the plant dictates. At least once a year give the entire installation a thorough inspection: dismantle the motors, gearboxes, etc., and check for structural weakness while tightening the bolts.

Do not permit the tower to remain out of service (drained) for any length of time if avoidable. To avoid fires in the tower, keep it wet. If the tower is kept out of service "dry" for any length of time, fire extinguishers should be made available for emergency use and should be located for easy accessibility. In large plants, a fire-protection system should be operational.

The best guarantee of trouble-free performance is the adoption of a preventive-maintenance schedule with a continuous record of performance. The life of a cooling tower depends on the care and attention it is given. Good operating and maintenance procedures should incorporate the recommendations of the manufacturer.

10.4   Auxiliaries

10.4.1   Traveling screens

When a natural source of cooling water is used from a lake, river, or ocean, a *traveling screen* is placed in the intake tunnel that supplies water to the plant and is operated at variable speed. This permits the screen to be rotated fast or slowly, continuously or intermittently. Operating costs can be reduced if the screen is run only when necessary. The frequency of operation can best be determined by installing an indicating or recording gauge to notify the operator when the pressure drop (differential head or water-level drop) across the screen has increased, indicating that the screen needs to be cleaned. Allowance must be made, however, for the head of water on the inlet side because the pressure drop across the screen will vary with the area exposed to the inlet water.

The usual operation is to run the revolving screen at intervals frequent enough to expose a new set of screens or buckets to the water flowing to the tunnel to avoid pressure drop across the screen. Since the pressure drop occurs because of foreign material on the screen, the simplest way to determine whether screens are clean or dirty is to operate as follows:

Assuming the water is 20 ft deep and each screen is 1 ft wide,

1. Turn the screens over slowly until all 20 ft (20 buckets) have been exposed. Check the screens for foreign material.

2. If the screens are dirty, continue to run them until they begin to run clean.

3. Determine the frequency of running necessary to keep them in this condition. Without a gauge, this is best determined by operating experience.

4. Once the frequency of rotation has been determined, a schedule can be set up to rotate or raise the screens to the necessary height once each hour or less frequently, depending on the condition of the water.

When starting the screen, operate at the slowest speed so that if interference exists in the form of logs, etc., fouling will not occur with resulting damage, thus requiring pulling the assembly from the well or sending a diver down to make an inspection.

In certain seasons of the year more difficulty is experienced than at others, due to seasonal rains, floods, etc., which increase the foreign matter carried along by the moving stream. It is best under these conditions to run the screen continuously.

Care should be exercised to prevent freezing and damage to the screen if ice should form in the intake during cold weather. To prevent this, screens are run continuously. In some plants, all or part of the circulating water is discharged at the mouth of the intake after leaving the condenser. This maintains the water above the freezing temperature and prevents ice formation on the screens.

Most of these drives are equipped with automatic lubricators. If they are not so equipped, lubrication must be applied while the screen is running. The motor should be properly lubricated, and a coating of grease should be applied to the rollers on the rear of each basket.

Preventive maintenance is the first step toward lower maintenance costs.

Weekly: Check the entire installation for rubbing, holes in the baskets, overloaded motor, etc. Lubricate when necessary.

Monthly: Check the bearings on the sprocket drive, rollers, and bearings. Replace damaged baskets.

Yearly: Replace all worn bushings and bearings and make a thorough and complete check of all moving parts. Overhaul the motor, gear reducer, etc.

10.4.2 Pumps

Of all the auxiliaries, the circulating pump requires the most power because of the large quantity of water that it handles. Most condensers are equipped with two circulating pumps (see Fig. 9.55), both operated at maximum load, while at partial load only one is required to maintain the vacuum desired. Some circulating pumps are provided with two-speed or variable-speed motors to reduce operating cost to a minimum.

Circulating pumps (see Fig. 9.56) are operated at relatively low speed and normally require little attention once alignment has been made correctly. Properly lubricated, operated, and maintained, circulating to equalize pumps will run for extended periods of time without difficulty.

Some pumps are set above the normal water level and have to be primed to be placed in service. Priming is accomplished by evacuating the air in the pump until water is ejected from the siphon. At this point the motor is started, and subsequently the discharge valve is opened slowly so as not to lose the vacuum. The priming is then discontinued.

If strainers are provided ahead of the pump, they must be inspected at frequent intervals. If twin strainers are provided, the cleaning can be accomplished without taking the pump out of service. If not, one condenser must be taken out of service for this inspection. The cleanliness of the strainer can be determined by placing a set of pressure gauges before and after the strainer and noting the pressure drop (loss) across the strainer.

Condensate (hot-well) pumps and other auxiliary pumps used on heater drains and sumps require the same type of service and maintenance as that outlined for pumps in Chap. 8.

10.4.3 Steam-jet ejectors and vacuum pumps

In order to maintain the highest possible vacuum, air and noncondensable gases must be removed from the condenser. Sometimes the pump that removes the condensate also removes the air. This type is called a *wet vacuum pump*. More often, condensers are provided with ejectors (*dry vacuum pumps*) to remove the air or noncondensable gases and a separate pump to remove the water. The design of wet vacuum pumps is similar to that of reciprocating pumps. Dry vacuum pumps may be either of the hydraulic type or steam ejectors.

With the hydraulic type, jets of water are hurled at high velocity and are made to pass through a revolving wheel that rotates because of the water passing through it. The water rushing through the discharge cone and diffuser in the form of a helix encloses the vapors that enter around the rotary transforming wheel between the suction and discharge of the pump. The air and vapors, in addition to the hurling water, are discharged into the hurling-water tanks, from which the gases are liberated by means of a vent. In operation, the hurling-water temperature should be maintained below the temperature in the condenser. This is accomplished by adding sufficient makeup water or by using a heat exchanger to reduce the hurling-water temperature.

The steam ejector (see Fig. 9.57) entrains the air and noncondensable gases in very much the same way as the hydraulic vacuum pump. Ejectors are designed for the vacuum and steam economy desired. The type of steam ejector usually used for large steam-driven turbine generators consists of two-stage ejectors in series provided with surface intercondensers and aftercondensers, with the steam being returned to the main condenser.

10.4.4 Vacuum gauges

Pressures below atmospheric are determined by means of a vacuum gauge. The gauge may be of the dial type actuated by a bourdon tube similar to the conventional pressure gauge. Or the guage may consist of a mercury tube, with the height of the mercury column indicating the difference in pressure between the condenser and the atmosphere. In either case, the gauge is calibrated to read in inches of mercury. Every inch of mercury corresponds to a pressure of 0.491 psi. A vacuum, then, of 27.25 is equivalent to $27.25 \times 0.491 = 13.38$ psi pressure or 1.32 psi below the atmospheric pressure of 14.7 psi (at sea level).

Since pressures in the steam tables are given in pounds per square inch absolute, it becomes necessary to convert gauge pressure into terms absolute. In detailed steam tables, pressure is identified in both pounds per square inch absolute (psia) and inches of mercury (in Hg) for pressures below atmospheric (i.e., a vacuum).

Example Assuming a barometer with a pressure of 29.50 in Hg and a condenser vacuum of 27.25 in, determine the absolute pressure difference in pounds per square inch.

Solution

$$\text{Pressure of atmosphere} = 29.50 \times 0.491 = 14.48 \text{ psi}$$

$$\text{Pressure in condenser} = 27.25 \times 0.491 = 13.38 \text{ psi}$$

$$\text{Pressure differential in condenser} = 14.48 - 13.38 = 1.10 \text{ psi}$$

Questions and Problems

10.1 For large electric utility power plants that burn fossil fuels, what are the system cycles that are most often designed?

10.2 For smaller utility or IPP plant sizes, why are significantly lower steam pressures and temperatures used?

10.3 Describe by use of a sketch the flow of steam through a tandem compound turbine consisting of a high-pressure turbine and a double-flow low-pressure turbine with the steam from the low-pressure turbine exhausting to a condenser.

10.4 When would a noncondensing turbine be selected?

10.5 Other than to drive a generator for electric production, what types of equipment are suitable to be driven by steam turbines?

10.6 Why is superheated steam used in turbine operation?

10.7 What are the disadvantages of using wet steam in a turbine?

10.8 If a higher superheat temperature offers higher turbine efficiency, why is there a steam temperature limit of approximately 1000°F?

10.9 If a turbine produces more power by operating at higher steam pressures or by operating at a vacuum at the exhaust, why not just increase the steam pressure?

10.10 For modern power plants, what are common exhaust pressures in inches of mercury and in pounds per square inch? Why are these pressures considered to be vacuums?

10.11 Although the use of condensers offers improvements in power output and efficiency, additional systems are necessary. These increase the capital costs of the plant and also operating costs. List these required systems and added operating costs, and provide a general evaluation on how they can be justified.

10.12 What are the more important things to watch for in bringing a turbine on line? After it is on line and operating?

10.13 How often should a turbine be dismantled to receive a thorough inspection and overhaul?

10.14 What types of operation have resulted in significant maintenance and operating problems on large turbines?

10.15 What are some of the common causes of turbine vibration?

10.16   What affects the proper operation of a condenser?

10.17   What is a common problem that affects performance of a condenser? Why is this minimized when a cooling tower is part of the system?

10.18   What maintenance problems can occur on a surface condenser? What is the impact on the performance of the plant?

10.19   How would you proceed to determine whether a condenser was leaking? How would you repair a tube leak? Would this have a significant impact on the performance of the condenser?

10.20   How would you determine the reason for low vacuum that had suddenly occurred on a condensing turbine?

10.21   List the operating advantages and disadvantages of an air-cooled steam condenser.

10.22   When are cooling towers selected to provide circulating water to condensers?

10.23   How does the relative humidity of the outside air affect the operation of a cooling tower?

10.24   What is the advantage gained by having two-speed motors on a mechanical-draft cooling tower?

10.25   In cold weather climates, what operational methods can be used on mechanical-draft cooling towers to minimize icing problems?

10.26   Explain the operation of a traveling screen when it is used with a natural source of cooling water.

10.27   What is the purpose of a steam-jet ejector and a vacuum pump on a condenser?

10.28   The vacuum gauge reads 26.3 in Hg and the barometer reads 30.03 in Hg. What is the absolute pressure in pounds per square inch?

10.29   The pressure of the atmosphere is 14.36 psi and the vacuum gauge reads 28.23 in Hg. What is the backpressure in pounds per square inch absolute (psia)?

10.30   A turbine operates with a backpressure of 1.5 in Hg. What is the pressure in pounds per square inch (psi)?

11

Auxiliary Steam-Plant Equipment

This chapter considers some of the more important general auxiliaries found in steam power plants. These devices are required to make it possible for the major equipment to perform its function with the highest efficiency that is practical for the particular plant. Satisfactory boiler operation depends on the heating and conditioning of the feedwater. It is essential that steam and water flow from one part of the plant to the other through pipelines of adequate strength and size. Condensate must be removed automatically from steam lines to prevent the possibility of waterhammer and from steam-heating coils to keep the water from blanking the heating surface and decreasing the rate of heat transfer. Pumps, filters, and various feeding devices must be used in the process of lubricating machinery. This auxiliary equipment requires careful attention and the necessary maintenance to keep it in good working condition.

The impurities in water affect boiler performance, and if they are not removed, they will cause excessive costly maintenance and the loss of revenues during the outage period necessary to perform that maintenance. These impurities can be either suspended or dissolved matter and consist of atmospheric gases (O_2, N_2, and CO_2) and minerals and organic materials. Suspended solids do not dissolve and can be removed by filtration, while dissolved solids are in solution and cannot be removed by filtration. These dissolved solids include iron, calcium, magnesium, and sodium salts. Hardness of water is caused by calcium or magnesium compounds, and if they are not removed, they will produce a scale inside the boiler tubes that reduces the effective heat transfer.

11.1 Feedwater Heaters

Feedwater heaters have two primary functions in power plants: (1) to provide the means for increasing the feedwater temperature, which improves the overall plant efficiency, and (2) to minimize the thermal effects in the boiler. Feedwater heaters use steam from selected turbine extraction points to preheat the feedwater from the condenser prior to it entering the economizer or boiler drum.

The number and type of feedwater heaters used depend on the steam cycle, the operating pressure of the cycle, and the plant economics, i.e., where lower operating costs can offset the additional capital cost expenditure. In general, smaller plants have fewer units. In utility and large industrial plants, five to seven stages of feedwater heaters are often part of the design. Feedwater heaters are classified as either closed or open designs and are designed for operating at low or high pressure.

11.1.1 Closed feedwater heaters

Closed feedwater heaters are specialized shell and tube heat exchangers. The steam flows from an extraction stage of the turbine and condenses on the shell side of the feedwater heater, while the feedwater flows inside the tubes and absorbs heat and thereby increases its temperature.

Most closed feedwater heaters are composed of bundles of a large number of tubes that are bent in the form of a U, and therefore, this type of design is called a *U-tube heat exchanger* or *feedwater heater.* The tubes are either expanded or welded into tube sheets at one end of the shell. A series of baffles and tube support plates are used to direct flow, minimize tube vibration, reduce erosion, and promote high heat transfer. The lowest-cost closed feedwater heaters are typically long, horizontal, two-pass designs with high water velocities.

Low-pressure feedwater heaters are located prior to (upstream) the boiler feed pump (see Fig. 1.1). They are generally designed for tube side pressures of less than 900 psig for utility boiler designs. The location of the feedwater heater relative to the boiler feed pump generally defines whether it is called a low- or high-pressure heater no matter what the actual pressure is. High-pressure heaters of a plant cycle are those heaters which are located after (downstream) the boiler feed pump.

The closed feedwater heater shown in Fig. 11.1 is different from a U-tube design in that straight tubes are used between two tube sheets. Feedwater heaters of this type are classified as one-, two-, three-, or four-pass designs depending on the number of times the

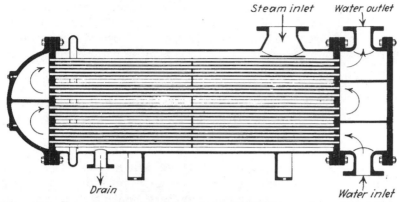

Figure 11.1 Four-pass closed feedwater heater. (*API Ketema, Inc., a subsidiary of American Precision Industries, Inc.*)

water passes the length of the unit before it is discharged. The design shown in Fig. 11.1 is baffled and provides four passes for the water.

As noted previously, the U-tube design of a feedwater heater is the preferred choice of heaters in today's power plants. This design has minimized the expansion problems of tubes being connected to two tube sheets. The tubes are bent into the form of the letter U, and these tubes are frequently referred to as *hairpin tubes*. Both ends of these tubes are expanded into the same tube sheet, and the water box is baffled to direct the flow of water through the tubes from one pass to the other. Although both ends of these tubes are rigidly attached to the tube sheets, their shape permits free expansion. However, because of their U shape, these tubes cannot be cleaned easily by mechanical means. But by having good water quality, the need for this cleaning has been reduced significantly. Also, because of concerns about potential transport of copper from the feedwater heaters into the boiler and turbine, stainless steel and carbon steel tubes generally are used in feedwater heaters.

The tube sections of closed feedwater heaters must withstand the pressure of the water, and the shell must withstand the pressure of the steam. The water is forced through one or more heaters by a single pump. High-pressure steam plants use heaters between the boiler feed pumps and the boilers, exposing the water side of the heater to the full boiler feed pump pressure. These are called *high-pressure heaters*. The low-pressure heaters are located between the condensate pumps and the boiler feed pumps. Safety measures in the form of relief valves must be installed to prevent the possibility that either the water or the steam pressure will exceed that for which the heater was built.

Closed heaters may be arranged to utilize steam at any pressure, either above or below that of the atmosphere. Some feedwater heaters are actually located in the condenser neck at the turbine exhaust outlet. When they are functioning properly, the outlet water temperature will be 2 to 5°F below that of the steam. The effectiveness of the heater is reduced by deposits on the tubes, the accumulation of noncondensable gases in the shell, or flooding of the shell with condensate. These heaters are not satisfactory for use with hard water because of the rapid accumulation of scale on the tube surfaces. When they are operating with a positive pressure in the shell, noncondensable gases may be vented to the atmosphere. When the shell is below atmospheric pressure, it can be vented either to a condenser or by a steam jet or other vacuum-producing auxiliary. Venting of the noncondensable gases is essential, but opening the vents more than is required results in a waste of steam. Condensate may be discharged through a trap to a vessel having a pressure sufficiently lower than that of the shell. If a low-pressure vessel is not available, a condensate pump is required. Gauge glasses located at the bottom of the heater provide a means of checking the performance of the drainage system.

11.1.2   Open feedwater heaters

These feedwater heaters are also called *deaerators,* and they serve the dual purpose of heating the feedwater to improve plant efficiency and deaerating the feedwater to remove gases that could cause corrosion of equipment and piping systems. Deaerators also provide the storage of high-quality feedwater for the boiler feed pump. Several deaerator arrangements are shown in Figs. 11.2 and 11.3.

Heat transfer in deaerators is by direct contact between the feedwater and the turbine extraction steam, and various design techniques are used such as bubbling, tray, spray, or various combinations of these. The drains from the high-pressure heaters usually flow into the deaerator, and noncondensable gases are vented to the atmosphere.

Originally, simple heaters were used to preheat boiler feedwater, which would improve the steam cycle efficiency. As boiler pressures and temperatures increased, it became necessary to mechanically strip dissolved oxygen from the water before it entered the boiler and other equipment and thus reduce corrosive attack. A similar need existed to remove CO_2 (which enters the system as part of any air infiltration) from feedwater, and, generally, equipment that was adequate to remove O_2 also removed CO_2.

Figure 11.2 Horizontal deaerator arrangement with storage tank. (*The Graver Company, Graver Water Division.*)

Two types of deaerating heaters are used most often in today's power plants: (1) the spray-tray type and (2) the spray-scrubber type. Schematics of these designs are shown in Fig. 11.4.

The feedwater that must be treated by the deaerator can vary significantly depending on the plant's application. A closed-loop system found in a typical utility has feedwater makeup requirements of only a few percent, while an industrial application such as a pulp and paper plant has requirements for feedwater makeup of 40 to 70 percent of the total flow.

Condensate that returns from a turbine is usually high in temperature and low in O_2 and CO_2. For industrial plants where makeup water may come from a number of sources, this water is usually much cooler and can contain high levels of dissolved O_2 and CO_2. Each situation requires a careful design and selection of the proper equipment.

Spray-type deaerator. The spray-type deaerator has three main sections, as shown in Fig. 11.4a:

1. A water box at the top, where water enters the unit through valves or nozzles

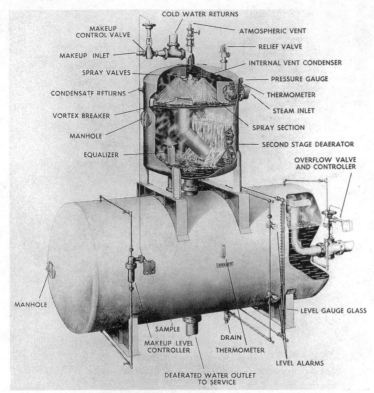

Figure 11.3 Vertical deaerator arrangement with storage tank. (*The Graver Company, Graver Water Division.*)

2. A spray area immediately below, where sprayed water interacts with steam to do 95 percent of the deaerating and heating of the feedwater

3. A tray section, where the final 5 percent of the deaeration and heating takes place

Below the tray tank is a temporary storage area for oxygen-free water. In most large units, the entire heater is mounted above a storage tank for volume retention. In some smaller units, the bottom of the heater serves as the storage area. The water box holds incoming water so that it may be evenly sprayed to the chamber below. Because of the potential for corrosion, the water box is constructed of stainless steel or lined with stainless steel.

The heater valves are held open by the pressure of water from above. Increasing feedwater flow forces the valves open further until full capacity is reached. These valves are also stainless steel. The spray chamber is the large area directly below the water box, where water meets the upward moving steam for the first time.

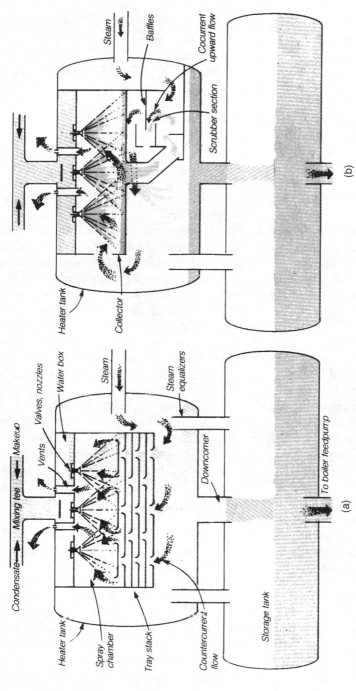

Figure 11.4 Deaerator designs. (*a*) Spray-tray deaerator uses trays to produce filmlike water flow to enhance contact with steam for maximum stripping of residual oxygen. (*b*) Spray-scrubber deaerator produces turbulent upward flow of steam and water in scrubber section for removal of oxygen. (*The Graver Company, Graver Water Division.*)

The operation of the tray area may seem of minor importance, since only 5 percent of the deaeration and heating takes place in the trays. The importance of the trays becomes apparent when it is realized that a deaerator heater must ensure minimum O_2 levels in the feedwater of about 7 parts per billion (ppb). It is in these trays that the cleanest steam meets the almost completely cleaned water for final polishing. These trays are also made from stainless steel for corrosion protection.

Below the tray station, at the bottom of the heater tank, there is a section where deaerated water is briefly retained before it drops into a separate storage tank. In one-piece heaters, the heater storage volume must be sufficient to permit water to be drawn off and directed to the boiler feed pumps.

Spray-scrubber deaerator. In this deaerator design, a scrubber section is used in place of trays, as shown in Fig. 11.4*b*. This design accomplishes the same thing as the tray design by removing residual dissolved gases from water that has passed through the spray chamber. While the steam can be counterflow, cross flow, or parallel flow to the water in tray-type deaerators, it is parallel flow in the scrubber section of spray scrubbers, as shown in Fig. 11.4*b*.

Turbulent mixing and scrubbing take place as water and steam are in intimate contact and rise through this region. Assisted by the turbulence caused by steam condensation, this results in a highly effective gas stripping action.

Spray-scrubber units are generally less expensive than tray types. However, they do have a disadvantage when compared with tray types in that their turndown range (design flow/low flow) is about 5:1, whereas a tray type can operate efficiently in a flow range of 10:1. Selection of the type of deaerator is often based on the load variations that are expected at the plant.

The deaerator is elevated to provide the necessary suction head on the boiler feed pumps. Regardless of the type of deaerator used, its location within the plant can vary with the plant design. Figure 11.5 shows a deaerator that is located outside on the roof of a power plant. Large power plants that have a number of boiler and turbine systems also require multiple deaerators. These deaerators are often designed for outdoor locations and are located on the roof of the boiler building in a series, as shown in Fig. 11.6. Often a deaerator is located in a boiler building, as shown in Fig. 11.7. In every case the deaerator is located prior to the boiler feed pump because it is considered to be a low-pressure feedwater heater. Sodium sulfite or hydrazine is frequently added to the water in the storage tank to remove the last trace of dissolved oxygen.

Open feedwater heaters must be provided with safety devices and instrumentation. Steam is supplied to these heaters from turbine

Figure 11.5 Deaerator outside location on roof of boiler building. (*The Graver Company, Graver Water Division.*)

Figure 11.6 Series of deaerators located on roof of boiler building for large utility plant. (*The Graver Company, Graver Water Division.*)

Figure 11.7 Deaerator located within boiler building. (*The Graver Company, Graver Water Division.*)

exhaust or other low-pressure systems. Pressure-reducing valves provide steam from high-pressure sources to supplement the low-pressure steam. Feedwater heaters are designed to operate within the pressure of the available exhaust steam.

Relief valves are provided to make sure the safe pressure limits are not exceeded. The storage section must have an overflow valve or loop seal to limit the level. Because of the critical nature of the level in the storage section, it is advisable to have a level indicator on the control panel and an annunciator alarm to warn the operator when the level is too high or too low. Refer to Fig. 11.3, where a storage tank level gauge and level alarms are shown.

The steam pressure in the heater and the water level in the storage tank are regulated automatically. Nevertheless, the operator should keep a close surveillance and know what action to take in event of an emergency. The available quantity of feedwater in the storage section normally will last only a few minutes if the supply fails. It is generally assumed that if the steam pressure is maintained in the heater, the water will be deaerated satisfactorily. However, it is good practice to occasionally run a dissolved oxygen analysis on the effluent water. Should the oxygen content be found excessive, the necessary corrective measures can be taken to prevent the possibility of corrosion in the boilers.

It might, at first, appear that an open feedwater heater is a cure-all, but this is not the case. The water and steam mix, and a part of the impurities that appear in the makeup water is introduced into the boiler. Therefore, the makeup water must be properly conditioned to ensure high-quality water.

The deaerating-type open heater shown in Fig. 11.8 consists of an external shell of welded low-carbon steel plate designed for the operating pressure. This steel shell is protected from corrosion by stainless steel components. Steam enters the heater and flows upward to the preheater compartment. The incoming water enters through the spray nozzles in the preheater compartment. The heating that takes place here releases most of the noncondensable gases, which, along with a small amount of steam, are discharged through the vent. The water flows from the preheater to the tray compartment. The steam and water come into further contact as the water flows downward through the distribution trays. The heated and deaerated water falls into the storage space.

Feedwater systems are available in factory-assembled package units, as shown in Fig. 11.9. These systems include deaerating heaters, condensate receivers, and booster and boiler feed pumps. The components are shop assembled on skids and include piping, motors, wiring, and controls. The use of these units reduces the fieldwork of installation to connecting the supply and discharge piping and providing the electric power; however, they are limited to small-capacity applications.

The heating and conditioning of feedwater are performed by a hot-process water softener. A typical installation is shown diagrammatically in Fig. 11.10a, and a section through the heater and softener unit is shown in Fig. 11.10b. There are no provisions for the introduction of condensate into this unit. It is intended for use where there is 100 percent makeup water or where the condensate returns are delivered to an independent deaerator. The installation includes a system that proportions the chemical feed to the rate of flow of makeup water; a combination heater, softener, and settling tank; a sludge-recirculating pump; filters; and a wash pump. The flow of cold makeup water to the softener is regulated by means of a flow-control valve to maintain a constant level in the treated-water chamber. An orifice-type flowmeter actuates the chemical feed in proportion to the rate of water flow. The raw water is sprayed into the steam-filled sections at the top of the softener. The oxygen and other noncondensable gases are discharged through the vent to the atmosphere. The sludge-recirculation pump transfers the sludge from the bottom of the softener into the chemical reaction portion of the softener. This recirculation of sludge accelerates the chemical action, thereby increasing the effectiveness of the reaction chamber. Treated water is discharged from the inverted-funnel-shaped baffle to the pressure filters and then to the boiler feed pumps. The filter backwash pump suction line

(a)

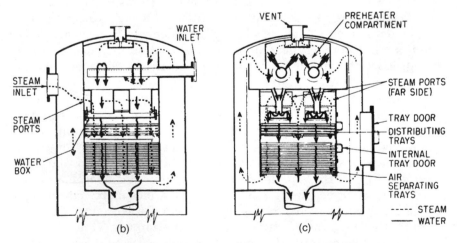

(b)   (c)

Figure 11.8 Deaerating heater. (*a*) Section view. (*b*, *c*) Flow diagrams. (*Crane Company, Cochrane Environmental Systems Division.*)

Figure 11.9 Package feedwater unit, including deaerator heater and pumps. (*Crane Co., Cochrane Environmental Systems Division.*)

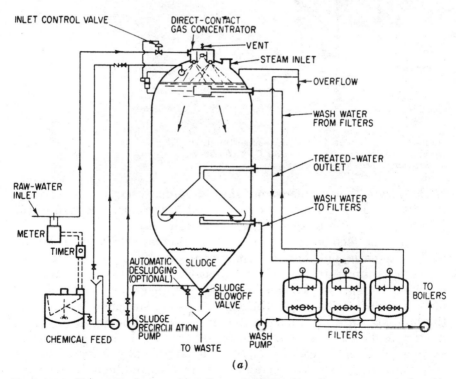

(*a*)

Figure 11.10 Deaerating hot process softener. (*a*) Diagrammatic arrangement of equipment.

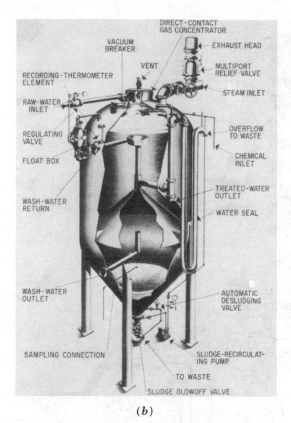

DIRECT-CONTACT
GAS CONCENTRATOR

VACUUM
BREAKER

EXHAUST HEAD

VENT

MULTIPORT
RELIEF VALVE

RECORDING-THERMOMETER
ELEMENT

STEAM INLET

RAW-WATER
INLET

REGULATING
VALVE

OVERFLOW
TO WASTE

CHEMICAL
INLET

FLOAT BOX

TREATED-WATER
OUTLET

WASH-WATER
RETURN

WATER SEAL

WASH-WATER
OUTLET

AUTOMATIC
DESLUDGING
VALVE

SAMPLING CONNECTION

SLUDGE-RECIRCULAT-
ING PUMP

TO WASTE

SLUDGE BLOWOFF VALVE

(b)

Figure 11.10 Deaerating hot process softener. (b) Downflow softener unit. (*Crane Co., Cochrane Environmental Systems Division.*)

is also located below this funnel-shaped baffle. After passing through the filters, the wash water is discharged to the upper section of the softener, thus reclaiming the filter backwash water.

These open feedwater heaters and hot-process softeners are essentially automatic, but there are some points that require the attention of the operator. Adequate steam must be supplied to maintain the specified pressure; this steam is usually obtained from the exhaust of plant auxiliaries, and when this exhaust is not sufficient, steam from the high-pressure lines is admitted through a reducing valve. These heaters and softeners may be designed for any desired steam pressure to satisfy plant conditions, but they are usually operated at 3 to 10 psig. Some softeners are equipped to receive and blend condensate with the makeup water. When the condensate returns are not sufficient to meet the demands for boiler feedwater, the level in the heater drops and makeup is added. The controls should be adjusted to supply a proportional amount of makeup and condensate to ensure uniform boiler feedwater conditions. The water-level control is important because failure to maintain the level will result in the plant's running out of feedwater, while flooding may cause the water to enter the steam supply lines. Because this water level is critical, alarms are frequently installed as an aid to the operator.

Noncondensable gases are liberated in open heaters and hot-process water softeners. These gases must be vented to prevent them from accumulating and partially blocking the flow of steam to the heater.

The softener-type open heaters provide a means of sludge removal during operation. Materials known as *coagulants* are added to the softener to speed up the settling action and thus decrease the load on the filters.

The control of the chemical treatment of water in a lime-soda softener is accomplished by testing or titrating[1] a sample of the effluent water for P (phenolphthalein) alkalinity and M (methyl orange) alkalinity, and for H (hardness). These values are expressed as parts per million (abbreviated ppm)[2] as calcium carbonate.

The lime control of a hot-process softener, $2P - M$, should normally be between 5 and 10 ppm as calcium carbonate. If this value is too low, increase the lime; if it is too high, decrease the lime.

The soda ash control of a hot-process softener, $M - H$, should normally be between 30 and 40 ppm as calcium carbonate. If this value is too low, increase the soda ash; if it is too high, decrease the soda ash.

This control should result in a hardness H between 15 and 25 ppm as calcium carbonate. Solids deposited in the bottom of the softener

---

[1]Titration for alkalinity consists of adding a few drops of a dye, phenolphthalein, to the sample. The sample turns pink, indicating the presence of hydroxides and carbonates. A standard-strength acid solution is then added in sufficient quantity to just remove the pink color. The quantity of acid used is the P alkalinity. Then another dye methyl orange, is added, and the sample becomes a straw color, indicating the presence of hydroxides, carbonates, or bicarbonates. As more standard-strength acid is added, the sample color will change from straw to pink. The quantity of acid added during this second titration is the M alkalinity.

One method of determining hardness is to add a standard-strength soap solution to the sample and record the amount required to produce a lather. A second method is by titration using an organic dye that causes the sample of hard water to turn red. Then a standard-strength sequestering (separating) agent is added in sufficient amount to change the color of the sample from red to blue. The milliliters of the titrating agent added multiplied by a constant gives the hardness H of the sample. The constant corrects for the size of the sample and strength of the titrating solution.

[2]Parts per million refers to a concentration of 1 in 1 million, that is to say, 1 lb of salts in 1,000,000 lb of water. Another way of expressing solids in water is grains per gallon, a grain being 1/7000 lb. To convert from grains per gallon to parts per million, multiply by 17.1. The P and M readings denote the alkalinity of the water. The M quantity is known as the *total alkalinity*. The alkalinity and hardness are expressed as calcium carbonate so that they may be added and subtracted directly. This procedure simplifies the calculations used in determining the necessary changes in treatment as indicated by the control test.

sedimentation compartment must be removed by blowing down once each 8-h shift. The filters must be backwashed before the pressure drop through them becomes excessive. A pump is usually provided, and the water used to backwash the filters is recirculated through the softener, resulting in a savings in both water and treating chemicals. The chemical mixing tank must be charged with lime, soda ash, and coagulants as indicated by the control analysis. A record should be kept of the total weight of chemicals and the amount in pounds per inch of water in the mixing tanks. The rate of chemical feed can be varied by changing the strength of the solution in the mixing tank or the rate at which the solution is fed to the softener in proportion to the water.

The treatment of water has become a sophisticated science to ensure that the highest-quality water is used in power plants. Many different techniques have been developed and used successfully to provide this quality water. In addition, many specialized testing procedures are used to ensure that this quality is maintained during operation. Water-treatment consultants and equipment suppliers should review the plant's specific water requirements and water sources and provide recommendations to the plant operating staff.

11.1.3 Operation and maintenance of feedwater heaters

Damage to feedwater heaters has involved primarily tube failures, which have been caused by the following:

1. Erosion from steam impingement

2. Tube vibration

3. Erosion and corrosion on the inlet tube end

4. Oxygen pitting

5. Stress corrosion cracking

In addition, failures of tube joints, improper plugging, and poor maintenance all lead to downtime and repairs of feedwater heaters. Generally, feedwater heaters require no daily maintenance; however, associated valves need to be given attention. During scheduled outages, nondestructive examination (NDE) techniques are used for tube-side and shell-side inspections. Since tube conditions are critical, eddy current and ultrasonic testing (UT) are used to evaluate tube integrity. Eddy current testing determines wall thinning and identifies cracks that have occurred.

When leaks are detected, tubes are usually plugged with tapered or mechanical plugs and are expanded in the tubesheet. Explosive plugs are also used. Units can have 10 to 30 percent of their tubes plugged and still meet thermal performance, although the pressure drop does increase and the condensate and boiler feed pumps must be designed to handle this additional pressure requirement.

The feedwater heater system has a significant impact on steam system performance. Much of the deposits found in steam-generating systems come from corrosion products and contaminants whose source is the feedwater heater system. Copper-based tubing is being replaced by other materials in order to reduce the carry-over of dissolved copper. Proper chemistry control and deaeration operation can minimize corrosion in the feedwater heater system and reduce corrosion product carry-over into the boiler. The proper control of oxygen is very important so that corrosion of the boiler system is minimized.

Feedwater heaters may be taken out of service for maintenance during plant operation by bypassing the feedwater around them and shutting off the extraction steam to them. However, this requires additional piping and valves, and the system must be designed to accommodate the bypassing.

11.2 Condensate Polishing Systems

In many cases, condensate does not require treatment prior to reuse. Properly treated makeup water is added directly to the condensate to form boiler feedwater. However, in cases where steam is used in industrial processes, the steam condensate is contaminated with corrosion products. Also, in both utility and industrial applications, condenser leaks can allow cooling water to contaminate the condensate.

Demineralized systems that are installed to purify condensate are known as *condensate polishing systems*. They are generally used in plants where boiler water chemical treatment and blowdown of solids will not provide consistent water quality, especially at pressures above 2000 psi.

A condensate polishing system will remove contaminates continuously in the cycle to provide proper boiler water chemistry. Silica, iron oxides, and other contaminants are effectively removed by the polisher. By preventing these contaminants from entering the boiler, their deposition on the turbine blades and other surfaces in the cycle is minimized. The frequency of acid cleaning in the boiler also will be reduced.

Figure 11.11 shows a schematic diagram of a condensate polishing system that uses precoat materials designed to provide the combination of filtration and ion exchange to recondition the condensate. The dual-vessel arrangement provides continuous capability, with one unit in service while the other is maintained in a standby mode,

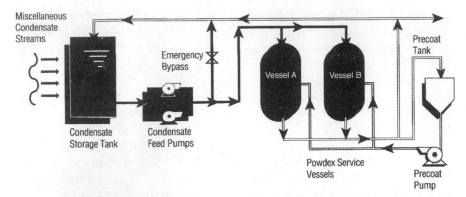

Figure 11.11 Schematic of Powdex condensate polishing system. (*The Graver Company, Graver Water Division.*)

ready to be placed into service when the on-line unit is exhausted to a differential pressure or conductivity endpoint. The exhausted unit is backwashed to waste, and a new precoat is applied by pumped recirculation.

This modern system combines the operation of filtration and ion exchange in a single system. Other designs use separate precoat filters and ion-exchange columns to obtain the required filtration efficiency and dissolved solids removal. When the concentration of soluble or particulate impurities to be removed is relatively low, precoat systems are preferred over deep-bed demineralizer systems because of their simplicity and lower initial cost. In addition, the operating costs associated with disposable precoats are often more economical than regenerable-bed systems.

Figure 11.12 is an illustration of a skid-mounted condensate polishing system including its control panel. The unit is shop assembled, which minimizes its field installation and startup period.

Another type of condensate polisher is the deep- or mixed-bed type, which consists of a service vessel with a bed that is approximately 3 ft high composed of mixed cation and anion exchange resins, usually in a 1:1 ratio. The mixed-bed system requires external regeneration of the cation and anion resins, and the cation and anion resins must be separated completely prior to regeneration. The cation resin is regenerated with hydrochloric or sulfuric acid. The anion resin is usually regenerated with a solution of sodium hydroxide.

A condensate polisher can be located in several portions of the plant cycle. For maximum benefits of condensate polishing, a full-flow system with a standby vessel is installed in the condensate line at the discharge of the condensate pumps. This provides high-quality con-

Figure 11.12 Condensate polisher, skid-mounted Powdex system. (*The Graver Company, Graver Water Division.*)

densate during operation and also provides maximum protection during a condenser leak.

Another arrangement is to treat a portion of the condensate either as a side stream off the main condensate line or as a recirculation line off the hot well of the condenser. The partial-flow polisher could be used during startup in a preboiler recycle cleanup mode, which will remove corrosion products before firing of the boiler.

The advantages of condensate polishing are reduced startup time, reduced blowdown, and an overall improvement in plant operations. Erosion, corrosion, and scale will be reduced, which results in fewer chemical and mechanical cleanings of the boiler and increased plant availability.

11.3   Raw Water Treatment

Makeup water for the plant must come from a natural source such as a lake, a river, or a well, and seldom, if ever, does pure water exist. All natural waters contain amounts of matter that are both dissolved and suspended, and these amounts vary depending on the source.

Gases in the atmosphere such as oxygen, nitrogen, and carbon dioxide are brought into the water supply with rain. On the ground, water dissolves and picks up minerals that would be detrimental to the boiler unless removed. Organic matter also must be removed from the boiler water, and raw water contains both dissolved and suspended solids.

Suspended solids are those which do not dissolve in water and which can be removed by filtration. Mud and silt are examples of suspended solids found in raw water. Those solids which are in solution and cannot be removed by filtration are *dissolved solids*. Dissolved materials found in water are silica, iron, calcium, magnesium, and sodium. There are also metallic constituents present with bicarbonate, carbonate, sulfate, and chloride.

While in solution many of these impurities are divided into component parts called *ions,* which can be either positively or negatively charged. The negatively charged ions include bicarbonate, carbonate, sulfate, and chloride, and these are called *anions*. Sodium and ammonium are included with positively charged ions, and these are called *cations*.

When calcium and magnesium compounds are dissolved in the water, the water is considered *hard water,* and scaling can occur when their impurities precipitate out and adhere to boiler tubes. This problem becomes more severe as water temperature increases, as it would in high-pressure boilers. This scaling can result in the overheating of the tubes, loss of heat-transfer surface, and eventual failure of the boiler tube.

Therefore, water treatment is necessary to ensure that high-water quality is maintained in the boiler, and various processes are used for this purpose.

11.3.1   Ion-exchange water conditioners

One of the simplest methods of removing hardness from a water supply is by use of a sodium zeolite water softener (Fig. 11.13). Softeners of this type use zeolite, an insoluble granular material having the ability to exchange hardness, in the form of calcium and magnesium ions, for sodium ions, which do not produce hardness. The total amount of chemicals dissolved in the water is not reduced, but they are changed to non-hardness-producing chemicals, thus softening the water.

The exchange material is contained in a steel tank adequate to withstand the pressure of the water system. The raw water is introduced above the bed, and the exchange takes place as the water flows downward through the zeolite. These softeners are equipped with meters and the necessary valves and accessories for controlling the regeneration procedure.

Normally the outlet water from these softeners shows zero hard-

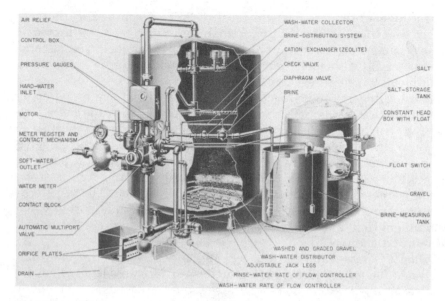

AIR RELIEF
CONTROL BOX
PRESSURE GAUGES
HARD-WATER INLET
MOTOR
METER REGISTER AND CONTACT MECHANISM
SOFT-WATER OUTLET
WATER METER
CONTACT BLOCK
AUTOMATIC MULTIPORT VALVE
ORIFICE PLATES
DRAIN

WASH-WATER COLLECTOR
BRINE-DISTRIBUTING SYSTEM
CATION EXCHANGER (ZEOLITE)
CHECK VALVE
DIAPHRAGM VALVE
BRINE
SALT
SALT-STORAGE TANK
CONSTANT HEAD BOX WITH FLOAT
FLOAT SWITCH
GRAVEL
BRINE-MEASURING TANK

WASHED AND GRADED GRAVEL
WASH-WATER DISTRIBUTOR
ADJUSTABLE JACK LEGS
RINSE-WATER RATE OF FLOW CONTROLLER
WASH-WATER RATE OF FLOW CONTROLLER

Figure 11.13 Zeolite water softener installation.

ness. When a sample of the outlet water shows hardness, the capacity of the exchange material has been exhausted, and it must be restored by regenerating with a brine solution (sodium chloride).

The regeneration procedure involves three steps:

1. The softener is taken out of service, and a swift flow of water is passed upward through the zeolite bed. This flow of water agitates and regrades the zeolite and at the same time washes away dirt that may have been deposited during the softening operation.

2. After the backwashing has been completed, the brine is introduced by means of a water-actuated ejector. The brine is evenly distributed above the zeolite bed by a system of piping, ensuring an even flow through the bed and regeneration with a minimum of salt. During the regeneration process, calcium and magnesium are removed as soluble chlorides, and the zeolite sodium content is replenished

3. After the salt has been introduced, the calcium and magnesium chloride together with the excess salts must be washed from the zeolite bed by means of a relatively slow flow of rinse water. After this rinsing operation has been completed, the softener may be returned to service. The hard water enters the top of the softener tank and flows downward through the bed, then through the meter, and out of service.

The regeneration requires a total of 35 to 65 min, and the soft water requirement plus the backwash and rinse must be provided either from a stored water supply or by another softener. It is customary to install two or more units so that adequate water is available to the boiler during the regenerative cycle.

These softeners can be regenerated either by hand manipulation or by automatic control. Softeners may be equipped with the necessary hand valves for backwashing, supplying the salt solution, and rinsing. However, the operation is simplified by the use of a multiported valve with a single control handle that is moved from the run position through the backwash, salt, and rinse positions and back to run, allowing the correct amount of time in each position. The regeneration is performed automatically by a power-driven multiported valve and timers that allow the valve to remain in each position the predetermined amount of time.

When the raw water hardness remains nearly constant, or when the variation takes place slowly, a meter provides an adequate means of determining the time to regenerate. After a predetermined amount of water has passed through the meter, a signal warns the operator that it is time to regenerate. In the case of an automatic unit, the signal from the meter may be used either to alert the operator or actually to initiate the regeneration cycle.

When the hardness of the raw water varies widely, the effluent must be analyzed for hardness to determine when regeneration is required. This determination may be made with an automatic hardness tester and the resulting signal used to warn the operator or initiate the regeneration cycle. When two or more automatically controlled softeners are installed, they are interlocked so that only one can be out of service for regeneration at any given time.

The dissolved solids are not removed by sodium zeolite softening; therefore, when the makeup water contains large quantities of impurities, the concentration of dissolved solids and alkalinity in the boiler water may become excessive. The concentration of the boiler water can be reduced by increasing the amount of blowdown, but supplemental treatment of the makeup water provides a more satisfactory solution. The alkalinity can be reduced by the addition of acid, but its use is limited, due to the close control that must be maintained and the precautions necessary in storage and handling. Another option is the use of a dealkalizer, which is physically like a zeolite softener but contains chloride anion resin. The effluent from the sodium zeolite softener flows through the dealkalizer. The alkaline sodium compounds formed in the zeolite softener are converted to chlorides in the dealkalizer. The alkalinity is reduced, but the solids remain in the water to build up the concentration in the boiler water. The dealkalizer, like the zeolite softener, is regenerated with sodium chloride and operated in the same manner.

Hot-lime-soda softeners are used in connection with sodium zeolite softeners. The makeup water first enters the lime-soda softener where it is heated; the dissolved solids, including silica, are partly removed, and the hardness is reduced. This hot treated water is then pumped through a sodium zeolite softener, where the hardness is reduced to zero. The ion-exchange material in sodium zeolite softeners must be selected for use with hot water.

Sodium zeolite softeners are relatively low in first cost and are simple to operate. The regeneration cycle including backwash admission of brine and rinse can be made fully automatic. The salt requirement varies from 0.3 to 0.5 lb per thousand grains of hardness removed. The salt can be stored in a tank into which water is added to provide a ready supply of brine for regeneration. Where the raw water supply pressure is adequate, it may be utilized to force the water through the softener, thus eliminating the use of extra pumps.

The use of the ion-exchange principle is not limited to the removal of calcium and magnesium hardness from water. By selection of exchange material and the solution used for regeneration, these units can be used to provide makeup for use in high-pressure boiler plants and in applications requiring water having a high degree of purity.

An arrangement of ion exchangers that removes practically all the dissolved solids in water is referred to as a *demineralizer* or *deionizer* (Fig. 11.14). The raw water passes through the cation unit, where the calcium, magnesium, and sodium ions are removed. The cation unit is regenerated with acid. From the cation unit the water cascades over fill material in the degasifier tank. Air is blown

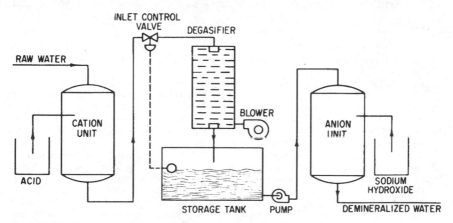

Figure 11.14 Diagrammatic arrangement of a demineralizer.

up through the degasifier tank in counterflow to the water. By this action the carbon dioxide gas is removed from the water and discharged from the tank with the air. The water discharged from the degasifier tank is next pumped through the anion exchanger and into the main water line, where it is stored in a demineralized water tank until it is needed. The anion unit contains an exchange material that will remove silica, carbon dioxide, chloride, and sulfate. The anion unit is regenerated with sodium hydroxide. The demineralizer will operate without the degasifier tank, but use of the latter reduces the size of the required anion unit and the amount of sodium hydroxide needed for regeneration.[3] Demineralizers with their ability to remove silica are essential to treating feedwater for use in high-pressure boilers.

When the supply of water to a demineralizer is obtained from city water mains, precaution must be taken because municipal water frequently contains residual chlorine. Chlorine deteriorates the exchange material, necessitating costly replacement. The chlorine can be removed by passing the water through a carbon filter before it reaches the demineralizer.

This ion-exchange principle is adaptable to many possible arrangements for different types of raw water and conditioned water requirements. The suppliers of water-conditioning equipment should be consulted so that the proper equipment is made available for the specific application.

11.3.2 Evaporators

Impurities in water are effectively removed by evaporators. Heat is applied, and the steam or vapor produced leaves the impurities to concentrate in the remaining water. When the steam or vapor is condensed, the resulting condensate has a high degree of purity.

---

[3]Pure water has a high resistance to the flow of electricity; therefore, its conductivity, the reciprocal of resistance, is low. The purity of the water is frequently determined by measuring its conductivity, using a dip cell and conductivity bridge. The results are expressed in microsiemens of specific conductivity. The relation between microsiemens of specific conductivity and parts per million of dissolved solids varies to some extent with the kind of dissolved solids in the water. The presence of carbon dioxide and ammonia in the water also affects the conductivity, and the readings must be corrected accordingly in order to determine the total dissolved solids.

In samples that do not contain carbon dioxide or ammonia, or when corrections for them have been made, the approximate purity in parts per million of dissolved solids can be determined by multiplying the conductivity in microsiemens by 0.55. In addition to checking relatively pure water, this method may, with suitably calibrated dip cells and conductivity bridges, be used to determine the concentration of boiler water. In the past, micromhos (mho) was the term for the basic unit of electrical conductivity instead of siemens (S). Thus 1 micromho = 1 microsiemens (μS).

The typical evaporator consists of a steel shell with the necessary outlets and a bank of tubes that constitute the heating surface. Steam supplied to the inside of the tubes gives up heat to the surrounding water and is thereby condensed. The condensate is removed through a trap that prevents the tubes from becoming flooded with condensate and at the same time prevents steam from being discharged and wasted. The heat received from the steam causes the water surrounding the tubes to evaporate, forming vapor, which leaves the unit through the outlet while the impurities remain in the shell of the evaporator. The vapor is at a lower pressure and hence a lower temperature than the steam supplied to the evaporator. The rate of evaporation depends on this temperature difference between the steam in the coils and the water surrounding them.

In the design and operation of an evaporator, consideration must be given to several details to ensure service and performance. Both the tube bundle and the shell must be designed to be strong enough to withstand the pressure involved. The shell must have a safety valve to prevent damage as a result of accidental overpressure. Priming and foaming, as experienced in boiler operation, also may occur in evaporators. This fault is overcome by designing the unit for sufficient water surface to prevent excessive agitation by steam bubbles and by installing moisture eliminators in the top of the shell at the steam outlet. The operator must keep the concentration of the liquid low enough to prevent foaming and to see that the water is at the specified level in the gauge glass. Careless operation results in a carry-over of solids with the vapor, thus diminishing the advantages derived from the use of the evaporator.

Except for radiation and blowdown losses, the heat supplied by the steam is returned to the condensate by the vapor condenser. The operator must remove the solids from the water in the evaporator by blowing down in order to prevent priming and foaming, since this would cause solids to be carried over with the vapor, thus defeating the purpose of the evaporator. The vapor must be checked to make sure that the water being produced is of the desired quality. The output of the evaporator may be controlled by varying the steam pressure to provide the necessary quality of distilled makeup water. Scale deposits on the evaporator tubes lower the maximum capacity but do not affect the efficiency or cause tube failures. Some scale is permissible as long as sufficient distilled water can be produced. Evaporators should be large enough to provide the necessary water with some scale on the tubes. Some plants prevent the formation of an excessive amount of scale by pretreating the water or by adding chemicals to the water in the evaporator. Another method of control consists of allowing the scale to deposit for several hours and then taking the evaporator out of service and subjecting the tubes to a rapid temperature change, which causes the hard scale to chip off the tubes. Evaporators are built to give a considerable tube movement with a change of temperature to cause the scale to chip off with this method of operation. When cold water is supplied to evaporators, oxygen, carbon dioxide, and other gases are driven off in the vapor, causing it to be corrosive. This difficulty has been overcome by the installation of hot-process softeners and deaerators to condition the water before it is fed to the evaporators.

When it is necessary to obtain a large quantity of distilled water, multiple-effect evaporators are used. These units consist of a number of successive stages, or effects, through which the vapor generated in one effect flows to the tubes of the next. This multistage operation increases the amount of distilled water that can be obtained from a given amount of steam.

11.4 Boiler Blowdown

In order to limit the concentration of solids in boiler water, it is necessary to discharge (blow down) some of the concentrated boiler water and replace it with makeup water (see Sec. 7.2). This procedure results in a loss of heat because the water blown down has been heated to the temperature corresponding to the boiler pressure. When blown down, it must be replaced by makeup water at a lower temperature. Furthermore, the cost of treating this added makeup must be charged against the blowdown operation.

Several arrangements of flash tanks and heat exchangers are available for reclaiming the heat from boiler blowdown water. Figure 11.15 is a schematic of a typical blowdown heat-recovery system. The flash tank removes heat from the blowdown water by decreasing its pressure to that of the heater steam supply and then using the resulting flash steam in the heater. Boiler blowdown water flows from the flash tank through the heat exchanger (closed type) to the sewer or, in today's zero-discharge facilities, to a liquid waste disposal system, where solids are removed and the water is recycled. Heat is transferred from the blowdown water to the makeup water as it flows through the heat exchanger to the feedwater heater. The system provides valves for controlling the rate of blowdown and cocks for obtaining samples of this water. Consider the following example, to estimate the possible saving that can be made by application of blowdown heat recovery.

Example   A plant generates steam at an average rate of 150,000 lb/h; pressure, 400 psig; blowdown, 10 percent of the boiler feedwater; fuel consumption, 8 tons of coal per hour; heating value of coal, 12,000 Btu/lb. What fuel saving in tons of coal per year will result from the installation of a blowdown heat-recovery system that will reduce the blowdown water temperature to 100°F?

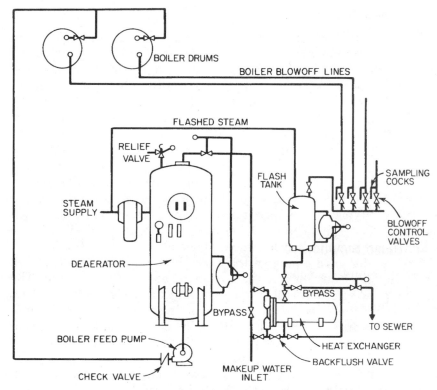

Figure 11.15 Continuous blowdown heat-recovery equipment. (*Crane Co., Cochrane Environmental Systems Division.*)

Solution

$$\text{Boiler feedwater} = \frac{150,000}{1.00-0.10} = 166,700 \text{ lb/h}$$

$$\text{Blowdown water} = 166,700-150,000 = 16,700 \text{ lb/h}$$

Enthalpy in blowdown water leaving the boiler
$$= 427.88 \text{ Btu/lb (400 psig, 414.7 psia)}$$

Enthalpy in water to sewer (100°F) = 67.97 Btu/lb

Heat reclaimed per lb of blowdown water = $427.88-67.97 = 359.91$ Btu/lb

Savings:

$$359.91 \times 16,700 = 6,010,500 \text{ Btu/h}$$

$$6,010,500/12,000 = 500.9 \text{ lb coal per hour}$$

$$500.9 \times 24 \times 365/2000 = 2194 \text{ tons per year}$$

If cost per ton of coal is known, annual fuel cost savings can be determined.

11.5 Piping System

Piping is used to transmit water, steam, air, gas, oil, and vapor from one piece of equipment to another. In this way the piping provides an essential steam plant function. An adequate piping system must provide for the necessary size to carry the required flow with the minimum amount of pressure drop, sufficient strength to withstand the pressure and temperature, expansion resulting from temperature change, the necessary support to hold the pipeline in place, drainage of water from steam and compressed-air lines, and insulation to reduce heat losses.

In determining the size of a pipeline, one must consider the velocity and pressure drop that will occur at the designated rate of flow. High-pressure systems that supply steam turbines are designed for velocities of 8000 to 15,000 ft/min depending on the pressure. Water-line velocities are significantly lower, ranging from 500 to 750 ft/min.

The best designed piping systems use the minimum sizes that will result in velocities and pressure drops that can be tolerated. The smaller pipes are lower in initial cost and have less radiation loss. However, it is false economy to select piping that is too small, since the pressure drop incurred may reduce the capacity of the equipment served by the piping or the increased design pressure may be cost prohibitive.

The diameter of pipe required for a given velocity and rate of steam flow may be determined from the following formula:

$$D = \sqrt{\frac{144Q}{0.7854V}}$$

where D = diameter of the pipe (in)
Q = ft^3/min of steam
V = velocity of steam (ft/min)

Example   Consider 10,000 lb of steam per hour flowing through a pipe at 100 psia pressure. Assume a velocity of 8000 ft/min. What size of pipe is required?

Solution   Saturated steam at 100 psia has a volume of 4.432 ft^3/lb (see steam tables in App. C).

$$Q = 4.432 \times \frac{10,000}{60} = 738.7 \text{ ft}^3 \text{ of steam per minute}$$

$$D = \sqrt{\frac{144 \times 738.7}{8000 \times 0.7854}} = 4.11$$

A 4.5-in pipe is therefore required.

Example Assume 400,000 lb/h of steam at 800 psia and 900°F flowing through the main steam line to a turbine. Assume a design velocity of 10,000 ft/min. What is the size of the main steam line?

Solution

$$v \text{ at 800 psia and 900°F} = 0.9633 \text{ ft}^3/\text{lb}$$

$$Q = 0.9633 \times \frac{400,000}{60} = 6422 \text{ ft}^3/\text{min}$$

$$D = \sqrt{\frac{144 \times 6422}{10,000 \times 0.7854}} = 10.85 \text{ in}$$

where v = specific volume, ft^3/lb

Since piping is fabricated in standard sizes, a 12-in pipe would be selected. Note that the designer could evaluate using a higher design velocity, which results in a smaller pipe size, 10 in. However, this must be evaluated against the additional pressure drop and higher design pressure of the system.

Piping must be selected to withstand the internal pressure, the strains resulting from expansion, and the external forces to which it will be subjected. The safe design pressure of a piping system may be calculated similar to that for a boiler drum (see Chap. 3). The method is included in the *ASME Code for Pressure Piping*. One method of designating the wall thickness of piping is by the terms *standard, extrastrong,* and *double extrastrong* (see Table 11.1). As steam pressures and temperatures increased, these designations became inadequate, and they have been supplemented by a system that uses schedule numbers to designate the wall thickness of pipe.

These schedule numbers are related to the steam pressure and strength by the following:

$$\text{Schedule numbers} = 1000 \times \frac{P}{S}$$

where P = steam pressure (lb/in^2)
S = working stress of pipe material (usually taken as 10 to 15 percent of ultimate strength)

Pipe having a schedule number of 40 corresponds approximately to standard pipe and a schedule number of 80 to extraheavy. Schedule 160 pipe is also available and is equivalent to double extrastrong pipe. The recommended values for S for steel and alloys at various temperatures are given in the *ASME Code for Pressure Piping*.

Pipe sizes 12 in and under are designated by their nominal inside diameter. Since the outside diameter of any nominal size of pipe remains the same, the extrastrong pipe has a smaller inside diameter

TABLE 11.1   Steel Pipe Data

Size, in	Standard				Extrastrong			
	External diameter, in	Internal diameter, in	Wall thick- ness, in	Weight, lb/ft	External diameter, in	Internal diameter, in	Wall thick- ness, in	Weight, lb/ft
⅛	0.405	0.269	0.068	0.244	0.405	0.215	0.095	0.314
¼	0.540	0.364	0.088	0.424	0.540	0.302	0.119	0.535
⅜	0.675	0.493	0.091	0.567	0.675	0.423	0.126	0.738
½	0.840	0.622	0.109	0.850	0.840	0.546	0.147	1.087
¾	1.050	0.824	0.113	1.130	1.050	0.742	0.154	1.473
1	1.315	1.049	0.133	1.678	1.315	0.957	0.179	2.171
1¼	1.660	1.380	0.140	2.272	1.660	1.278	0.191	2.996
1½	1.900	1.610	0.145	2.717	1.900	1.500	0.200	3.631
2	2.375	2.067	0.154	3.652	2.375	1.939	0.218	5.022
2½	2.875	2.469	0.203	5.793	2.875	2.323	0.276	7.661
3	3.500	3.068	0.216	7.575	3.500	2.900	0.300	10.252
3½	4.000	3.548	0.226	9.109	4.000	3.364	0.318	12.505
4	4.500	4.026	0.237	10.790	4.500	3.826	0.337	14.983
4½	5.000	4.506	0.247	12.538	5.000	4.290	0.355	17.611
5	5.563	5.047	0.258	14.617	5.563	4.813	0.375	20.778
6	6.625	6.065	0.280	18.974	6.625	5.761	0.432	28.573
7	7.625	7.023	0.301	23.544	7.625	6.625	0.500	38.048
8	8.625	8.071	0.277	24.696	—	—	—	—
8	8.625	7.981	0.322	28.554	8.625	7.625	0.500	43.388
10	10.750	10.192	0.279	31.201	—	—	—	—
10	10.750	10.136	0.307	34.240	—	—	—	—
10	10.750	10.020	0.365	40.483	10.750	9.750	0.500	54.735
12	12.750	12.090	0.330	43.773	—	—	—	—
12	12.750	12.000	0.375	49.562	12.750	11.750	0.500	65.415

than a standard pipe having the same normal size designation (see Table 11.1). This also applies to piping when the thickness is specified by schedule numbers.

There is no essential difference between piping and tubing from the standpoint of geometric shape and physical construction. However, pipe sizes generally are designated by their nominal inside diameter for a particular type of service, and tube sizes usually are specified in terms of outside diameter and a minimum wall thickness.

Piping is made in a relatively few standard sizes and manufactured in large quantities. Tubing is generally made in smaller quantities and must meet relatively strict specification requirements for such areas as tolerances, finish, chemical composition, and physical properties.

In addition to wall thickness, piping materials and method of manufacture must be considered. The material used in the manufacture of piping is designated by American Society of Testing Materials (ASTM) numbers.[4]

In order to assemble piping systems and make connections to equipment, it is necessary to have joints and use fittings of various types. Screwed, flanged, and welded joints are used for steel piping.

Screwed fittings are made of cast or malleable iron. It is advisable to limit the use of screwed fittings to sizes of 2 in and smaller. The larger sizes of screwed fittings are bulky, and the joints are difficult to align. Unions must be used in connection with screwed pipe assemblies to facilitate maintenance and replacement.

Standard fittings are satisfactory for pressures of 125 psi and temperatures to 450°F. Extraheavy fittings are required for pressures from 125 to 250 psi and temperatures to 450°F. For pressures above 250 psi and temperatures above 450°F, steel fittings are selected in accordance with specific code requirements for the design conditions.

Flanges and flanged fittings are used in connection with larger piping. The flanges may be attached to the pipe by screwed joints, slipped over the pipe and seal-welded (slip-on flanges), or by butt welding (weld-neck flanges). The use of screwed joints is restricted to low pressures. Welding is preferred for medium pressures and required for high pressures. Composition gaskets are used between the pipe flanges for water and saturated steam, while steel or stainless steel gaskets are used for superheated steam. Flanged joints are subject to leaks due to failure of the gaskets but provide accessibility for replacement and maintenance.

Piping systems can be assembled by butt or socket welding. Butt welding is used successfully for all piping above $1\frac{1}{2}$ or 2 in in diameter. Fittings are available with chamfered ends for butt welding, suitable for all pressures and made of materials for all temperatures.

Backing rings must be used or other precautions taken to prevent the welding material from entering the pipe and reducing the cross-

---

[4]The chemical analysis and manufacturing procedure for steel to meet these code requirements are fully explained in *ASTM Standard for Ferrous Metal.* These ASTM specifications cover pipe for all applications, from the ordinary low pressure to steels containing various percentages of molybdenum, chrome, nickel, and other materials for high-temperature applications.

sectional area. This work must be performed by a qualified welder in compliance with local codes. Steel having a thickness of $\frac{1}{2}$ in or less and containing neither carbon in excess of 0.35 percent nor molybdenum need not be preheated before welding or stress relieved after welding (see Sec. 3.4 for a discussion of stress relieving of welded joints).

Socket-welded fittings are used on pipe less than $1\frac{1}{2}$ or 2 in in diameter in all high-temperature water systems and other applications where tight joints are necessary. Socket welding involves a special steel fitting into which the full-sized end of the pipe may be inserted. The pipe is welded in place to hold against the pressure and seal against leakage.

Various types of valves, including gate, globe, angle, quick-opening, ball, butterfly, plug-type, check, nonreturn, reducing, etc., are used in pipelines to stop or regulate the flow. Valves are available for all pressures and are arranged for use with screwed-, flanged-, or welded-joint piping.

Cast or ductile iron pipe is used underground for water services and drainage systems and in other places to resist corrosion. (Ductile iron is made by adding other metals in the molten gray iron; the resulting change in the shape of the graphite particles results in a stronger, tougher, and more ductile material.) The sections are connected together, and fittings are attached either by use of bell and spigot or by mechanical joints. A bell-and-spigot joint has one end enlarged (bell) to receive the neck (spigot) of the other. The space between the bell and spigot is caulked with lead or provided with a rubber gasket to form a tight joint.

The mechanical joint likewise has one end enlarged to receive the other, but packing compressed by bolts is used to make a tight joint. Mechanical joints are more costly than bell-and-spigot joints but are preferred especially when the line is under a structure or roadway, where a leak would result in great damage and be difficult and costly to repair. Valves and fittings are available for use in cast iron piping systems using either mechanical or bell-and-spigot joints. These joints do not restrain the pipe from longitudinal movement, and adequate anchors or ties must be provided when there is a change in direction or other cause for longitudinal thrust.

Where there is a high corrosion potential, and also for underground service, many modern plants utilize fiberglass or plastic piping.

As the temperature of a pipeline changes, its length also changes, as a result of the expansion and contraction characteristics of the metal. This expansion results in a surprisingly large change in the length of the pipe and, if not taken care of, creates strains that might result in leaks or rupture. The amount of expansion can be calculated from the following:

$$E \text{ (expansion, in)} = (T_2 - T_1) \times L \times \text{C.E.} \times 12$$

where T_2 = final temperature of steam pipe (°F)
  T_1 = initial temperature of steam pipe (°F)
  L = length of pipe (ft)
  C.E. = coefficient of expansion for 1°F (for steel, 0.00000734)

Note: Refer to App. A for coefficients for materials other than steel.

Example Assume a steel pipe carrying steam at a pressure of 100 psia and a total temperature of 500°F. The header is 200 ft long and the room temperature is 100°F. Find the amount of expansion when heated from room to steam temperature.

Solution

$$E = (500 - 100) \times 200 \times 0.00000734 \times 12 = 7.0464 \text{ in}$$

In order to reduce the pressure drop in a pipeline to the minimum, the length of run should be as short as possible, and bends and fittings should be kept to a minimum. However, this practice frequently leads to a pipeline that is too rigid to provide for expansion. Arranging the piping with several bends frequently will provide adequate flexibility, but when this method fails, special devices to provide for expansion must be used.

In high-pressure installations, the U-bend is universally used as an expansion joint. This consists of a section of pipe formed with a long-radius bend to allow for considerable variation in the length of the pipe without imposing undue strain (Fig. 11.16a). The size of the pipe and the radius of the bend determine the maximum permissible pipe movement that the pipe bend will accommodate. These U-bends have the disadvantage of requiring considerable space, but when installed correctly, they require no maintenance.

For low and medium pressures, the packed expansion joint is used to advantage (Fig. 11.16b). This consists of a sleeve that can move inside a larger sleeve with packing to prevent leakage. Joints of this type require less space than U-bends and therefore may be installed in pipe tunnels. They can be made to adjust for a considerable amount of expansion. Stops are provided to prevent overtravel in both directions. The joints must be inspected and the packing tightened and occasionally replaced. Increased life of the packing may be obtained by applying a heavy heat-resistant grease to the packing gland with a grease gun.

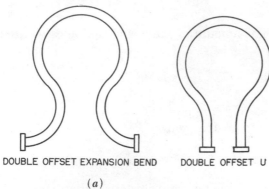

DOUBLE OFFSET EXPANSION BEND   DOUBLE OFFSET U

(a)

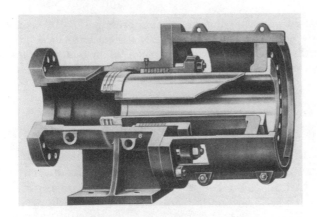

(b)

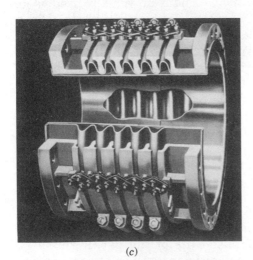

(c)

Figure 11.16 Piping expansion. (a) Expansion bends for pipelines. (b) Sliding sleeve, internally guided. (c) Corrugated expansion joint.

Another compact expansion joint consists of a corrugated-metal section installed in the pipeline (Fig. 11.16c). The circular corrugations provide for a limited amount of endwise movement of the pipe. The number of corrugations can be varied to meet the maximum expansion requirements. When this type of joint is used for high pressure, metal support bands are clamped to the outside of the corrugations to prevent distortion. The sidebar arrangement on these bands equalizes the movement of the corrugations, thus preventing some from being overstressed.

The support of piping systems presents a difficult problem because the pipeline must be held in place, and at the same time allowance must be made for movement resulting from expansion. The pipeline must be so anchored that the expansion will move the pipe toward the expansion joint. This is accomplished by securely anchoring the pipeline between the expansion joints and supporting it at the other points on rollers or spring hangers. In this way the pipe will be held in line, and as the temperature increases, the movement will be toward the expansion joint, allowing it to function as intended. The magnitude of the end thrust on the fixed anchors can be shown by an example. A 12-in pipe carrying steam at 150 psig pressure is to be anchored at an ell so that the expansion will extend toward an expansion joint. The thrust on the anchor will be

$$150 \times 12 \times 12 \times 0.7854 = 16,964 \text{ lb}$$

This shows that the thrust on the anchors will be approximately $8\frac{1}{2}$ tons. If the anchors should fail, the expansion joint would be expanded to the limit of its travel and the pipeline subjected to as much strain as if the expansion joint had not been installed.

Piping systems must be designed to limit the stress in the piping and the thrust imposed on the equipment to acceptable values. Overstressed pipe will result in failure. When piping exerts too much thrust or torque on the nozzles of a pump or a turbine, it will be forced out of alignment. Manufacturers of this type of equipment state the allowable forces that can be imposed on the nozzles by the piping. After initial design of a piping system, the stress and forces at all major points are calculated by a computer. If the pipe is overstressed, or if the forces are too great on the nozzle of the equipment to which the piping is attached, additional flexibility must be provided and the stresses and forces recalculated by the computer. It is essential that fixed hangers, spring supports, and anchors be maintained to prevent pipe failures and misalignment of equipment. When repairs or modifications are made to a piping system, consideration must be given to possible change in the flexibility.

Vibration is sometimes encountered in piping connected to air compressors and reciprocating pumps. This difficulty can be overcome by reducing the pulsating effects and by supporting the pipe more securely. In addition to the conventional air chambers for pumps, snubbers are used for reducing the pulsating effect produced by reciprocating machinery. Depending on the pressure, flexible piping is often used where there is the potential for vibration and resulting leaks.

Steam piping systems must be installed to prevent an accumulation of water while steam is flowing and to make it easy to drain the line when it is being placed in service. To accomplish this, the horizontal runs must slope in the direction of steam flow, with steam traps at all points where water may accumulate. This precaution lessens the possibility of water collecting in the line and being carried along with the steam to produce waterhammer.

The temperature of the substance in the pipe is frequently different from that of the surrounding air, making it advisable to insulate the line. When the temperature of the substance in the pipe is higher than that of the surrounding air, there is a heat loss from the pipe and a rise in the surrounding air temperature. This not only constitutes a loss of heat but frequently results in a high room temperature objectionable to personnel and injurious to equipment and to personnel if they come into contact with the hot piping. When the temperature of the substance in the pipe is lower than that of the surrounding air, moisture will condense, resulting in corrosion and dripping of water from the line. This condition is especially prevalent when the air is warm and moist (i.e., when there is high relative humidity).

High-temperature lines are effectively insulated by the use of 85 percent magnesium blocks in varying thickness, depending on the temperature. Actually, the economical thickness of insulation should be based on a balance of heat loss against the cost of insulation, with consideration to the comfort and safety of personnel. Heat loss is given in handbooks as Btu per hour per linear foot of pipe for various steam temperatures and thicknesses of insulation.

Example Consider 8-in pipe carrying steam at 160 psig, with a loss per linear foot of bare pipe of 2140 Btu/h. Loss per linear foot of 2-in high-grade insulation is 190 Btu/h. The saving per linear foot of pipe resulting from insulation is 1950 Btu/h. With 13,500 Btu per pound of coal and 70 percent boiler efficiency, the insulation on 1000 ft of 8-in pipe would save (using 30 days per month and 24 hours per day)

$$\frac{1950 \times 1000 \times 30 \times 24}{13,500 \times 0.70 \times 2000} = 74.3 \text{ tons of coal per month as compared with bare pipe}$$

For most power plants, the insulation is designed to provide safety for personnel and to minimize heat loss. For units that are located indoors, ventilation is required both for the comfort of plant personnel and for the changing of room air.

The materials used predominantly for insulation are

1. Mineral wool block and blankets, which are used to insulate membrane tube walls and boiler casings
2. Calcium silicate block, which is used on enclosures and piping
3. High-temperature plastic, which is used on irregularly shaped valves and fittings and to fill gaps between block insulation

Underground steam lines and return condensate lines are available in common insulated ducts. These lines are insulated and covered with a waterproof metal shield. Water must not come into contact with underground steam lines, since it will produce steam and give a false indication of leakage, as well as create a nuisance.

To prevent the accumulation of moisture, hair, felt, fiberglass, and cork are used to insulate lines that carry substances at a temperature lower than the surrounding atmosphere. These insulations must be applied with a vapor barrier to exclude the air, which, if allowed to enter the voids, will deposit moisture. When this occurs, the insulation will become saturated with water, and mold will form, creating an unsightly, unsanitary condition and causing rapid deterioration of the insulation.

The operator must recognize the piping system as an essential part of the steam plant. Steam and compressed air lines must be operated to avoid waterhammer by preventing steam and water or air and water from mixing in the pipelines. Steam lines must be placed in service slowly to allow the water to drain out and expansion to take place without creating strain in isolated sections. Large steam line valves should be provided with small bypass valves for this purpose. It is best to bypass the traps to ensure complete drainage during the time the steam line is being placed in service.

The cause of leakage should be determined immediately and repairs made as soon as possible. Leakage may be a warning of weakness that will cause serious rupture. Continual leakage of flanged joints results in cutting of the metal, making it necessary to replace or reface the flanges.

In time, insulation deteriorates as a result of maintenance, leakage, and mechanical damage, and sections of the pipe and fittings become exposed. This condition should not be tolerated, since it exposes personnel to burns, creates a source of heat loss, and makes the plant unsightly.

11.6 Steam Traps

A *steam trap* is a device attached to the lower portion of a steam-filled line or vessel that will pass condensate but will not allow the escape of steam. Traps are used on steam piping, headers, separators, and purifiers, where they remove the water formed as the result of unavoidable condensation or carry-over from the boilers. They are also used on all kinds of steam-heating equipment in which the steam gives up heat and is converted into condensate. Coils used in heating buildings, in water heaters, and in a wide range of industrial processing equipment are included in this classification.

Whether a trap is used to keep condensate from accumulating in a steam line or to discharge water from a steam-heated machine, its operation is important. If it leaks, steam will be wasted; if it fails to operate, water will accumulate. A satisfactory trap installation must pass all the water that flows to it without discharging steam, must not be rendered inoperative by particles of dirt or by an accumulation of air, and must be rugged in construction with few moving parts so that it will remain operative with a minimum of attention. Several principles are used in the operation of traps.

Figure 11.17 shows the operation of an inverted-bucket-type trap. The water and steam enter at the bottom and flow upward into the inverted bucket. As long as the bucket contains steam, it is buoyed up in the same way that an inverted empty bucket is buoyed up in water. While in this position, the valve is closed and there is no discharge of water or steam from the trap. As water enters the bucket, it displaces the steam, and the bucket loses its buoyancy and drops, causing the valve to open. After the water has been discharged, the bucket again

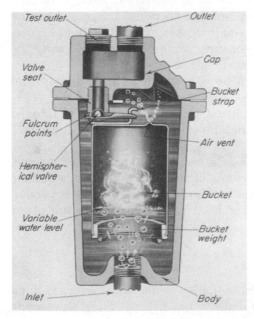

Figure 11.17 Inverted-bucket trap. (*Armstrong International, Inc.*)

fills with steam, the buoyancy is restored, and the valve closes. A small vent in the top of the bucket allows air to escape, thus preventing it from interfering with trap operation. In some models, the release of air is controlled by a thermostatic vent, which opens when the temperature decreases and allows the air to be discharged from the trap. Under normal operating conditions with steam in the trap, the vent remains closed. The valve seats and disks are made of stainless steel or other alloys to resist corrosion and wear. The moving parts, including the seats and disks, are easily replaced.

Float-actuated traps may be adapted to a wide range of operating conditions (Fig. 11.18). When water fills the body of the trap, the float rises, opening the valve. After the water has been discharged, the float drops, closing the valve and preventing the escape of steam.

The trap (Fig. 11.18a) is intended for use in removing condensate from steam systems. It has a thermovent valve that prevents air from interfering with the trap operation. When the temperature of the trap decreases, this vent opens automatically, allowing the air to escape. The trap (Fig. 11.18b) is suitable for removing water from compressed-air systems. The strainer (Fig. 11.18c) is installed in the line, ahead of the trap, to prevent sediment from stopping up the trap orifice. When selected for the correct pressure and provided with strain-

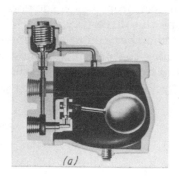
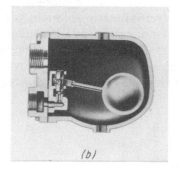

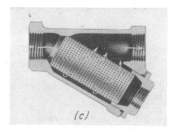

Figure 11.18 Float-actuated traps and pipeline strainers. (a) Steam trap with thermostatic air vent. (b) Trap for removing water from compressed-air lines. (c) Pipeline strainers.

ers, these traps will give satisfactory service. The valve assembly, including the float, float arm, and linkage, may be readily replaced.

The impulse steam trap (Fig. 11.19) operates on the principle that hot water under pressure tends to flash into vapor when the pressure is reduced. When hot water is flowing to the trap, the pressure in the control chamber (K) is reduced, causing the valve (F) to rise from its seat and discharge the water. As steam enters the trap, the pressure in the chamber (K) increases, causing the valve (F) to close, thus reducing the flow of steam to that which can pass through the small center orifice in the valve. It requires 3 to 5 percent of full condensate capacity to prevent steam from being discharged through this small orifice. These traps are available for all pressures, they are compact, and the few parts subject to wear can be readily replaced. It is necessary to install a strainer, since small particles of foreign matter will interfere with the operation of the valve.

Condensate is effectively returned from the point of steam usage to the boiler plant by the trap system shown in Fig. 11.20. The condensate flows into the receiver and then through the check valve into the condensate chamber. When the chamber fills to make contact with the short electrode, a solenoid positions the three-way steam valve to apply steam pressure to the chamber. This pressure forces the condensate through the discharge check valve and back to the boiler plant. When the level in the condensate chamber drops below the end of the long electrode, the solenoid positions the three-way valve to shut off the steam supply to the condensate chamber and at the same time vent this chamber to the receiver. Since the pressure is now equalized, the condensate will flow by gravity from the receiver to the chamber. During the condensate discharge cycle, when the chamber is

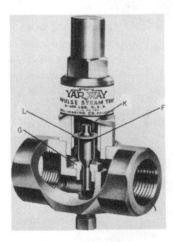

Figure 11.19 Impulse steam trap. F = valve. G = valve seat. K = control chamber. L = control disk. (*Yarway Corp.*)

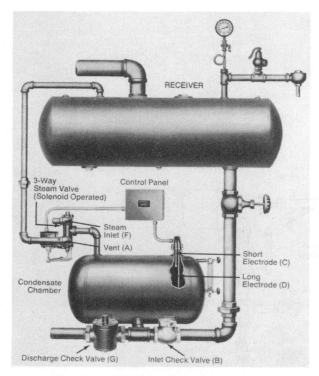

Figure 11.20 Condensate-handling system. (*The Johnson Corp.*)

pressurized, condensate accumulates in the receiver, providing an uninterrupted flow from the steam-utilizing equipment.

This system delivers the condensate to the feedwater heater under pressure. In this way the heat in the condensate, as discharged from the steam-utilizing equipment, is retained. This is in contrast to conventional condensate pumping systems, in which either the condensate tank must be vented to the atmosphere or cooling must be provided to obtain satisfactory operation of the pumps.

Traps are termed *nonreturn* when the condensate is discharged into a receiver, or heater, rather than directly to the boiler. A *return* trap delivers the condensate directly to the boiler. Return traps are located above the boiler; when they are filled with water, a valve automatically opens and admits steam at boiler pressure. This equalizes the pressure, and the water flows into the boiler as a result of hydrostatic head caused by the elevation of the trap. These traps are sometimes used in connection with low-pressure heating boilers.

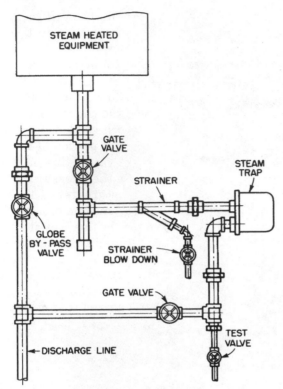

Figure 11.21 Typical steam trap piping arrangement.

Figure 11.21 shows a trap piping arrangement on a vital piece of equipment that must remain in continuous service. The two gate valves provide a means of shutting off the trap so that it can be removed for replacement or repairs. During this time, the globe bypass valve is used to regulate the flow of condensate. Bypass on steam header traps may be opened when the header is being placed in service. This procedure ensures against the possibility of water-hammer if the trap should fail to drain the condensate from the steam line. The strainer is placed in the line ahead of the trap. A test valve is useful for checking trap operation by observing the discharge.

Traps correctly sized and installed should require very little attention or maintenance, but they should be inspected periodically. The frequency of inspection will depend on the operating conditions. Gauge glasses provide an effective means of observing trap performance. When the trap is functioning satisfactorily, the water level will be near the centerline of the body and will continue to show some fluctuations in level. If the water is above the glass (as observed by opening the gauge-glass drain), the trap is too small, the orifice is stopped up, the chamber contains air, there is an obstruction in the line, or the backpressure is too high. If the water is low or out of sight in the glass, the shutoff mechanism in the trap is not functioning and steam is blowing through and being wasted. The strainer should be blown out periodically to remove the accumulation of dirt. A test valve (see Fig. 11.21) may be opened and the discharge of the trap observed to determine how often it is operating and how much condensate is being discharged. Do not be confused by the fact that a portion of the condensate discharged from the trap will appear as vapor; this happens because the water is flashing into steam as the pressure is reduced. The higher the pressure in the trap, the greater is the amount of flashing. Traps can be checked by placing one end of a rod (sounding rod) to the ear and the other to the trap. In this way it is possible to detect by sound if the trap is blowing straight through or failing to operate and discharge the condensate.

Traps, strainers, and associated piping installed in areas that experience freezing must be adequately protected. This is accomplished by installing either steam tubing or electric cable tracing and then insulating the system.

11.7 Pipeline Separators and Strainers

The steam, air, and gas flowing in a pipeline often contain particles of moisture, oil, and other foreign matter. These impurities must be removed by the use of separators or strainers located in the pipeline. Although there are many types of separators, they all use the same principle. The moving stream is directed against the baffle or obstruction, which suddenly changes the direction of flow of the steam, air, or gas. The particles of moisture or oil (being heavier than the steam, air, or gas) are thrown out of the main stream and are retained in the separator.

The separators shown in Fig. 11.22 have a ribbed baffle area extending outside the jet of steam or compressed air projected from the inlet pipe. On entering the separator, the flow is directed against this ribbed baffle. The particles of moisture or oil are first deposited on the baffle and then drip into the collecting chamber. The ribs in the baffle retain the film of water or oil while the flow of steam or air passes through ports at the sides of the baffle. This arrangement prevents the flow from coming in contact with the water or oil drops as they fall into the chamber, thereby decreasing the tendency of water and oil to be picked up by the flow of steam or compressed air. The collecting chamber is provided with a trap for drainage and a gauge glass to check the liquid level.

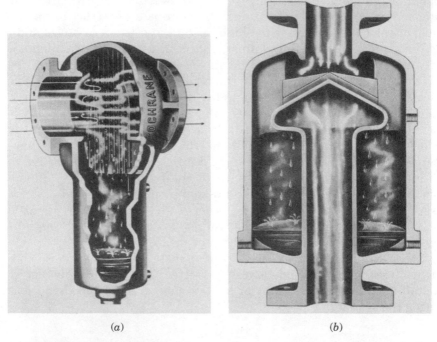

(a)   (b)

Figure 11.22 Steam and compressed-air separators. (a) For horizontal lines. (b) For downward flow in vertical lines. (*Crane Co., Cochrane Environmental Systems Division.*)

A live-steam separator (Fig. 11.23) is suitable for use in a vertical steam line between the boiler and a pump or turbine. This is the baffle-type separator in which both the baffle and the walls contain slanting corrugations. The moisture is collected by these corrugations, deposited in the well, and drained off by use of a steam trap. The complete reversal of flow at the lower edge of the baffle separates the steam from the moisture.

Centrifugal separators are used to remove oil, moisture, and solids from steam, compressed air, and gas. Figure 11.24 shows how these separators function to remove the entrainment.

The exhaust head used when steam is discharged to the atmosphere above the roof of a building is a special adaptation of this type of separator. When exhaust steam is discharged to the atmosphere, the exhaust head (separator) is installed on the discharge end of the pipe. In this way the oil and moisture are collected to prevent them from becoming an environmental problem.

Because oil separators are not effective in removing all the oil from steam, air, or gas, secondary filtering is often necessary. Condensate is often passed through a filter to remove the last trace of oil before

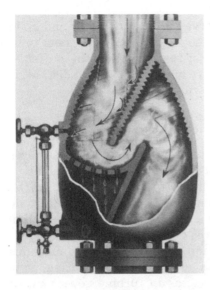

Figure 11.23 Live steam separator. (*Hayward Industrial Products, Inc.*)

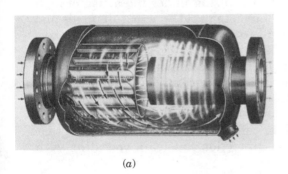

(*a*)

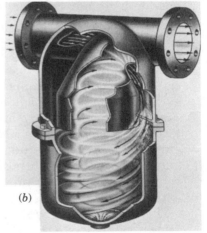

(*b*)

Figure 11.24 Centrifugal separators for steam, air, and gas. (*a*) Line-type for use in supply lines ahead of equipment. (*b*) T-type for compact in-line mounting. (*Hayward Industrial Products, Inc.*)

the condensate is returned to the boilers. Filters are available for cleaning compressed air when small amounts of oil are objectionable.

Strainers are available for removing foreign matter from liquids flowing in a pipeline. They are made in various sizes for installation in the pipeline. The two-element arrangement provides for continuous service, since the shutoff valves place one element in operation while the other is being cleaned. The strainers or baskets may be made of any suitable material and mesh, depending on the material in the line and the particle size of the impurities to be removed.

The difficulty most likely to be encountered with pipeline separators is failure of the trap to discharge the oil or water removed. Some separators have gauge glasses to observe the level of water and oil in the bowl, but it is difficult to keep these clean enough for them to be of any value. It is advisable to have a test valve on the outlet or bypass of the trap. By opening this valve and observing the discharge, it is possible to tell whether or not the trap is functioning.

11.8   Lubricants and Lubricating Devices

The primary purpose of a lubricant is to prevent two bearing surfaces from coming into direct contact. When one lubricated surface slides or rolls over another, the lubricant adheres to each surface and the motion takes place within the lubricant. The two metallic surfaces do not come into contact, and wear is therefore reduced. Fluid friction occurs as one film of the lubricant moves over another.

In addition to this primary function, lubricants are sometimes required to carry away heat developed, as in a turbine bearing, or to operate hydraulic cylinders and devices, as in turbine governors. In performing these functions, the oil must resist mixing with water (emulsifying), thinning out and carbonizing under the action of high temperature, and oxidizing and becoming acid on exposure to air and other contamination. It is evident from this that care should be exercised in selecting lubricants.

Progress has been made in the manufacture of lubricants to meet the needs of the industry, and various lubricants for all purposes are available. Detergents are added to oil to prevent formation of dirt and sludge in crankcases and gearboxes. Wetting agents cause oil to adhere to the surfaces being lubricated. Lubricants are available that contain additives that provide protection for gears operating with extreme pressure between the teeth.

Oils are thickened by the addition of soap to form plastic lubricants known as *greases*. Greases are used as lubricants where oil would leak out of the bearings, where water and dirt cannot be excluded, where it is difficult to give the bearings frequent attention, or where the operating speed is low and the bearings are heavily loaded.

Plant operators can obtain the benefits of advanced developments by presenting their problems to the manufacturers of lubricants or of equipment. Each application should receive separate consideration. The request presented by the operator should include a description of the machine, the size and speed of the shaft, the types of bearings, the possibility of leaking oil or grease, whether water and grit are excluded from the bearings, the method of applying the lubricant, whether service is continuous or intermittent, and the manufacturer's lubrication recommendations. From this information, the type of lubricant and frequency of application can be determined. For simplicity and economy, the number of types of lubricants used in the plant should be kept at a minimum. In the majority of cases, equipment manufacturers provide recommendations on lubricants based on the operating experience of their equipment in many plants.

Viscosity is one of the qualifications mentioned in equipment manufacturers' specifications of lubricants. It may be defined as a measure of the resistance to flow. A high-viscosity lubricant is one that has high internal fluid friction. This means that a high-viscosity lubricant has a tendency to develop heat, especially when applied to high-speed machinery. It follows that, in general, light or low-viscosity lubricants are adaptable to high-speed machines and heavy or high-viscosity lubricants to low-speed machines.

The viscosity of oil can be measured by a standard saybolt universal (SSU) viscosimeter. This test consists of noting the time in seconds required for 60 ml of an oil at standard temperatures of 70, 100, 130, and 210°F to flow through the orifice. Thus an SSU viscosity of 80 at 210°F would mean that it required 80 s for the 60 ml of oil at 210°F to pass the orifice. The familiar Society of Automotive Engineers (SAE) viscosity number takes into account the SSU viscosity range of an oil. For example, an SAE 30 oil has an SSU viscosity of no less than 185 and no more than 255 at 130°F. Additives are available that reduce the effect of temperature changes on the viscosity of lubricating oil. They are widely used where the machinery operates under varying temperatures. The automobile engine is a typical example.

Other lubrication specifications include *specific gravity,* which can be expressed either as the weight of oil compared with an equal volume of water or in API degrees (see Chap. 4); *pour point,* or lowest temperature at which the oil can be poured; *flash point* or *fire point,* meaning, respectively, the temperature at which the vapors given off will flash or burn continuously when exposed to an open flame; *demulsibility,* or ability of the oil to separate from water; and *acidity,* as measured by the amount of caustic of known strength required to neutralize a given quantity of oil.

The method of applying lubricants varies from frequent hand operation to automatic methods that require little attention. The method of application and the type of bearing must be considered when determining the frequency of application. The oil cups shown in Fig. 11.25a, b, c, and d are used in various ways to apply oil to bearings. The ring oil bearing is self-lubricating, but the level in the reservoir must be checked. An oil cup such as that shown in Fig. 11.25a may be attached to the side of these bearings to provide a means of observing

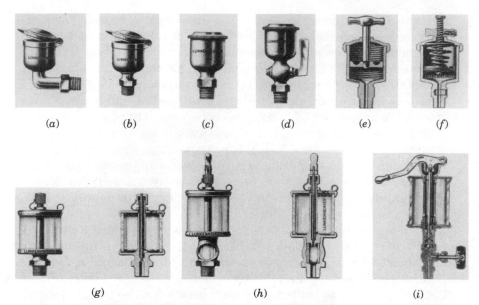

(a)   (b)   (c)   (d)   (e)   (f)

(g)   (h)   (i)

Figure 11.25 Lubricators, oil and grease cups. (*a–d*) Hinge-spring and screw-lid cups. (*e*) Plunger screw-feed grease cup. (*f*) Automatic-feed grease cup. (*g*) Oil cup without sight feed. (*h*) Oil cup with sight feed. (*i*) Hand-operated oil pump. (*The Lunkenheimer Co.*)

the oil level. Wool waste is sometimes placed in the oil cups to act as a feed and to decrease the necessary frequency of oiling. The oil cups in Fig. 11.25*g* and *h* have needle valves for regulating the flow of oil to the bearings. When the machine is stopped, the valve can be shut off without changing the needle-valve adjustment. The oiler shown in Fig. 11.25*i* is a hand pump for injecting oil into a system that is under pressure. The grease cup shown in Fig. 11.25*e* can be filled—and the grease can be applied as required—by turning the handle; this method of grease application has been almost completely replaced by the high-pressure fitting and detachable grease gun. After the grease cup shown in Fig. 11.25*f* has been filled, the spring exerts a pressure that slowly feeds the grease into the bearing.

Multipoint lubricators are available for supplying grease to a large number of points automatically. Each lubrication point is provided with a lubricator. These lubricators are connected in series and supplied by one grease pump. The pump may be operated by hand or by power with a timer set for the desired interval. When pressure is applied to the system by the pump, grease is delivered to each point. The individual lubricators are adjustable so that the required amount of grease will be supplied.

In pressure systems, the oil is circulated by pumps to the different points requiring lubrication. This oil accumulates impurities that must be removed, or serious damage to the equipment will result.

These impurities cause undue wear of the bearings, clogging of pipes in the lubrication system, coating of cooler tubes, and subsequent higher oil temperatures as a result of reduced heat transfer. When oil comes into contact with water, there is a tendency to form emulsions, usually referred to as *sludge*. By removing these impurities from the oil, it is rendered satisfactory for service, and the plant lubrication costs are thereby reduced.

Oil is conditioned by separators, filters, or a combination of both. In some instances, the entire batch of oil is removed from the machine and reconditioned; in others, a portion of the oil is continually circulated through the reconditioning apparatus.

If contaminated oil is allowed to stand for a long period of time, the water and sludge, being heavier than the oil, will settle to the bottom of the container. This process may be accelerated by passing the oil through a mechanical centrifuge that subjects it to a centrifugal force several thousand times that of gravity. The centrifuge

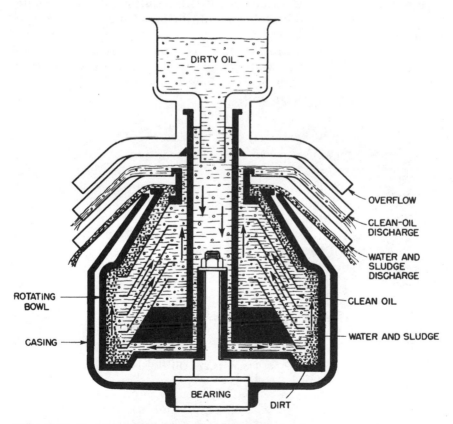

Figure 11.26 Cross section of centrifugal oil separator bowl.

shown in Fig. 11.26 consists of a frame that supports a bowl on a vertical shaft. This shaft and bowl constitute the rotating elements, which are power driven at high speed. The dirty oil is fed into the top center of the separator, flows down through the hollow shaft, and is discharged into the bottom of the bowl by the centrifugal force. The water and sediment, being heavier than the oil, are forced to the outside of the bowl and travel up its sides and overflow through the lower discharge tube. A number of conical plates or baffles direct the movement of the oil toward the center of the bowl, where it flows upward and is discharged through the center spout. The top spout acts as an emergency overflow. Sediment collects on the conical plates and in the bowl, making it necessary to clean them frequently.

Conditioners that use a combination of settling and stages of filtration are useful in maintaining the lubricating properties of low-viscosity oil. These conditioner units remove small particles of solids and water from the oil, thus restoring its lubricating value. Care must be exercised in selecting this equipment, since some filter media remove additives and inhibitors, thus changing the characteristics of the oil.

The oil reclaimer shown in Fig. 11.27 is capable of removing contamination from a lubricating system and thus restoring the lubricating qualities of the oil. Figure 11.27a shows how the components are assembled to form a package unit.

Figure 11.27 Oil reclaimer. (a) Package unit.

(a)

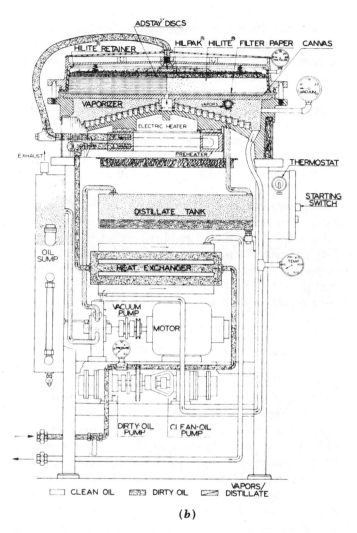

Figure 11.27   Oil reclaimer. (*b*) Flow diagram. (*The Hilliard Corp.*)

The operation can be explained by reference to the flow diagram in Fig. 11.27*b*. The dirty-oil pump forces the incoming oil through the heat exchanger, where it receives heat from the clean oil. The dirty oil next passes through the electric preheater. Then it is drawn by vacuum through a filter that removes carbon, sludge, abrasives, and other solids. This filter also neutralizes the acidity and restores the color. From the filter, the oil enters the vaporizer. Here, moisture, solvents, dissolved gases, and other volatile impurities are removed by the combined action of the elevated temperature and the vacuum. The volatile substances that are removed from the oil may be discharged to the atmosphere or they may be condensed and reclaimed. The clean-oil pump removes the oil from the vaporizer and forces it through the heat exchanger into a clean-oil storage tank or back into service.

The type of filter medium is varied depending on the characteristics of the oil and the severity of the service. The filter media must be changed when the gauges show that the pressure drop has exceeded the established limits.

The expendable or throwaway type of cartridge filter is effective in removing small particles of sediment from low-viscosity oil. Filters of this type consist of a case designed to hold a cartridge that has a large filter area. The cartridge is made of cellulose material and can be designed to remove extremely small particles of micron size. When the filter becomes clogged, as indicated by an increase in pressure drop, the cartridge is replaced with a new one to restore the unit to its original performance.

After the operator has determined the correct type of lubricant and the frequency of application and has made a study of the method of application, the next step is to use this information in normal operating procedures. Written procedures will serve as instructions and reduce the possibility of misunderstandings and neglect. Figure 11.28 is a sample of a lubrication-schedule sheet for a small boiler plant. Such sheets are issued to the personnel responsible for lubricating the equipment at the various specified intervals. Form 10 is issued each week and informs operating personnel of the equipment involved, the points of application, and the lubrication to be used and provides a space to initial to indicate that the work has been done. The type of lubricant can be referred to by the manufacturer's specification number. At the end of the week the sheet is returned to the supervisor with comments on any difficulties that have been encountered in carrying out the instructions. In like manner, Form 15 is issued every 120 days and covers the lubrication work that is to be done at that frequency. Adequate lubrication is essential to continued operation and low maintenance costs.

In today's modern plants, computerized systems are used for record keeping and immediate access to information by plant operating personnel. These systems also alert operators on the maintenance schedule for all equipment.

Figure 11.28 Lubrication schedules. (*a*) Form 10. (*b*) Form 15.

Form 10 BOILER-PLANT LUBRICATION SCHEDULE Frequency: daily									
Equipment and parts	Lubricant to be used	Work to be done	Work done by						
			M	T	W	T	F	S	S
Spreader stoker: Motor sheave pulley	Light grease	1 fitting with gun							
Motor slide	Oil	Apply oil with can							
Drive shaft	Heavy grease	4 fittings with gun							
Coal-feed regulator	Heavy grease	6 fittings with gun							
Stoker-feed mechanism	Heavy grease	5 fittings with gun							
Boiler feed pump and turbine: Pump shaft	Light turbine oil	Check oil level; 2 wells							
Turbine shaft	Light turbine oil	Check oil level; 2 wells							
Governor shaft	Light grease	1 grease cup							

(*a*)

Form 15 BOILER-PLANT LUBRICATION SCHEDULE Frequency: 120 days			
Equipment and parts	Lubricant to be used	Work to be done	Work done by
Spreader stoker: Motors	Light grease	2 motors; 2 fittings each	
Boiler feed pump: Pump shaft	Light turbine oil	Flush and refill bearing	
Turbine shaft	Light turbine oil	Flush and refill bearing	

(*b*)

Questions and Problems

11.1 What is the purpose of feedwater heater? How is it classified?

11.2 What is a low-pressure feedwater heater? A high-pressure feedwater heater?

11.3 Describe the design for a closed feedwater heater. Why is the U-tube type the preferred design in today's power plants?

11.4 Describe the operation of an open feedwater heater. What is it commonly called? Other than feedwater heating, what function does it serve?

11.5 Name some of the common locations where deaerators are found in a power plant.

11.6 What are the primary problems associated with feedwater heaters? What is a common correction for a tube leak?

11.7 Compare the quality of the condensate that returns from a condensing turbine with that of water from an industrial power plant where a larger quantity of makeup water is common.

11.8 What is a condensate polishing system?

11.9 What are the advantages of condensate polishing? List possible arrangements for this system.

11.10 Why does raw water for power plant use have to be treated?

11.11 What are suspended solids? Dissolved solids?

11.12 What are anions and cations?

11.13 When is water considered hard water, and what problems occur with hard water?

11.14 What is the purpose of a sodium zeolite water softener? How does it operate?

11.15 Explain the procedure for regenerating a sodium zeolite water softener.

11.16 What difficulties are likely to be encountered when using a sodium zeolite softener to condition boiler makeup water? What system is often used in connection with sodium zeolite softeners?

11.17 What is the purpose of a demineralizer? Under what condition would a demineralizer be used to condition boiler makeup water?

11.18 What is the function of an evaporator?

11.19 List the precautions to be taken when operating evaporators?

11.20 What is the purpose of blowdown from a boiler? What are the disadvantages of a blowdown system?

11.21 A boiler produces 450,000 lb/h of steam at 750 psia and 800°F when burning coal that has a heating value of 12,800 Btu/lb. The percentage of blowdown is 5 percent. The drum pressure is 775 psia. What annual fuel savings would result if the cost of coal were $75 per ton and a blowdown heat-recovery system were installed that reduced the blowdown water temperature to 100°F? Assume that the plant operates at this capacity for 95 percent of the year.

11.22 A boiler produces 225,000 lb/h of steam at 685 psig and 700°F to a turbine. The design velocity of the steam line is 9000 ft/min. Calculate and select the diameter of the main steam line.

11.23 What is the basic difference between piping and tubing?

11.24 On the admission of steam to the main steam line, the temperature changes from 80 to 650°F. If the line is 135 ft long, what is its expansion?

11.25 What are two methods used to account for the expansion in steam lines?

11.26 What are the reasons for insulating hot and cold piping?

11.27 What materials are commonly used for insulation?

11.28 What is a steam trap, and how does it operate?

11.29 What is a steam separator, and how does it work?

11.30 What is meant by *lubrication?* Describe the qualities that are essential to a good lubricant.

11.31 What is grease? When is it used in preference to oil?

11.32 What is meant by the *viscosity* of oil? When would you use a low- or a high-viscosity oil?

11.33 How would you proceed to establish a lubrication program for a steam power plant?

12

Environmental Control Systems

12.1 Introduction

The operators of today's industrial and utility power plants not only must be concerned with the production of steam for process or for the generation of electricity that will meet the demands of the plant, but also they must meet these demands within the confines of environmental restrictions. Thus the operators must be well aware of restrictions associated with their operating permit as related to air pollution, water pollution, noise, and other restrictive areas.

Operators must be well aware of the need for meeting all environmental requirements or their plants will face possible shutdowns and/or severe monetary penalties for not adhering to imposed regulations.

This chapter will concentrate on air pollution requirements and the equipment available to ensure compliance.

12.2 Environmental Considerations

Environmental control is driven primarily by government legislation and the resulting regulations at the local, national, and international levels. These emission restrictions have evolved because of the public consensus that has determined that the costs of environmental protection are worth the tangible and intangible benefits now and in the future. The design philosophy of energy-conversion systems, such as steam generators, has evolved from providing the lowest-cost energy to providing low-cost energy with an acceptable impact on the environment.

12.3   Sources of Plant Emissions

Figure 12.1 shows the significant waste streams from a modern coal-fired power plant. For a new 500-MW coal-fired boiler, the typical discharge rates for the significant emissions with and without control equipment are shown in Table 12.1.

Atmospheric emissions result primarily from the by-products of the combustion process (SO_2, NO_x, and particulates) and are exhausted from the stack. Another source of particulates is fugitive dust from coal piles, fuel-handling equipment, and ash-handling equipment. A final source of air emissions is from the cooling towers and the thermal rise plume that contains heat and water vapor.

The primary sources of solid waste are from the collection of the coal ash from the bottom of the boiler, its economizer, and air heater hoppers and from the electrostatic precipitator or fabric filter. The ash is transported to an ash settling pond, where it settles out and is prepared for either landfill or other use. At some locations it is transported directly to a landfill that is designed specifically for the handling of this waste.

Another major source of solids is the by-product that results from the flue gas wet scrubbing process. This is generally a mixture of calcium sulfate or calcium sulfite. After dewatering, processing, and treatment, the by-product may be sold as gypsum or put directly into a landfill. Other sources of solids include the sludge from cooling tower basins, wastes from water-treatment systems, and wastes that result from periodic boiler chemical cleaning.

TABLE 12.1   Emissions from a Typical 500-MW Coal-Fired Power Plant (with Coal at 2.5 Percent Sulfur, 16 Percent Ash, 12,360 Btu/lb)

Emission	Typical control equipment	Discharge rate, tons/h	
		Uncontrolled	Controlled
SO_x as SO_2	Wet limestone scrubber	9.3	0.9
NO_x as NO_2	Low NO_x burners	2.9	0.7
Fly ash to air	Electrostatic precipitator or baghouse	22.9	0.05
Thermal discharge to water stream	Natural-draft cooling tower	2.8×10^9 Btu/h	0
Ash to landfill*	Controlled landfill	9.1	32.0
Scrubber sludge	Controlled landfill	0	25.0

*As fly ash emissions to air decline because of collection of precipitator or baghouse, ash shipped to landfill increases.
SOURCE: Courtesy of Babcock & Wilcox, a McDermott company.

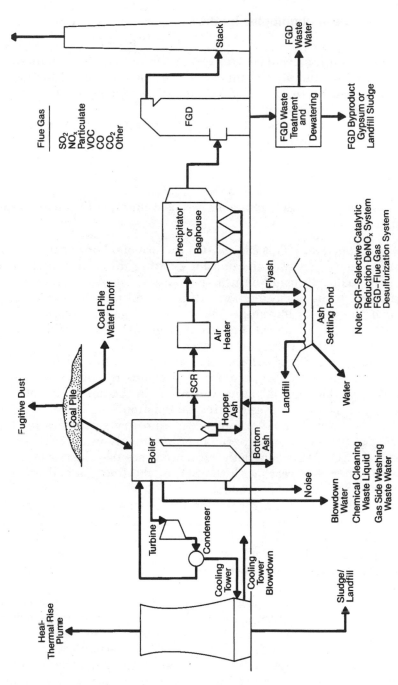

Figure 12.1 Steam power plant emissions. (*Babcock & Wilcox, a McDermott company.*)

12.4   Available Technologies for the Control of Emissions

The major environmental control systems found in power plants are discussed in this chapter. Combustion methods for the reduction of emissions such as fluidized bed combustion and low-NO_x burners are discussed elsewhere in this book.

Atmospheric emissions of SO_2, NO_x, and particulates can be controlled by the following major technologies.

12.4.1 SO_2 control

For coal-fired plants, SO_2 emissions can be reduced with the following techniques:

1. *Precombustion.*   This can include the use of oil or natural gas or low-sulfur coals in new units and for existing units by changing to low-sulfur coals. By using natural gas, sulfur emissions are reduced to near zero, while the use of low-sulfur oil will minimize emissions. Although the use of low-sulfur oil or natural gas is advantageous from the standpoint of SO_2 emissions, the cost and availability of these fuels can make them less attractive.

2. *Combustion modification.*   The use of a fluidized bed boiler where limestone is the bed material is a means to reduce SO_2 emissions. This method of combustion can result in the absorption of over 90 percent of the SO_2 released during combustion.

3. *Wet and dry scrubbers.*   The wet and dry scrubbers for the reduction of SO_2 emissions are the most commonly used technologies for reducing SO_2 emissions. Both these processes use slurries of a sorbent and water that react with the SO_2 in the flue gas and produce either a wet or a dry waste product.

In the wet scrubbers, a sorbent slurry consisting of water mixed with primarily lime, limestone, or sodium carbonate (Na_2CO_3) is placed in contact with flue gas. This process can absorb over 90 percent of the SO_2 in the flue gas, and it is a proven technology that also can produce a usable by-product, gypsum, for commercial drywall production.

Dry scrubbing is also a proven technology and involves the spraying of a sorbent slurry in a manner where the slurry droplets dry as they contact the flue gas. The SO_2 reaction occurs during the drying process and results in a dry particulate. This particulate, together with the fly ash, is captured downstream in the particulate-control equipment, most often a fabric filter as compared with an electrostatic precipitator.

12.4.2 NO$_x$ control

NO$_x$ can be controlled by the use of the following technologies:

1. *Combustion modification.*   The use of a fluidized bed boiler where combustion occurs at relatively low temperatures results in low NO$_x$ emissions. Emissions are generally so low that postcombustion techniques are not required to meet requirements. The use of low-NO$_x$ burners and two-stage combustion techniques is also an effective means to reduce NO$_x$.

2. *Postcombustion.*   The two major postcombustion technologies for NO$_x$ control are (a) selective noncatalytic reduction (SNCR) and (b) selective catalytic reduction (SCR). SNCR is a process in which ammonia or other compounds, such as urea, which thermally decomposes to form ammonia, are injected into the furnace downstream of the combustion zone, where the furnace temperatures are in the range of 1400 to 2000°F. At these temperatures, NO$_x$ is removed from the flue gas by its reaction with ammonia. SCR technology is the most effective method of reducing NO$_x$ emissions, especially where high removal efficiencies are required. It removes NO$_x$ from the flue gas by a reaction with ammonia in the presence of a catalyst.

12.4.3 Particulate control

The particulate emissions from boilers come from the noncombustible portion of the fuel, called *ash,* that is released during the process of combustion. Another source of particulates comes from the incomplete combustion of fuel, which is in the form of unburned carbon particles.

The primary means of controlling particulates are as follows:

1. *Mechanical collectors.*   These are used primarily on small boilers, where there are fewer emission limits, but often in conjunction with a precipitator or a baghouse. Mechanical dust collectors are limited to a collection efficiency of about 90 percent and are not good at collecting small particles. These units also require high draft loss to operate effectively.

2. *Fabric filters.*   These units also have high draft loss, but they can achieve collection efficiencies greater than 99.9 percent. They also have the ability to enhance SO$_2$ capture in dry scrubbing systems.

3. *Electrostatic precipitators.*   These units also can achieve collection efficiencies greater than 99.9 percent and with very low draft loss.

However, they require high electric power usage and therefore are equivalent to fabric filters in overall power requirements.

12.5 Regulatory Requirements

Emissions of air pollutants from utility and industrial steam plants in the United States came under broad government regulation with the passage of the first Clean Air Act in 1963. Since then, that law has been amended to establish national health and air quality standards, as well as regulations limiting discharges of specific air pollutants from power plants, and has given the federal government the leading role in air pollution control. Implementation of the act has required the formation of federal and state agencies and the expenditure of billions of dollars by industry to comply with its standards and rules.

Prior to 1963, industry practice with respect to air pollution control generally was governed by the motivation of individual companies to be good neighbors and, therefore, air pollution control varied widely from region to region. Only particulate emissions were of concern then, and little attempt was made to reduce discharges of acid gases such as sulfur dioxide (SO_2) and of nitrogen oxides (NO_x).

Except possibly for plants in urban areas, mechanical collectors generally were used on coal-fired boilers until the mid-1950s to provide a degree of emission control and to protect induced-draft fans from erosion. As boilers became larger and flue gas flows greater, electrostatic precipitators gained in popularity because of their lower power and maintenance costs and higher collection efficiencies as compared with mechanical collectors. Particulate removal efficiencies typically averaged between 90 and 95 percent for precipitators.

This beginning has led to plants that incorporate high-efficiency electrostatic precipitators or bag filterhouses that have a particulate collection efficiency as high as 99.95 percent, depending on the ash content in the coal, and sulfur dioxide scrubbing systems that must remove over 90 percent of the SO_2 in the flue gas, depending on the sulfur content in the fuel.

A typical example of a modern utility power plant that incorporates such high-efficiency environmental control equipment is shown in Figs. 12.2 and 12.3. These illustrations show the 400-MW San Miguel plant, which is located in Texas.

The need for high-efficiency particulate and sulfur dioxide scrubbing systems has not only created a new set of operating criteria for plant operators but also has created a new set of problems associated with the disposal of the collected particulate from precipitators or bag filterhouses and of the sludge produced from the sulfur dioxide removal. These also have their associated environmental problems,

Figure 12.2   San Miguel Power Project. (*Babcock & Wilcox, a McDermott company.*)

since suitable landfill and other disposal means have to be created to satisfy environmental restrictions. This book will not attempt to cover the various ways of handling waste disposal.

Nevertheless, it is important to know that waste disposal is expensive whether the waste is used as landfill, roadbeds, or cement blocks or for making by-products such as gypsum. This disposal has its own unique environmental restrictions.

12.6   Particulate and Sulfur Dioxide Removal Requirements

12.6.1   Particulate removal

For new utility boilers with heat inputs greater than 250 million Btu/h (250×10^6 Btu/h), the permissible particulate emission is 0.03 lb per million Btu of heat input for coal-fired boilers.

> **Example**   A boiler plant generates 225,000 lb of steam per hour and burns 13.9 tons of coal per hour. The coal has a heating value of 11,400 Btu/lb. A test of the particulates leaving the boiler shows that 3804 lb of particulate is being discharged per hour.
>
> 1. What is the particulate discharged per million Btu heat input to the furnace?

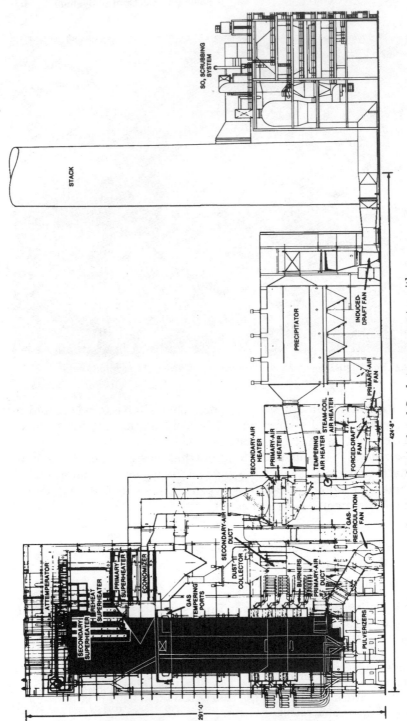

Figure 12.3 San Miguel Power Project showing pulverized-coal-fired steam generator with electrostatic precipitator and sulfur dioxide (SO_2) wet scrubbing system. (*Babcock & Wilcox, a McDermott company.*)

SO₂ SCRUBBING SYSTEM

STACK

PRECIPITATOR

INDUCED-DRAFT FAN

SECONDARY-AIR HEATER

PRIMARY-AIR HEATER

TEMPERING AIR HEATER STEAM-COIL AIR HEATER

FORCED-DRAFT FAN

PRIMARY-AIR FAN

SECONDARY-AIR DUCT

DUST COLLECTOR

GAS-RECIRCULATION FAN

BURNERS

PRIMARY-AIR DUCT

PULVERIZERS

ATTEMPERATOR

SECONDARY SUPERHEATER

REHEAT SUPERHEATER

PRIMARY SUPERHEATER

ECONOMIZER

GAS TEMPERING PORTS

424'-9"

29'-0"

2. What must the efficiency of the precipitator or bag filterhouse be to reduce this emission to 0.03 lb per million Btu?

Solution

$$1. \text{ Furnace heat input} = 13.9 \times 2000 \times 11{,}400$$

$$= 317 \times 10^6 \text{ Btu/h}$$

$$\text{Particulate released from boiler} = \frac{3804}{317 \times 10^6} = \frac{12 \text{ lb}}{10^6 \text{ Btu}}$$

$$2. \text{ Collection efficiency} = \frac{12 - 0.03}{12} \times 100 = 99.75 \text{ percent}$$

Solid particles carried by flue gases consist essentially of a portion of the ash that was in the fuel with some unburned carbon particles. The amount and nature of this material depend on the fuel, combustion equipment, capacity, and operating skill. Restrictive ordinances make it necessary to provide some type of collection equipment with solid fuel and in most instances with liquid fuels such as fuel oil.

The quantity and particle size must both be considered when evaluating and controlling particulate emission. Particle size is designated in micrometers, or microns. (A micrometer, or micron, equals one-millionth of a meter, or 1 in = 25,400 μm.) A size distribution analysis of a particulate sample states the percentage of the material that is less than the specified number of microns. For example, an analysis showed that 30 percent of the particulate from a pulverized-coal-fired boiler was less than 10 μm. These data are required in the selection of collection equipment.

12.6.2 Sulfur dioxide removal

Again, for utility boilers with heat inputs greater than 250×10^6 Btu/h, the removal requirements for sulfur dioxide on coal-fired units are as follows:

1. For coal with a sulfur content that would result in an emission of 0.6 lb of sulfur dioxide per million Btu from the boiler, 70 percent of the sulfur dioxide must be removed from the flue gas.

2. For coal with a sulfur content that would result in an emission of 2 lb of sulfur dioxide per million Btu from the boiler, 70 percent of the sulfur dioxide must be removed from the flue gas, the remaining 0.6 lb per million Btu being released to the atmosphere.

3. For coal with a sulfur content that would result in an emission of 2 to 6 lb of sulfur dioxide per million Btu from the boiler, 70 to 90 percent of the sulfur dioxide must be removed from the flue gas, with the remaining 0.6 lb per million Btu being released to the atmosphere.

4. For coal with a sulfur content that would result in an emission of 6 to 12 lb of sulfur dioxide per million Btu from the boiler, 90 percent of the sulfur dioxide must be removed from the flue gas, with a maximum of 1.2 lb per million Btu being released to the atmosphere.

5. For very high sulfur coals with a sulfur content that would result in an emission of 12 lb or more of sulfur dioxide per million Btu from the boiler, sulfur dioxide removal must be sufficient to limit the amount released to the atmosphere to a maximum of 1.2 lb per million Btu, a removal efficiency of greater than 90 percent.

Table 12.2 summarizes the preceding and shows typical coals that are affected by these requirements.

These regulations require a sulfur dioxide removal system on every new utility boiler, even when it is burning the lowest-sulfur coal available, but they allow the use of a less expensive dry sulfur dioxide scrubbing system, requiring a 70 percent removal efficiency, if the fuel's sulfur content is low.

These regulations represent the minimum acceptable emission levels established by federal standards. State and local requirements may impose stricter limits to meet their own air-quality standards.

12.6.3 Industrial boilers

The emission limits for new industrial boilers has been established by the Environmental Protection Agency (EPA). The particulate emission limit on coal-fired units is 0.05 lb per million Btu for all boilers having a heat input of 30 million Btu/h (30×10^6 Btu/h) or greater.

For SO_2 removal on coal-fired units, the requirements are as follows:

TABLE 12.2   Sulfur Dioxide Removal Requirements

Uncontrolled SO_2 emissions, lb/10^6 Btu	SO_2 removal (%) and discharge to atmosphere (lb/10^6 Btu)	Typical coals
0.6	70%	Very little such coal exists.
2.0	70%; 0.6 lb/10^6 Btu	Wyoming subbituminous, North Dakota lignite
6.0	90%; 0.6 lb/10^6 Btu	Illinois high-volatile bituminous
12.0	90% or more; 1.2 lb/10^6 Btu	High-sulfur Eastern coals

1. For heat inputs between 10 and 75×10^6 Btu/h, there is no specific removal efficiency that has to be achieved; however, a maximum release of 1.2 lb of SO_2 per million Btu from the boiler is allowed.

2. For heat inputs greater than 75×10^6 Btu/h, 90 percent of the sulfur dioxide must be removed from the flue gas, with a maximum of 1.2 lb per million Btu being released to the atmosphere.

Because of these requirements for SO_2 removal, many industrial-sized boilers that burn solid fuels are using the fluidized bed boiler combustion technology.

12.7 Equipment for Particulate Emission Control

There are three principal types of equipment used to remove ash from the flue gases before they are discharged from the stack: (1) mechanical collectors, (2) electrostatic precipitators, and (3) bag filterhouses.

12.7.1 Mechanical collectors

Mechanical collectors use a combination of centrifugal, inertial, and gravitational forces to remove fly ash from the flue gas. Mechanical collectors, which are often called *cyclones* or *multiclones,* are used primarily in the industrial market to remove material having carbon from the flue gas (e.g., as a result of wood or bark burning) and to reinject it into the boiler. This not only improves the emissions but also improves the boiler efficiency.

Mechanical collectors consist of multitube units that remove the fly ash by centrifugal action (Fig. 12.4). The gases flow downward through the spinner vanes in the annular space between the outer and inner tubes. Then the gases reverse direction and flow up through the inner tube and out of the collector. The combination of centrifugal force created by the spinning of the gases and the action of gravity when the flow is reversed separates the dust particles, and they fall through the bottom end of the outer tube into the hopper. These collectors can be designed to remove 92 percent of the fly ash from flue gases on coal-fired spreader-stoker boilers. However, their efficiency decreases rapidly on that portion of the fly ash with a particle size less than 10 μm. A draft loss of 3.0 in of water (gauge) is required to obtain high efficiency. The induced-draft fan must be designed for this added resistance. Operating costs are increased by the added power required to operate the fan.

As the load on the boiler decreases, the velocity of gases through the collector also decreases, lowering its efficiency. This loss is overcome by providing two or more sections in the collector. All but one of

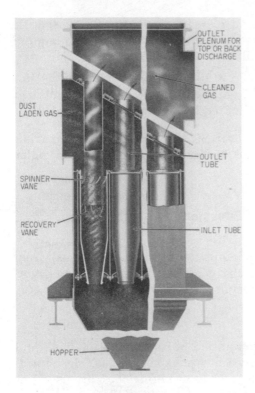

Figure 12.4 Mechanical dust collector. (*American Standard, Inc.*)

these sections have dampers that shut off the flow of gases. As the load decreases on the boiler, the collector dampers are closed progressively to maintain the design velocity and therefore the desired efficiency at the reduced capacity.

The ash hoppers must be emptied on a regular schedule and frequently enough to prevent the ash from building up and sealing off the bottoms of the tubes. Should this occur, the tubes may remain plugged even after the ash has been removed from the hopper. Failure to remove the ash also may result in a smoldering fire in the hopper. Continual discharge of the hoppers through the use of automatically operated valves should be strongly considered as a design and operating practice.

Mechanical dust collectors can be utilized if the emission requirements are 0.3 lb per million Btu or greater and if fewer than 25 percent of the particles are under 10 μm.

For this reason, mechanical dust collectors have limited application when emission requirements must be met, and, therefore, an electrostatic precipitator or a bag filterhouse must be used to meet the more stringent requirements. However, where applicable, the mechanical dust collector should receive careful consideration, since its capital cost is about one-eighth that of an electrostatic precipitator or bag filterhouse.

On fluidized bed boilers, mechanical collectors are used extensively to capture and recirculate the bed material. A high-efficiency collector, such as a baghouse, is then used in series with the mechanical collector to meet the particulate emission requirement.

12.7.2 Electrostatic precipitators

Approximately 80 percent of the ash from a pulverized-coal-fired boiler is carried through the boiler as fly ash, and about 50 percent of that is less than 10 μm in size. This requires the use of electrostatic precipitators to capture the fly ash.

In an *electrostatic precipitator* (ESP), dust-laden flue gas is distributed uniformly between rows of discharge electrodes and grounded collecting plates (Fig. 12.5). A high-voltage dc current is applied to the electrodes, which causes the dust particles to become ionized and then to be attracted to the grounded collecting plate. These collected particles are removed periodically from the plates by a rapping system that generates vibrations and causes the collected dust to fall into the hoppers.

Maximum efficiency is obtained by automatic control of the high voltage. The voltage is maintained at the maximum value without excessive sparking between the discharge electrodes and collecting plates. When sized and operated correctly, these collectors can remove more than 99.9

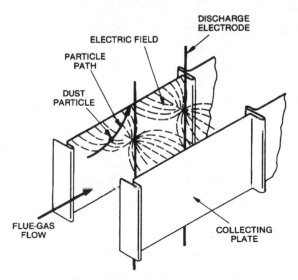

Figure 12.5 Flue gas distribution between discharge electrodes and collecting plates. (*Power Magazine, a McGraw-Hill publication.*)

percent of the fly ash from the flue gases. However, the size of an ESP depends on a characteristic of ash known as the *resistivity*. This determines the susceptibility of the fly ash particles to the influence of the electrostatic field. Fly ash resulting from burning coal with a high sulfur content has a low resistivity and therefore is more easily collected, requiring a smaller precipitator than one designed to collect fly ash from a coal having a low sulfur content. Therefore, the proper sizing of a precipitator depends on knowing the various types of coal that will be used over the life of the plant. Otherwise, the precipitator might be too small if it were designed for high-sulfur coal, and in the future low-sulfur coal were used. Other fuels such as wood, bark, or municipal solid waste (MSW) have similar resistivity characteristics, and these must be used for the proper sizing of an ESP.

Because the gas velocities through an ESP are very low (approximately 5 ft/s), the draft loss is minimal, at about 0.5 in of water (gauge) through the precipitator, as compared with mechanical collectors and, as we will see later, bag filterhouses.

In general, there are two types of electrostatic precipitators currently in use, with variations in each of the designs. These designs can be classified as (1) weighted wire and (2) rigid frame.

Perhaps the most important feature that differentiates precipitator designs is the manner in which the discharge electrodes are supported; the various types are illustrated in Fig. 12.6. The weighted-wire design shown in Fig. 12.7 attempts to hold the discharge electrode

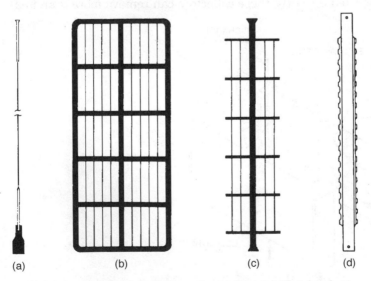

(a)   (b)   (c)   (d)

Figure 12.6 Types of discharge electrodes. (*a*) Weighted wire (shrouded). (*b*) Rigid frame (bedspring). (*c*) Rigid frame (strung) mast. (*d*) Rigid electrode.

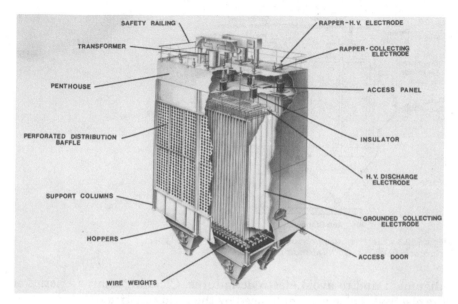

SAFETY RAILING

TRANSFORMER

PENTHOUSE

PERFORATED DISTRIBUTION BAFFLE

SUPPORT COLUMNS

HOPPERS

WIRE WEIGHTS

RAPPER-H.V. ELECTRODE

RAPPER-COLLECTING ELECTRODE

ACCESS PANEL

INSULATOR

H.V. DISCHARGE ELECTRODE

GROUNDED COLLECTING ELECTRODE

ACCESS DOOR

Figure 12.7   Electrostatic precipitator, weighted-wire design.

wires in a position uniform between the collecting plates by means of wire weights at the bottom of the individual wires. This design is also characterized by the two-point suspension system of the discharge electrodes for each bus section and by the top rapping of both the discharge electrodes and the collecting plates.

With the arrival of large power plants requiring larger and taller precipitators to meet the more stringent emission requirements, the weighted-wire precipitator was found to be inadequate in many cases to provide the performance and reliability necessary. Long weighted wires had a high incidence of failure, and subsequent bus section outages resulted, deteriorating performance and requiring plant outages for repair.

Clear stacks were not only a public relations requirement but also a regulatory requirement. Therefore, power plant operators had to be more concerned with the reliability of their precipitators, since their poor performance meant reduced loads or shutdowns for repair that resulted in added costs for replacement power.

The rigid-frame design used in Europe for decades was introduced into the United States to overcome this problem. Several designs were developed to provide the performance necessary to meet the more stringent performance and reliability requirements. The designs emphasized rigidly affixing the discharge electrodes to provide better

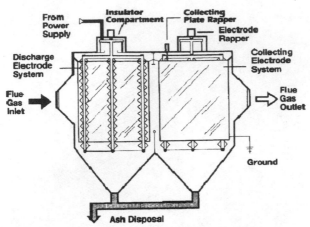

Figure 12.8 Electrostatic precipitator, rigid-mast-plate type. (*Babcock & Wilcox, a McDermott company.*)

alignment and to avoid electrode failures. Cleaning characteristics of the electrodes also were enhanced by the rigid construction.

Figures 12.8, 12.9, and 12.10 show different types of rigid-frame precipitators, providing more rigidity and therefore, it is hoped, more reliability.

Figure 12.8 shows a schematic of a rigid mast type of electrostatic precipitator that uses alternating vertical collecting plates and dis-charge electrodes that are hung in rows to form gas passages. A precipitator is designed for uniform gas distribution, and, therefore, each precipitator includes a variety of gas-distribution devices such as a gas-distribution plate at the inlet and turning/straightening vanes to ensure uniform velocity through the precipitator.

Its operation is similar to that of all electrostatic precipitators. A high dc voltage to the discharge electrode system (i.e., the rigid mast) generates a corona. As the particulate matter that is entrained in the flue gas passes through the corona, an electrical field is formed that ionizes the particles and creates an electron flow from the discharge to the collecting plate. The dust particulate becomes negatively charged and is attracted to the collecting plates, which are positively ground. The particulate adheres to the plate and forms a dust layer.

At periodic intervals, both the discharge electrode and the collecting plates are rapped, which shears the accumulated layer of fly ash particulate. The released fly ash is collected in hoppers located beneath the precipitator. The collected ash is then removed by an ash-removal system.

Figure 12.9 shows a design that provides a pipe framework to support the discharge electrodes at the top, the bottom, and several inter-mediate locations within the frame. Individual frames are restrained

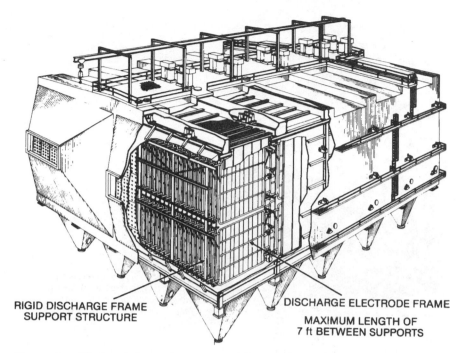

RIGID DISCHARGE FRAME
SUPPORT STRUCTURE

DISCHARGE ELECTRODE FRAME

MAXIMUM LENGTH OF
7 ft BETWEEN SUPPORTS

Figure 12.9 Electrostatic precipitator, rigid-frame design using discharge electrode frame. (*Babcock & Wilcox, a McDermott company.*)

and supported at both the front and rear to maintain electrical clearances; this design has virtually eliminated wire breakage. Among its design features, a four-point suspension system of each bus section provides a more rigid design than a two-point system.

The design in Fig. 12.10 uses a rigid discharge electrode design that combines both the electrical and mechanical functions in a single structural member, thus providing a more reliable design.

A major factor in good precipitator operation is to have a rapping system that can dislodge hard-to-remove ash and prevent the buildup of ash, which can cause sparking and subsequent loss of collection efficiency. Undoubtedly, if the internals cannot be cleaned properly, deterioration in performance will result. There are many precipitators in operation that have been sized correctly to meet performance but whose performance has deteriorated because of an inadequate rapping system, which makes it difficult to keep the precipitator clean.

The precipitator designs shown in Figs. 12.8, 12.9, and 12.10 illustrate different philosophies in rapping designs. The design shown in Fig. 12.9 incorporates individual rappers for both discharge electrode frames and collecting plates, as shown in Fig. 12.11. This design

Figure 12.10 Hi-R electrostatic precipitator featuring rigid discharge electrodes (*Research-Cottrell, Air Pollution Control Division.*)

brings the rapping force to the point where the blow can be utilized effectively with one hammer at the bottom edge of each collecting plate curtain and one hammer on each discharge frame section. This rapping system is effective because

1. Rapping one collector plate curtain at a time means that a smaller proportion of the precipitator is disturbed and stack puffing is eliminated.

2. In-plane rapping results in a uniform acceleration from the bottom to the top of the plates, thereby cleaning all surfaces.

The rapping design shown in Figs. 12.8 and 12.10 is a top rapping design incorporating magnetic-impulse gravity-impact rappers, which provide multiple rapping to both the discharge electrodes and the collecting plates. They utilize one moving part and are located outside the dirty gas environment.

The type of design selected must be evaluated carefully for the type of application required, since each design has its advantages.

The reliability and collection efficiency of precipitators have been improved dramatically by use of rigid discharge electrodes instead of weighted wires. The problem with wires is that, being thin (about 0.1

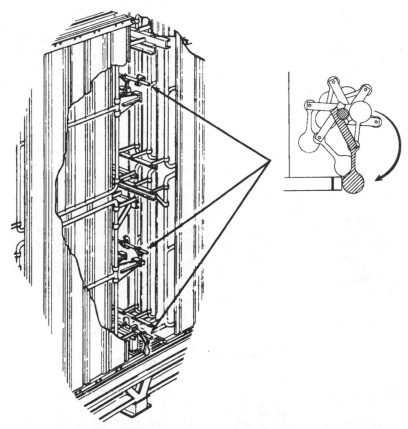

Figure 12.11 Rapping system for electrostatic precipitator. (*Babcock & Wilcox, a McDermott company.*)

in in diameter), they are very susceptible to fatigue failure from cold working caused by rapping vibration, to electrical failure from repeated sparkover at one spot caused by swinging wires and close tolerances, and to chemical corrosion. This is not to say that wire generally makes a poor discharge electrode. In industrial plants where precipitators are smaller, the reliability of properly designed wire electrodes is good.

Maintenance of precipitators is generally at a minimum when flue gas temperatures are kept above the acid dew point and fly ash is not stored in hoppers. Hoppers are simply funnels, not storage bins, and continuous hopper evacuation systems should be considered. In many designs, the following are integral parts of the hoppers to ensure good fly ash removal: hopper heaters, vibrators, poke holes, anvil bars, and level detectors.

12.7.3 Bag filterhouses

The bag filterhouse (often called *baghouse* or *fabric filter*) is another method of removing particulates from flue gases. The decision on whether to have a fabric filter or an electrostatic precipitator is one of the most important decisions that a plant designer and operator must make, since the consequences have to be lived with for the life of the plant.

Bag filterhouses are characterized according to the method used to remove the fly ash from the filter medium (Fig. 12.12). To illustrate, in reverse-air (actually cleaned flue gas) collectors, the type preferred for large power plants, cleaned flue gas is pushed through the bags in the direction opposite to filtration after the compartment containing the bags is removed from service. Damper valves isolate the compartment, and the cleaned flue gas is drawn from the plenum and is reversed. The reversed cleaned flue gas dislodges the fly ash from the bags and causes the fly ash to fall into the hopper. The compartment is then returned to service.

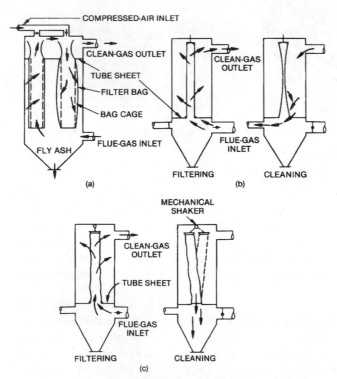

Figure 12.12 Types of bag filterhouses. (*a*) Pulse jet. (*b*) Reverse air. (*c*) Shaker. (*Power Magazine, a McGraw-Hill publication.*)

Sometimes a mechanical shaker is installed on these units to help remove fly ash. Pulse-jet baghouses often are used on industrial boilers and rely on a short burst of compressed air directed down through the filter bag to dislodge fly ash into the hopper below. A compressed-air pulse is activated through a solenoid valve and is injected into the filter bags, flexing them and removing the ash. An advantage of this method is on-line cleaning, which means the baghouse can remain in continuous operation while the bags are cleaned.

Bag filterhouses remove more than 99.9 percent of the fly ash from the flue gas, and for low-sulfur coals, where the ash has a high resistivity, they are preferred in many cases over electrostatic precipitators. Of course, careful evaluation of each collecting device is required to determine the proper equipment for the application. This evaluation would consider such areas as maintenance costs, power costs, draft loss, bag replacement costs, etc. Fabric filters operate at high pressure drops as compared with precipitators (6 in of water as opposed to 0.5 in), which is a highly important consideration as it increases power costs.

The principal advantage of filters over precipitators is that coal sulfur content (resistivity) does not influence the collection efficiency. Another benefit of reverse-air filters is that their compartmented design permits maintenance while the unit is operating. Fabric filters also can capture the very fine particulates and, when used with a dry scrubber, assist in the capture of acid gases because of the accumulation of a dust cake through which the flue gas passes.

Figure 12.13 shows a typical reverse-air fabric filter system. In operation, fly ash–laden flue gas from the boiler enters the system through the ash hopper below the bag compartment. The flue gas flows into the bags from the bottom, and particulate is collected on the inside surface of the fabric, forming a filter cake. It is the filter cake that performs the cleaning, not the fabric itself. The gas passes through the filter cake and the fabric to the stack.

Figure 12.14 shows another reverse-air baghouse design that is applicable to both utility and industrial boilers. These units provide gentle cleaning of the bags, to long bag life. It is relatively simple to operate and has few moving parts, and its performance results in low emission levels. For utility boiler applications, these baghouses can become quite large, as shown in Fig. 12.15. The baghouse shown is on a 350-MW boiler designed to burn low-sulfur western coal, and it handles 1.7 million ft³/min of flue gases at 270°F.

To clean the bags, the appropriate compartment is isolated from the gas stream, and airflow (clean gas) is reversed so that the bags collapse gently. This collapsing action breaks the filter cake, which falls from the bag into the hopper below. Too little cleaning can reduce per-

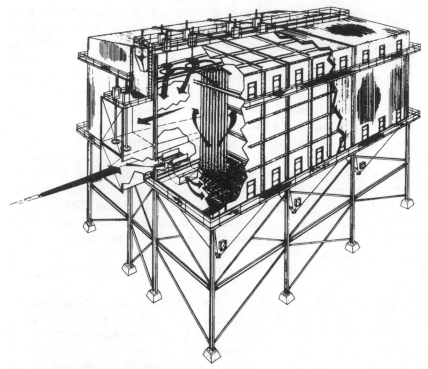

Figure 12.13   Reverse-air fabric filter system. (*Research-Cottrell, Air Pollution Control Division.*)

formance, while overcleaning the bags strips away the residue needed to collect fine particulate.

In order to prevent complete bag collapse when the airflow (cleaned gas) is reversed, the fabric is supported with anticollapse rings. This prevents the bags from creasing and the fabric from working loose and breaking. Careful attention should be made to the manufacturer's operating instructions so that optimal operating procedures are maintained.

The air-to-cloth ratio establishes the size of the baghouse; for the woven glass fabrics used on reverse-air filters, this ratio is about 2:1. Although these materials can operate at much higher ratios, system designers generally agree that the 2:1 ratio provides good balance among bag life, pressure drop, and time between cleaning cycles.

If operating conditions are such that the bags do not clean properly, the high pressure drop would force a reduction in plant output or even a plant shutdown. Another problem is the failure of individual bags, which could quickly put the plant out of compliance if a signifi-

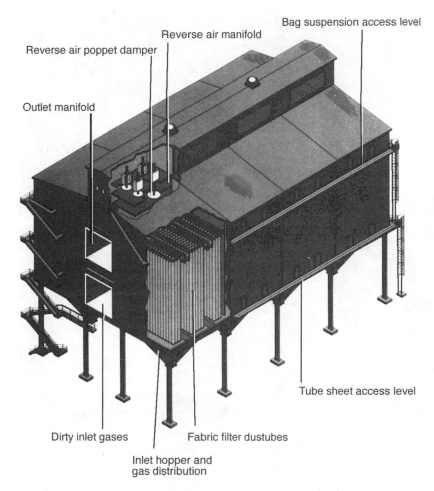

Reverse air poppet damper

Reverse air manifold

Bag suspension access level

Outlet manifold

Tube sheet access level

Dirty inlet gases

Fabric filter dustubes

Inlet hopper and
gas distribution

Figure 12.14 Reverse-air fabric filter system showing internal arrangement of filter bags. (*Wheelabrator Clean Air System, Inc.*)

cant number of bags failed. However, designs incorporate spare compartments for both maintenance and spare capacity. There is also concern about the use of fabric filters in cycling plants because of the potential for corrosion problems during startup and low-load operation. On stoker-fired coal-burning boilers, fires are a possibility in fabric filters if glowing char is carried over into the collector, but the installation of a mechanical collector ahead of the filter can prevent this from occurring.

Bag failures and subsequent replacement costs present a very high potential maintenance cost to the bag filterhouse. Various bag materi-

Figure 12.15 Reverse-air fabric filter for 350-MW pulverized-coal-fired boiler. (*Wheelabrator Clean Air System, Inc.*)

als are continually being examined and tested to develop a long-life bag that can withstand high temperatures and provide both acid and alkaline resistance.

Operation below a dew point may attack the bag fibers and shorten bag life. Factors that affect this action are the specific compounds condensed, the fiber materials and the influence of protective coatings on the filters, and the neutralizing materials present in the collected dust. Regardless of chemical attack, water alone may allow salt recrystallization within the fiber structure, resulting in stiffening and failure of the bags by physical action.

Dew point protection is designed into the bag filterhouse by careful insulation procedures and careful control of air in-leakage that could cause localized cold zones.

Various types of filter bags and bag coatings are available. Their selection will vary from one application to another according to dust loading, particle size, stickiness of dust, available time for complete cleaning cycle, and type of dust. Teflon-B coatings and acid-resistant finishes are popular, as are fiberglass bags.

Cleaning cycles of the baghouse can be controlled by timed cycles, pressure drop across the fabric filter, or pressure drop across individual compartments. Pressure drop across the fabric filter is generally used because it provides longer bag life by minimizing cleaning cycles.

A pulse-jet type of baghouse is often used for industrial-sized power plants, and it is gaining rapid acceptance on utility-sized boilers. A schematic of this type of baghouse is shown in Fig. 12.16, and this system uses vertically suspended filter bags that are designed for outside collection of ash. The bags are arranged in rows inside gastight compartments that permit cleaning of the bags while the unit is operational.

Fly ash–laden flue gas enters the hoppers through an inlet manifold and is distributed over the outer surfaces of the filter bags. Heavier fly ash particles that are entrained in the flue gas settle in the hoppers by gravity because of the very low gas velocity. Finer fly ash is separated from the flue gas as it passes from the outside to the inside of the bag, and the fly ash particulates are deposited on the outside of the bag in the form of a filter cake.

At prescribed intervals set by operating duration or usually by pressure drop, the bag compartments are cleaned. Compressed-air headers that are mounted above the bag rows provide compressed-air

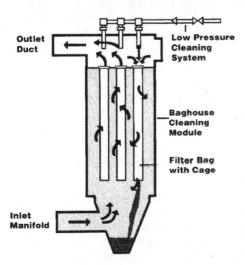

Figure 12.16 Pulse-jet-type baghouse system where filter bags are cleaned by low-pressure compressed-air pulses into the bags that causes the bags to expand and dislodge the dust filter cake. (*Babcock & Wilcox, a McDermott company.*)

pulses into the bags, which cause the bags to expand. As the bags expand, the dust filter cake is dislodged and falls into the hopper.

Because of their high collection efficiency, of their ability to collect fine particulates, and of their immunity of resistivity of fly ash, bag filterhouses have gained acceptance in many applications as an alternative design to electrostatic precipitators. Thus selection of the proper equipment must carefully consider all the fuels that are planned for use and the operating features and operating costs that are associated with each technology.

12.8   Equipment for Sulfur Dioxide Scrubber Systems

In the early 1970s, utilities were faced with the requirement of reducing sulfur dioxide emissions that resulted from fossil fuel combustion. As a result, utilities selected *flue gas desulfurization* (FGD) systems (sulfur dioxide scrubbers) to meet these requirements. However, many of these systems were treated as a separate and exclusive process and segregated from the rest of the power plant.

Power plant operators and plant designers soon realized that the FGD system was an integral part of the complete environmentally compatible power generation process. Today, as regulatory agencies become less tolerant of flue gas bypassing the cleaning system, the power plant becomes directly dependent on a reliable FGD system or some other form of SO_2 reduction, such as fluidized bed combustion, whether it is for a large utility system or an industrial-sized power plant.

Sulfur dioxide is a gas formed when the sulfur in the fuel is burned. All but a small percentage of the sulfur is converted to sulfur dioxide during the combustion process. Wet scrubbers using lime or limestone, dry scrubbers, coal gasification, fluidized bed combustion, and reducing the sulfur content of the fuel are proven methods of reducing the amount of this pollutant released to the atmosphere. For scrubber systems, over 80 percent of those in operation are wet scrubbers, with the remaining being dry scrubbers.

There are many designs for wet scrubbing systems, such as lime, dual-alkali, and double-loop processes, the last of which can produce a commercial by-product, gypsum. This book will discuss the single-loop wet limestone process and the double-loop process.

Another method of sulfur dioxide scrubbing, dry scrubber systems, also will be discussed as an alternative to wet scrubbing when the sulfur content of the coal permits its use. Dry scrubbing is also a popular choice on smaller industrial power plants and for combined HCl and SO_2 control on waste-to-energy plants.

12.8.1 Wet scrubbers

Wet scrubbing, by its name, means that a wet product is produced after the sulfur dioxide has been removed from the flue gas. Today, the wet limestone throwaway type of FGD system is a popular choice. The system is based on a closed-loop waste-disposal system, which means that solids are removed from the FGD system in the form of a sludge, while an effort is made to recover the water and use it again within the system.

One of the first items to consider in an FGD system is the layout of the plant (refer to Fig. 12.3). Layout is site-dependent and is constrained by such considerations as available space, future unit requirements, and waste-disposal methods. Note that the particulate-control equipment (precipitator or baghouse) is located immediately downstream of the air heaters, followed by the induced-draft fans. Ductwork from the fans runs past the stack to the sulfur dioxide scrubbing system. Flue gas then leaves the scrubber and flows to the stack.

A typical wet scrubbing system is shown in Fig. 12.17. Flue gas enters the absorber towers at the bottom and passes up through the tower through a series of sprays where the flue gas comes in contact with the lime or limestone slurry. Depending on the design, more than 90 percent of the sulfur dioxide can be removed, and the scrubbed flue gas continues upward through a mist eliminator to the stack.

The absorption-loop system consists of the tower, the hold or reaction tank and mixers, and the absorber pumps and piping. The tower design will vary by manufacturer, but it basically consists of three sections: quencher, absorption chamber, and mist eliminator.

The first section is the quencher, which is the area where the flue gas enters the tower and comes into contact with the first series of sprays. This section can be totally separate, as with two-loop systems, or integrated with the tower, as in spray-tower designs.

The second section is the absorption section, where incoming sulfur dioxide is removed by bringing the flue gas into contact with a lime or limestone slurry. Several types of absorption sections exist, such as the spray tower, with its open design, and other designs that incorporate internal components such as packed beds or trays

The third section is the mist eliminator, which is a fiberglass or plastic chevron-type design.

One of the major considerations in absorber design is materials of construction. Normal operation involves slurries with low pH, which create both corrosive and abrasive conditions for the absorbers. In addition, operation with zero discharge or operation in a closed loop causes a significant buildup of chlorides. Towers must be constructed

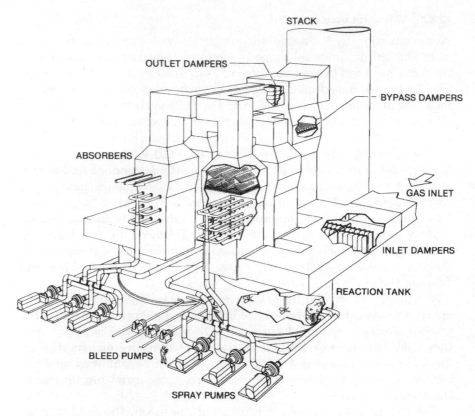

Figure 12.17 Sulfur dioxide (SO_2) wet scrubber system. (*ABB Environmental Systems.*)

of materials to withstand the hostile environment for long periods. Thus equipment made of stainless steel or rubber-lined carbon steel has been predominant.

The absorption loop is supported by a variety of subsystems that can be divided into two broad groupings, slurry and nonslurry. The major slurry subsystems are (1) reagent or limestone preparation, (2) waste slurry, (3) dewatering, and (4) solids disposal. The nonslurry subsystems include (1) water reclamation, (2) mist eliminator wash, (3) seal water, and (4) makeup or potable water. These systems vary in design and are not discussed in this book.

The sulfur dioxide removed in the absorber is converted to calcium sulfite ($CaSO_3$) and calcium sulfate ($CaSO_4$). These are the waste products that must be removed continuously and sent to disposal.

Another type of wet scrubber design is the double-loop wet limestone FGD system. This wet scrubbing method uses limestone to reduce SO_2 emission levels while producing quality gypsum, an environmentally compatible and reusable product used in commercial dry wall. The system is contained within a single tower, and it operates as follows:

1. *Lower loop.* Flue gas exiting from an electrostatic precipitator or fabric filter is quenched with a recirculating limestone slurry in the prescrubber and quencher or the lower loop. The reaction occurs at a pH of about 4.0. Conditions are optimum for oxidation and the formation of gypsum.

2. *Upper loop.* In the absorber or upper loop, flue gas contacts a recirculating limestone slurry at a pH of 5.8 to 6.0. Optimal SO_2 removal rates are ensured by efficient contact between slurry and flue gas in the tower as well as precise limestone residence time in the scrubbing loop system.

3. *Flue gas exit.* After passing through an additional spray zone and undergoing final treatment through a wet-film contactor, the cleaned flue gas leaves the absorber tower. Up to 98 percent of the SO_2 can be removed, depending on the initial sulfur content of the flue gas.

4. *Slurry dewatering.* The quencher slurry that contains approximately 15 percent solids passes through a dewatering system. The system uses hydroclones for primary dewatering and is followed by additional dewatering and gypsum processing as necessary. Calcium sulfite is the predominant end product of a SO_2 scrubber system and can be changed to calcium sulfate through forced oxidation to produce a salable wallboard-grade gypsum.

Wet scrubbing systems can use limestone, lime slurry, or soda ash liquor and are a combination of many subsystems ranging from reagent-preparation equipment to spent-slurry-disposal equipment. The actual SO_2 absorption takes place in the absorber tower.

Figure 12.18 shows another type of wet scrubber design. With this design, flue gas first flows into the absorber tower's quench section, where spray nozzles inject a lime slurry into the gas stream, resulting in the initial stage of SO_2 absorption. The walls of the quench section are washed continuously with slurry to eliminate any wet-dry zones so that deposition and scaling are prevented.

Intimate contact between slurry and flue gas is important for operating stability and for efficient SO_2 absorption. Therefore, the slurry and the flue gas must be equally distributed across the absorber tower's cross section.

The flue gas than passes through a perforated tray. Further SO_2 absorption takes place in this section because of the violent, frothing action that takes place on the tray.

The flue gas passing upward through the absorbing tower carries droplets with it. Therefore, droplet separation is essential to prevent

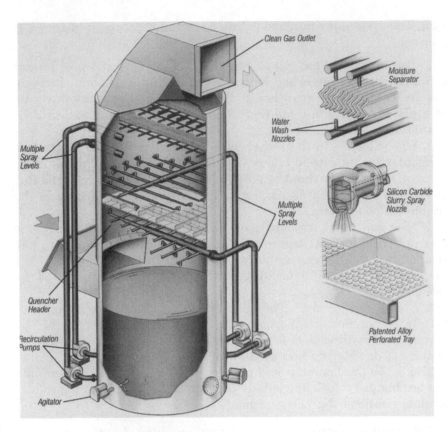

Figure 12.18 Wet flue gas desulfurization system with single-loop absorber spray tower. (*Babcock & Wilcox, a McDermott company.*)

slurry carry-over as well as corrosion and deposition in the flue downstream of the absorber. This separation occurs in chevron moisture separators. (Chevrons are closely spaced corrugated plates that collect slurry deposits.)

There are a variety of methods for sludge disposal. They include landfill, mine disposal, ocean disposal, and sludge utilization such as gypsum extraction, as described above. Selection of any of these methods is very site specific and economic dependent.

A common method for disposal is landfill. In this process, the dewatered sludge is mixed with fly ash. The lime in the fly ash creates a pozzolanic (cement-like) reaction in the mixture. If lime in the fly ash is insufficient, quicklime can be added.

With any method of disposal, the plant operator must face an entirely new set of problems with the handling of the material that results from collection of sulfur dioxide from the flue gas.

The use of scrubbers, whether wet or dry, for the removal of SO_2 is a costly addition to a power plant and results in an increase to the overall cost of electricity. For wet scrubbers, there are advantages and disadvantages to the technology. Some of these are as follows:

Advantages

A developed and proven technology.

Can handle flue gas from the burning of coals with high sulfur content.

Reagents (e.g., limestone) are inexpensive and readily available.

Scrubbing process is relatively simple.

High removal efficiencies, over 90 percent.

Disadvantages

Produces a large amount of waste sludge.

Sludges can be difficult to pump and handle.

Sludges are difficult to dewater.

Significant landfill area is required for disposal.

Reheating of flue gas to stack may be necessary for proper flue gas discharge.

Large quantity of water is required.

12.8.2 Dry scrubbers

Spray drying is a flue gas scrubbing technique that has proven operational acceptance. The rapid development of dry scrubbing for removing sulfur dioxide from the flue gas of coal-fired boilers has taken place because of its basic simplicity when compared with the wet scrubbers. Dry scrubbing means that a dry product, similar to ordinary fly ash, is produced and that the flue gas is always maintained above the moisture dew point. For waste fuels such as municipal solid waste, dry scrubbers are used to remove other acid gases from the flue gas in addition to sulfur dioxide. These include hydrogen chloride (HCl).

In the dry scrubbing process, the heat in the flue gas at the air heater outlet is used to dry and condition a finely atomized slurry of alkaline reactants. As the drying occurs, the majority of the sulfur dioxide is removed from the flue gases. The reactant material, along with the normal fly ash in the flue gas, is then collected in a conventional precipita-

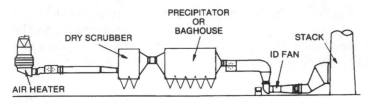

Figure 12.19 General arrangement of environmental control equipment with a dry scrubber system.

tor or baghouse. Handling of the collected material is then accomplished with conventional ash-handling equipment. Figure 12.19 shows the general arrangement of plant equipment in the dry scrubbing process. This arrangement requires the scrubber to be located prior to the precipitator or baghouse as compared with the wet scrubber, which is located after the particulate collector, as shown in Fig. 12.3.

Dry scrubbers offer the following advantages over wet scrubbers: (1) They are less complex than wet scrubbers and have a lower initial cost. (2) Their simplified design offers the prospect of greater reliability at lower maintenance costs than wet scrubbers. (3) Dry systems use less water than wet scrubbers, which is an important consideration where water is limited, such as the western part of the United States. (4) The amount of energy required to operate dry systems is less than that required for wet systems. (5) The resulting dry waste product is more easily disposed of than wet sludge.

Nevertheless, dry scrubbers are generally limited to low-sulfur coals because of high reagent costs, which is their major disadvantage. However, design improvements and operating experience have proved that dry scrubbing can be effective even with medium-sulfur coals. Lime is generally used as the reagent with dry scrubbers, and it is considerably more costly than the limestone used with wet scrubbers. Thus, even though the capital cost may be lower for dry systems, economics must still be determined from the operating costs of the two types of scrubbers by comparing the costs of using lime versus limestone. For applicable utility and industrial plants, the relative simplicity and lower capital costs of dry scrubbers have made them attractive alternatives to wet scrubbers.

There are basically two types of designs used in the dry scrubber systems: (1) rotary atomizers and (2) dual-fluid nozzles utilizing nozzles with fluid atomization, similar to that used in the combustion of liquid fuels.

Rotary atomizers.   The design of a spray-dryer absorption system with a rotary atomizer is shown schematically in Fig. 12.20. The

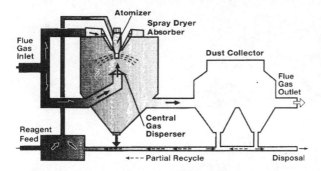

Figure 12.20 Flue gas desulfurization with spray-dryer absorption system. (*Babcock & Wilcox, a McDermott company.*)

reagent is generally a lime slurry that is introduced into the spray dryer by a rotary atomizer. The fine droplets (by atomization) of lime slurry mix and react with hot flue gas and a dry powder is formed. During this process the sulfur dioxide (or other acid gas) contained in the flue gas is captured. The dry powder falls to the bottom of the spray dryer for removal or is entrained with the flue gas, where it is collected in the dust-collection system, either a baghouse or an electrostatic precipitator.

When flue gas conditions dictate, and where benefits of reduced lime usage can be economically justified, some of the powder, containing unused lime that has been collected, is recirculated into the reagent mixer to reduce lime consumption. The remaining dry powder is conveyed to the ash-disposal system.

The rotary atomizer is centrally positioned in the spray dryer and is driven by a motor as large as 1300 hp depending on the size of the system. In order to reduce slurry wear, since the atomizer rotates at speeds up to 12,000 rpm, it has an abrasion-resistant stainless steel wheel with removable silicon carbide inserts.

The spray-dryer absorber module incorporating a rotary atomizer (Fig. 12.21) ensures both gas and reagent flow-pattern symmetry and provides a flue gas disperser system to adequately handle and mix the large gas volumes involved. The rotary atomizer is on the vertical centerline of the spray-dryer absorber, and the flue gas is spirally introduced completely around the atomizer from both above and below.

A schematic of the spray-dryer system is shown in Fig. 12.22. The rotary atomizer is shown in Fig. 12.23, and its use with a flue gas disperser within the spray-dryer assembly is shown in Fig. 12.24.

Since the wet-dry interface thus occurs in space and not on any surface, scaling and plugging are not problems. Sulfur dioxide monitors control the amount of reactant to the feed slurry.

Figure 12.21 Sulfur dioxide (SO$_2$) dry scrubber system, rotary atomizer design. (*Babcock & Wilcox, a McDermott company.*)

The atomizer is powered by a vertically mounted motor, as shown in Fig. 12.23. By varying the design of the gear drive, atomizer spindle speed can be varied to suit the particular operating conditions.

The liquid slurry is fed onto a rotating wheel, where it is accelerated in the wheel under centrifugal force and is atomized at the wheel's multiple nozzles to form a spray of droplets. The spray leaves the nozzles of the wheel in a horizontal direction with a spray angle of 180°. Since flue gas enters the spray-dryer absorber at a relatively low temperature, approximately 350°F, a high volume of flue gas must be thoroughly treated with a relatively small amount of reactant slurry to ensure complete drying. This means that as the plant size increases, the dimensions of the spray-dryer absorber increase faster than the dimensions of the atomizer itself, and gas contact with the cloud of reagent droplets occurs farther away from the atomizer wheel. As a result, the inlet flue gas is divided into two streams to

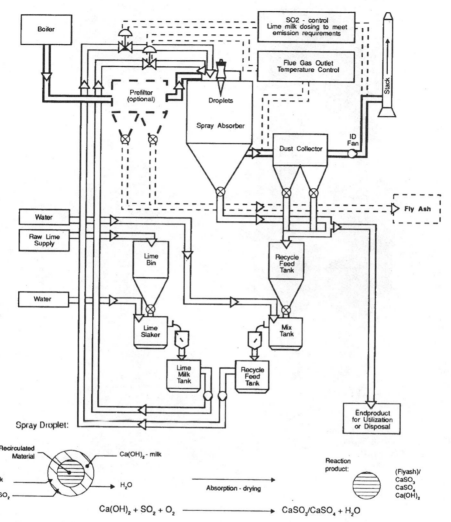

Figure 12.22 Schematic of spray-dryer absorber. (*Babcock & Wilcox, a McDermott company.*)

balance and shape the cloud of droplets. This helps prevent deposits on the spray-dryer absorber walls or wet-bottom operation. For smaller industrial applications or utility plants, the flue gas is generally introduced in one stream. The equipment designer determines the configuration based on research and operating experience.

As the flue gas comes in contact with the cloud of fine lime slurry droplets, the flue gas dries the droplets as they react with the sulfur dioxide in the flue gas to form dry calcium sulfate and calcium sulfite powders.

1. Upper part
2. Intermediate part
3. Lower part
4. Center Cone
5. Insert
6. Bottom plate
7. Top plate

Figure 12.23 Rotary atomizer and abrasion-resistant atomizer wheel design. 1 = upper part. 2 = intermediate part. 3 = lower part. 4 = center cone. 5 = insert. 6 = bottom plate. 7 = top plate. (*Babcock & Wilcox, a McDermott company.*)

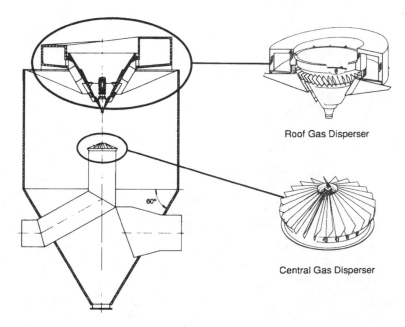

Roof Gas Disperser

Central Gas Disperser

Figure 12.24 Spray-dryer assembly module with gas disperser and rotary atomizer. (*Babcock & Wilcox, a McDermott company.*)

The flue gas exits from the spray-dryer absorbers carrying some of the dry-reacted chemicals to a baghouse or precipitator, as shown in Fig. 12.19. Additional reaction between the dry chemicals and flue gas takes place during removal of the particulates. From the precipitator or the baghouse, the cleaned flue gas flows to the induced-draft fan and then to the stack.

The waste-product material and fly ash received from the spray-dryer absorbers and the baghouse or precipitator hoppers have the appearance of being totally dry. The material is pneumatically or mechanically transported to the ash-disposal bins for disposal in a landfill. Based on economics, a portion of the collected material may be recirculated to mix with the reagent feed to provide improved lime slurry utilization.

Nozzle atomizers. Dual-fluid nozzle designs are also effective as dry scrubbers. Both horizontal and vertical flow designs are utilized and are based on the manufacturer's design and operating experience. The horizontal-flow reactor (Fig. 12.25) utilizes a number of systematically arranged nozzles where a fluid such as air provides the energy for atomization, similar to that found in liquid-fuel burner nozzles.

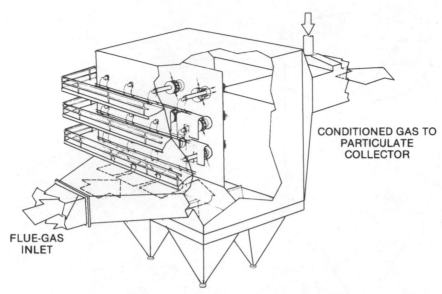

CONDITIONED GAS TO PARTICULATE COLLECTOR

FLUE-GAS INLET

Figure 12.25 Sulfur dioxide (SO₂) dry scrubber system, horizontal-flow reactor utilizing pneumatic nozzle atomization. (*Babcock & Wilcox, a McDermott company.*)

With this system, flue gas from the boiler enters a reaction chamber directly ahead of the precipitator or baghouse. At the reactor inlet a spray of finely atomized lime slurry is mixed uniformly with the hot flue gas. The sulfur dioxide in the gas reacts with the lime in the droplets and forms calcium sulfite and calcium sulfate. Flue gas heat dries the reaction products and any unreacted lime into a powder.

The system controls water input so that the gas temperature remains above the dew point as it passes into the precipitator or baghouse. The waste product produced from this type of design is handled in a similar manner to that described above for the rotary atomizer design.

Another design of a dual-fluid spray dryer is shown schematically in Fig. 12.26. This is a vertical flow design and is used on both industrial and utility boilers as well as boilers that burn municipal solid waste (MSW). In this system, an alkaline slurry is atomized vertically downward into the flue gas in the spray-dryer absorber (scrubber). The finely atomized droplets absorb SO_2 and other acid gases (such as HCL on MSW units), while the heat of the flue gas evaporates the droplets. The solid particulates and the dried reaction products that are contained in the flue gas are then collected in a fabric filter or an electrostatic precipitator. Some products do fall into the hopper below the spray dryer.

Flue Gas

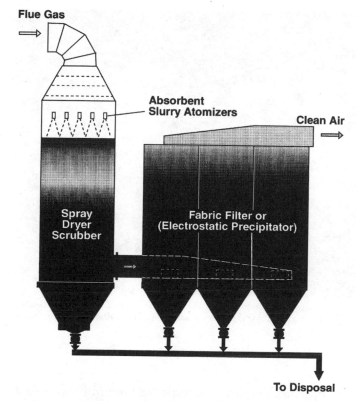

Absorbent Slurry Atomizers

Clean Air

Spray Dryer Scrubber

Fabric Filter or (Electrostatic Precipitator)

To Disposal

Figure 12.26 Spray-dryer scrubber, vertical flow of flue gas in absorber with slurry atomizers in the upper region of scrubber. (*Wheelabrator Clean Air System, Inc.*)

Figure 12.27 is an illustration of three spray dryers and fabric filters that control emissions from three boilers that burn 2250 tons per day of municipal solid waste.

12.9 Systems for Control of Nitrogen Oxides (NO_x)

Nitrogen oxide (NO_x) is one of the primary pollutants released during the combustion process. The primary sources of NO_x are from internal combustion engines and from the burning of fossil fuels. In the United States, as well as in many other countries, NO_x generated by transportation (cars, trucks, etc.) contributes about the same percentage as that contributed by the burning of fossil fuels, each about 45 percent of the total. Together with the other major emissions, SO_2 and particulates, NO_x emissions have been identified as contributors to

Figure 12.27 Vertical spray-dryer scrubbers and fabric filters for control of emissions from three MSW-fired boilers at a 2250 tons/day waste-to-energy plant. (*Wheelabrator Clean Air System, Inc.*)

acid rain, visibility degradation, and concerns for human health. As a result, NO_x emissions from most combustion sources are regulated and require control of some type.

NO_x emissions from a combustion process are approximately 90 percent NO, with the remainder being NO_2. However, after the flue gas leaves the stack, most of the NO is oxidized in the atmosphere to NO_2. It is the NO_2 in the flue gas that creates the brownish plume often seen in a power plant's discharge from the stack. When the NO_2 is in the air, it becomes involved in a series of reactions that form secondary pollutants. The NO_2 can react with sunlight to produce smog and acid rain constituents.

Nitrogen oxides that exist in flue gases are formed by two phenomena: (1) the thermal dissociation and recombination of nitrogen and oxygen from the combustion of air into nitrogen oxides (thermal NO_x) and (2) the release of nitrogen atoms from the fuel during the combustion process and subsequent re-formation with oxygen atoms to form nitrogen oxides (fuel NO_x).

The staging of the combustion process results in initial combustion with a deficiency of oxygen. The intent is to force the freed nitrogen atoms to recombine with themselves to form molecular nitrogen, N_2.

Low excess air combustion and burners that produce staged flame patterns have proved effective in reducing NO_x levels relative to uncontrolled levels. While these measures are generally adequate for meeting the current emission standard, it is doubtful that more stringent limits can be met. For this reason, research in burner technology continues with the established goal of designing effective NO_x control methods.

Recently, the use of thermal denitrogen oxide (de-NO_x) systems that inject ammonia into the furnace of a boiler have proven to be effective in the control of NO_x production.

Low NO_x emissions also result with the use of fluidized bed combustion technology, where there are low combustion temperatures.

With emission regulations becoming more stringent, more emphasis is placed on NO_x emission reduction, and NO_x control systems have been developed for this purpose. One such system employs the selective catalytic reduction (SCR) technology, which is used when burning natural gas, oil, coal, or waste fuels. When combined with low-NO_x burners, the SCR method has resulted in an effective means in reducing NO_x emissions.

The major component of the system is the catalyst that changes the NO_x-containing flue gas properties. This system is a dry process in which ammonia (NH_3) is used as a reducing agent, and the NO_x contained in the flue gas is decomposed into harmless N_2 and H_2O. Ammonia is injected and mixed with the flue gas prior to entering the SCR reactor. The flue gas then passes to the reactor and through the catalyst bed, with the NH_3 and NO_x reacting to form N_2 and H_2O.

Questions and Problems

12.1 In addition to the production of steam or the generation of electricity, what operating restrictions must the plant operator be concerned with meeting?

12.2 If environmental restrictions are not met, what are the potential consequences of such action?

12.3 List the emissions that must be controlled from a coal-fired power plant.

12.4 As more particulate and SO_2 emissions are collected from power plants, what additional disposal problem does this create? Does this solution present an environmental situation that also must be dealt with? If so, what problems must be considered?

12.5 What methods are available for the control of SO_2?

12.6   How can NO_x emissions be controlled?

12.7   What are the primary means for the control of particulates? Describe the operation of each, and identify the advantages and disadvantages of each.

12.8   For coal-fired boilers with heat inputs over 250×10^6 Btu/h, what are the sulfur dioxide removal requirements?

12.9   What do the solid particles in the flue gas consist of? Why does the nature of these particles vary? What type of firing would result in particles having a small micron size?

12.10   What are the permissible particulate emissions for utility boilers and industrial boilers? What are the heat-input limitations for these restrictions?

12.11   For a pulverized-coal-fired boiler, approximately what percentage of the ash in the coal is carried through the boiler as part of the flue gas? What is the size of the particulates?

12.12   Explain the operation of an electrostatic precipitator.

12.13   What is ash resistivity? How does it vary for high- and low-sulfur coals, and why is this important in the sizing of an electrostatic precipitator?

12.14   Is resistivity an important criterion with a baghouse? Why?

12.15   How does the draft loss through an electrostatic precipitator compare with that of a mechanical collector and a bag filterhouse? Why is this of concern to an operator?

12.16   Explain the differences between a weighted-wire precipitator and a rigid-frame precipitator. What design appears to be more reliable? Why?

12.17   What system ensures cleanliness of the precipitator? Why is this system important?

12.18   Describe the various types of bag filterhouses.

12.19   What are the advantages of baghouses over precipitators? What are the disadvantages?

12.20   How are bags cleaned with a reverse-air type of fabric filter? With a pulse-jet filter? What device prevents the bags from collapsing?

12.21 What portion of the bag filterhouse has the highest potential for maintenance? Why?

12.22 What is sulfur dioxide, and how is it formed? Name two types of scrubber systems used for SO_2 removal. What other techniques can be used for reducing SO_2 emissions?

12.23 Describe the operation of a wet scrubber system. Identify its location relative to the arrangement of the particulate collector of a power plant. Why is its location different from a dry scrubber?

12.24 What reagents are commonly used to absorb sulfur dioxide in a wet scrubber system? Which is the most commonly used? Why?

12.25 What are the three major sections of a wet scrubber spray tower? Describe their functions.

12.26 What materials of construction are commonly used in wet scrubbers? Why is material selection important?

12.27 Briefly describe the waste-disposal methods of wet scrubbers. Which is the most common?

12.28 Provide the advantages and disadvantages of wet scrubbing. What would be the major detriment to a wet scrubber system for a plant located in an arid or semiarid region?

12.29 What is a dry scrubber? Describe its operation. Identify the location of a dry scrubber system relative to the arrangement of the particulate collector of a power plant. Why is its location different from a wet scrubber?

12.30 What are the advantages and disadvantages of a dry scrubber system?

12.31 Briefly describe the operating characteristics of the two types of dry scrubbing systems.

12.32 What reagent is commonly used with dry scrubbers? Why do some systems recirculate a portion of the collected dry powder from the particulate collector such as a baghouse?

12.33 Explain the formation of NO_x from a combustion process.

12.34 Describe an SCR system for the reduction of NO_x emissions. What is the major operating cost for this system?

13

Waste-to-Energy Plants

13.1 Introduction

Waste-to-energy facilities are part of the solution of the worldwide solid waste disposal problem. These facilities, when combined with recycling of critical materials, composting, and landfilling, will be a long-term economic solution as long as they are designed and operated in an environmentally acceptable manner. All these options have problems, since they involve considerations such as politics, siting, public acceptance, and financing. However, these problems must be overcome because vast quantities of solid wastes are produced each day, and this waste has to be disposed of in some manner.

Resource-recovery facilities (waste-to-energy facilities) are not a new idea; in the 1890s the city of Hamburg, Germany, incinerated municipal refuse and used the resulting energy to produce steam and electricity. This technology was created in Europe because of the lack of available land for landfilling, and today Europe is a world leader in waste-to-energy facilities. Many portions of the United States now face this same problem. In 1903, the first solid-waste-fired plant that produced electricity in the United States was installed in New York City.

Refuse incinerators, as they were originally called, were basically refractory-lined furnaces that were designed to reduce waste. However, most of these incinerators did not recover the heat energy resulting from the combustion. A few installations did use separate waste-heat boilers for steam production. However, it was not until the early 1950s that waterwall incinerators (boilers) started to be designed and built.

In the 1970s, nearly all the municipal solid waste (MSW) generated in the United States was disposed of in landfills. Some areas used incinerators for waste reduction, but these plants did not recover the generated heat that resulted from combustion. Incinerators had little, if any, pollution-control equipment, and they became unacceptable because of the air pollution requirements that were imposed and also because of the availability of inexpensive landfills.

In the United States, MSW is high in paper and plastic and typically has less moisture content and therefore a higher heating value than the MSW found elsewhere in the world. The design of MSW-fired boilers is strongly dependent on the contents of the refuse, and this content changes over time and with the seasons of the year.

With the use of more convenience foods, plastics, and packaging and with less food refuse due to residential garbage disposals, the moisture content of MSW has decreased, with an associated increase in heating value. As more recycling programs are implemented, the change in MSW content will continue.

As old landfills have closed, new landfills have become more difficult and costly to open, and they tend to be located further from the source of the refuse. This not only has increased transportation costs, because of the increased distance and additional requirement for more trucks, but also has added to the air pollution and traffic problems in the area. Concerns about groundwater contamination have resulted in more expensive landfill designs that require several containment layers and added systems for leachate and methane gas monitoring and control.

In the United States there has been a very slow growth and acceptance of waste-to-energy facilities. Certainly the availability of inexpensive land for landfilling is a primary reason. In addition, the availability of other fuels such as coal, oil, and natural gas and the relative low cost of electricity have been factors. The rapid filling of the present landfills, the problems with contamination from existing landfills, and the difficulties in siting new landfills, such as public opposition, have all led to evaluating other means of handling and reducing solid wastes. Waste-to-energy facilities play an important role in the solution of these problems, although it is still a relatively minor role, as we will see. This role should be expanded in the coming years as it becomes an accepted means of solid waste management and as a complementary process with recycling.

Waste-to-energy facilities have demonstrated their ability to reduce dependence on landfills and their reliability as a renewable energy source. The large number of successfully operating plants has proven that they are a dependable part of the waste-management solution. The major factors that have limited the rate of growth of waste-to-energy facilities include their cost relative to other options, such as landfilling, public perception of environmental risks, and an uncertain support from regulatory agencies. These factors will be overcome because of increasing unavailability of landfills, increasing successful operation of existing waste-to-energy facilities, and an increasing public understanding of waste disposal problems and of the available solutions.

In the mid-1990s, there were 162 municipal waste combustion (MWC) plants in the United States, including all waste-to-energy facilities, refuse-derived fuel (RDF) processing plants, and incinerator operations (Table 13.1). Of this total, 148 are combusting MSW, and these facilities handle over 31 million tons of MSW annually, which is approximately 15 percent of the 207 million tons of MSW generated annually in the United States. When incineration is not considered, 121 waste-to-energy plants handle over 30 million tons annually.

By the year 2000, the United States is expected to generate more than 218 million tons of MSW annually. Estimates project that one-third of this total will be recycled or composted, which will leave 150 million tons per year to be managed either by landfilling or by waste-to-energy plants. In the mid-1990s, just over 30 million tons per year were handled by waste-to-energy plants, which leaves a significant potential for this technology.

TABLE 13.1   Municipal Waste Combustion (MWC) Plants in the United States

Technology	No. of operating plants	Daily design capacity, tons/day	Annual capacity, million tons*	Percentage of total
Incineration	27	4,346	1.3	3.9
Modular	26	3,221	1.0	3.0
Mass burn	68	70,278	21.8	66.3
RDF	15	20,570	6.4	19.5
RDF, processing	14	4,713	1.5	4.6
RDF, combustion	12	2,908	0.9	2.7
Total plants	162	106,036	32.0	100.0
MWC facilities†	148	101,323	31.4	
WTE facilities‡	121	96,977	30.1	

*Assumes operation at 85 percent design capacity.
†Excludes RDF processing plants.
‡Excludes RDF processing and incinerators.
SOURCE: Integrated Waste Services Association.

The political situation, however, could divert this large potential to landfills, which will delay the waste problem for future generations to solve. Waste-to-energy plants effectively reduce MSW by using safe, environmentally friendly, and economical technologies. Waste-to-energy plants reduce the MSW volume by approximately 90 percent, which significantly reduces landfill requirements.

The 121 waste-to-energy plants currently in operation generate enough electricity for over 1 million homes, and they serve the disposal needs of over 37 million people. This energy production replaces the need for over 30 million barrels of oil each year and reduces the demand for valuable landfill space.

The solution to this solid-waste-disposal problem is not to simply add more waste-to-energy plants. The management of MSW involves a variety of options that will minimize the effect on the environment while maximizing the conservation and recovery of energy and materials. Each option also should minimize costs and protect the taxpayer. The solution to this problem must be able to integrate the four options available for the management of MSW:

1. Source reduction

2. Recycling, including composting

3. Conversion to energy

4. Landfilling

The potential for increasing the production of energy from waste is high. However, the decision makers often let politics stand in the way, and current thinking is that recycling is the answer, and, as a result, new waste-to-energy plants are delayed. But it is not only one option that solves this problem. The best solution is one that integrates a combination of recycling, waste-to-energy plants, and landfilling into a satisfactory economic and environmental solution.

This chapter will discuss some of the various options available as a solution to this solid waste disposal problem, and steam power plants play a very important role in this process.

13.2   Waste Disposal—A Worldwide Problem

In accordance with figures from the National Renewable Energy Laboratory, in the mid-1990s, the annual quantity of MSW in the United States that required some form of disposal was approximately 207 million tons. The means of disposal is divided approximately as follows:

	Million tons per year	Percentage
Landfill	129	62.4
Combusted*	33	15.9
Recycled (including composting)	45	21.7
TOTAL	207	100.0%

*With and without energy recovery.

Table 13.2 shows the composition of the preceding MSW and the percentage recovered by recycling and composting. Note how consistent this MSW composition analysis is as compared with that in Table 13.5, which was based on a Franklin Associates study. MSW can change significantly at various locations in the world and in individual countries, as well as at different times of the year.

By analyzing the preceding methods of disposal and Table 13.2, it is obvious that a significant amount of combustible material is disposed of in landfills. Not only do landfills reach their capacity, but this pre-

TABLE 13.2   Typical MSW Composition and Percentage Recovered by Recycling and Composting per Year

Material	MSW quantity, millions of tons	Percent of total MSW	Percent recovered
Paper and paperboard	77.8	37.6	34.0
Glass	13.7	6.7	22.0
Metals			
Ferrous	12.9	6.2	26.1
Aluminum	3.0	1.4	35.4
Other nonferrous	1.2	0.6	62.9
Plastics	19.3	9.3	3.5
Rubber and leather	6.2	3.0	5.9
Textiles	6.1	2.9	11.7
Wood	13.7	6.6	9.6
Other materials	3.3	1.6	22.1
Other wastes			
Food wastes	13.8	6.7	0
Yard wastes	32.8	15.9	19.8
Miscellaneous inorganic wastes	3.2	1.5	0
Total	207.0	100.0%	21.7%

SOURCE: National Renewable Energy Laboratory.

vents the use of this energy source in waste-to-energy plants, whose use can conserve valuable fossil fuel assets.

In 1960, approximately 31 percent of the MSW was combusted with no energy recovery and without air pollution controls. In 1970, only 20 percent of the MSW was combusted, but less than 1 percent was converted to energy recovery.

By 1990, approximately 17 percent of the MSW was combusted, with 16 percent of it for energy recovery, and in 1993, nearly all combustion was for energy recovery, and all the plants required air pollution controls.

By the year 2000 it is expected that only about 19 percent of the total MSW will be handled by waste-to-energy plants. Another factor to consider is that since 1960 the waste that each person produces has increased dramatically, which naturally has increased the annual waste disposal requirements. The recovery of materials from MSW for recycling and composting was less than 10 percent until about 1985. This increased to about 22 percent by 1993.

The increase in recycling over the years is attributable to several factors, including people's belief that it is the right thing to do and, perhaps more important, the mandate from states and local communities to specific rates of recycling. Unfortunately, this latter political decision is based on the belief that the solid waste disposal problem can be solved only by recycling, and, therefore, new waste-to-energy plants, as part of an integrated plan, have been delayed indefinitely in most areas of the United States.

As a result, there has been a significant increase in facilities to receive and process recyclables. In addition to an increase in these material recycling facilities (MRFs), there also has been a significant increase in buy-back centers, dropoff centers, curbside recycling, and other methods for the public to participate. However, the key to success of any recycling program is to have the necessary markets for the recovered materials. Unfortunately, when market prices are depressed, these recovered materials have found their way into the landfill, thus compounding the problem of the landfill being filled and of the costs of recovery being absorbed, with no benefit of the actual recovery of materials.

The combustion of MSW has developed into a technology that has demonstrated reliable and environmentally sound operation. It forms an integral part of waste management for many industrialized nations.

The United States, however, disposes of most of its waste in landfills and has one of the lowest uses of waste-to-energy plants for disposal as compared with many industrialized nations. Refer to Table 13.3. A major reason for this is political, since the impression that the

TABLE 13.3   Comparison of MSW Management in Various Countries of the World

Country	Percentage of MSW disposal		
	Landfill	WTE	Recycling*
United States	62	16	22
Germany	46	36	18
Switzerland	12	59	29
France	45	42	13
Sweden	34	47	19
Canada	74	5	21
Japan	23	72	5
United Kingdom	90	7	3
Average	51	33	16

*Includes composting.
SOURCE: Energy Technology Support Unit for the United Kingdom Department of Trade and Industry.

general public has is that a waste-to-energy plant is damaging to the environment. However, as will be shown later, each waste disposal option has its costs and its risks, and landfilling has significant potential problems for both air and ground pollution.

13.3   Potential Energy from MSW

The potential energy from *municipal solid waste* (MSW) is enormous. In the Untied States alone, each person produces approximately 5 pounds of refuse each day. This equates to over 200 million tons of refuse each year, which requires some means of disposal such as recycling, composting, combustion, or landfilling. Refuse has a heating value of approximately 4500 Btu/lb, and thus there exists a potential energy supply of 2×10^{15} Btu/year. It is estimated that over 50 percent of the waste generated is contributed from residential sources, the remaining 50 percent is from commercial and industrial sources. The amount of waste generated is also expected to increase by 2 percent per year even with expanded recycling efforts.

To put this potential energy into proper perspective, this amount of energy from refuse represents approximately 3 percent of the total energy consumption in the United States and 10 to 15 percent of the oil imports. If only half this potential energy were recovered, it still would represent a significant energy source, which not only would reduce the dependence on other fuels, such as imported oil, but also would significantly reduce the solid waste disposal problem. However, currently, only approximately 16 percent of the solid waste is handled in waste-to-energy facilities. Therefore, a tremendous opportunity for expansion presents itself because these facilities are a necessary part of an integrated waste-management solution that includes recycling, composting, and landfilling.

13.4   Landfills

Today the strongest competition to a waste-to-energy facility is the landfill. The landfills designed today are often called *sanitary land-fills* because of the care taken in using multiple liners that prevent leaching of contaminated liquids to surrounding soil and groundwater. Such designs are very different from those landfills designed without synthetic liners and where leaching has resulted in the contamination of groundwater. The first sanitary landfills were unlined pits covered with dirt. The disposal fee often covered only the cost of putting a daily layer of dirt over the trash.

Landfills originally represented a low-cost solution to the solid waste problem because they often occupied land that was considered worthless, such as swamps, estuaries, and even open water. Now state, local, and federal laws have imposed strict environmental pollution laws because of surface and groundwater pollution that resulted from improperly located and poorly designed and operated landfills. The sanitary landfill design of today requires important practices and systems that include leachate collection and recovery systems, extensive methane gas monitoring and collection systems, groundwater monitoring systems, and, finally, elaborate capping, closure, and postclosure requirements.

With or without waste-to-energy plants, landfills are a necessity. The problem is the rapid filling of existing landfills and the difficulty in siting new ones. Thus disposal problems will become overwhelming if critical decisions are not made soon.

The need for properly designed landfills can be illustrated by an analysis of residential waste. Modern household wastes deposited in sanitary landfills contain many of the same hazardous chemicals that require special treatment at hazardous waste facilities. Toxic pollutants in household garbage come from chemicals in cleaning products, paints, oils, insecticides, and solvents in addition to other products. The primary problem with these wastes is their high potential for mobility from the disposal site to the surrounding soil and groundwater. Therefore, synthetic lining systems, rather than conventional soil liners, are required for solid waste containment. In addition, methane collection systems are incorporated into the landfill to protect the air.

If a waste-to-energy plant is not part of the waste disposal solution, the majority of waste would eventually be landfilled. In landfills, a putrescible waste degrades and produces methane gas. In addition to methane, raw waste in a landfill emits volatile organic compounds (VOC), which are a contributor to the formation of smog. Therefore, the combustion of MSW in modern plants helps to mitigate the emissions of gases that contribute to the greenhouse effect as well as being air pollutants.

Thus, not only is the availability of properly designed landfills a problem, but also the costs of landfills will increase dramatically. Waste-to-energy plants not only extend the life of landfills but also produce energy that conserves other valuable fuel resources. Waste-to-energy plants can reduce the solid waste volume by over 90 percent, and the resulting ash from combustion is a much better material to landfill than raw MSW.

13.5 Solid Waste Composition

The composition of municipal solid waste obviously depends on what is thrown away. Although this varies between different sections of the country and different countries in the world and is a function of such variables as seasonal climate and the degree of recycling, generally MSW has a composition that is very similar to that of wood. Table 13.4 shows a comparison of municipal refuse with two different types of wood on the basis of their ultimate analyses[1] and their ash- and

TABLE 13.4 Comparison of Ultimate Analysis and Heating Value of Wood and Municipal Refuse

	Pine bark, %	Hardwood bark, %	Municipal refuse, %
Ultimate analysis			
Ash	1.5	5.3	14.4
Sulfur	0.1	0.1	0.2
Hydrogen	5.5	5.4	5.7
Carbon	55.3	49.7	42.5
Oxygen and nitrogen	37.6	39.5	37.3
Total	100.0	100.0	100.0
HHV on ash- and moisture-free basis, Btu/lb	9300	8830	8600

SOURCE: Courtesy of Babcock & Wilcox, a McDermott company.

---

[1]The ultimate analysis is the chemical analysis of a fuel determining carbon, hydrogen, sulfur, nitrogen, chlorine, oxygen, and ash as percentages of the total weight of the sample.

TABLE 13.5 Solid Waste Composition

Material	Percent by weight
Paper and paperboard	40.1
Textiles	2.1
Plastics	8.0
Rubber and leather	2.5
Wood	3.6
Glass	7.0
Iron	6.5
Aluminum	1.4
Nonferrous	0.6
Yard waste	17.6
Food waste	7.4
Other	3.2
Total	100.0

SOURCE: Courtesy of Westinghouse Electric Corp.

moisture-free heating values. The heating values of both woods and of MSW are quite similar and vary only between 8600 and 9300 Btu/lb.

Based on a Franklin Associates study of the waste stream in the United States, the waste generally can be characterized as shown in Table 13.5. Note the high percentage of paper, plastics, wood, and other combustibles that are the primary contributors to the heating value. As noted previously, Table 13.2 shows another study of solid waste composition, and the results are very similar for such a heterogeneous material.

Table 13.6 shows the composition of waste for a waste-to-energy plant in West Palm Beach, Florida. As stated previously, waste varies by location, and when Tables 13.2, 13.5, and 13.6 are compared, several important similarities and differences should be noted. The percentages of paper and plastics are very comparable. The percentage of yard wastes, however, are significantly different. The reason for this may be a source-separation program used by the local community for possible composting. The ultimate analysis for the MSW in Table 13.6 shows a high moisture content (25.3 percent) and a relatively high ash content (23.65 percent). The actual heating value for this MSW is 4728 Btu/lb.

Knowledge of the MSW's composition is of critical importance to the operating performance of a waste-to-energy facility. Various

TABLE 13.6   Waste Composition and Refuse-Derived Fuel Analysis in Palm Beach County, Florida

	MSW	Refuse-derived fuel
	Component analysis, % by weight	
Corrugated board	5.46	—
Newspapers	17.16	—
Magazines	3.44	—
Other paper	19.46	—
Plastics	7.24	—
Rubber and leather	1.94	—
Wood	0.83	—
Textiles	3.07	—
Yard waste	1.11	—
Food waste	3.71	—
Mixed combustibles	17.52	—
Ferrous	5.43	—
Aluminum	1.80	—
Other nonferrous	0.32	—
Glass	11.51	—
Total	100.00	
	Ultimate analysis, %	
Carbon	26.65	31.00
Hydrogen	3.61	4.17
Sulfur	0.17 (max. 0.3)	0.19 (max. 0.36)
Nitrogen	0.46	0.49
Oxygen	19.61	22.72
Chlorine	0.55 (max. 1.0)	0.66 (max. 1.2)
Water	25.30	27.14
Ash	23.65	13.63
Total	100.00	100.00
	Heating value, Btu/lb	
	4728	5500

Fuel value recovery, percent of MSW   96
Mass yield, percent (lb RDF/lb MSW)   83
SOURCE: Courtesy of Babcock & Wilcox, a McDermott company.

constituents of the waste stream have different heating values and also cause varying degrees of problems in the operating plant.

The composition of MSW can vary greatly because the definition of solid waste can vary significantly. In Florida, for example, *solid waste* means garbage, refuse, yard trash, clean debris, white goods (dishwashers, clothes dryers, etc.), ashes, and sludge or other discarded material (including solid, liquid, semisolid, or contained gaseous material resulting from domestic, industrial, commercial, mining, agricultural, or governmental operations). With such a variety of solid waste, a number of solid waste solutions must be utilized. The refuse coming to a waste-to-energy facility obviously can vary significantly, based on the waste-management plan used in the region.

Perhaps the two most important properties of waste that most directly affect the operating performance of a waste-to-energy facility are moisture content and heating value. These properties are interrelated in that the higher the moisture content, the lower is the heating value. This is best illustrated with yard waste. Yard waste typically has a high moisture content as compared with other waste. Operators of waste-to-energy plants constantly seek ways to reduce or eliminate yard waste from the waste stream. The handling of yard waste is often part of a local composting program.

Even without a waste-to-energy plant, yard waste takes up significant landfill volume, and composting this material will both reduce the need for landfills and contribute to the recycling effort by producing a marketable compost. The recycling of other reusable materials in the waste stream, such as plastics, glass, ferrous metal, aluminum cans, and some paper products, is also an important part of an integrated waste management plan, as will be discussed later in this chapter.

A unique feature of a design for a waste-to-energy plant is that it must be sized to handle the quantity of MSW that is delivered to that plant, regardless of the heating value of the MSW. The boiler, however, is a heat-input device that must be sized for the maximum heat input expected. In the design of the boiler, the tons per day of MSW and the typical range of expected heating value for the MSW are required. As an example, a plant could be expected to handle 1000 tons per day of MSW. But depending on the plant's location in the world, where MSW composition varies greatly, and the time of the year (dry versus wet season), the heating value of the MSW could vary from 3500 to 6500 Btu/lb. This significantly affects the heat input to the boiler and, when the MSW has a high heating value, could actually limit the quantity of MSW that could be handled because of heat-input limits on the boiler.

13.6 Types of Energy-Recovery Technologies

Over the years, a number of technologies have been used in an attempt to recover energy from solid waste. These have included

- Pyrolysis
- Fluidized bed combustion
- Methane recovery from sanitary landfills
- Mass burning
- Conversion of municipal solid waste to refuse-derived fuel (RDF) and then burning of the RDF

Pyrolysis. In this method, MSW is heated in an oxygen-starved atmosphere to break down the solids into a gas that may be burned directly in a boiler.

Fluidized bed combustion. This method involves the combustion of MSW, which is processed as RDF, in a fluidized bed boiler. (Refer to Chap. 2.) The hot gases from combustion transfer their heat to the production of steam.

Methane recovery. This process utilizes the methane gas produced from the decay of organic matter in MSW. By correctly tapping the landfill, the escaping methane gas can be collected and burned in a boiler for steam production or utilized in a gas turbine.

Mass burning. In this process MSW is burned in the form that it is delivered to the facility. This process can be further categorized as follows:

1. *Combustion without energy recovery.* This is an alternative to landfilling but is currently unacceptable. These units have been referred to as *incinerators.*
2. *Combustion using refractory furnaces with heat-recovery boilers.* These have been in operation for many years but are now replaced by more efficient designs for the recovery of the energy produced.
3. *Combustion using modular furnaces.* A popular means of energy recovery from MSW that is generally used by smaller communities with capacity requirements ranging between 5 and 200 tons per day of MSW. Modular systems are assembled at the factory and shipped to the site for easy assembly. These systems offer the flexi-

bility of being able to group several modules to obtain the needed capacity.

4. *Combustion using waterwall furnaces.* This technology was successfully developed in Europe and is the most efficient use of mass-burning technology. These systems range in size from approximately 200 to 1000 tons per day.

Refuse-derived fuel burning. This requires the MSW to undergo front-end processing for the removal of noncombustible materials, for the recovery of acceptable materials, and for the development of a more homogeneous-type fuel with a fairly high heating value. The recovered noncombustibles are landfilled, some materials are recovered, and the combustible material is processed and subsequently burned in a dedicated boiler or burned as a supplemental fuel with pulverized coal. A dedicated RDF plant is generally designed for a plant capacity of 1000 tons per day or greater.

Of these technologies, the two technologies that are most prominent, mass burning using waterwall furnaces and refuse-derived-fuel burning also using waterwall furnaces, are described here in greater detail.

13.7 Mass Burning

MSW combustion is dominated by systems that use mass-burn technology. These are systems that can accept solid waste that has undergone little or no preprocessing other than the removal of oversized items or materials that were diverted by source-separation programs.

The early mass-burn systems recovered energy primarily by the addition of waste heat boilers to refractory-lined furnaces. This resulted in relying totally on convection for heat transfer with no recovery of the radiant heat produced.

The modern waterwall mass-burn designs have become a major method for the recovery of energy from solid waste. When the method is combined with proper environmental control equipment, it becomes an effective and efficient part of integrated waste management.

Figure 13.1 shows a typical mass-burn facility. The refuse collection vehicles enter the facility on the tipping floor and dump the MSW directly into a large pit area. The pit is usually sized to handle several times the daily processing capacity of the steam-producing plant. This is necessary because refuse collection is often only a 5- or 6-day per-week, 12-hours-per-day activity, and most plants are in operation 7 days a week, 24 hours per day. Storage is also required because there are periods when the boilers are out of service because of maintenance. In addition, when refuse (garbage) is very wet from a heavy rain, it will not burn properly, and a larger pit allows the operator to use the drier MSW first or to mix the dry and wet MSW.

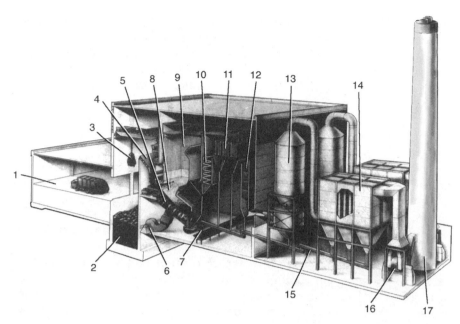

Figure 13.1 Typical mass-burn waste-to-energy plant. (*Ogden Projects, Inc.*) 1 = tipping floor. 2 = refuse holding pit. 3 = feed crane. 4 = feed chute. 5 = Martin stoker grate. 6 = combustion air fan. 7 = Martin residue discharger and handling system. 8 = combustion chamber. 9 = radiant zone (furnace) 10 = convection zone. 11 = superheater. 12 = economizer. 13 = dry gas scrubber. 14 = baghouse or electrostatic precipitator. 15 = fly ash handling system. 16 = induced draft fan. 17 = stack.

MSW is loaded from the pit by means of a feed crane and is fed into a feed chute where the MSW is loaded uniformly onto a stoker for burning. The stokers used to burn MSW are of various designs and generally have a moving grate with a typical residence time in the furnace of 25 to 40 min.

This mass-burn grate system uses a ram-type feeder that distributes the MSW evenly onto the grate. The grate is a step-grate-burning stoker where the design causes the refuse to be churned and redistributed on basically four levels: feeder, drying grate, combustion grate, and burn-out grate. Following complete combustion, the resulting ash falls into an ash-handling system, where it is transported to a landfill. Prior to the ash being carried to the landfill, the iron and steel contained in the ash can be removed by magnets for recycling.

The furnaces and boilers that are used vary based on the experience and criteria established by the designers. The principles used and the equipment selected are similar to those described in previous chapters. Flue gas cleaning equipment, similar to that described in Chap. 12, is utilized to ensure that the facility meets all the air permit regulations. Figure 13.1 shows a typical layout in this area with the use of a dry scrubber for the removal of acid gas, such as hydrogen chloride (HCl) and sulfur dioxide (SO_2), and a collector for particulates, such as a baghouse or an electrostatic precipitator.

The steam produced in the boiler can be sold to an industrial customer for process, or it can be used for heating. It is very common for the steam to be used in the production of electricity by way of a steam turbine.

The boilers in waste-to-energy plants are subjected to a much more hostile operating environment than other types of boilers. They therefore require certain features.

Because fuel for a mass-burn plant consists of solid waste, objects in the MSW can vary greatly in size. Large noncombustibles are common in the waste stream, and since they will not burn, they will come out of the boiler with the ash into the ash-handling system. Larger ash openings with a more complex structure than that found in a fossil-fuel-fired boiler are required at the back end of the boiler.

MSW normally has a high moisture content and is often very wet from rains. Thus the combustion air is vented through a number of underfire and overfire windboxes or compartments. The air from these compartments helps dry the waste and creates turbulence so that the waste is burned more efficiently.

Because of the plastics and other corrosive materials in the MSW, boilers must be designed so that their size and materials will minimize any accelerated corrosion. This corrosion will cause costly outages, resulting in high maintenance and loss of revenue.

Corrosion in refuse-fired boilers, whether they be for mass or RDF firing, is usually caused by the chlorides that deposit on the tubes in the furnace, superheater, and boiler bank. The rate of tube metal loss due to corrosion is temperature-dependent, with high rates of metal loss directly related to high metal temperatures. Refuse boilers that operate at high steam pressures have higher-temperature saturated water in the furnace tubes, and, therefore, these tubes have higher metal temperatures. Superheater tube metal temperatures are directly related to steam temperature inside the tubes. It is the temperature of the water or steam inside the tube that is dominant in the tube metal temperature rather than the temperature of the flue gas outside the tube.

Furnace-side corrosion can be aggravated by poor water chemistry. If water-side deposits are permitted to form, tube wall metal temperatures will increase, and furnace corrosion will be accelerated. Therefore, a waste-to-energy plant requires good water chemistry, just like any well-operated power plant that burns conventional fossil fuels. This results in lower maintenance costs and higher availability.

Figures 13.2 and 13.3 show another arrangement of a mass-burn facility using another type of stoker design. The feed of MSW to the stoker is similar to that described previously, and the stoker is again an inclined arrangement. With this design, a roller-grate stoker is utilized.

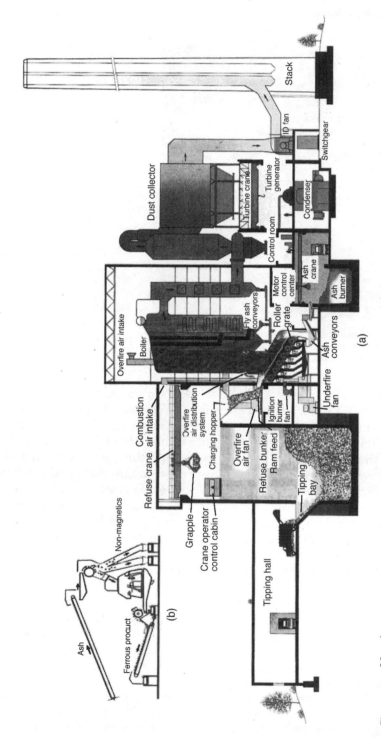

Figure 13.2 Mass-burn waste-to-energy plant using roller-grate stoker design. (*American REF-Fuel.*)

745

Figure 13.3 Duesseldorf roller-grate stoker system, viewing grate from ash discharge upward to stoker feed area. (*American REF-Fuel.*)

The ram feeder pushes the solid waste onto the uppermost roller of the Duesseldorf roller grate. The constant rotation of the set of rollers tumbles and distributes the waste evenly along the downward slope of the roller grate to promote complete combustion of the refuse. The speed of the rollers and of the ram feeder and the quantity of combustion air provided are individually controlled to maintain optimal furnace conditions. The facility includes a flue gas cleaning system consisting of a dry scrubber and either a baghouse or an electrostatic precipitator for the removal of acid gases and particulates.

In Fig. 13.2*b*, note the ash-handling system. It incorporates a magnet to capture the ferrous material that is in the ash from the boiler. This material is recycled into the metals market for reuse into new ferrous products.

Although similar in plant layouts, mass-burn facilities do vary based on the available space, the designer's equipment and experience, and the owner's requirements. Figure 13.4 shows an arrangement that incorporates the features described previously, including a flue gas cleaning system consisting of a dry scrubber with either an electrostatic precipitator or a fabric filter.

The boiler designs for every mass-burn application vary and are based on the operating experience of the designer and on the design philosophy. Figure 13.5 shows a mass-burn boiler designed to handle

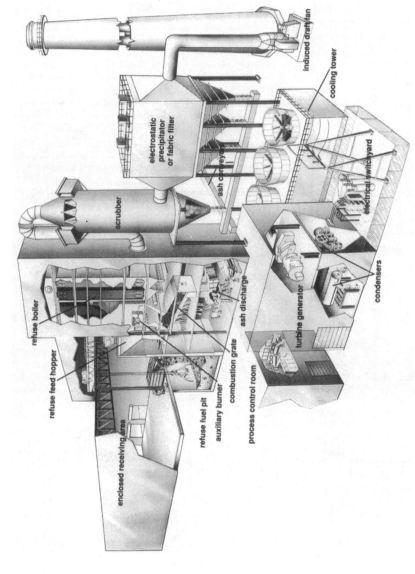

Figure 13.4 Typical arrangement of major power plant equipment in a mass-burn waste-to-energy plant. (*Wheelabrator Environmental Systems, Inc.*)

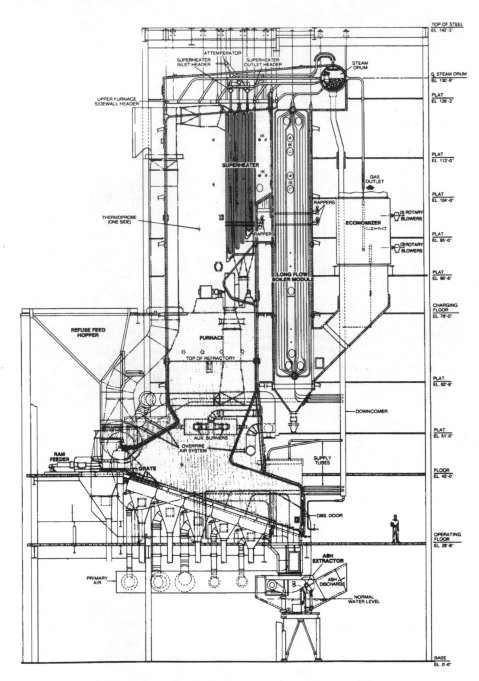

Figure 13.5 Side view of 750-ton/day mass-burn boiler. (*Babcock & Wilcox, a McDermott company.*)

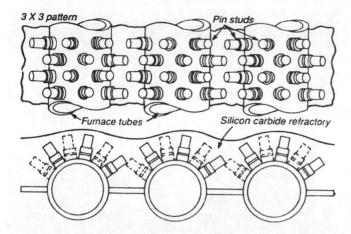

Figure 13.6 Pin stud and refractory arrangement for protection on lower furnace waterwalls in a mass-burn boiler. (*Babcock & Wilcox, a McDermott company.*)

750 tons/day of MSW with a heating value of 5000 Btu/lb. This unit produces 192,000 lb/h of steam at 900 psig and 830°F. The steam in this application is used in a turbine generator for the production of electricity. The lower furnace of this boiler design is protected with pin studs and silicon carbide refractory, as shown in Fig. 13.6. Nearly all mass-fired boilers have a similar arrangement for protection of the lower furnace from corrosion.

The pin stud and refractory design of the lower furnace should provide maximum cooling so that the lowest possible refractory temperature is maintained. This low refractory surface temperature results in a longer refractory life, less fouling of the furnace walls, and lower maintenance costs. In addition, by having more heat removed in the lower furnace, less upper furnace heating surface is required to achieve the required furnace exit gas temperature. This results in a less costly boiler.

Waste-to-energy facilities not only must contribute to the solution of the solid waste problem but also must be environmentally acceptable. This requires that the necessary environmental equipment to protect the air, water, and ground environments be used. In addition, the facility must be accepted by the public, which requires a design having the proper aesthetics for the surrounding area. Figure 13.7 shows the Westchester County, N.Y., waste-to-energy mass-burn facility, which has a plant capacity of 2250 tons of refuse per day.

All waste-to-energy plants, whether they use the mass-burn or RDF-burn technology, must be economically sound operating facilities and meet all environmental requirements. These facilities require

Figure 13.7 Westchester County, N.Y., waste-to-energy plant. (*Wheelabrator Environmental Systems, Inc.*)

scrubbers and particulate removal systems for acid gas and fly ash removal from the flue gas.

In addition to meeting air pollution requirements, waste-to-energy plants must meet the requirements for liquid discharges. These plants, as well as all power plants, produce liquid discharges from ash quenching, boiler operation, and flue gas cleaning equipment. Liquid discharges are controlled and treated prior to any release. It is common to see new waste-to-energy plants that are designed to have no water discharge. Liquids are evaporated or recycled within the plant, and the remaining solids are disposed in properly designed landfills.

The advantage of the mass-burn designs is that they require negligible front-end processing of waste unless there is a recycling requirement. However, because of the bulky heterogeneous nature of MSW, high quantities of excess air are required for proper combustion, using a larger boiler and a larger flue gas cleaning system when compared with a facility using RDF technology and for the identical quantity of incoming MSW. Nevertheless, because of the experience obtained from the extended operations in Europe and now in the United States, mass-burn facilities are dependable and relatively easy to maintain and operate and are the predominant choice of technologies today.

13.8 Refused-Derived Fuel (RDF) Burning

Refuse-derived fuel technology was developed in the United States as an alternative to mass burning of refuse that was common in Europe.

RDF firing is believed by many to be a more efficient and economical means of converting MSW into usable energy such as steam and electricity. RDF technology is often considered to be more compatible with the current trend to conserve resources by recycling.

RDF is the combustible fraction of MSW that has been prepared by a mechanical processing method for use as a fuel in a boiler. RDF can be used as a supplemental fuel to a primary fuel such as pulverized coal or as the only fuel in a boiler designed and dedicated to the burning of RDF. Table 13.6 shows the ultimate analysis of RDF with an expected heating value of 5500 Btu/lb for a facility located in Florida. Although the moisture content remains quite high at 27 percent, the quantity of ash is greatly reduced (23.63 percent for MSW versus 13.63 percent for RDF) because a significant quantity of the noncombustibles have been removed in the processing of the RDF.

The RDF preparation system includes a combination of conveyance, size separation, shredding, material recovery, and storage. The type of processing system used depends on the philosophy of the plant designer. The variation in processing produces a fuel (RDF) of significantly different composition.

In one type of processing system that is often called a *shred and burn* type of RDF, the MSW is fed to a shredder, where the MSW is reduced in size. Following this size reduction, the shredded waste is conveyed to magnetic separators, where ferrous metal is removed. The remaining waste, now called RDF, is then fed to the boiler.

By subjecting the incoming MSW to size reduction, fuel size is reduced, which provides advantages in the combustion process. The mixing of air with the reduced-in-size fuel particles is improved as compared with mass burning. In addition, by removing a large percentage of ferrous materials from the MSW, an abrasive and high-impact component is removed from the fuel-handling system, grate, and ash-removal system. Therefore, the combustible fraction of shred-and-burn-type RDF is improved; however, it still retains the noncombustible fraction, including sand, glass, grit, and dirt.

Figure 13.8 shows a boiler designed to utilize shred-and-burn-type RDF. Multiple boilers are utilized at the SEMASS waste-to-energy facility in Rochester, Mass., with each boiler designed to produce approximately 280,000 lb/h of steam at 650 psig and 750°F.

In Palm Beach County, Fla., a waste-to-energy facility incorporates a more sophisticated RDF processing system that produces a more homogeneous fuel by removing a significant amount of noncombustibles (Figs. 13.9 and 13.10). At the same time, the processing system recovers nearly 96 percent of the energy from the incoming MSW.

This RDF processing system consists of a flail mill, magnetic separator, two-stage trommel, secondary shredder, and disk screen.

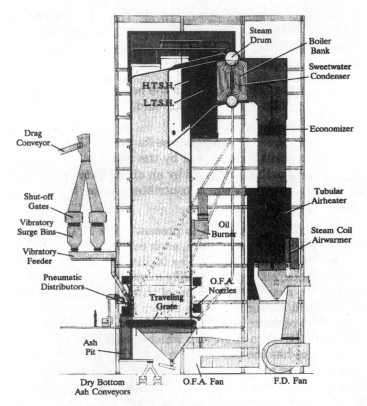

Figure 13.8 RDF boiler for SEMASS waste-to-energy project. (*DB Riley, Inc.*)

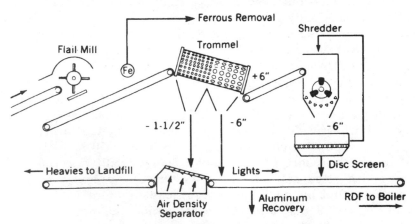

Figure 13.9 RDF processing system, Palm Beach County, Florida. (*Babcock & Wilcox, a McDermott company.*)

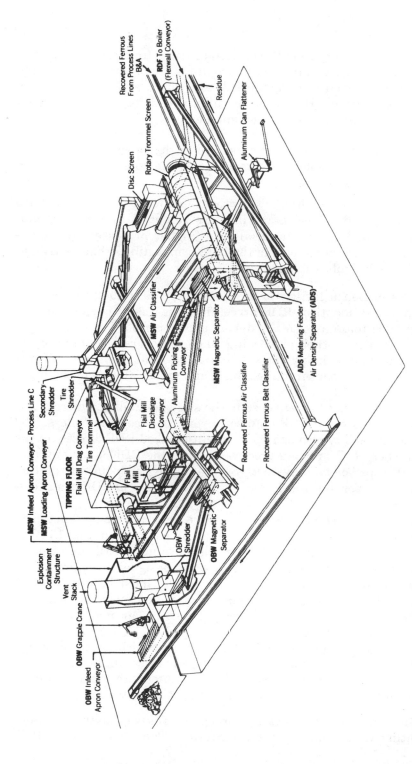

Figure 13.10 Arrangement of equipment for RDF processing of 2000 tons/day of municipal solid waste (MSW), Palm Beach County, Florida. (*Babcock & Wilcox, a McDermott company.*)

Combustible recovery from the material coming from the first stage of the trommel is also obtained as shown in Fig. 13.9. The purpose of the flail mill is to open the bags, break any glass, and provide a very coarse size reduction. Having a flail mill instead of a shredder located prior to a trommel has several advantages:

1. Glass is not pulverized and imbedded in the paper and cardboard. Thus noncombustibles are removed, and the potential for slagging in the boiler is reduced.

2. Materials that are already the proper size for use as RDF fuel are not shredded. These materials are removed by the trommel. Trommel-undersized materials are further processed by an air-density separator to reclaim the light combustibles that contribute to the high combustible recovery.

The Palm Beach facility is designed to process 2000 tons/day. It contains three identical RDF processing lines, each capable of processing 1000 tons/day of MSW during a 16-h period. The facility also incorporates two other processing lines, an *oversized-bulky waste* (OBW) line and a tire-processing line. An overall illustration of this facility's layout is shown in Fig. 13.11, and an aerial view of the plant is shown in Fig. 13.12.

The RDF produced is conveyed to the storage building, where front-end loaders move the RDF onto conveyers that feed the boilers. The Palm Beach facility also has the flexibility to feed RDF directly to the boilers when RDF is being processed. Since the RDF is usually produced 16 h per day, 5 days per week, and the boiler-turbine portion operates 24 h per day and 7 days per week, storage of RDF is a necessity.

RDF is introduced via feeders to a traveling-grate stoker (similar to coal firing on a stoker described in Chap. 5), as shown in Fig. 13.13. A combination of on-grate and suspension burning of RDF takes place with under-grate and over-fire air systems designed to optimize the combustion process. The over-fire air system must be designed for broad flexibility to accommodate changes in fuel moisture, ash content, heating value, etc. This is accomplished by varying the over-fire and under-grate air ratios.

RDF facilities require front-end processing, which is the most significant difference from mass-burn units because of its associated initial cost and subsequent operating and maintenance costs. However, RDF technology does offer several significant advantages:

1. RDF is a better fuel than unprocessed refuse, having a higher heating value than unprocessed refuse. The combustion of RDF

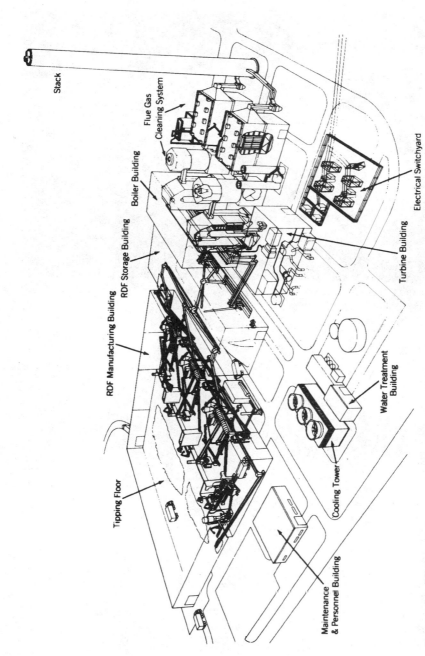

Figure 13.11 RDF waste-to-energy plant, Palm Beach County, Florida. (*Babcock & Wilcox, a McDermott company.*)

Stack

Flue Gas Cleaning System

Boiler Building

RDF Storage Building

RDF Manufacturing Building

Tipping Floor

Maintenance & Personnel Building

Cooling Tower

Water Treatment Building

Turbine Building

Electrical Switchyard

Figure 13.12 Aerial view of 2000-ton/day RDF waste-to-energy plant, Palm Beach County, Florida. (*Babcock & Wilcox, a McDermott company.*)

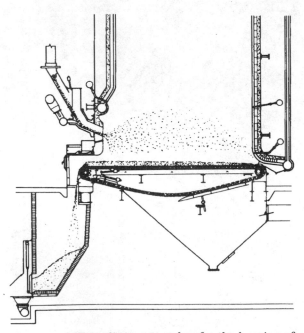

Figure 13.13 Traveling-grate stoker for the burning of RDF. (*Babcock & Wilcox, a McDermott company.*)

requires less excess air and is therefore more thermally efficient. This translates into smaller boilers and flue gas cleaning equipment for a given quantity of incoming refuse.

2. Lower excess air in the furnace means less flue gas effluents are released into the atmosphere even after these effluents pass through a flue gas cleaning system.

3. Recyclable materials that are recovered prior to combustion have more value than those recovered after combustion.

4. Burning a more homogeneous fuel results in lower maintenance costs and higher availability for the boiler. This higher availability leads to extended landfill life because less waste has to be bypassed to the landfill during boiler outages that are required for maintenance.

The air pollution control equipment for RDF plants is similar to that required for mass-burn plants. Dry scrubbers with either a baghouse or an electrostatic precipitator are used; however, because the layer of ash and lime that collects on the bags of fabric filters improves the lime utilization, the dry scrubber-baghouse combination is currently the preferred system in many plants.

13.9 Operation and Maintenance of Refuse Boilers

In addition to being a corrosive, hard-to-burn fuel, refuse is a fuel that can cause slagging in the furnace and fouling in the convection sections of the boiler (such as the superheater and boiler bank) if the boiler is not designed and operated properly. The potential for slagging in the lower furnace can be minimized by a well-designed combustion system, especially the over-fire air system. The removal of glass and metal prior to the introduction of refuse into the furnace, as part of the mass-burn process or as a necessary part of the RDF process system, definitely reduces this potential for slagging.

MSW is a corrosive fuel. Paper and plastic products in the MSW contain chlorine which, during combustion, forms acid gases that can corrode the tubes in the boiler's furnace walls, superheater, and boiler bank. Refuse-fired boilers, whether they are mass fired or RDF fired, are subject to some amount of corrosion. It is important to have a design that will prevent accelerated corrosion. The corrosion in refuse boilers is a function of the relatively high amount of chlorides in the MSW. A protective coating on the lower furnace tubes is an effective means of minimizing the effect of corrosion.

In mass-fired boilers, it is necessary to have the lower furnace covered with refractory material to ensure higher temperatures in this area for proper combustion of MSW (Fig. 13.6). The pin stud and refractory design not only enhances combustion but also protects the furnace tubes from excessive wear from the MSW and provides protection from lower furnace corrosion.

However, combustion of RDF results in higher furnace temperatures than a mass-fired boiler. Using a refractory-coated furnace would result in even higher temperatures, which would lead to slagging problems. Therefore, for corrosion protection in the lower furnace when burning RDF, a properly designed unit should use either an alloy-weld overlay on the lower furnace wall tubes or should incorporate composite tubing, i.e., an Inconel-clad tube. The amount of protection that is needed in the furnace can only be determined from operating experience. MSW composition varies significantly from region to region as well as seasonally. Therefore, a conservatively designed boiler is extremely important to ensure high availability, reduce maintenance costs and extended outages, optimize steam and revenue outputs, and reduce the need for the landfilling of refuse, which is the ultimate objective.

Aluminum metal is also part of the MSW. This metal melts at 1200°F, a melting temperature lower than glass, which melts at 2000°F and forms slag. The melted aluminum plugs the air ports on the stoker, clogs the stoker drive mechanisms, reduces combustion efficiency, and increases the need for maintenance and resulting plant outages. The aluminum can only be removed manually during costly maintenance outages.

Of all the maintenance concerns in a waste-to-energy plant, the largest concern is the rapid corrosion of boiler components caused by the severe furnace environment and the acid flue gases that result from burning MSW. For a coal-fired boiler, boiler tubes typically last 20 to 30 years, while in many waste-to-energy plants tube replacement is often required within 3 to 5 years. In addition, tube failures in these plants are accelerated in some cases when the boilers were designed to operate at higher pressures and temperatures. Special design features and the use of special materials have minimized these effects.

Although corrosion in refuse-fired boilers can adversely affect almost any boiler component, such as stokers and flues, the greatest concern has been the effect of corrosion on the boiler tubes. Tube leaks are critical because they take the boiler out of service, while other failures may require operating at a reduced capacity. Corrosion (and erosion) reduces the wall thickness of tubes, and, eventually, the tube cannot withstand the internal pressure and it bursts.

Chlorides are the principal cause of boiler tube corrosion. MSW has a relatively high chloride content of 0.5 to 1.0 percent, and most of it is contributed by polyvinyl chloride, which is found in plastics and paper. When released during combustion, the chlorides attack steel boiler tubes by various means:

1. Corrosion by solid chlorides that form on the tubes
2. Corrosion by molten salts
3. Corrosion by hydrochloric or sulfuric acid in the flue gas

In the superheater, the corrosion is often accelerated by higher tube metal temperatures and erosion.

Erosion is another factor that exacerbates the corrosion problem. The erosion may be caused by the use of sootblowers that direct steam or air directly at the tubes, baring the tube totally to the flue gas. Erosion shields are often used to minimize this problem. Some designs use rappers instead of sootblowers to remove the major accumulations of fly ash because this maintains a light coating of fly ash for protection of the tube against corrosion.

Erosion also may be caused by the movement of flue gases along the tubes. The flue gas carries solid particles, and these act as an abrasive.

Changes in both the boiler design and the boiler operation have been used to reduce corrosion. Some of these include the following:

1. The use of refractory can protect tubes in the lower waterwall section of the furnace. However, the benefits of refractory for corrosion resistance are offset by a reduction in heat transfer. This reduced heat transfer lowers the boiler's efficiency and accelerates furnace fouling due to slagging, which leads to increased maintenance costs. The use of refractory also can lead to higher initial costs for the boiler because a larger boiler is required to meet the same furnace exit gas temperature.

2. Increasing the tube spacing in the convection sections of the boiler will decrease flue gas velocity. This, however, increases heat-transfer surface requirements.

3. The use of higher-grade materials such as stainless steel and nickel-based alloys, such as Inconel, instead of carbon steel and low-alloy steel.

4. The most effective operational corrosion-control strategy is more frequent inspections. Inspections of the stoker grate can detect overloaded grate sections, where operational corrections could be made to avoid grate section failures, and to identify where combus-

tion is incomplete, thus minimizing corrosive reducing atmospheres, i.e., where insufficient air exists. Inspections to detect slagging, particularly in areas around secondary air nozzles, will ensure proper airflow and good combustion conditions. Frequent monitoring of the tubing, erosion shields, rappers, and other components for deterioration will allow their timely repair or replacement before a more expensive failure occurs. In many facilities, scheduled maintenance periods bring boilers off line once a year for 2 weeks to conduct thorough inspections and to complete routine maintenance.

Composite tubes (a coaxial tube having carbon steel inside and a corrosion-resistant metal outside) are often used in WTE boilers, and they offer the following advantages:

1. The pressure boundary is carbon steel, and plant operators can use normal water chemistry controls. The use of stainless steel or Inconel tubing would require more costly water chemistry.

2. In addition to corrosion and erosion resistance that is superior to carbon or low-alloy steel tubes, the outer tubing can be much thinner than the equivalent Inconel or stainless steel tubing because it is not part of the pressure wall. This results in improved heat transfer.

The most significant disadvantage of composite tubing is its material cost, since its cost is approximately three times that of carbon steel. However, as it is with other higher-grade material selections, increased life and lower maintenance costs can offset the initial material cost, and this option should be evaluated carefully.

Continuous emission monitors (CEMs) have become a standard piece of equipment on waste-to-energy facilities to ensure compliance with regulatory requirements. Previously, only particulates needed to be measured, but most plants now require, and all new plants will require, the monitoring and recording of most, if not all, of the following: O_2, CO, CO_2, NO_x, HCl, volatile organic compounds (VOCs), temperature of the gases, and opacity. CEM systems collect and record the emissions data required by the facility permit. Many of these plants are directly linked via computer to a regulatory enforcement agency, thus enabling the continuous monitoring of emission printouts and ensuring compliance with the operating permit.

All waste-to-energy plants require flue gas cleaning systems. For acid gas removal, a dry scrubber system is the choice of design, with either a baghouse or an electrostatic precipitator used for particulate removal. Because of the heterogeneous composition of MSW, some toxics, such as mercury and dioxins, are formed during the combustion process. Although the quantity of these toxics is small, when required, systems are installed for their removal.

One such system uses activated carbon, which is a highly adsorbent form of carbon used to remove odors and toxic substances from gaseous emissions. The carbon is injected upstream of the spray dryer and is intimately mixed with the flue gas within the spray dryer. The turbulent mixing within the spray dryer ensures contact between the carbon and the toxics in the flue gas, and the mercury and dioxins are captured on the carbon particles.

In the United States, all waste-to-energy plants represent less than 1 percent of the known sources of dioxin. All the sources of dioxins are not known, and the Environmental Protection Agency (EPA) knows the source of only about half these dioxins. Yet the waste-to-energy industry seems to attract the major attention on controlling this emission. This is another example of how the political scene has hampered the solution to the solid waste problem.

13.10   Recycling

The compatibility of recycling with waste-to-energy plants is a clear trend in some areas and one that should be adopted in all parts of the world. As regards MSW as a fuel, rather than reducing the availability of combustible refuse, recycling actually improves the quality of the waste as a fuel.

The removal of grass and leaves for possible composting eliminates materials that have high moisture content and low heating value. The removal of metal and glass improves the fuel because it becomes more homogeneous, which results in higher quality with a higher heat content. The removal of aluminum cans improves the revenues by their sale and also reduces maintenance and improves the availability of the boiler. Aluminum melts in the furnace, flows through the stoker grate bars, and then resolidifies, which causes serious maintenance problems with extended outages for repairs.

Tables 13.2 and 13.5 show examples of the composition of typical solid wastes in the United States. This composition can and does vary by season and even by the day of the week, as well as the region of the world where it is collected.

13.10.1   Advantages and disadvantages of recycling

Everything cannot be recycled easily, nor is there a consistent market for these materials. The following are typical examples of different materials that are recycled. It shows the approximate amount of each that is contained in the MSW and approximately how much of that amount is recycled.

For each of these materials, the advantages and disadvantages of recycling are identified, as well as the general outlook for future recycling possibilities.

1. Paper and paperboard. Percent in MSW, 37.6 percent; percent recycled, 34 percent.
   a. Advantages
      (1) Saves more landfill space than recycling any other material.
      (2) Reduces air and water pollution.
      (3) Large supply of newspaper and cardboard.
      (4) Least expensive of all materials to sort.
   b. Disadvantages
      (1) Weak market for mixed paper.
      (2) Recycled paper is lower in quality than virgin paper for some uses.
      (3) Cannot be recycled indefinitely.
      (4) Difficult and expensive to deink.
   c. Outlook: Very good material for recycling a relatively high percentage.
2. Plastic packaging. Percent in MSW, 9.3 percent; percent recycled, 3.5 percent.
   a. Advantages
      (1) Reduces air pollution.
      (2) Conserves oil and natural gas.
   b. Disadvantages
      (1) New plastic packaging is rarely recycled.
      (2) Only PETE (polyethylene terephthalate, e.g., dark plastic cola bottles) and HDPE (high-density polyethylene, e.g., white plastic milk containers) are recycled in quantity.
      (3) Cannot be recycled indefinitely.
      (4) Generally not recycled into food containers.
      (5) Pickup expensive due to light weight and high volume.
      (6) Difficult to sort different types.
      (7) Some virgin plastics are available cheaply.
   c. Outlook: Most difficult to achieve a high recycling percentage.
3. Container glass. Percent in MSW, 6.7 percent; percent recycled, 22 percent.
   a. Advantages
      (1) Recyclable containers make up 90 percent of discarded glass.
      (2) Can be recycled indefinitely.

(3) Can be recycled into food containers.

(4) Labels and food residue burn off in furnaces.

(5) Steady markets for clear and brown glass.

  b. Disadvantages

(1) Bottles break during sorting.

(2) Broken glass is hard to reuse.

(3) Must be hand sorted by color.

(4) Poor market for green glass.

(5) Often contaminated with unusable glass.

  c. Outlook: New uses and markets are needed for mixed-color and broken glass.

4. Steel cans. Percent in MSW, 1.5 percent; percent recycled, 41 percent.

  a. Advantages

(1) Recycling reduces pollution and conserves iron ore.

(2) Can be recycled indefinitely.

(3) Can be recycled into food containers.

(4) Dirt and contaminants burn off in furnaces.

(5) Easy to separate with magnets.

(6) Steel mills are organized to use scrap steel.

(7) Good market for recycled cans

  b. Disadvantages: Generally none; however, when part of mixed waste (MSW), efficiency of recovery declines.

  c. Outlook: No solid waste disposal problem if all material were as easy to recycle as steel cans.

5. Aluminum cans and foil. Percent in MSW, 1 percent; percent recycled, 68 percent.

  a. Advantages

(1) Recycling uses over 90 percent less energy than virgin production.

(2) Recycling reduces pollution and conserves ore.

(3) Can be recycled indefinitely.

(4) Can be recycled into food containers.

(5) Dirt and contaminants burn off in furnaces.

(6) System for collection and processing in place.

(7) Strong market for recycled cans.

  b. Disadvantages

(1) Light weight makes collection expensive.

(2) When part of mixed waste (MSW), efficiency of recovery declines.

  c. Outlook: Energy savings made aluminum the first large-scale recyclable material and the most valuable.

13.10.2   Economics and quality products from recycling

The obvious incentive for recycled materials is for them to have increasing value. However, in general, prices for recycled material have been depressed because of worldwide lower prices for raw materials and because the increase in recycling programs has flooded the market with recycled materials.

Most manufacturing industries are still organized to operate on virgin material rather than recycled material because of the relative abundance and low cost of virgin resources. And for some recycled materials, the retooling requirements are expensive.

The use of virgin materials often has another clear advantage over recycled material in the area of consistent quality. Although contamination is not a problem for aluminum or steel, it can compromise the quality and affect the marketability of glass, plastic, and paper.

Glass-making shops that accept recycled bottles must expend a great deal of effort to exclude contaminants because they have different melting points and they can ruin entire furnaceloads of glass. Some of these glass contaminants include clear Pyrex baking dishes, window panes, and light bulbs, all of which cannot be distinguished from container glass after they have been broken and mixed together. By contrast, sand and limestone, which are the raw materials of glass, are pure.

Despite the problems, recycling efforts will continue for those materials which have marketability or have an impact on the environment, and all these materials must be evaluated in determining a viable solution to the solid waste disposal problem.

Recycling is not without its additional costs, environmental impacts, and social problems. Recycling, by demanding source separation, requires many additional trucks and operating personnel. These additional trucks will produce additional pollution as well as added traffic problems. All these problems must be evaluated carefully in determining the overall benefits of recycling.

And finally, recycling can be such an emotional subject that it becomes the thing to do. However, for it to be successful, people must purchase products made from recycled materials, even though at times these products may be more expensive than those made from virgin material. Again, the overall benefits must be evaluated, and not everyone looks at the overall picture, but only the initial cost and perceived benefits.

13.11   Material Recovery Facilities (MRFs)

The development of recycling programs of various types has led to the requirement for *material recovery facilities* (MRFs). A MRF receives waste, and then workers and machinery are used to remove and separate recyclable materials from the waste stream. The recovered materials then leave the facility and are destined for some use other than occupying space in a landfill.

The design of a MRF must be based on the type of waste that will enter the facility. This can be of three types: (1) mixed waste (MSW), (2) source-separated material, or (3) comingled waste. Also of significance are the following:

1. Is the source-separated material arriving at the facility in dedicated trucks, or is it mixed with other MSW?
2. How is the waste packaged: loose, bundled, or in plastic bags?
3. What is the content of the separated waste: paper, types of plastics, glass (colors), etc.?

Each community can have a different arrangement based on the recycling programs instituted, and each concept could necessitate a different design of a MRF. A MRF design must include the following factors:

1. The size of the community and the amount of waste that can be expected to be delivered to the facility
2. Any state or local regulations that require certain materials to be collected or certain volumes to be diverted from landfills
3. The financial aspects of the project

The most important factor, however, is determining whether there is a market for the recovered material. The MRF must provide a recovered material that the market needs, or else it can be left holding a warehouse full of unneeded materials such as plastic bottles, bales of paper, or containers of glass. Eventually, these recovered materials could end up in a landfill if a market is not available, with the added cost of material recovery.

Perhaps the most important question is how should the waste be collected. Therefore, the answer does not occur with the MRF but at the curbside where the waste is collected, and this means an even earlier decision at the residence or business where the waste is actually generated. The issue is how much separation residents and businesses should do versus how much to leave for the processing facility (MRF). At issue are the increased costs associated with curbside collection of source-separated materials (additional trucks, personnel, etc.) and the ability and willingness of residents to correctly separate their recyclables in the home or at their place of business. The collection decision determines the design of a MRF because it determines whether the facility will be designed to receive mixed waste, separated waste, or a combination of waste streams (comingled waste). The decision helps determine the size of the facility, equipment needs, and degree of automation.

The primary argument against mixed-waste processing is the lower quality and quantity of the final material because of its difficulty in being separated, the additional handling, and more contamination. However, mixed waste may be the proper option if it is expected that the community will not source separate to a degree that is beneficial.

When the separation is performed at the source by residents and by businesses, the material stays cleaner throughout the process, which contributes to a cleaner and higher-quality material. Collection costs increase, but these can be offset by reduced processing costs and higher revenues for a better product.

It is a common concept to have a MRF in combination with a waste-to-energy facility. This not only would retain the materials that have recyclable market value but also could remove some materials that can cause operation problems in the boiler such as aluminum, steel, and other noncombustible materials. In addition, up-front recycling has the potential for reducing air pollutants and ash emissions from the boiler.

MRFs vary in size and design based on the community being served, the expected constituents of the MSW, and the recyclables desired based on the expected recycle market for those products. Figure 13.14 is a mass-flow diagram of a typical MRF designed to handle 80,000 tons per year of bulky waste and 350,000 tons per year of residential and commercial waste. Based on a 5-day-per-week operation, the plant could handle approximately 1600 tons/day of MSW.

This plant is designed to recover both ferrous and nonferrous metals, aluminum cans, plastic bottles, and various paper products. With reference to Fig. 13.14, the acronyms are as follows:

OCC   old corrugated containers

ONP   old newsprint

PETE   polyethylene terephthalate (e.g., dark plastic cola bottles)

HDPE   high-density polyethylene (e.g., white plastic milk containers)

The design of the MRF must be consistent with the methods by which the waste is picked up at its source. This facility has separate trucks that deliver bulky waste (called type 13 in the diagram) to a specific location on the tipping floor. A grapple sorts out portions of the waste containing metals and cardboard and places them in small

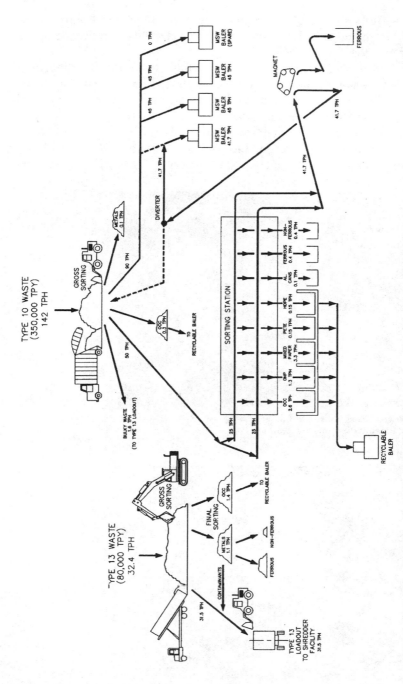

Figure 13.14 Mass-flow diagram for material processing and recovery facility (MRF), Monmouth County, N.J. (*National Ecology, a Babcock & Wilcox company.*)

piles for final sorting by operating personnel. Recovered materials are stored for subsequent load out. Unrecovered waste is placed into trailers for disposal to a landfill.

Residential and commercial waste (called type 10 in the diagram) is unloaded on the tipping floor similar to that shown in Fig. 13.15. This MRF design performs gross sorting on the tipping floor, where the recyclable-rich waste proceeds to the sorting station for recovery and the remaining MSW is baled and sent to a landfill.

The recyclable-rich waste is fed into two steel apron conveyors by a front-end loader similar to that shown in Fig. 13.16. These conveyors

Figure 13.15 Truck dumping MSW onto tipping floor. (*National Ecology, a Babcock & Wilcox company.*)

Figure 13.16 Front-end loader moves MSW from tipping floor to processing lines. (*National Ecology, a Babcock & Wilcox company.*)

Figure 13.17 Sorting stations where certain recoverable materials are removed from the processing lines by hand sorting. Paper, plastics, and glass are often hand sorted. (*National Ecology, a Babcock & Wilcox company.*)

feed two parallel sorting belt conveyors. Workers at each sorting conveyor rip open any bags and expose materials for subsequent recovery. In some MRF designs, the bag-opening process is performed mechanically by a flail mill, a trommel, or similar piece of equipment.

Sorters stand on each side of the belts and manually remove the targeted recyclables such as cardboard, newspaper, mixed paper, and plastic bottles, similar to that shown in Fig. 13.17. These collected materials drop down chutes into storage bunkers where they are eventually put into a baler feed conveyer and baled as shown in Figs. 13.18 and 13.19. Ferrous metals, aluminum cans, and mixed nonferrous metals are dropped down chutes into containers, where they also can be baled, if desired, or delivered as unbaled material.

Ferrous metals are often removed from the waste stream by the use of magnets, where the expected quantity of ferrous material can justify the expense of this equipment. Figure 13.20 shows an overhead magnet that is often used for this purpose.

A control system similar to that shown in Fig. 13.21 uses a programmable logic controller as the main controlling device and also allows rapid and continuous reporting of the status of all equipment in the MRF.

13.12 Waste-to-Energy Plants and Recycling

There is a growing acceptance of waste-to-energy plants as a viable part of the solution to the waste disposal problem. This acceptance is

Figure 13.18 Corrugated cardboard, newsprint, and mixed paper are baled for reuse and recycling. (*National Ecology, a Babcock & Wilcox company.*)

Figure 13.19 Aluminum cans are baled for recycling. (*National Ecology, a Babcock & Wilcox company.*)

Figure 13.20 Magnetic separator for recovery of ferrous materials. (*National Ecology, a Babcock & Wilcox company.*)

Figure 13.21 Control room for material recovery facility (MRF). (*National Ecology, a Babcock & Wilcox company.*)

partly due to the use of more sophisticated air quality control equipment, i e , dry scrubbers and particulate control systems. However, the influence of politics has delayed the addition of many waste-to-energy projects, since many people want to believe that recycling alone is the solution to the waste disposal problem.

The size of waste-to-energy plants generally has fallen into two categories: smaller plants of 400 tons/day or less and larger plants of 1000 tons/day or more. However, the trend toward solving the waste problems in a general area has led to the possibility of regional plants that would require larger plant sizes.

There also has been the trend that these plants produce only electricity rather than only steam or a combination of steam and electricity. Energy revenues are important to the economics of a project, and steam typically provides less revenues than electricity. In addition, the needed long-term contracts are more likely from electric utilities rather than from steam customers.

However, it will be the limitations associated with landfilling and recycling that will create the opportunity and necessity for waste-to-energy facilities. This coupled with successfully operating waste-to-energy plants will increase the public's confidence in this technology.

The combination of recycling and waste-to-energy plants enables communities to become, or remain, self-sufficient by reducing their reliance on others to manage their MSW. Many communities would have to export their MSW, often to out-of-state destinations, if it were not for their successful implementation of recycling and waste-to-energy programs.

Properly sized waste-to-energy plants can help ensure that a community's recycling objectives can be achieved. Accurate waste-management planning will enable communities to establish meaningful recycling goals while also implementing sufficient disposal capacity. If recycling activities reduce the amount of MSW going to the waste-to-energy plant to the point where capacity becomes available, the service area can be broadened to help solve regional waste-management problems.

Even after a community has recycled to the maximum extent, there is still MSW remaining for waste-to-energy programs. Currently, the average rate of recovery for recycling is approximately 22 percent. Refer to Table 13.2. Even if this were increased to 30 percent, the majority of waste has to be managed in other ways. Therefore, recycling and waste-to-energy plants do not compete for the same waste, and both programs can ensure that a community's entire MSW will be managed.

The removal of most types of recyclable materials before combustion results in fewer operation and maintenance problems. The removal of ferrous, aluminum, and other metals helps prevent clogging of the system. This reduces maintenance on the stoker and ash-handling systems. The removal of recyclables also provides a cleaner fuel with potentially higher energy content, since many recyclables are noncombustible. O&M costs as a result will be lower. Therefore, it will take a well-managed integrated solution of recycling, waste-to-energy programs, and landfilling to properly solve the waste disposal problem.

Questions and Problems

13.1 What are the four parts of an effective waste-management plan that lead to a practical solution of the solid waste disposal problem?

13.2 What is unique about the municipal solid waste (MSW) found in the United States as compared with other places in the world? How does this affect a boiler design? How do the seasons of the year affect MSW?

13.3 Why are waste-to-energy plants so important to the solution of the solid waste disposal problem?

13.4 In the United States, approximately what percentage of the annual quantity of MSW is handled in waste-to-energy plants? How much is landfilled?

13.5 Provide the primary reasons for recycling. What happens if there is little or no market for the recycled materials?

13.6 Approximately how much daily waste is generated by each person in the United States, and how much annual waste does this equate to?

13.7 What are the advantages of putting this potential energy to use?

13.8 What is the importance of landfills? What are some of their major problems? With a waste-to-energy plant, what is the potential volume reduction of solid waste?

13.9 Identify some of the major characteristics of the composition of solid waste.

13.10 What two properties of MSW affect the operating performance of a waste-to-energy facility? Why?

13.11 What contents of solid waste have a significant impact on its heating value?

13.12 In analyzing the waste composition of two different locations, you notice that there is a significant difference in the percentage of yard waste. What would be some reasons for this?

13.13 In the sizing of a boiler for the combustion of MSW, how must the variation of heating value and MSW quantity, in tons per day, be evaluated?

13.14 Name and describe the various types of technologies that have been used to recover energy from solid waste. Which are the types used primarily today?

13.15 Describe a mass-burn facility. What is its unique feature? Name some advantages and disadvantages of this technology.

13.16 What is a significant operation problem encountered in the burning of MSW? How can this problem be accelerated, and what must a well-operated plant include?

13.17 In a mass-fired plant, if ferrous material cannot be removed prior to MSW combustion, what method is used to recover this material for recycling?

13.18 What is refuse-derived fuel? What is the significance and importance of the required processing?

13.19 Name the advantages and disadvantages of RDF technology.

13.20 When operating a boiler designed for burning MSW, name the major operating problems encountered. What can be done to minimize these potential problems?

13.21 What is a composite tube, and what are its benefits?

13.22 How are continuous-emission-monitoring (CEM) systems used to support the operation of a waste-to-energy plant?

13.23 Explain how recycling can complement the efficient operation of a waste-to-energy plant.

13.24 What are the primary disadvantages of recycling? How can a recycling program have a detrimental impact on the environment?

13.25 In order for a recycling program to be successful, what must happen to the products made from recycled materials? What happens to the program if this does not happen?

13.26 What is a material recovery facility (MRF), and for what types of waste can such a facility be designed?

13.27 Why does the method of waste collection have an impact on the MRF design?

13.28 If source separation of waste is considered advantageous from the standpoint of the quality of the recycled material and the relative design simplicity of a MRF, why is this concept often proven unfeasible?

13.29 Why does the combination of a recycling program and a waste-to-energy plant provide a good solution to the solid waste disposal problem?

Unit Conversions

Length and Area

$$1 \text{ inch (in)} = 2.54 \text{ centimeters (cm)}$$

$$= 25.4 \text{ millimeters (mm)}$$

$$1 \text{ foot (ft)} = 12 \text{ inches (in)}$$

$$= 30.48 \text{ centimeters (cm)}$$

$$1 \text{ yard (yd)} = 3 \text{ feet (ft)}$$

$$= 0.914 \text{ meter (m)}$$

$$1 \text{ mile (mi)} = 5280 \text{ feet (ft)}$$

$$= 1760 \text{ yards (yd)}$$

$$= 1.609 \text{ kilometers (km)}$$

$$1 \text{ meter (m)} = 100 \text{ centimeters (cm)}$$

$$= 1000 \text{ millimeters (mm)}$$

$$= 1.094 \text{ yards (yd)}$$

$$= 3.28 \text{ feet (ft)}$$

$$= 39.37 \text{ inches (in)}$$

$$1 \text{ kilometer (km)} = 1000 \text{ meters (m)}$$

$$= 0.621 \text{ mile (mi)}$$

$$1 \text{ millimeter} = 1000 \text{ micrometers } (\mu\text{m, or microns})$$

$$1 \text{ square inch (in}^2) = 6.45 \text{ square centimeters (cm}^2)$$

$$1 \text{ square foot (ft}^2) = 144 \text{ square inches (in}^2)$$

$$= 0.0929 \text{ square meter (m}^2)$$

$$1 \text{ square yard (yd}^2) = 9 \text{ square feet (ft}^2)$$

$$= 0.836 \text{ square meter (m}^2)$$

$$1 \text{ square mile (mi}^2) = 640 \text{ acres}$$

$$1 \text{ acre} = 43,560 \text{ square feet (ft}^2)$$

$$= 4840 \text{ square yards (yd}^2)$$

$$1 \text{ square meter (m}^2) = 10,000 \text{ square centimeters (cm}^2)$$

$$= 11.196 \text{ square yards (yd}^2)$$

$$= 10.76 \text{ square feet (ft}^2)$$

$$1 \text{ square centimeter (cm}^2) = 100 \text{ square millimeters (mm}^2)$$

$$= 0.155 \text{ square inches (in}^2)$$

$$1 \text{ nautical mile} = 6080 \text{ feet (ft)}$$

$$= 1.853 \text{ kilometers (km)}$$

$$1 \text{ nautical mile per hour} = 1 \text{ knot}$$

Volume

$$1 \text{ cubic inch (in}^3) = 16.39 \text{ cubic centimeters (cm}^3)$$

$$1 \text{ cubic foot (ft}^3) = 1728 \text{ cubic inches (in}^3)$$

$$= 28.32 \text{ liters}$$

$$1 \text{ cubic yard (yd}^3) = 27 \text{ cubic feet (ft}^3)$$

$$= 0.765 \text{ cubic meter (m}^3)$$

$$1 \text{ cubic meter (m}^3) = 1000 \text{ liters}$$

$$= 1.308 \text{ cubic yards (yd}^3)$$

$$= 35.31 \text{ cubic feet (ft}^3)$$

$$1 \text{ imperial gallon} = 277.4 \text{ cubic inches (in}^3)$$

$$= 4.55 \text{ liters}$$

$$1 \text{ U.S. gallon} = 0.833 \text{ imperial gallon}$$

$$= 3.785 \text{ liters}$$

$$= 231 \text{ cubic inches (in}^3)$$

$$1 \text{ U.S. barrel (petroleum)} = 42 \text{ U.S. gallons}$$

$$= 35 \text{ imperial gallons}$$

$$1 \text{ liter} = 1000 \text{ cubic centimeters (cm}^3)$$

$$= 0.22 \text{ imperial gallon}$$

$$= 0.2642 \text{ U.S. gallon}$$

$$= 61 \text{ cubic inches (in}^3)$$

Weight

$$1 \text{ pound (lb)} = 16 \text{ ounces (oz)}$$

$$= 7000 \text{ grains (gr)}$$

$$= 454 \text{ grams (g)}$$

$$= 0.454 \text{ kilogram (kg)}$$

$$1 \text{ grain (gr)} = 64.8 \text{ milligrams (mg)}$$

$$= 0.0648 \text{ gram (g)}$$

$$= 0.0023 \text{ ounce (oz)}$$

$$1 \text{ gram (g)} = 1000 \text{ milligrams (mg)}$$

$$= 0.03527 \text{ ounce (oz)}$$

$$= 15.43 \text{ grains (gr)}$$

$$1 \text{ kilogram (kg)} = 1000 \text{ grams (g)}$$

$$= 2.205 \text{ pounds (lb)}$$

$$1 \text{ U.S. short ton} = 2000 \text{ pounds (lb)}$$

$$= 907 \text{ kilograms (kg)}$$

$$1 \text{ U.S. long ton} = 2240 \text{ pounds (lb)}$$

$$= 1016 \text{ kilograms (kg)}$$

$$1 \text{ metric ton} = 1000 \text{ kilograms (kg)}$$

$$= 0.984 \text{ U.S. long ton}$$

$$= 1.102 \text{ U.S. short tons}$$

$$= 2205 \text{ pounds (lb)}$$

Density of Water (at 62°F)

$$1 \text{ cubic foot } (ft^3) = 62.5 \text{ pounds (lb)}$$

$$1 \text{ pound (lb)} = 0.01604 \text{ cubic foot } (ft^3)$$

$$1 \text{ gallon} = 8.33 \text{ pounds (lb)}$$

Pressure

$$1 \text{ atmosphere} = 14.696 \text{ pounds per square inch}$$
$$(lb/in^2; \text{ psi})$$

$$1 \text{ inch of water (gauge) at } 62°F = 0.0361 \text{ psi}$$

$$= 5.20 \text{ pounds per square foot } (lb/ft^2)$$

$$1 \text{ foot head of water at } 62°F = 0.433 \text{ psi}$$

$$1 \text{ inch of mercury (inHg)} = 0.491 \text{ psi}$$

Temperature—Comparison of Fahrenheit and Celsius (Centigrade) Scales

$$°F = (\% \times °C) + 32$$

$$°F_{absolute} = °F + 460$$

$$°C = \tfrac{5}{9}(°F-32)$$

$$°C_{absolute} = °C + 273$$

Power and Heat

1 British thermal unit (Btu) = 778 foot-pounds (ft · lb)

= 0.252 calorie (cal)

1 calorie (cal) = 3088 foot-pounds (ft · lb)

= 3.968 Btu

1 kilowatt (kW) = 1000 watts (W)

= 738 foot-pounds per second (ft · lb/s)

= 1.341 horsepower (hp)

1 megawatt (MW) = 1000 kilowatts (kW)

1 horsepower = 33,000 foot-pounds per minute (ft · lb/min)

= 0.746 kilowatts (kW)

1 kilowatt-hour (kWh) = 3413 Btu

= 860 calories (cal)

1 horsepower-hour (hp · h) = 2544 Btu

1 boiler horsepower (bhp) = 10 square feet of boiler heating surface (water-tube boiler)

= 12 square feet of boiler heating surface (fire-tube boiler)

= 34.5 pounds per hour of evaporation from a temperature of 212°F into dry saturated steam at the same temperature

Linear Coefficient of Expansion per Degree Fahrenheit

Substance	Coefficient of expansion	Substance	Coefficient of expansion
Bronze	0.00001024	Cast iron	0.00000589
Copper	0.00000926	Steel	0.00000734
Glass (flint)	0.00000438	Zinc	0.00001653

Weights of Common Substances

Substance	lb/ft^3	Substance	lb/ft^3
Air*	0.0763	Ice	57.5
Brass	525	Cast iron	450
Brick	125	Steel	490
Coal, bituminous	47–56	Zinc	438

*At 60°F, barometer 29.92 in of mercury.

Specific Heat of Various Substances

Substance	Specific heat
Air (constant pressure)	0.24
Dry flue gases (constant pressure)	0.24
Water vapor (atmospheric pressure)	0.48
Ice (0–32°F)	0.50
Steel	0.117

Geometric Formulas

1. *Fractional equivalents*

   $\frac{1}{4} = 0.25$ $\frac{1}{8} = 0.125$ $\frac{1}{16} = 0.0625$

   $\frac{2}{4} = 0.50$ $\frac{3}{8} = 0.375$ $\frac{3}{16} = 0.1875$

   $\frac{3}{4} = 0.75$ $\frac{5}{8} = 0.625$ $\frac{5}{16} = 0.3125$

   $\frac{7}{8} = 0.875$ $\frac{7}{16} = 0.4375$

   $\frac{9}{16} = 0.5625$

   $\frac{11}{16} = 0.6875$

   $\frac{13}{16} = 0.8125$

   $\frac{15}{16} = 0.9375$

2. *Area of a rectangle*

   $$\text{Area} = A \times B$$

   where A = height
   B = width

3. *Volume of a rectangular solid*

   $$\text{Volume} = A \times B \times C$$

where A = height
  B = width
  C = depth

4. *Circumference and area of a circle*

$$\text{Circumference} = \pi D$$

$$\text{Area} = \frac{\pi D^2}{4} = \pi r^2$$

where $\pi(\text{pi}) = 3.1416$
  D = diameter
  r = radius

5. *Surface and volume of a cylinder*

$$\text{Surface} = \pi D \times H + \frac{2\pi D^2}{4}$$

$$\text{Volume} = \frac{\pi D^2}{4} \times H$$

where D = diameter
  H = height

6. *Area of a ring*

$$\text{Area} = \frac{\pi D^2}{4} - \frac{\pi d^2}{4} \qquad \text{or} \qquad \frac{\pi}{4}(D^2 - d^2)$$

where D = diameter of outer ring
  d = diameter of inner ring

Steam Tables and Charts

Steam Tables

The properties of steam, including pressure, temperature, specific volume, total heat (enthalpy), entropy, and superheat, are given in steam tables for use in solving problems (see Chap. 3). Table C.1 gives the properties of saturated steam with reference to the absolute pressure in pounds per square inch, shown in the left column. Table C.2 gives the same data except that the data correspond to the temperature (left column). When the pressure of saturated steam is known, use Table C.1 to find the other properties. When the temperature of saturated steam is known, use Table C. 2 to find the other properties.

The properties of wet steam are not given directly in the steam tables but may be calculated from the data given in these tables when the moisture content is known. (See Chap. 3 for the method used in making wet steam calculations.)

When steam is heated above the saturation temperature, it is said to be *superheated*; it contains more heat per pound and has a greater volume than shown in the saturated steam tables. Table C.3 gives volume in cubic feet per pound of steam (v), total heat (enthalpy) in Btus per pound (h), and entropy (s)[*] for superheated steam at various absolute pressures and temperatures in degrees Fahrenheit. The saturated temperatures for the respective absolute pressures are given by the numbers in parentheses directly under the pressures in the left column. The temperature above saturation (amount of superheat) is found by subtracting the saturation temperature from the total

---

[*]Entropy is used in detailed thermodynamic studies, and not included in this book.

temperature. For example, at 400 psia and 800°F total temperature, the degrees of superheat are

$$800 - 444.59 = 355.41°F$$

since the saturated temperature at 400 psia is 444.59°F.

Table C.4 shows the saturation temperature at steam pressures varying from 0.20 inHg to the critical pressure of 3206.2 psia. For example, as known, the saturation temperatures at 14.7 psia is 212°F. At 1 inHg, the saturation temperature is 79.03°F.

Steam Charts

The properties of steam may be arranged graphically in the form of charts for convenience in solving problems. The enthalpy-entropy diagram for steam (Chart C.1 and Table C.4), known as the *Mollier chart,* has a wide application. The vertical axis represents total heat (enthalpy) in Btus per pound; the horizontal axis, entropy per pound. The curved lines plotted on the chart represent the pressure in pounds per square inch absolute, the steam temperature and superheat in degrees Fahrenheit, and the percentage of moisture in the wet steam range. When two properties of steam are known, the others may be read directly from the chart. This applies to wet, saturated, and superheated steam.

The chart shows that steam at 100 psia and 600°F will have a heat content of 1329 Btu as compared with 1329.1 as given in Table C.3.

Steam having an absolute pressure of 50 psia and containing 1100 Btu/lb is found from the chart to contain 8.0 percent moisture. Many problems involving the use of steam may be solved directly by use of the Mollier chart.

TABLE C.1 Dry Saturated Steam Pressure

Abs press., psi p	Temp., °F t	Specific volume, ft³/lb Sat. liquid v_f	Sat. vapor v_g	Enthalpy, Btu/lb Sat. liquid h_f	Evap. h_{fg}	Sat. vapor h_g	Entropy Sat. liquid s_f	Evap. s_{fg}	Sat. vapor s_g
1.0	101.74	0.01614	333.6	69.70	1036.3	1106.0	0.1326	1.8456	1.9782
2.0	126.08	0.01623	173.73	93.99	1022.2	1116.2	0.1749	1.7451	1.9200
3.0	141.48	0.01630	118.71	109.37	1013.2	1122.6	0.2008	1.6855	1.8863
4.0	152.97	0.01636	90.63	120.86	1006.4	1127.3	0.2198	1.6427	1.8625
5.0	164.24	0.01640	73.52	130.13	1001.0	1131.1	0.2347	1.6094	1.8441
6.0	170.06	0.01645	61.98	137.96	996.2	1134.2	0.2472	1.5820	1.8292
7.0	176.85	0.01649	53.64	144.76	992.1	1136.9	0.2581	1.5586	1.8167
8.0	182.86	0.01653	47.34	150.79	998.5	1139.3	0.2674	1.5383	1.8057
9.0	188.28	0.01656	42.40	156.22	985.2	1141.4	0.2759	1.5203	1.7962
10	193.21	0.01659	38.42	161.17	982.1	1143.3	0.2835	1.5041	1.7876
14.696	212.00	0.01672	26.80	180.07	970.3	1150.4	0.3120	1.4446	1.7566
15	213.03	0.01672	26.29	181.11	969.7	1150.8	0.3135	1.4415	1.7549
20	227.96	0.01683	20.089	196.16	960.1	1156.3	0.3356	1.3962	1.7319
25	240.07	0.01692	16.303	208.42	952.1	1160.6	0.3533	1.3606	1.7139
30	250.33	0.01701	13.746	218.82	945.3	1164.1	0.3680	1.3313	1.6993
35	259.28	0.01708	11.898	227.91	939.2	1167.1	0.3807	1.3063	1.6870
40	267.25	0.01715	10.498	236.03	933.7	1169.7	0.3919	1.2844	1.6763
45	274.44	0.01721	9.401	243.36	928.6	1172.0	0.4019	1.2650	1.6669
50	281.01	0.01727	8.515	250.09	924.0	1174.1	0.4110	1.2474	1.6585
55	287.07	0.01732	7.787	256.30	919.6	1175.9	0.4193	1.2316	1.6509
60	292.71	0.01738	7.175	262.09	915.5	1177.6	0.4270	1.2168	1.6438
65	297.97	0.01743	6.655	267.50	911.6	1179.1	0.4342	1.2032	1.6374
70	302.92	0.01748	6.206	272.61	907.9	1180.6	0.4409	1.1906	1.6315
75	307.60	0.01753	5.816	277.43	904.5	1181.9	0.4472	1.1787	1.6259
80	312.03	0.01757	5.472	282.02	901.1	1183.1	0.4531	1.1676	1.6207
85	316.25	0.01761	5.168	286.39	897.8	1184.2	0.4587	1.1571	1.6158
90	320.27	0.01766	4.896	290.56	894.7	1185.3	0.4641	1.1471	1.6112
95	324.12	0.01770	4.652	294.56	891.7	1186.2	0.4692	1.1376	1.6068
100	327.81	0.01774	4.432	298.40	888.8	1187.2	0.4740	1.1286	1.6026
110	334.77	0.01782	4.049	305.66	883.2	1188.9	0.4832	1.1117	1.5948
120	341.25	0.01789	3.728	312.44	877.9	1190.4	0.4916	1.0962	1.5878
130	347.32	0.01796	3.455	318.81	872.9	1191.7	0.4995	1.0817	1.5812
140	353.02	0.01802	3.220	324.82	868.2	1193.0	0.5069	1.0682	1.5751
150	358.42	0.01809	3.015	330.51	863.6	1194.1	0.5138	1.0556	1.5694
160	363.53	0.01815	2.834	335.93	859.2	1195.1	0.5204	1.0436	1.5640
170	368.41	0.01822	2.675	341.09	854.9	1196.0	0.5266	1.0324	1.5590
180	373.06	0.01827	2.532	346.03	850.8	1196.9	0.5325	1.0217	1.5542
190	377.51	0.01833	2.404	350.79	846.8	1197.6	0.5381	1.0116	1.5497
200	381.79	0.01839	2.288	355.36	843.0	1198.4	0.5435	1.0018	1.5453
250	400.95	0.01865	1.8438	376.00	825.1	1201.1	0.5675	0.9588	1.5263
300	417.33	0.01890	1.5433	393.84	809.0	1202.8	0.5879	0.9225	1.5104
350	431.72	0.01913	1.3260	409.69	794.2	1203.9	0.6056	0.8910	1.4966
400	444.59	0.0193	1.1613	424.0	780.5	1204.5	0.6214	0.8630	1.4844
450	456.28	0.0195	1.0320	437.2	767.4	1204.6	0.6356	0.8378	1.4734
500	467.01	0.0197	0.9278	449.4	755.0	1204.4	0.6487	0.8147	1.4634
550	476.94	0.0199	0.8424	460.8	743.1	1203.9	0.6608	0.7934	1.4542

TABLE C.1   Dry Saturated Steam Pressure (*Continued*)

Abs press., psi p	Temp., °F t	Specific volume, ft³/lb		Enthalpy, Btu/lb			Entropy		
		Sat. liquid v_f	Sat. vapor v_g	Sat. liquid h_f	Evap. h_{fg}	Sat. vapor h_g	Sat. liquid s_f	Evap. s_{fg}	Sat. vapor s_g
600	486.21	0.0201	0.7698	471.6	731.6	1203.2	0.6720	0.7734	1.4454
650	494.90	0.0203	0.7083	481.8	720.5	1202.3	0.6826	0.7548	1.4374
700	503.10	0.0205	0.6554	491.5	709.7	1201.2	0.6925	0.7371	1.4296
750	510.86	0.0207	0.6092	500.8	699.2	1200.0	0.7019	0.7204	1.4223
800	518.23	0.0209	0.5687	509.7	688.9	1198.6	0.7108	0.7045	1.4153
850	525.26	0.0210	0.5327	518.3	678.8	1197.1	0.7194	0.6891	1.4085
900	531.98	0.0212	0.5006	526.6	668.8	1195.4	0.7275	0.6744	1.4020
950	538.43	0.0214	0.4717	534.6	659.1	1193.7	0.7355	0.6602	1.3957
1000	544.61	0.0216	0.4456	542.4	649.4	1191.8	0.7430	0.6467	1.3897
1100	556.31	0.0220	0.4001	557.4	630.4	1187.8	0.7575	0.6205	1.3780
1200	567.22	0.0223	0.3619	571.7	611.7	1183.4	0.7711	0.5956	1.3667
1300	577.46	0.0227	0.3293	585.4	593.2	1178.6	0.7840	0.5719	1.3559
1400	587.10	0.0231	0.3012	598.7	574.7	1173.4	0.7963	0.5491	1.3454
1500	596.23	0.0235	0.2765	611.6	556.3	1167.9	0.8082	0.5269	1.3351
2000	635.82	0.0257	0.1878	671.7	463.4	1135.1	0.8619	0.4230	1.2849
2500	668.13	0.0287	0.1307	730.6	360.5	1091.1	0.9126	0.3197	1.2322
3000	695.36	0.0346	0.0858	802.5	217.8	1020.3	0.9731	0.1885	1.1615
3206.2	705.40	0.0503	0.0503	902.7	0	902.7	1.0580	0	1.0580

SOURCE: Abridged from Joseph H. Keenan and Frederick G. Keyes, *Thermodynamic Properties of Steam,* Wiley, New York.

TABLE C.2   Dry Saturated Steam Temperature

Temp., °F t	Abs press., psi p	Specific volume, ft³/lb Sat. liquid v_f	Evap. v_{fg}	Sat. vapor v_g	Enthalpy, Btu/lb Sat. liquid h_f	Evap. h_{fg}	Sat. vapor h_g	Entropy Sat. liquid s_f	Evap. s_{fg}	Sat. vapor s_g
32	0.08854	0.01602	3306	3306	0.00	1075.8	1075.8	0.0000	2.1877	2.1877
35	0.09995	0.01602	2947	2947	3.02	1074.1	1077.1	0.0061	2.1709	2.1770
40	0.12170	0.01602	2444	2444	8.05	1071.3	1079.3	0.0162	2.1435	2.1597
45	0.14752	0.01602	2036.4	2036.4	13.06	1068.4	1081.5	0.0262	2.1167	2.1429
50	0.17811	0.01603	1703.2	1703.2	18.07	1065.6	1083.7	0.0361	2.0903	2.1264
60	0.2563	0.01604	1206.6	1205.7	28.06	1059.9	1088.0	0.0555	2.0393	2.0948
70	0.3631	0.01606	867.8	867.9	38.04	1054.3	1092.3	0.0745	1.9902	2.0647
80	0.5069	0.01608	633.1	633.1	48.02	1048.6	1096.6	0.0932	1.9428	2.0360
90	0.6982	0.01610	468.0	468.0	57.99	1042.9	1100.9	0.1115	1.8972	2.0087
100	0.9492	0.01613	350.3	350.4	67.97	1037.2	1105.2	0.1295	1.8531	1.9826
110	1.2748	0.01617	265.3	265.4	77.94	1031.6	1109.5	0.1471	1.8106	1.9577
120	1.6924	0.01620	203.25	203.27	87.92	1025.8	1113.7	0.1645	1.7694	1.9339
130	2.2225	0.01625	157.32	157.34	97.90	1020.0	1117.9	0.1816	1.7296	1.9112
140	2.8886	0.01629	122.99	123.01	107.89	1014.1	1122.0	0.1984	1.6910	1.8894
150	3.718	0.01634	97.06	97.07	117.89	1008.2	1126.1	0.2149	1.6537	1.8685
160	4.741	0.01639	77.27	77.29	127.89	1002.3	1130.2	0.2311	1.6174	1.8485
170	5.992	0.01645	62.04	62.06	137.90	996.3	1134.2	0.2472	1.5822	1.8293
180	7.510	0.01651	50.21	50.23	147.92	990.2	1138.1	0.2630	1.5480	1.8109
190	9.339	0.01657	40.94	40.96	157.95	984.1	1142.0	0.2785	1.5147	1.7932
200	11.526	0.01663	33.62	33.64	167.99	977.9	1145.9	0.2938	1.4824	1.7762
210	14.123	0.01670	27.80	27.82	178.05	971.6	1149.7	0.3090	1.4508	1.7598
212	14.696	0.01672	26.78	26.80	180.07	970.3	1150.4	0.3120	1.4446	1.7566
220	17.186	0.01677	23.13	23.15	188.13	965.2	1153.4	0.3239	1.4201	1.7440
230	20.780	0.01684	19.365	19.382	198.23	958.8	1157.0	0.3387	1.3901	1.7288
240	24.969	0.01692	16.306	16.323	208.34	952.2	1160.5	0.3531	1.3609	1.7140
250	29.825	0.01700	13.804	13.821	216.48	945.5	1164.0	0.3675	1.3323	1.6998
260	35.429	0.01709	11.746	11.763	228.64	938.7	1167.3	0.3817	1.3043	1.6860
270	41.858	0.01717	10.044	10.061	238.84	931.8	1170.6	0.3958	1.2769	1.6727
280	49.203	0.01726	8.628	8.645	249.06	924.7	1173.8	0.4096	1.2501	1.6597
290	57.556	0.01735	7.444	7.461	259.31	917.5	1176.8	0.4234	1.2238	1.6472
300	67.013	0.01745	6.449	6.466	269.59	910.1	1179.7	0.4369	1.1980	1.6350
310	77.68	0.01755	5.609	5.626	279.92	902.6	1182.5	0.4504	1.1727	1.6231
320	89.66	0.01765	4.896	4.914	290.28	894.9	1185.2	0.4637	1.1478	1.6115
330	103.06	0.01776	4.289	4.307	300.68	887.0	1187.7	0.4769	1.1233	1.6002
340	118.01	0.01787	3.770	3.788	311.13	879.0	1190.1	0.4900	1.0992	1.5891
350	134.63	0.01799	3.324	3.342	321.63	870.7	1192.3	0.5029	1.0754	1.5783
360	153.04	0.01811	2.939	2.957	332.18	862.2	1194.4	0.5158	1.0519	1.5677
370	173.37	0.01823	2.606	2.625	342.79	853.5	1196.3	0.5286	1.0287	1.5573
380	195.77	0.01836	2.317	2.335	353.45	844.6	1198.1	0.5413	1.0059	1.5471
390	220.37	0.01850	2.0651	2.0836	364.17	835.4	1199.6	0.5539	0.9832	1.5371
400	247.31	0.01864	1.8447	1.8633	374.97	826.0	1201.0	0.5664	0.9608	1.5272
410	276.75	0.01878	1.6512	1.6700	385.83	816.3	1202.1	0.5788	0.9386	1.5174
420	308.83	0.01894	1.4811	1.5000	396.77	806.3	1203.1	0.5912	0.9166	1.5078
430	343.72	0.01910	1.3308	1.3499	407.79	796.0	1203.8	0.6035	0.8947	1.4982
440	381.59	0.01926	1.1979	1.2171	418.90	785.4	1204.3	0.6158	0.8730	1.4887

TABLE C.2   Dry Saturated Steam Temperature (*Continued*)

Temp., °F t	Abs press., psi p	Specific volume, ft³/lb			Enthalpy, Btu/lb			Entropy		
		Sat. liquid v_f	Evap. v_{fg}	Sat. vapor v_g	Sat. liquid h_f	Evap. h_{fg}	Sat. vapor h_g	Sat. liquid s_f	Evap. s_{fg}	Sat. vapor s_g
450	422.6	0.0194	1.0799	1.0993	430.1	774.5	1204.6	0.6280	0.8513	1.4793
460	466.9	0.0196	0.9748	0.9944	441.4	763.2	1204.6	0.6402	0.8298	1.4700
470	514.7	0.0198	0.8811	0.9009	452.8	751.5	1204.3	0.6523	0.8083	1.4606
480	566.1	0.0200	0.7972	0.8172	464.4	739.4	1203.7	0.6645	0.7868	1.4513
490	621.4	0.0202	0.7221	0.7423	476.0	726.8	1202.8	0.6766	0.7653	1.4419
500	680.8	0.0204	0.6545	0.6749	487.8	713.9	1201.7	0.6887	0.7438	1.4325
520	812.4	0.0209	0.5385	0.5594	511.9	686.4	1198.2	0.7130	0.7006	1.4136
540	962.5	0.0215	0.4434	0.4649	536.6	656.6	1193.2	0.7374	0.6568	1.3942
560	1133.1	0.0221	0.3647	0.3868	562.2	624.2	1186.4	0.7621	0.6121	1.3742
580	1325.8	0.0228	0.2989	0.3217	588.9	588.4	1177.3	0.7872	0.5659	1.3532
600	1542.9	0.0236	0.2432	0.2668	617.0	548.5	1165.5	0.8131	0.5176	1.3307
620	1786.6	0.0247	0.1955	0.2201	646.7	503.6	1150.3	0.8398	0.4664	1.3062
640	2059.7	0.0260	0.1538	0.1798	678.6	452.0	1130.5	0.8679	0.4110	1.2789
660	2365.4	0.0278	0.1165	0.1442	714.2	390.2	1104.4	0.8987	0.3485	1.2472
680	2708.1	0.0305	0.0810	0.1115	757.3	309.9	1067.2	0.9351	0.2719	1.2071
700	3093.7	0.0369	0.0392	0.0761	823.3	172.1	995.4	0.9905	0.1484	1.1389
705.4	3206.2	0.0503	0	0.0503	902.7	0	902.7	1.0580	0	1.0580

SOURCE: Abridged from Joseph H. Keenan and Frederick G. Keyes, *Thermodynamic Properties of Steam,* Wiley, New York.

TABLE C.3 Properties of Superheated Steam*

Abs press., psi (sat. temp)		400	500	600	700	800	900	1000	1100	1200	1400
1 (101.74)	v	512.0	571.6	631.2	690.8	750.4	809.9	869.5	929.1	988.7	1107.8
	h	1241.7	1288.3	1335.8	1383.8	1432.8	1482.7	1533.5	1585.2	1637.7	1745.7
	s	2.1720	2.2233	2.2702	2.3137	2.3542	2.3932	2.4283	2.4625	2.4952	2.5566
5 (162.24)	v	102.26	114.22	126.16	138.10	150.03	161.95	173.87	185.79	197.71	221.6
	h	1241.8	1288.0	1335.4	1383.6	1432.7	1482.6	1533.4	1585.1	1637.7	1745.7
	s	1.9942	2.0456	2.0927	2.1361	2.1767	2.2148	2.2509	2.2851	2.3178	2.3792
10 (193.21)	v	51.04	57.05	63.03	69.01	74.98	80.95	86.92	92.88	98.84	110.77
	h	1240.6	1287.5	1335.1	1383.4	1432.5	1482.4	1533.2	1585.0	1637.6	1745.6
	s	1.9172	1.9689	2.0160	2.0596	2.1002	2.1383	2.1744	2.2086	2.2413	2.3028
14.696 (212.00)	v	34.68	38.78	42.86	46.94	51.00	55.07	59.13	63.19	67.25	75.37
	h	1239.9	1287.1	1334.8	1383.2	1432.3	1482.3	1533.1	1584.8	1637.5	1745.5
	s	1.8743	1.9261	1.9734	2.0170	2.0576	2.0958	2.1319	2.1662	2.1989	2.2603
20 (227.96)	v	25.43	28.46	31.47	34.47	37.46	40.45	43.44	46.42	49.41	55.37
	h	1239.2	1286.6	1334.4	1382.9	1432.1	1482.1	1533.0	1584.7	1637.4	1745.4
	s	1.8396	1.8918	1.9392	1.9829	2.0235	2.0618	2.0978	2.1321	2.1648	2.2263
40 (267.25)	v	12.628	14.168	15.688	17.198	18.702	20.20	21.70	23.20	24.69	27.68
	h	1236.5	1284.8	1333.1	1381.9	1431.3	1481.4	1532.4	1584.3	1637.0	1745.1
	s	1.7608	1.8140	1.8619	1.9058	1.9467	1.9850	2.0212	2.0555	2.0883	2.1498
60 (292.71)	v	8.357	9.403	10.427	11.441	12.449	13.452	14.454	15.453	16.451	18.446
	h	1233.6	1283.0	1331.8	1380.9	1430.5	1480.8	1531.9	1583.8	1636.6	1744.8
	s	1.7135	1.7678	1.8162	1.8305	1.9015	1.9400	1.9762	2.0106	2.0434	2.1049
80 (312.03)	v	6.220	7.020	7.797	8.562	9.322	10.077	10.830	11.582	12.332	13.830
	h	1230.7	1281.1	1330.5	1379.9	1429.7	1480.1	1531.3	1583.4	1636.2	1744.5
	s	1.6791	1.7346	1.7836	1.8281	1.8694	1.9079	1.9442	1.9787	2.0115	2.0731
100 (327.81)	v	4.937	5.589	6.218	6.835	7.446	8.052	8.656	9.259	9.860	11.060
	h	1227.6	1279.1	1329.1	1378.9	1428.9	1479.5	1530.8	1582.9	1635.7	1744.2
	s	1.6518	1.7085	1.7581	1.8029	1.8443	1.8829	1.9193	1.9538	1.9867	2.0484

Temperature °F

v = specific volume, ft³/lb; h = enthalpy, Btu/lb; s = entropy.
SOURCE: Abridged from Joseph H. Keenan and Frederick G. Keyes, *Thermodynamic Properties of Steam*, Wiley, New York.

TABLE C.3 Properties of Superheated Steam* (Continued)

Abs press, psi (sat. temp)		Temperature °F									
		400	500	600	700	800	900	1000	1100	1200	1400
120 (341.25)	v	4.081	4.636	5.165	5.683	6.195	6.702	7.207	7.710	8.212	9.214
	h	1224.4	1277.2	1327.7	1377.8	1428.1	1478.8	1530.2	1582.4	1635.3	1743.9
	s	1.6287	1.6869	1.7370	1.7822	1.8237	1.8625	1.8990	1.9335	1.9664	2.0281
140 (353.02)	v	3.468	3.954	4.413	4.861	5.301	5.738	6.172	6.604	7.035	7.895
	h	1221.1	1275.2	1326.4	1376.8	1427.3	1478.2	1529.7	1581.9	1634.9	1743.5
	s	1.6087	1.6683	1.7190	1.7645	1.8063	1.8451	1.8817	1.9163	1.9493	2.0110
160 (363.53)	v	3.008	3.443	3.849	4.244	4.631	5.015	5.396	5.775	6.152	6.906
	h	1217.6	1273.1	1325.0	1375.7	1426.4	1477.5	1529.1	1581.4	1634.5	1743.2
	s	1.5908	1.6519	1.7033	1.7491	1.7911	1.8301	1.8667	1.9014	1.9344	1.9962
180 (373.06)	v	2.649	3.044	3.411	3.764	4.110	4.452	4.792	5.129	5.466	6.136
	h	1214.0	1271.0	1323.5	1374.7	1425.6	1476.8	1528.6	1581.0	1634.1	1742.9
	s	1.5745	1.6373	1.6894	1.7355	1.7776	1.8167	1.8534	1.8882	1.9212	1.9831
200 (381.79)	v	2.361	2.726	3.060	3.380	3.693	4.002	4.309	4.613	4.917	5.521
	h	1210.3	1268.9	1322.1	1373.6	1424.8	1476.2	1528.0	1580.5	1633.7	1742.6
	s	1.5594	1.6240	1.6767	1.7232	1.7655	1.8048	1.8415	1.8763	1.9094	1.9713
220 (389.86)	v	2.125	2.465	2.772	3.066	3.352	3.634	3.913	4.191	4.467	5.017
	h	1206.5	1266.7	1320.7	1372.6	1424.0	1475.5	1527.5	1580.0	1633.3	1742.3
	s	1.5453	1.6117	1.6652	1.7120	1.7545	1.7939	1.8308	1.8656	1.8987	1.9607
240 (397.37)	v	1.9276	2.247	2.533	2.804	3.068	3.327	3.584	3.839	4.093	4.597
	h	1202.5	1264.5	1319.2	1371.5	1423.2	1474.8	1526.9	1579.6	1632.9	1742.0
	s	1.5319	1.6003	1.6546	1.7017	1.7444	1.7839	1.8209	1.8558	1.8889	1.9510
260 (404.42)	v	—	2.063	2.330	2.582	2.827	3.067	3.305	3.541	3.776	4.242
	h	—	1262.3	1317.7	1370.4	1422.3	1474.2	1526.3	1579.1	1632.5	1741.7
	s	—	1.5897	1.6447	1.6922	1.7352	1.7748	1.8118	1.8467	1.8799	1.9420
280 (411.05)	v	—	1.9047	2.156	2.392	2.621	2.845	3.066	3.286	3.504	3.938
	h	—	1260.0	1316.2	1369.4	1421.5	1473.5	1525.8	1578.6	1632.1	1741.4
	s	—	1.5796	1.6354	1.6834	1.7264	1.7662	1.8033	1.8383	1.8716	1.9337

300	v	—	—	1.7675	2.005	2.227	2.442	2.652	2.859	3.065	3.269	3.674
(417.33)	h	—	—	1257.6	1314.7	1368.3	1420.6	1472.8	1525.2	1578.1	1631.7	1741.0
	s	—	—	1.5701	1.6268	1.6751	1.7184	1.7582	1.7954	1.8305	1.8638	1.9260
350	v	—	—	1.4923	1.7036	1.8980	2.084	2.266	2.445	2.622	2.798	3.147
(431.72)	h	—	—	1251.5	1310.9	1365.5	1418.5	1471.1	1523.8	1577.0	1630.7	1740.3
	s	—	—	1.5481	1.6070	1.6563	1.7002	1.7403	1.7777	1.8130	1.8463	1.9086
400	v	—	—	1.2851	1.4770	1.6508	1.8161	1.9767	2.134	2.290	2.445	2.751
(444.59)	h	—	—	1245.1	1306.9	1362.7	1416.4	1469.4	1522.4	1575.8	1629.6	1739.5
	s	—	—	1.5281	1.5894	1.6398	1.6842	1.7247	1.7623	1.7977	1.8311	1.8936

v = specific volume, ft^{3}lb; h = enthalpy, Btu/lb; s = entropy.

SOURCE: Abridged from Joseph H. Keenan and Frederick G. Keyes, *Thermodynamic Properties of Steam*, Wiley, New York.

TABLE C.3 Properties of Superheated Steam* (Continued)

Abs press., psi (sat. temp)		Temperature °F											
		500	600	620	640	660	680	700	800	900	1000	1200	1400
450 (456.28)	v	1.1231	1.3005	1.3332	1.3652	1.3967	1.4278	1.4584	1.6074	1.7516	1.8928	2.170	2.443
	h	1238.4	1302.8	1314.6	1326.5	1337.5	1348.8	1359.9	1414.3	1467.7	1521.0	1628.6	1738.7
	s	1.5095	1.5735	1.5845	1.5951	1.6054	1.6153	1.6250	1.6699	1.7108	1.7486	1.8177	1.8803
500 (467.01)	v	0.9927	1.1591	1.1893	1.2188	1.2478	1.2763	1.3044	1.4405	1.5715	1.6996	1.9504	2.197
	h	1231.3	1298.6	1310.7	1322.6	1334.2	1345.7	1357.0	1412.1	1466.0	1519.6	1627.6	1737.9
	s	1.4919	1.5588	1.5701	1.5810	1.5915	1.6016	1.6115	1.6571	1.6982	1.7363	1.8056	1.8683
550 (476.94)	v	0.8852	1.0431	1.0714	1.0989	1.1259	1.1523	1.1783	1.3038	1.4241	1.5414	1.7706	1.9957
	h	1223.7	1294.3	1306.8	1318.9	1330.8	1342.5	1354.0	1409.9	1464.3	1518.2	1626.6	1737.1
	s	1.4751	1.5451	1.5568	1.5680	1.5787	1.5890	1.5991	1.6452	1.6868	1.7250	1.7946	1.8575
600 (486.21)	v	0.7947	0.9463	0.9729	0.9988	1.0241	1.0489	1.0732	1.1899	1.3013	1.4096	1.6208	1.8279
	h	1215.7	1289.9	1302.7	1315.2	1327.4	1339.3	1351.1	1407.7	1462.5	1516.7	1625.5	1736.3
	s	1.4586	1.5323	1.5443	1.5558	1.5667	1.5773	1.5875	1.6343	1.6762	1.7147	1.7846	1.8476
700 (503.10)	v	—	0.7934	0.8177	0.8411	0.8639	0.8860	0.9077	1.0108	1.1082	1.2024	1.3853	1.5641
	h	—	1280.6	1294.3	1307.5	1320.3	1332.8	1345.0	1403.2	1459.0	1513.9	1623.5	1734.8
	s	—	1.5084	1.5212	1.5333	1.5449	1.5559	1.5665	1.6147	1.6573	1.6963	1.7666	1.8299
800 (518.23)	v	—	0.6779	0.7006	0.7223	0.7433	0.7635	0.7833	0.8763	0.9633	1.0470	1.2088	1.3662
	h	—	1270.7	1285.4	1299.4	1312.9	1325.9	1338.6	1398.6	1455.4	1511.0	1621.4	1733.2
	s	—	1.4863	1.5000	1.5129	1.5250	1.5366	1.5476	1.5972	1.6407	1.6801	1.7510	1.8146
900 (531.98)	v	—	0.5873	0.6089	0.6294	0.6491	0.6680	0.6863	0.7716	0.8506	0.9262	1.0714	1.2124
	h	—	1260.1	1275.9	1290.9	1305.1	1318.8	1332.1	1393.9	1451.8	1508.1	1619.3	1731.6
	s	—	1.4653	1.4800	1.4938	1.5066	1.5187	1.5303	1.5814	1.6257	1.6656	1.7371	1.8009
1000 (544.61)	v	—	0.5140	0.5350	0.5546	0.5733	0.5912	0.6084	0.6878	0.7604	0.8294	0.9615	1.0893
	h	—	1248.8	1265.9	1281.9	1297.0	1311.4	1325.3	1389.2	1448.2	1505.1	1617.3	1730.0
	s	—	1.4450	1.4610	1.4757	1.4893	1.5021	1.5141	1.5670	1.6121	1.6525	1.7245	1.7886

1100 (556.31)	v	—	—	0.4532	0.4738	0.4929	0.5110	0.5281	0.5445	0.6191	0.6866	0.7503	0.8716	0.9885
	h	—	—	1236.7	1255.3	1272.4	1288.5	1303.7	1318.3	1384.3	1444.5	1502.2	1615.2	1728.4
	s	—	—	1.4251	1.4425	1.4583	1.4728	1.4862	1.4989	1.5535	1.5995	1.6405	1.7130	1.7775
1200 (567.22)	v	—	—	0.4016	0.4222	0.4410	0.4586	0.4752	0.4909	0.5617	0.6250	0.6843	0.7967	0.9046
	h	—	—	1223.5	1243.9	1262.4	1279.6	1295.7	1311.0	1379.3	1440.7	1499.2	1613.1	1726.9
	s	—	—	1.4052	1.4243	1.4413	1.4568	1.4710	1.4843	1.5409	1.5879	1.6293	1.7025	1.7672
1400 (587.10)	v	—	—	0.3174	0.3390	0.3580	0.3753	0.3912	0.4062	0.4714	0.5281	0.5805	0.6789	0.7727
	h	—	—	1193.0	1218.4	1240.4	1260.3	1278.5	1295.5	1369.1	1433.1	1493.2	1608.9	1723.7
	s	—	—	1.3639	1.3877	1.4079	1.4258	1.4419	1.4567	1.5177	1.5666	1.6093	1.6836	1.7489
1600 (604.90)	v	—	—	—	0.2733	0.2936	0.3112	0.3271	0.3417	0.4034	0.4553	0.5027	0.5906	0.6738
	h	—	—	—	1187.8	1215.2	1238.7	1259.6	1278.7	1358.4	1425.3	1487.0	1604.6	1720.5
	s	—	—	—	1.3489	1.3741	1.3952	1.4137	1.4303	1.4964	1.5476	1.5914	1.6669	1.7328
1800 (621.03)	v	—	—	—	—	0.2407	0.2597	0.2760	0.2907	0.3502	0.3986	0.4421	0.5218	0.5968
	h	—	—	—	—	1185.1	1214.0	1238.5	1260.3	1347.2	1417.4	1480.8	1600.4	1717.3
	s	—	—	—	—	1.3377	1.3638	1.3855	1.4044	1.4765	1.5301	1.5752	1.6520	1.7185
2000 (635.82)	v	—	—	—	—	0.1936	0.2161	0.2337	0.2489	0.3074	0.3532	0.3935	0.4668	0.5352
	h	—	—	—	—	1145.6	1184.9	1214.8	1240.0	1335.5	1409.2	1474.5	1596.1	1714.1
	s	—	—	—	—	1.2945	1.3300	1.3564	1.3783	1.4576	1.5139	1.5603	1.6384	1.7055
2500 (668.13)	v	—	—	—	—	—	—	0.1484	0.1686	0.2294	0.2710	0.3061	0.3678	0.4244
	h	—	—	—	—	—	—	1132.3	1176.8	1303.6	1387.8	1458.4	1585.3	1706.1
	s	—	—	—	—	—	—	1.2687	1.3073	1.4127	1.4772	1.5273	1.6088	1.6775
3000 (695.36)	v	—	—	—	—	—	—	—	0.0984	0.1760	0.2159	0.2476	0.3018	0.3505
	h	—	—	—	—	—	—	—	1060.7	1267.2	1365.0	1441.8	1574.3	1698.0
	s	—	—	—	—	—	—	—	1.1966	1.3690	1.4439	1.4984	1.5837	1.6540
3206.2 (705.40)	v	—	—	—	—	—	—	—	—	0.1583	0.1981	0.2288	0.2806	0.3267
	h	—	—	—	—	—	—	—	—	1250.5	1355.2	1434.7	1569.8	1694.6
	s	—	—	—	—	—	—	—	—	1.3508	1.4309	1.4874	1.5742	1.6452

* v = specific volume, ft³/lb; h = enthalpy, Btu/lb; s = entropy.

SOURCE: Abridged from Joseph E. Keenan and Frederick G. Keyes, *Thermodynamic Properties of Steam*, Wiley, New York.

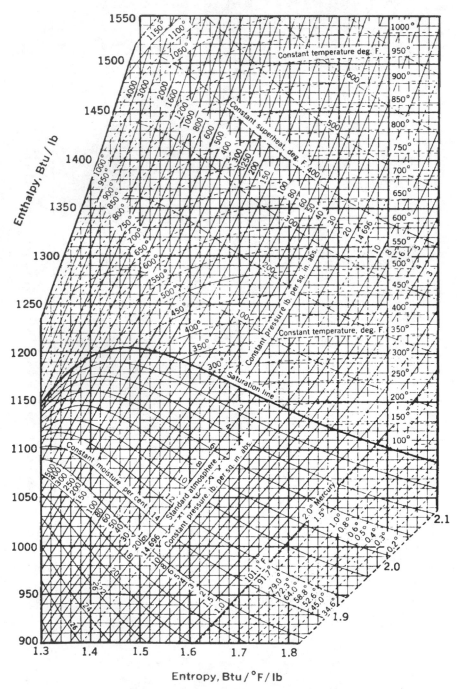

Chart C.1   A Mollier chart for steam. (Abstracted by permission from J. H. Keenan and F. G. Keyes, *Thermodynamic Properties of Steam*, Wiley, New York.)

TABLE C.4   Saturation Pressures and Temperature of Steam

Abs. press., in Hg	Sat. temp., °F	Abs. press., lb/in²	Sat. temp., °F	Abs. press., lb/in²	Sat. temp., °F
0.20	34.56	1.0	101.74	120	341.25
0.25	40.23	2	126.08	140	353.02
0.30	44.96	3	141.48	160	363.53
0.35	49.06	4	152.97	180	373.06
0.40	52.04	5	162.24	200	381.79
0.45	55.87	6	170.06	220	389.86
0.50	58.80	7	176.85	240	397.37
0.55	61.48	8	182.86	260	404.42
0.60	63.95	9	188.28	280	411.05
0.65	66.26	10	193.21	300	417.33
0.70	68.40	12	201.96	400	444.59
0.75	70.43	14	209.56	500	467.01
0.80	72.32	14.696	212.00	600	486.21
0.85	74.13	16	216.32	700	503.10
0.90	75.84	18	222.41	800	518.23
0.95	77.47	20	227.96	1000	544.61
1.00	79.03	25	240.07	1200	567.22
1.10	81.95	30	250.33	1400	587.10
1.20	84.65	35	259.28	1600	604.90
1.30	87.17	40	267.25	1800	621.03
1.40	89.51	45	274.44	2000	635.82
1.50	91.72	50	281.01	2200	649.46
1.60	93.80	60	292.71	2400	662.12
1.70	95.77	70	302.92	2600	673.94
1.80	97.65	80	312.03	2800	684.99
1.90	99.43	90	320.27	3000	695.36
2.00	101.14	100	327.81	3206.2	705.40

SOURCE: Abridged from Joseph H. Keenan and Frederick G. Keyes, *Thermodynamic Properties of Steam,* Wiley, New York.

Answers to Problems

Chapter 3

3.3 3.1416 in^2; 2546 psi

3.6 2827.4 in^2 (19.6 ft^2)

3.7 2,120,550 lb

3.8 $1\frac{5}{8}$ in, or 1.625 in

3.18 27,150 psi elastic limit
56,561 psi ultimate strength

3.19 610,725 lb

3.20 7,776,000 lb

3.21 1354 psi

3.22 $t = 3.21$ in; commercial size $t = 3.25$ in

3.26 1381 ft^2

3.27 115 boiler hp

3.29 269.4×10^6 Btu/h

3.33 h at 900 psia saturated = 1195.4 Btu/lb
h at 900 psia and 950°F = 1480 Btu/lb
Degrees superheat = 418.02°F

3.34 17,518, 13,841, and 3200 Btu/h per square foot (Btu/h/ft^2);
88,543 Btu/h per cubic foot (Btu/h/ft^3)

Chapter 4

4.10 2.09 lb of O$_2$/lb of coal; 9.02 lb of air/lb of coal

4.15 28.71 percent

4.17 13.53 lb of air/lb of coal

4.18 12,336 Btu/lb

4.30 22.3°API

Chapter 5

5.65 1.135 in of water theoretical
(0.908 in of water available)

Chapter 6

6.4 255.8 psi

6.5 6.5 psi low

6.20 3976 lb

Chapter 8

8.34 43.3 psi

8.35 288.7 ft

8.36 840 ft^3; 1,451,520 in^3; 52,500 lb; 6302.5 gal

8.37 5891 gal

8.38 48.4 gpm; 2906 gph

8.39 130 min

8.40 9.12 h

8.41 256.6 psi; 593 ft

8.42 3757 gal

8.43 33 ft

8.44 112.5 psi

8.45 3.53 in in diameter

8.46 102 strokes per minute

8.47 15.15 hp

8.48 485.4 hp

8.49 motor size, 550 hp
input, 393.4 kW

Chapter 10

10.10 1 in Hg = 0.491 psi
3.5 in Hg = 1.72 psi

10.28 1.83 psi

10.29 0.5 psi

10.30 0.74 psi

Chapter 11

11.21 $253,560/yr

11.22 7.74 in; use 8 in

11.24 6.78 in

Glossary

Absolute Pressure   The common gauge expresses a pressure in pounds per square inch called *gauge pressure*. When the gauge is open to the atmosphere, it reads zero. The gauge pressure plus atmospheric pressure is known as *absolute pressure*. Atmospheric pressure is 14.7 psi at sea level; it varies with location and atmospheric conditions. It is accurately indicated by a barometer.

Absolute Temperature   The temperature as read on the Fahrenheit scale plus 460 is the absolute temperature.

Absolute Zero of Pressure   The starting point of the absolute-pressure scale is absolute zero. It is lower than a *zero gauge* by an amount equal to the atmospheric pressure.

Absolute Zero of Temperature   The temperature 460° below zero Fahrenheit is *absolute zero*. At absolute zero there is a complete absence of heat. Absolute zero has never been attained, but it has been approached within a few degrees.

Acidity   Water is a chemical combination of hydrogen (H) and oxygen (O) that is represented by the formula H_2O. It may also be treated as a chemical combination of hydrogen ions (H^+) and hydroxyl ions (OH^-). If there is a greater number of hydrogen ions than hydroxyl ions as a result of the chemical action of impurities or solutes, the solution is acidic. A greater number of hydroxyl ions results in an alkaline solution. The degree of acidity or alkalinity of a substance is known as the hydrogen-ion concentration and is called the pH value. A pH value of 7.0 indicates neutral water; a value less than 7.0, acidity; and a value greater than 7.0, alkalinity.

Air Heater   A heat exchanger that transfers heat from a high-temperature medium, such as flue gas or steam, to incoming air.

Alkalinity   See *Acidity*.

Attemperator   This is also called a *desuperheater,* and it reduces the temperature of superheated steam at higher loads to permit a constant steam temperature over a defined load range.

Bag   A deep bulge in the shell of a fire-tube boiler.

Bag Filterhouse   A device consisting of multiple filter bags designed to remove dust particles from the flue-gas stream. It is also commonly called a *bag house* or a *fabric filter.*

Banking   The burning of solid fuel on a grate at a rate that is sufficient to maintain ignition only. It is also a combustion rate that is just sufficient to maintain normal operating pressure but with no steam capacity.

Boiler   A closed vessel in which water is heated and steam is generated and superheated, if desired, all performed under pressure by the application of heat.

Boiling Out   The process of boiling highly alkaline water in boiler pressure parts for the purpose of removing oils, greases, etc., prior to normal operation or after initial construction or after major repairs.

Boiling Point of Water   Water at atmospheric pressure boils at 212°F or 100°C.

Boyle's Law of Gases   When the temperature of a gas remains constant, the volume will be reduced to one-half if the absolute pressure is doubled; the absolute pressure will be reduced to one-half if the volume is doubled. (The volume of a gas varies inversely as the pressure: $p_1V_1 = p_2V_2$.)

British Thermal Unit (Btu)   A British thermal unit is used to measure heat energy. It is defined as the quantity of heat required to raise the temperature of 1 lb of water 1°F.

Caustic Embrittlement   This refers to the cracking of boiler steel occurring while the boiler is in service but not subjected to excessive pressure or temperature. Such failures are attributed to the boiler water's being too caustic, that is, too alkaline; hence the term *caustic embrittlement.*

Collection Efficiency of Particulates   The ratio of the weight of dust collected to the total weight of dust entering the collector.

Combustion   The chemical combination of oxygen with the combustible elements of a fuel, such as carbon and hydrogen, and this results in the production of heat.

Condensation   When steam or any other vapor is subjected to a change of state which reduces it to a liquid, it is said to be *condensed.* Steam is condensed in a condenser or heater by extracting heat. The water formed is called *condensate.*

Conduction   When heat is transmitted through a substance or from one substance in contact with another but without the bodies' themselves moving, the transfer is by conduction. Heat is conducted through the metal in the shell and tubes of a boiler. Substances differ widely in their ability to conduct heat. Metal is a good conductor; soot and boiler scale are very poor conductors. Heat transfer by conduction is a primary method used in fluidized bed boiler designs.

Continuous Blowdown   The continuous removal of concentrated boiler water for the purpose of controlling the total solids concentration in the remaining water.

Convection   When heat is carried by means of the movement of currents within a body, it is said to be transmitted by *convection*. The change in density of the substance, due to the heating, causes the movement. The circulation of water in a boiler carries heat from the tubes to the boiler drum.

Conversion of Heat Energy and Mechanical Energy   Heat energy and mechanical energy are convertible. There is a direct relation between heat energy and mechanical energy; 778 ft · lb is equivalent to 1 Btu.

Critical Pressure   The pressure at which there is no difference between the liquid and vapor states for water. This occurs at a pressure of 3206 psia.

Deaeration   The removal of air and gases from boiler feedwater prior to its introduction to a boiler. A deaerator serves this purpose and is part of the feedwater heating system.

Degrees of Superheat   The number of degrees between the superheated steam temperature and the saturated steam temperature at a particular pressure.

Demineralizer   An ion-exchange system that removes solids from water.

Density   The density of a substance is the number of units of weight that it contains per unit of volume. Water has a density of 62.5 lb/ft^3.

Design Pressure   The pressure used in the design of a boiler for the purpose of determining the minimum permissible thickness of pressure parts and also other characteristics of the boiler.

Dew Point   The temperature at which a vapor liquefies is called the *dew point*.

Downcomer   A tube or pipe in a boiler circulating water system through which water flows downward. On a boiler with a steam drum only, the downcomer is connected from the drum to the lower furnace waterwall headers. On a two-drum boiler, the downcomer is connected to the lower drum.

Draft   The difference between atmospheric pressure and a lower pressure that is present in a boiler.

Draft Loss   The drop in the pressure of flue gas between two points within a system that is caused by a resistance to flow.

Economizer   Heat-transfer surface that transfers heat from the flue gas to the boiler feedwater.

Efficiency   The efficiency of any system or piece of equipment is the output divided by the input, sometimes stated as the useful energy divided by the energy expended. The input and output may be expressed in any energy units. They must, however, be in the same units.

Electrostatic Precipitator   A device for collecting dust from a flue gas stream by placing an electric charge on the dust particle and removing that particle onto a collecting plate.

Energy   Energy is the ability to do work. Mechanical energy is expressed in footpounds or horsepowerhours; electric energy, in kilowatthours; and heat energy, in British thermal units.

Enthalpy   Enthalpy is the number of Btus that a substance contains above a specific datum. In the case of water and steam the reference condition is water at 32°F. The enthalpy values given in the steam tables are the Btus required to raise water or steam from 32°F to the specific temperature and pressure. These values are also referred to as *total heat*. It is expressed as Btus per pound of fluid (Btu/lb).

Evaporation   The process of changing a liquid into a vapor or a gas is known as *vaporization*. This is usually accomplished by the application of heat.

Excess Air   The combustion of fuel is primarily the combining of combustible substances of the fuel with oxygen of the air. A fuel requires a definite amount of oxygen, therefore air, to result in complete combustion. The amount of air used in excess of this amount is known as *excess air*. Excess air is necessary to result in complete combustion, but too much causes a decrease in efficiency.

Expansion   A change in temperature produces a change in the size of practically all substances. Each material changes a different amount for a given change in temperature. The change in length per degree of temperature change is known as the *linear coefficient of expansion*. Such coefficients for common materials may be found in handbooks. If the pressure of a gas is kept constant, the volume will change in proportion to the absolute temperature.

Factor of Evaporation   If 970.3 Btu is added to 1 lb of water that is at atmospheric pressure and 212°F, it will be converted into steam and the steam will be at atmospheric pressure and 212°F. This is termed *evaporation from and at 212°F*, or, briefly, *from and at*. The heat added to a pound of water by the boiler (from the time at which it enters until it leaves as steam) divided by 970.3 is the factor of evaporation.

Feedwater   This is the water that enters a boiler during operation, and it includes makeup water and condensate from the condenser.

Flame Impingement The contact on a surface by a flame from a burner that results in carbon deposits, incomplete combustion, and the high potential for failure of furnace wall tubes.

Fluidized Bed Combustion A process where a fuel is burned in a bed of granulated particles (such as sand or limestone) which are maintained in a mobile suspension by the upward flow of air and combustion products.

Fly Ash Suspended ash particles carried in the flue gas.

Force Force is that which produces, or tends to produce, motion. The force on the blade of a steam turbine produces motion. The force exerted on the head of a steam boiler does not produce motion, but it tends to; both are examples of force.

Freezing Point of Water Water freezes at 32 on the Fahrenheit scale of temperature measurement. This corresponds to 0 on the centigrade scale. When water freezes, its volume increases by about 9 percent.

Furnace Heat Release This is the number of Btus developed per hour in each cubic foot of furnace volume. It is usually assumed that all the heat available in the fuel burned is transformed into heat in the combustion process. Therefore, the furnace heat release equals the total fuel burned per hour times the Btu content divided by the furnace volume.

Fusion Temperature Fusion is the act or process of melting by heat or the state of being fused or melted. For coal, the fusion temperature is reported as initial-deformation temperature (I.D.T.), ash-softening temperature (A.S.T.), and ash-fusion temperature (A.F.T.), the test being made under a reducing atmosphere.

Grindability The characteristic of coal that identifies its ability to be pulverized and is used as a factor in determining the capacity of a pulverizer.

Head Head is the energy per pound of fluid.

Potential head: This refers to energy of position, measured by work possible in dropping a vertical distance.

Static pressure: This refers to energy per pound due to pressure; it is the height to which liquid can be raised by a given pressure.

Velocity: This refers to kinetic energy per pound; it is the vertical distance a liquid would have to fall to acquire the velocity V.

Total: This refers to the net difference between total suction and discharge heads.

Net Positive Suction (NPSH): This is the amount of energy in the liquid at the pump datum. It is the measure of the energy increase imparted to the liquid by the pump.

Heat   Heat is a form of energy.

Heat Content of Steam or Total Heat   This refers to the Btus that must be added to produce steam at the condition in question. Water at 32°F is usually taken as the starting point.

Horsepower   Horsepower (hp) is a unit of power, which is defined as the rate at which work is being performed. One horsepower is equal to 33,000 ft · lb/min.

Horsepowerhour   A horsepowerhour is 1 hp of energy expended continuously for 1 h. (1 hp · h = 2545 Btu.)

Hydrostatic Test   A strength and tightness test of a closed pressure vessel by water pressure.

Incomplete Combustion   The partial chemical combination of the combustible elements of a fuel with the oxygen in the air that results in unburned carbon loss.

Industrial Boiler   A boiler that produces steam or hot water primarily for process applications for industrial use with some use for heating. Industrial boilers cover a wide range of capacities, pressures, and temperatures and can burn a variety of solid, liquid, and gaseous fuels. Some units are designed for small electric utility applications.

Kilowatt   A kilowatt is a unit of electric power and is equal to 1000 watts. For direct current, watts equal amperes times volts ($P = I \times E$); for alternating current (single phase), watts equal amperes times volts times power factor ($P = I \times E \times \text{PF}$, where I = amperes, E = volts, P = watts, and PF = power factor).

Kilowatthour   A kilowatthour is 1 kW of energy expended continuously for 1 h.

Kinetic Energy   Kinetic energy is energy that a body has due to its motion.

Latent Heat of Evaporation   When a liquid is vaporized, a large amount of heat must be added to produce the change. This heat does not increase the temperature and is therefore called *latent heat*. The latent heat of vaporization of water at atmospheric pressure is 970.3 Btu.

Law of Conservation of Energy   The amount of energy in existence is constant. The machines that are built and operated do not produce energy; they merely change it from one form to another.

Mechanical Equivalent of Heat   This is sometimes called Joule's equivalent: 778 ft · lb of mechanical energy is equivalent to 1 Btu of heat energy.

Municipal Solid Waste (MSW)   Solid waste material as collected from households and commercial sources. It is highly heterogeneous in nature in that it varies in appearance and content in various parts of a country and the world and also varies from season to season. Its content also changes based on the recycling programs in the area.

Over-Fire Air Combustion air that is admitted into the furnace at a point above the fuel bed.

Package Boiler A boiler that is shipped complete with fuel-burning equipment (normally oil and gas fired), mechanical-draft equipment, automatic controls, and accessories. Shipment takes place in one or more major sections.

Power Power is the rate at which work is done. Footpounds express work, but the rate or time required determines the power: 33,000 ft · lb/min is one horsepower.

Pressure Drop The difference in pressure between two points in a system. For example, a superheater pressure drop is the difference in pressure from the boiler drum outlet, through the superheater tubes, to the superheater outlet.

Primary Air Combustion air that is introduced with the fuel at the burners.

Products of Combustion The gases, vapors, and solids that result from the combustion of fuel. It is often called *flue gas.*

Proximate Analysis The analysis of a solid fuel that determines the content of moisture, volatile matter, fixed carbon, and ash as percentages of the total weight of the fuel.

Radiation, Thermal Radiation is the transmission of heat without the use of a material carrier. The earth receives heat from the sun, and for most of that distance, the heat travels through a vacuum. When a furnace door is open, you can feel the heat even though air is being pulled into the furnace through the door. The lower rows of furnace and boiler tubes receive much heat by radiation. Radiated heat is very similar to visible light; they both travel at the same speed, namely, 186,000 miles per second.

Reheater A portion of a boiler that adds heat to steam to raise its temperature after the steam has performed part of its work. This is usually done on a utility boiler application after the initial superheated steam passes through the high-pressure section of a turbine. The lower-pressure steam from this section of the turbine returns to the boiler through the reheater, where the steam temperature is increased, and then returns to the low-pressure section of the turbine.

Safety Valve A valve that is spring loaded and automatically opens when pressure increases to the valve setting. It is used to prevent pressure in a vessel from exceeding the design pressure.

Saturated Steam Saturated steam is steam that contains no moisture. It is saturated with heat, since additional heat will raise the temperature above the boiling point and the removal of heat will result in the formation of water.

Secondary Air Combustion air that is supplied to a furnace and supplements the primary air.

Smoke Smoke refers to flue gases that contain enough unburned carbon and hydrocarbons to cause discoloration. The degree of coloration depends on the carbon present.

Specific Heat Different substances have different heat capacities. In fact the heat capacity of some substances changes as the temperature changes. By the definition of a Btu, the heat content of a pound of water per degree Fahrenheit is 1. The specific heat of any substance is the heat required to raise 1 lb of it 1°F.

Steam Generator A boiler to which water, fuel, and air or waste heat are supplied and in which steam is generated. It includes a furnace, heat-transfer heating surface, and fuel-burning equipment, and it also may include a superheater, reheater, economizer, air heater, flues and ducts, controls, and various auxiliaries.

Stoker Consists of a system that includes a fuel-feeding mechanism and a grate. It feeds a solid fuel into a furnace and distributes it over a grate where air is admitted to the fuel for its combustion. A means is provided for the removal of the ash that resulted from combustion.

Stoker Combustion Rate This rate is expressed in terms of the number of pounds of fuel or the Btu developed per square foot of stoker grate area per hour.

Sulfur Dioxide Scrubber A device for the removal of sulfur dioxide from flue gases. The process is often called *flue-gas desulfurization.*

Superheated Steam Steam that has been raised to a temperature higher than the boiling temperature corresponding to the boiling pressure is said to be *superheated.*

Superheater Heat-transfer surface that transfers heat from the flue gas to the steam for the purpose of raising the temperature of the steam above its saturation temperature.

Temperature The temperature of a substance must be carefully distinguished from the heat content. Temperature is thermal pressure and is a measure of the ability of a substance to give or receive heat from another substance.

Tertiary Air Combustion air that is supplied to a furnace and supplements both the primary and secondary air. This is often part of staged combustion to control NO_x emissions that result from the combustion process.

Thermal Efficiency The net work produced divided by the total heat input to the cycle.

Thermodynamics   The science that describes and defines the transformation of one form of energy into another: chemical to thermal, thermal to mechanical, mechanical to electrical.

Torque   Torque is a force that tends to produce rotation. It is measured by the product of the force and the applied distance from the center of rotation. A force of 100 lb applied at a distance of 2 ft will produce 200 ft · lb of torque.

Ultimate Analysis   The chemical analysis of a fuel that determines the contents of carbon, hydrogen, sulfur, nitrogen, chlorine, oxygen, and ash as percentages of the total weight of the fuel.

Utility Boiler   A boiler designed to produce steam for the production of electricity in the utility industry.

Vacuum   The word *vacuum* refers to pressures below atmospheric, units expressed in inches of mercury (14.7 psi atmospheric equals 30 in Hg).

Viscosity   Viscosity is the resistance that a fluid offers to flow. The viscosity of lubricating oil is an important characteristic. Viscosity varies with the temperature.

Wet Steam   When steam contains particles of water that have not been evaporated, it is said to be *wet*.

Work   Work is force exerted through a distance. No mention is made of the time required. Work is conveniently expressed in footpounds.

Index

ABOUT THE AUTHORS

Everett B. Woodruff was a project engineer at A. M.
Kinney, Inc., where he worked on the design of complete
plants, operation and maintenance, performance tests,
plant records, and survey reports.

Herbert B. Lammers was a consultant for a wide range of
industries. His experience covered all phases of industrial
plant engineering, including installation and operation and
maintenance of steam power plant equipment.

Thomas F. Lammers was Senior Project Manager with
Babcock and Wilcox, where he successfully managed
projects in engineering, marketing, and project
management for more than 35 years.